OECD Communications Outlook 2013

OECD
BETTER POLICIES FOR BETTER LIVES

Please cite this publication as:
OECD (2013), *OECD Communications Outlook 2013*, OECD Publishing.
http://dx.doi.org/10.1787/comms_outlook-2013-en

ISBN 978-92-64-19459-5 (print)
ISBN 978-92-64-19463-2 (PDF)

Annual: OECD Communications Outlook 2013
ISSN 1562-8795 (print)
ISSN 1999-1460 (online)

Corrigenda to OECD publications may be found on line at: *www.oecd.org/publishing/corrigenda*.

Foreword

This report, the twelfth in a biennial series, was prepared in the context of the OECD's work on the analysis of communication policy in member countries.

This edition of the OECD Communications Outlook *was drafted by the staff working in the OECD Directorate for Science, Technology and Industry, including Sam Paltridge, Agustín Díaz-Pínes, Alexia Lee González Fanfalone, Pierre Montagnier, Hajime Oiso, Jaesung Song, Rudolf van der Berg and Frédéric Bourassa. It benefited from valuable contributions from Andrea de Panizza, Kayoko Ido and Cristina Serra Vallejo. They are grateful for the contribution of information by telecommunication carriers and to national delegations which responded in 2012 to an OECD questionnaire relating to industry regulation and data.*

The assistance of Geoff Huston from APNIC, André Lange from the European Audiovisual Observatory, IHS Screen Digest, CISCO, Netcraft, the International Telecommunication Union, Akamai, CAIDA, M-Lab is gratefully acknowledged where they provided data. The pricing comparisons are undertaken in co-operation with Teligen, a division of Strategy Analytics Ltd. and quarterly updates of some pricing indicators using the OECD methodology are available directly from Teligen. Many of the other indicators in this report are available in electronic format from the OECD Telecommunications Database 2013, *covering the period 1980-2012.*

The draft of this report was presented to the OECD Working Party on Communication Infrastructure and Services Policy at its meeting on 10-11 December 2012. The Committee for Information, Computer and Communications Policy subsequently recommended that the report be made available to the general public.

Table of contents

Tables

Figures

Executive summary

Key findings

In 2011, the total number of OECD communication access paths was 2 066 million, or 166 subscriptions per 100 inhabitants. Mobile subscriptions represented 65.4% of paths, versus 64% in 2009, and traditional fixed telephony subscriptions continue to decline. Fibre broadband subscriptions grew at 16.61% year on year between 2009 and 2011. Greater use of mobile broadband access has been stimulated by the popularity of smartphones. The average subscription rate of mobile Internet access in OECD countries as a whole rose to 56.6% in June 2012, up from just 23.1% in 2009.

Prices for fixed telephony and, more markedly, for mobile voice services decreased from 2010 to 2012, showing significant declines across all consumption patterns, with the exception of fixed business services.

A laptop-based wireless broadband basket (offers within the 500 MB per month range) cost USD 13.04 on average across the OECD in PPP terms, although it reached USD 30 in some countries. Average expenditure was USD 37.15 for a 10 GB basket. A 250 MB tablet package cost USD 11.02 per month on average.A 5 GB basket for tablets cost USD 24.74 on average, but varied from USD 7.98 (Finland) to USD 61.84 (New Zealand).

Previously distinct communication services are converging rapidly, while digitalisation plus the rolling out of fixed and wireless infrastructures are expanding the bandwidth available for all types of communication services. Examples from the Internet include the quick uptake of the long-term evolution (LTE or 4G) standard for mobile networks based on Internet Protocol (IP)-only architecture and using Voice over LTE (VoLTE) as an application; and IP-based Video-on-Demand and live-streaming television services by cable companies, satellite providers, public broadcasters, and cloud-based and other "over the top" (OTT) providers.

Telecommunication revenues experienced a notable decline in 2009 but stabilised in 2010 and rebounded in 2011. This can be attributed to the strength of mobile communication markets and specifically to the rapid increase in smartphone penetration during this period. By far the greatest traffic generated by smartphones or tablets is linked to the use of Wi-Fi associated fixed networks, rather than cellular networks. Fixed networks have, in effect, become the backhaul for mobile and wireless devices with some studies claiming that 80% of data used on mobile devices is received via Wi-Fi connections to fixed networks.

Revenues corresponding to data services are growing at double-digit rates in most OECD countries, and transport of data is now the major source of growth for network operators. While there are significant opportunities in new services such as mobile payments, essentially these involve the transport of data in association with partners such as credit companies. Few expect growth in traditional services such as telephony or SMS as measured by their share of revenue.

The key to the success of the mobile ecosystem has been the presence of sufficient competition in the provision of network infrastructure and services. This competition drove some operators to open and share their access to customers with far more success than could have been achieved under the imposition of regulatory arrangements.

The Internet is still growing strongly, but relative growth has decreased compared to previous periods in some categories, as might be expected given widespread adoption of this technology. The Internet, together with analogue audio broadcasting, has become the primary distribution method for audio content. The conversion to digital television is almost completed in the OECD area. In many countries, broadcasters offer their content either live or via catch-up television over the Internet. Subscription video-on-demand services are seeing rapid adoption.

Emerging issues

Policy makers and regulators have a vital role to play in ensuring sufficient competition. This includes making sure there is adequate available spectrum, abundant IP addresses or other numbering resources for new market entry, and fair competition between operators and OTT providers.

Ensuring markets remain open to OTT and facilities-based providers is essential to innovation in broadband infrastructures, and critical to addressing major industry and broader economic and social challenges.

A growing number of industry leaders claim high prices for international mobile roaming are detrimental to their relationship with their customers, and a significant barrier to trade and travel in OECD economies.The OECD Recommendation of the Council on International Mobile Roaming Services (February 2012) recommends assessing and removing barriers that may prevent access by mobile virtual network operators to local wholesale mobile services to offer roaming services.

Limited spectrum and the increasing demand for data services mean that mobile networks will strive to offload traffic to fixed networks. Policy makers and regulators need to ensure enough supply to maintain sufficient backhaul for wireless networks, especially if there is insufficient fixed access network competition. While there is debate as regards the schedule for fibre-to-the-residence, all agree that network operators will continue to bring this technology closer to residences and end users. The challenge for regulators is that, regardless of the technology used, many parts of the OECD look likely to face monopolies or duopolies for fixed networks. Wireless canprovide competition, but spectrum availability will always impose limits that are not a constraint for fibre.

Since the 2011 *Communications* Outlook, the Asia Pacific Network Information Centre has run out of Internet Protocol version 4 (IPv4) addresses under normal procedures, as has Réseaux IP Europeéns Network Coordination Centre. Africa, North America and South America will use up their allocated address space in due time.The successor to IPv4, IPv6, allows 2^{128} addresses, a near unlimited amount, but has not been implemented to any significant extent. Although over half the equipment deployed on the wired Internet is capable of supporting IPv6 today, less than 1% of this equipment connects to a service that provides IPv6.

While industry levies or fees may be justified for specific purposes, such as funding the sector regulator or contributing to universal service goals, additional tax burdens on the telecommunication sector may harm both consumers and the industry itself.

Chapter 1

Main trends in the communications industry

The communications industry is an ever-evolving landscape – one in which stakeholders need to embrace dynamic technological innovation and rapid commercial development. Long-predicted trends such as the convergence of previously distinct communication services are now occurring at a fast pace across all sectors of industry and having profound and widespread impacts on economies and societies. This welcome process presents OECD countries with new opportunities to promote innovation and competitiveness and to address key challenge areas such as the promotion of greater equity. The data presented here show a return to revenue and investment growth in the sector following the global financial and economic crisis.

The statistical data for Israel are supplied by and under the responsibility of the relevant Israeli authorities. The use of such data by the OECD is without prejudice to the status of the Golan Heights, East Jerusalem and Israeli settlements in the West Bank under the terms of international law.

In 2013, communication infrastructure and service ecosystems continue to demonstrate tremendous innovation and increasing competitiveness, not only in the sector itself but also in the broader economy. OECD countries welcome these developments, recognising their potential to strengthen and sustain their economies as they emerge from the global financial crisis. They also view these developments as an opportunity to improve social welfare and to address challenges related to equity.

The rolling out of broadband infrastructures, whether fixed or wireless networks, continues across the OECD area, while digitalisation (e.g. in broadcasting) is reaching fruition in most countries. Taken together, these developments are expanding the bandwidth available for all types of communication services. Meanwhile, use of the Internet Protocol (IP) has gathered pace with convergence among communication networks and services. One such example is the rapid adoption of the long-term evolution (LTE) standard for mobile networks, based on IP-only architecture and using Voice over LTE (VoLTE) as an application. Further examples include IP-based Video-on-Demand and live-streaming television services by cable companies, satellite providers, public broadcasters, and cloud-based and other "over-the-top" (OTT) providers.*

The convergence of the Internet and traditional communication services first occurred with fixed telecommunication in areas such as telephony. The changes, while seemingly rapid at the time, took more than a decade to implement, and represented a combination of liberalisation and a shift from dial-up Internet access to broadband networks. The result was that consumers began to treat telephony as no different than any other application over IP networks. For users this was most notable in the elimination of tariffs based on factors such as distance, time of day/week and call duration.

Greater availability of bandwidth has heightened competition among communication services, but has also brought new opportunities. Services with lower bandwidth requirements in relative terms (e.g. radio) were the first affected; however, these changes are now evident in video and have similar implications for traditional broadcasting, cable or satellite services. These fields and others are now converging in the mobile sector, with the ability of users to consume these services "on the go" creating incentives to develop devices that better enable them to do so. The development of handheld devices that take advantage of new communication technologies has transformed markets such as book publishing – now effectively a market for e-books. New devices such as smartphones and tablet computers build on the nomadic capabilities made possible by expanded bandwidth and now enable entry, for an increasing range of different actors, into any area where the primary activity is the transfer of information.

Actors all along the value chain are exploiting the potential of mobile devices for payment or money transfer services. Mobile network operators such as Turkcell in Turkey, or Verizon, AT&T and T-Mobile through their joint venture "Isis" in the United States, have

* OTT refers to video, voice and other services provided over the Internet rather than solely over the provider's own managed network.

launched products to enable users to pay for goods and services via mobile devices. In Mexico, joint ventures between Telcel and financial institutions such as Banamex and Inbursa allow customers to transfer funds via text message from accounts that have no fees to open or maintain, as well as no minimum balance. Meanwhile, device or operating system manufacturers, as well as OTT service providers such as Apple and Google, have launched products conceived as the online equivalent of wallets or purses. Features range from direct payment to pre-purchased tickets, vouchers and loyalty cards. Other actors including banks, credit card companies, online payment providers or direct suppliers of goods and services, from airlines to coffee shops, are providing their own applications or aligning themselves with one or more providers to support better services for their customers. These companies also seek to benefit from leveraging products and services around "big data", a term that reflects the greater use of data in economies and societies.

Policy makers that take advantage of these new communication capabilities could make substantial gains towards policy objectives, such as promoting greater equity. In one well-known example, Kenya has pioneered the use of mobiles for making payments and providing services to people without bank accounts. Such uses can also be applied in OECD countries to address exclusion, where some proportion of the community may not have easy access to banking because of distance or other factors. These new capabilities can be provided to buyers and sellers across economies. This means that every constituent in a global value chain must strive to service its customers; otherwise an actor from a different part of that chain may enter their traditional market segment.

For policy makers and regulators a key to taking advantage of these developments is keeping the Internet economy open for innovation and new market entry. The previous *Communications Outlook* (2011) predicted the introduction of new mobile operators foursome countries with initial penetration rates much higher than those faced by incumbents. As expected, new entrants, for example, in France and Israel made markets significantly more competitive and encouraged new innovation. Meanwhile, regulatory authorities held firm in other countries, blocking developments to decrease the number of mobile network operators (MNOs). In the United States, regulatory authorities rejected the planned consolidation between AT&T and T-Mobile, thereby preventing a decrease in facilities-based competition. All major facilities-based mobile networks in the United States, including AT&T, Sprint, T-Mobile and Verizon, host mobile virtual network operators (MVNOs). In some markets, these account for a significant share of prepaid services that target users with lower incomes, thereby acting against digital exclusion.

Ensuring markets remain open to OTT and facilities-based providers is essential to innovation in broadband infrastructures, and critical to addressing major industry and broader economic and social challenges (Chapter 2). For example, a growing number of industry leaders recognise high prices for international mobile roaming as detrimental to their relationship with their customers, and a significant barrier to trade and travel in OECD economies. The prices on offer for consumers roaming from key partner economies, such as India and China, into the OECD are significantly lower than for business and consumers roaming from OECD countries to other OECD countries or key partner economies.

The OECD Recommendation of the Council on International Mobile Roaming Services (February 2012) included options to increase competition. It recommended assessing and removing barriers that may prevent access by mobile virtual network operators to local wholesale mobile services for the purpose of offering roaming services. In 2012, MVNOs on

T-Mobile's network in the United States provided some of the lowest prices for domestic unlimited voice, text and data services. In 2013, one of these MVNOs, Voyager Mobile, has said it plans to introduce a cloud-based approach to international mobile roaming in partnership with foreign networks as an innovative way to eliminate prohibitive pricing. Meanwhile, another US-based MVNO, FreedomPop, is selling sleeves with SIM cards for iPods with 500 MB per month of data included at no additional cost, and plans to launch services in 2013 on Sprint's LTE network. The European Commission is also widening the market for MVNOs in European Union member states by introducing provisions on the separate sale of international roaming services to boost competition. While the marketplace will constitute the ultimate test for such approaches, the key point for policy makers is that markets open to new ideas and competitive entry are a fundamental driver of innovation.

Competitive entry may derive from MVNOs adding capabilities to portable devices, or from consumers using "apps" combined with technologies such as near field communications (NFC). NFC is a technology that enables devices in very close proximity or placed in direct contact to exchange data. The number of smartphones or smaller tablets with NFC is increasing, as are the number of applications and services that make use of this technology. Turkcel and Isis both use NFC as does Google Wallet and any number of OTT providers. At present there is no international standard, so NFC will only function in the country in which a network operator provides service. However, over-the-top NFC services work independent of the network used (comparable with changing a SIM card or roaming). Traditional operators are reacting by creating standards around roaming. For example, KT in Korea and NTT DoCoMo in Japan are developing ways to use NFC to better serve their customers when they are roaming abroad. This includes the use of NFC in both Japan and Korea for shopping coupons, making payments and accessing information in different languages.

There is, of course, nothing to prevent OTT providers from offering certain types of services using NFC across borders. NFC itself would have been less open to competition if it had relied on SIM cards and international roaming to provide such services. However, some OTT providers need users to leave international mobile data roaming active for any service requiring access to the cloud, something many users may be reluctant to do. The data connection is necessary to provide authentication and user approval of transactions. Depending on the design of services offered by operators, a SIM card could provide such authentication, avoiding the charges that would otherwise be incurred by customers. Given the high costs of mobile data roaming many mobile users will turn off their data services while abroad. This may be one area presenting a challenge for OTT providers offering similar services using NFC across borders.

Over-the-top providers are challenging traditional telecommunication operators to improve their services. In 2012, in Japan, NTT DoCoMo became the first mobile company in the world to offer real-time telephony translation. A DoCoMo customer can make a call in Japanese and be heard in English, Mandarin or Korean. The recipient can speak in any of these languages and have their words translated back to the caller in Japanese. The service operates by accessing the cloud and would have obvious advantages for international roaming if made competitive with local substitutes.

Ensuring that there is sufficient spectrum available to existing and new entrants to wireless markets is essential. OECD countries pioneered the use of auctions and unlicensed spectrum: auctions led to more efficient and transparent allocation of spectrum and facilitated the entry of players into mobile markets; the success of unlicensed spectrum in areas such as

Wi-Fi have made markets more competitive by offering a further way for new players to enter markets and existing providers to better serve their customers. Policy makers and regulators are now striving to find new and more efficient ways to make spectrum available and utilised. In the United States, for example, the Federal Communications Commission (FCC) is introducing incentive auctions, a market-based means of repurposing spectrum by encouraging licensees to voluntarily relinquish spectrum usage rights in exchange for a share of the proceeds from an auction of new licenses to use that spectrum. Further examples in OECD countries include spectrum sharing through the use of "white spaces", which is being pursued in countries such as the United Kingdom, and adopting more flexible approaches to the allocation and use of spectrum.

A key way to enable new innovation and keep markets open is to take advantage of convergence. In the United States, for example, media companies are using spectrum originally allocated for broadcasting to offer television and video direct services to consumers that have appropriately equipped smartphones (Chapter 6). These services compete with those of traditional mobile companies. In France, the launch of a new mobile operator, in 2012, was associated with the innovative use of the Extensible Authentication Protocol (EAP-SIM) method to authenticate mobile users accessing Wi-Fi, provided by fixed networks, in an automatic and seamless way. Overnight, the mobile customers of the new provider could access more than four million hotspots without needing to log on to those networks or enter passwords.

These two examples from France and the United States show how innovation can be used to make markets more competitive to meet policy objectives, especially in areas such as promoting equity. In the example from the United States smartphone users can access television on their smartphones without incurring data charges. The service in France allows the new entrant to leverage its existing broadband network and offload traffic instead of burdening its 3G network or incurring a charge from the national roaming agreement it has with an incumbent operator. Both initiatives use different approaches to avoid the challenges posed by spectrum or backhaul constraints to offer more competitive outcomes. The emergence of smartphones has been fundamental to these developments and will be the basis for developing innovation for broader economic and social development.

In 2013, Apple celebrates the sixth anniversary of the launch of the iPhone. While not the first smartphone, and subsequently joined by an increasingly competitive group of rivals, it did represent a fundamental shift in the relationship between mobile operators and customers. The success of the ecosystem pioneered around the iPhone spread quickly across the industry, first in the United States then worldwide. In 2008, at the time of launch of Apple's "app store", approximately 20% of operating systems were headquartered in the United States and 80% offshore. By 2012, this figure had been reversed, primarily by Apple (IOS) and Google (Android).

The introduction of Android and, later, Windows Mobile further opened the market for competition between device manufacturers for smartphones. Manufacturers had greater flexibility to focus on specific groups and price points and to use readily available operating systems for their devices. This removed further barriers to entry and allowed new players into the market. Chipset manufacturers, for example, reacted by creating systems on a chip combined with reference designs, which eliminated much of the research and development from designing smartphones. This trend has opened up the entire market to new manufacturers and many Chinese firms, such as Huawei, Gionee, Oppo and ZTE, have gained substantial market shares with low-priced smartphones, mostly in non-OECD economies. For communication regulators, however, the key global change has been the introduction of intermediary relationships into the value chain.

The most valuable asset of telecommunication operators was direct customer access. Prior to the "app store" approach, mobile operators dominated the applications and services offered on feature phones and the first smartphones. These devices largely offered traditional applications (i.e. telephony) with minimal innovation (text messaging was an unplanned and unexpected success). Interoperability issues and a lack of compelling price plans hampered newer services such as Mobile Media Messaging (MMS). The possible exception was Japan's i-mode, a system controlled by an operator that had limited success outside the country due to the lack of applications developed in foreign markets. In contrast, the ecosystems that emerged around Apple and Google's operating systems have provided technological and commercial platforms for applications which created the demand for smartphones. In turn, mobile operators around the world have been driven by customer demand to adopt the new ecosystems around smartphones, even if this has meant relinquishing or sharing some of their traditional levers of control.

Today, market demand has resulted in services initially resisted by some mobile operators, for instance, OTT applications like Skype or WhatsApp (which offers substitutes for voice and text) as well as capabilities such as tethering being accepted as a regular and expected part of the features available to smartphone users. It is important to note that these developments were the outcome of a competitive market rather than the implementation of functional or structural separation. Where the latter two regulatory actions have been taken, the obvious difference was the degree of market competition.

In the United States, the presence of four major mobile providers drove one to reach a deal with Apple for the first generation iPhone and continues to drive other operators to reach agreements with MVNOs. In contrast, in OECD countries with fixed networks, there may be considerable monopoly or duopoly power. The presence of a sufficiently competitive number of players enables regulators to apply lighter oversight to market development around mobile communications. The pertinent question is whether one set of guardians has replaced another. Network operators were formally the arbiters of most available applications for use on mobile devices; these decisions are now sometimes dependent on approval for inclusion in a proprietary app store. Some consumers value such an approval process as being akin to an imprimatur from the device vendor in areas such as device compatibility or security. Others prefer a model where they alone select the applications available to them. The key point here is that different models are available and compete against each other. In this light, it is worth noting that at least one operating system enables customers to download applications directly from their authors as well as via their app store. Android has also divided into competing operating systems, supported by different companies such as Amazon and Bada. In addition, mobile operators can introduce applications on some operating systems that they can use to compete with OTT providers in areas such as mobile payments. Finally, a number of new operating systems, for smartphones and other devices, have been announced that will add to competitive choices available for manufactures and mobile operators such as Firefox OS, Ubuntu, Sailfish and Tizen.

The GSMA (GSM Association) has coordinated an initiative to promote inter-operator services among its members via an OTT app called joyn. The app, initially offered to mobile operator customers with Android or IOS, allows them to exchange messages enriched with images, voice and video. Three mobile carriers in Spain launched the service in November 2012, enabling communication between the networks and any other operators that introduce joyn. Meanwhile, in Sweden and the Netherlands, some operators have proposed price plans that would allow consumers to pay a surcharge to use OTT services. The Netherlands parliament reacted by adopting a network neutrality law, which came into force in January 2013 (Chapter 2).

The apparent benefits to business and consumers from the transformation of the mobile sector resulted from the application of innovation first introduced on fixed networks to mobile networks, enabling the emergence of many new applications built around features such as location and mobility. Today, smartphones and tablets are more than devices to communicate and gather information; they are the springboard for home automation, smart vehicles and other connected or smart devices. This area has tremendous potential for OECD economies to increase efficiency and meet broader policy objectives in fields such as health and energy.

In the next decade billions of devices will connect to the Internet. Some estimates place numbers as high as 50 billion connected devices between 2020 and 2025. Though seemingly high, upon evaluation these assessments do not appear unrealistic. In 2012, an OECD estimate concluded that a family with two teenagers may have 10 connected devices; by 2017 this number could reach 25, and by 2022 it could reach 50. Some of the functionality of machine-to-machine (M2M) communications will be built into machines such as automobiles, which will be controlled by and interact with devices such as smartphones. These devices will connect millions of sensors, providing ever-greater amounts of "big data" to enable the monitoring of environments and machines or the health of people.

The resources needed to provide identifiers for M2M technologies play as critical a role as spectrum does for wireless services. Further liberalisation in wireless markets could enable M2M users to buy wholesale access to mobile networks, change networks without switching SIMcards, and directly negotiate national and international roaming. However, this would involve changing current numbering policies both for international mobile subscriber identity (IMSI) numbers for SIMcards and telephone numbers, to enable access by M2M-users as well as traditional telecommunication companies. Such changes could lead to a more dynamic market for mobile wholesale access and mobile roaming and strengthen competition between mobile network operators.

The transition from IPv4 to IPv6 is a further critical requirement to ensure an innovative and competitive market. The "permission-less Internet" depends on new and innovative market entrants having access to IP numbers. This is as true today as it has been for any previous market in telecommunication services. Without such a development only existing players along the Internet economy's value chain will have access to IP addresses (i.e. IPv4). While such players can continue to use IPv4, the costs of implementing carrier grade network address translation (NAT) to stretch its use will increasingly rise and hamper the innovation that OECD economies need to ensure economic and social development. Innovators will have to assess the impacts that carrier grade NATs have on new applications instead of having an unencumbered path to the end customer.

Internet Protocol version 4 (IPv4), which uses 32-bit addressing, allows 4 billion devices to connect directly to the Internet. The impending shortage of IP addresses that was recognised in the 90s is now a reality. Since publication of the previous Communications

Outlook, the Asia Pacific Network Information Centre (APNIC) has run out of IPv4 addresses under normal procedures. A year later, Reseaux IP Europeénne Network Coordination Centre (RIPE NCC) also ran out of addresses under normal procedures. These Regional Internet Registries (RIRs) can now only allocate small blocks of IPv4 addresses for special groups of applications. The other three RIRs in Africa, North America and South America will use up their allocated address space in due time.

The successor to IPv4 is Internet Protocol version 6 (IPv6). This allows 2 128 addresses, a near unlimited amount. However, the transition to IPv6 has not been implemented to any significant extent. Meanwhile, the Internet has continued its inexorable growth with each year seeing greater levels of deployment of IP devices.

Some parts of the industry have recognized the issues relating to the depletion of addresses and have planned accordingly (Chapter 5). Microsoft's Windows operating system has incorporated an IPv6 protocol engine alongside the IPv4 engine since late 2001. Apple have made similar moves in their MAC OSX operating system, and support is also found in Linux systems and Linux-based derivative platforms such as Android. Much of the network infrastructure including backbone transmission paths and Domain Name System (DNS) infrastructure is also IPv6-capable.

Nonetheless, much more needs to be done to support the development of the Internet economy through greater take up and use of IPv6. While it appears that over half of the equipment deployed on the wired Internet is capable of supporting IPv6 today, less than 1% of this same equipment connects to a service that provides IPv6 (Figure 1.1). Only four countries, France, Luxembourg, Japan and the United States, are above the OECD average in this respect.

Figure 1.1. **IPv6 deployment by country, November 2012**

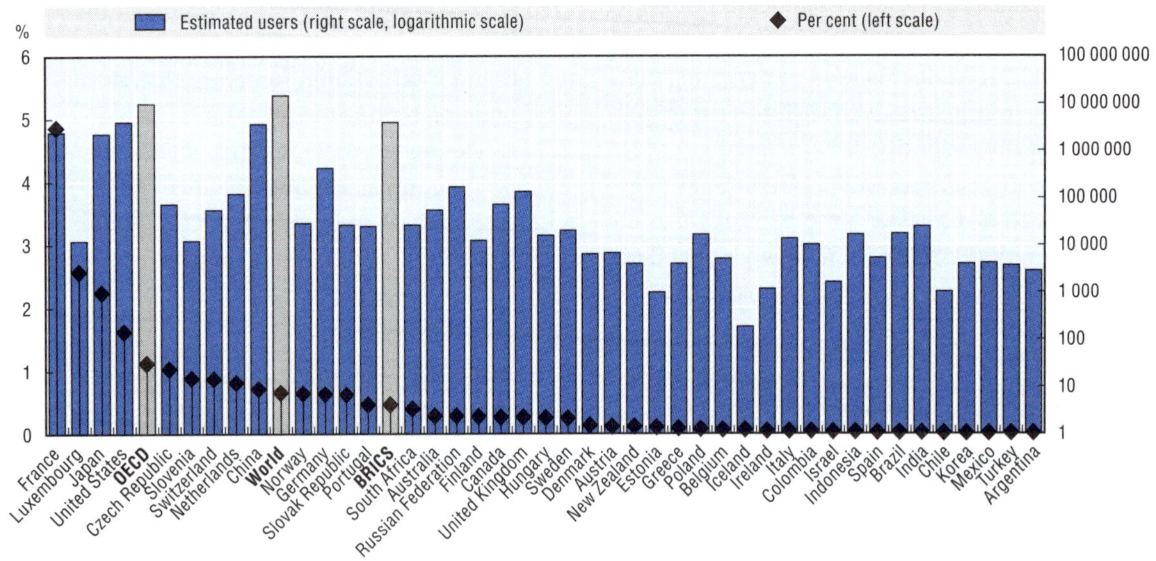

Note: Data collected on 19 November 2012.
Source: *www.potaroo.net/reports/oecd*, based on report files published by the RIRs.

StatLink http://dx.doi.org/10.1787/888932798354

On 6 June 2012, in a co-ordinated effort, some of the world's largest content providers made their services available via IPv6 ("IPv6 launch day"). This may assist with the "chicken and egg" conundrum in which there was no demand for IPv6 because there was no content available for users of that protocol. For their part, a number of providers in OECD countries have made substantial efforts to extend IPv6 services through the access network. Notable efforts include Free.fr in France, KDDI in Japan, XS4ALL in the Netherlands and Comcast in the United States. All these providers have recently made significant efforts to extend IPv6 access to consumers through their mass-market Internet access services. Several major ISPs such as AT&T, RCS, TWC and Verizon Mobile, among others, have committed to enabling IPv6 for their new users. Some equipment vendors such as Cisco and D-Link have also configured IPv6 as a default feature in their products.

Market growth

There were just over 2 billion communication access paths in OECD countries at the end of 2011 (Figure 1.2). Mobile access paths continue to make up the bulk with 1.35 billion prepaid and post-paid subscriptions. This was followed by a total of 379 million traditional analogue fixed-line connections and a further 175 million DSL lines. Cable networks (94 million) and fibre (43 million) make up most of the remainder.

Figure 1.2. **Access growth in the OECD, 1997-2011**

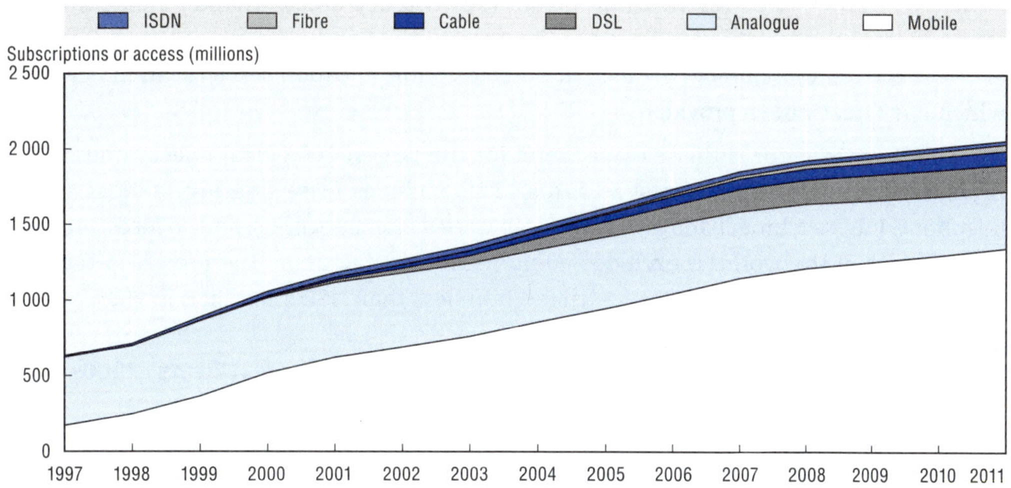

StatLink http://dx.doi.org/10.1787/888932798373

Fixed (wired) broadband subscriptions reached 314 million in the OECD area at the end of 2011, although growth slowed to 1.8% in the second half. The overall share of DSL subscriptions continues to decrease (55.8%) to the benefit of cable (30%) and, especially, fibre-to-the-home subscriptions, which now represent 13.7% of the total number of fixed broadband subscriptions. Between 2007 and 2011, OECD countries had a net increase of 279 million access paths. The relative increases were fibre (114%), DSL (20%), cable (37%) and mobile (18%) (Figure 1.3).

The influence of greater smartphone use is evident in the increasing use of broadband mobile services. Wireless broadband subscriptions showed healthy growth of over 13% in the second half of 2011. The global number of wireless broadband subscriptions in OECD countries totals 667 million, up from 590 million in June 2011. It is notable that during the

Figure 1.3. **Net access path growth, 2007-11**

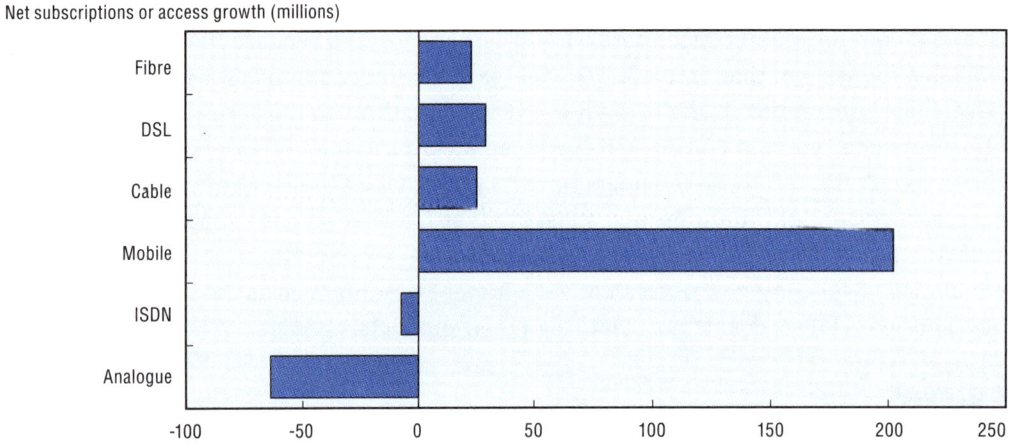

StatLink http://dx.doi.org/10.1787/888932798392

global financial crisis (which spilled over into the broader economy from 2008 onwards) the number of access paths to networks continued to grow year on year.

The previous *Communications Outlook* considered the unfolding of the financial crisis and the associated economic recession, and recorded a decrease in the size of total industry revenue. The present edition examines two key areas. Chapter 3 examines the recovery from the recession in terms of the communication industry's revenue. Chapter 4 explores the trend in number of subscriptions to communication services, an indication of which has already been provided.

One key indicator is the revenue trend for the largest 100 global telecommunication firms (Figure 1.4, Table 1.1 largest 50 firms). This includes firms offering services such as telephony, Internet, broadband access (e.g. cable television networks), and so forth, including all countries of the world. It excludes equipment manufacturers. Data for 2011 reveal an increase in global revenue to USD 1.6 trillion from less than USD 1.5 trillion in 2009.

Figure 1.4. **Communications revenue for global 100 largest firms, 2000-11**

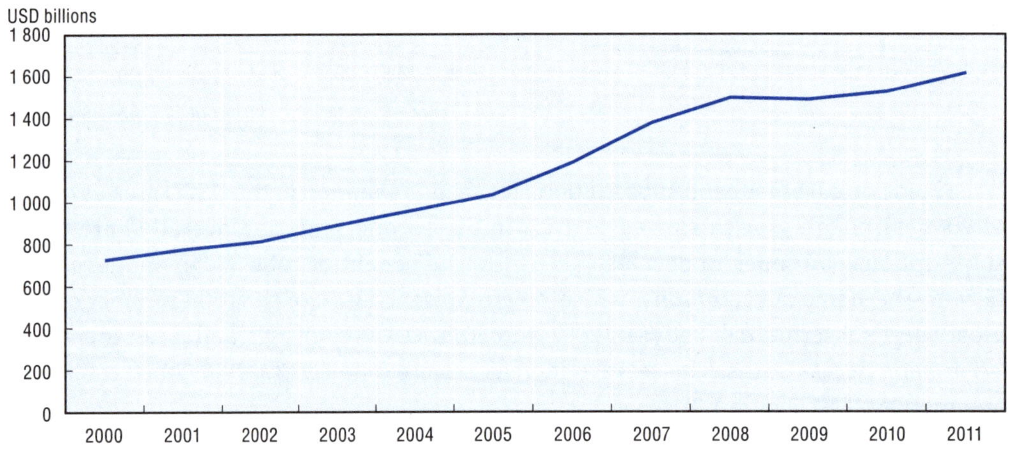

StatLink http://dx.doi.org/10.1787/888932798411

Total revenue for the leading 100 firms rose 5.9% between 2010 and 2011. Nonetheless, the long-term trend for net income was relatively flat between 2006 and 2011 (Figure 1.5). In 2011, net income decreased 6.4% from 2010, but based on the first three-quarters of 2012 is likely to increase for that year to pre-recession levels(of 2008) in many parts of the global economy.

Figure 1.5. **Communications net income for global 100 largest firms, 2000-2011**

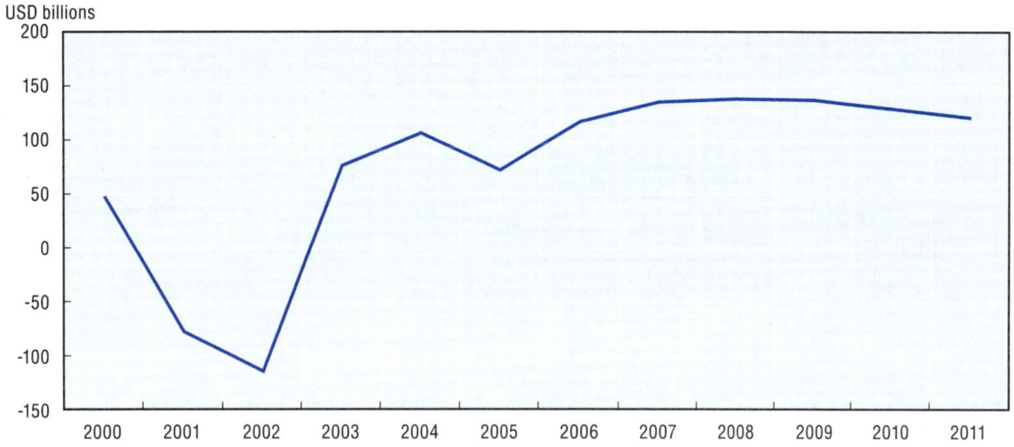

StatLink http://dx.doi.org/10.1787/888932798430

Chapter 3 explores the trends in OECD country revenue. It is noted that the smartphone effect has increased revenue with greater numbers of consumers adopting monthly post-paid services. They have done this for two main reasons. One was that initial smartphone offers in many countries were linked to long-term contracts and discounted equipment. It is also true that while operators record these sales of equipment as revenue they will have different effects on net-income depending on market conditions in individual countries. In 2012, a number of operators were examining or implementing measures to reduce the amount of up-front smartphone discounts and much of the financial community welcomed this move as one that would likely improve the margins of these operators.

Revenue and investment

Following a sharp decline during the economic downturn, levels of revenue and investment in OECD countries stabilised in 2010 and increased in 2011. In 2011, telecommunication industry revenue was USD 1.35 trillion, up from USD 1.27 trillion at the end of 2009 (Figure 1.6). Meanwhile, investment also increased between 2010 (USD 179 billion) and 2011 (USD 188 billion).

Increases in capital expenditure by operators contributed to a rise in international trade for ICT equipment (Figure 1.7). Between 2009 and 2011, OECD countries witnessed a rise in both imports and exports of telecommunication equipment.

Figure 1.6. **Subscriptions, revenue and investment growth, 1980-2011**

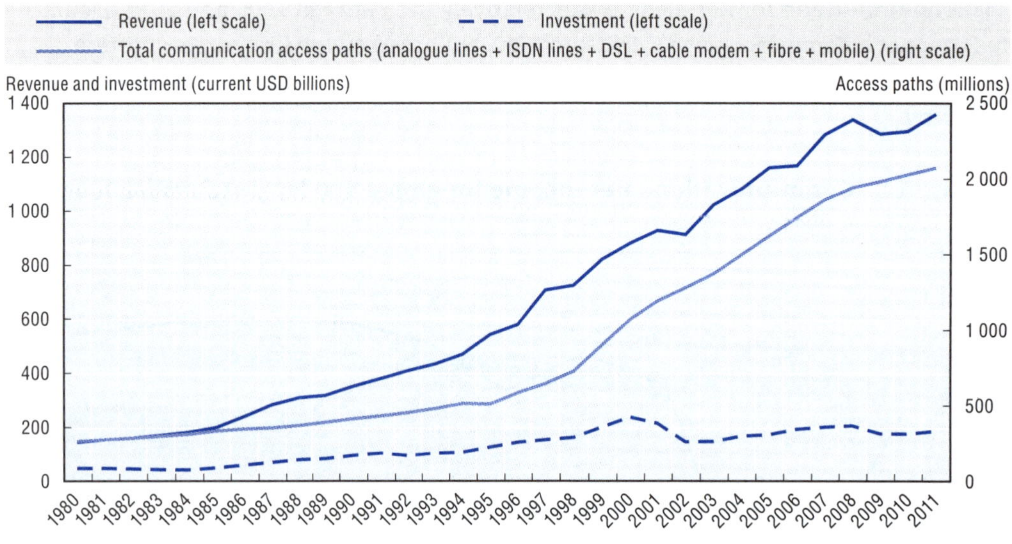

StatLink http://dx.doi.org/10.1787/888932798449

Figure 1.7. **Trade in telecommunication equipment and communication services for OECD countries**

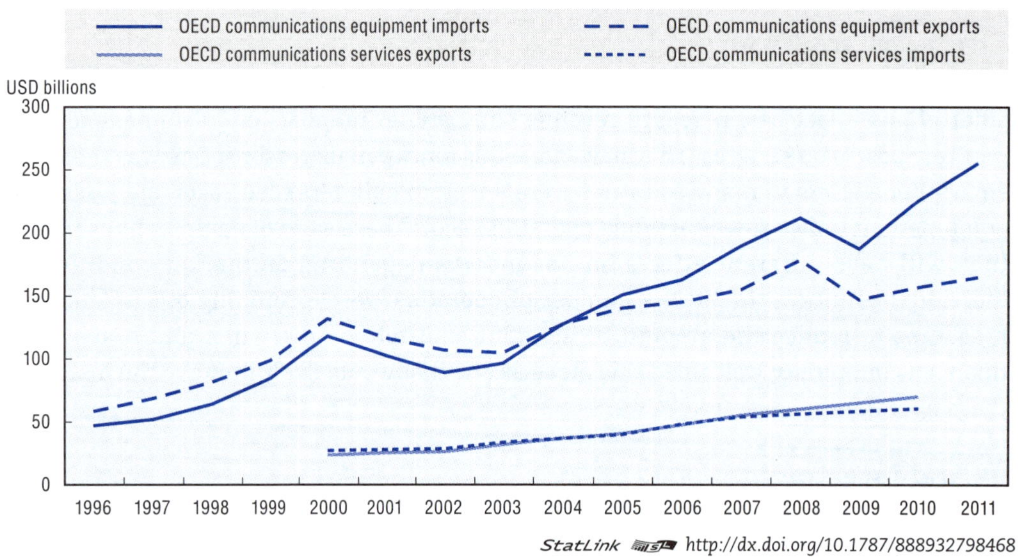

StatLink http://dx.doi.org/10.1787/888932798468

The outlook for traffic growth

Global Internet traffic continued to grow in 2011 (Figure 1.8). According to Cisco's VNI Forecast, global IP traffic volume has grown eightfold over the period 2006-11, but will only grow threefold over the period 2011-16, at a growth rate of 29% per year. In 2010, global volume of IP traffic was 20 exabytes per month according to Cisco, and rose to 30.7 exabytes per month in 2011, an increase of 40%. The various regions in the world showed different growth levels from 2010 to 2011. They also reflect the trend towards lower growth rates compared to the previous edition of *Communications Outlook*. With the exception of Latin America, no region of the world had a growth rate above 50% during this period. North America, Western Europe and Asia Pacific saw traffic grow by 40% over the two-year period.

Figure 1.8. **Global IP traffic, 2005-11**

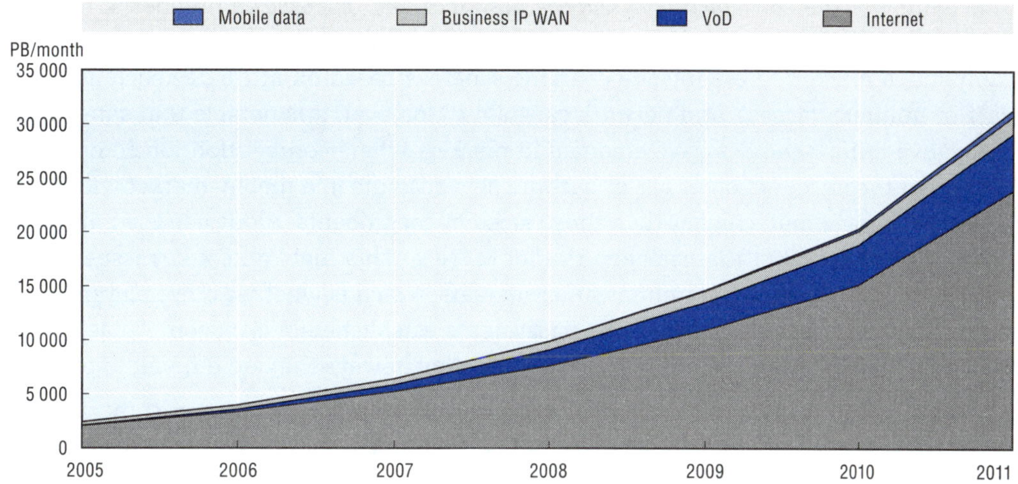

Source: Cisco VNI.

StatLink http://dx.doi.org/10.1787/888932798487

Mobile data showed the fastest growth from 2006 to 2011, growing by more than 15 000%. It is, however, still a small but growing part of total IP-data at 597 petabyte (PB) per month, which represents less than 2% of total IP traffic. In 2010, mobile data accounted for over 1% of IP traffic for the first time. Video-on-Demand, which is not delivered over the public Internet, grew by over 2 200% from 2007, and represents 17% of IP traffic, down from 18% the year before. Traffic over the public Internet has the largest share of IP data at 77%. It has grown by 716% since 2006 and become a larger part of total traffic, increasing from 74.5% to 77.8%. Business use of IP that does not traverse the public Internet grew 384% during that period, but decreased in its share of total traffic from 6.2% to 5.2%.

Networks and services: Smarter, broader and a return to growth

The communication infrastructure and services market has emerged stronger from the global economic recession, reflected once more in a long-term trend towards growth. This position is the result of market-driven competition, which has made the sector more relevant and useful to the demands of customers. The industry flagship has been the development and rapid growth of smartphones and other devices connected to networks. Whereas feature phones had only a handful of applications, there are now several million across different operating systems. The combined number of apps on IOS and Android nearly doubled between the end of 2011 (850 000) and 2012 (1 400 040).

The key to the success of the mobile ecosystem has been the presence of sufficient competition in the provision of network infrastructure and services. This competition drove some operators to open and share their access to customers with far more success than could have been achieved under the imposition of regulatory arrangements. However, policy makers and regulators do have a vital role to play in ensuring sufficient competition to maintain this momentum. This includes making sure that there is adequate available spectrum, abundant IP addresses or other numbering resources for new market entry, and fair competition between operators and OTT providers.

For their part, operators have the tools to take advantage of growth in mobile markets. This edition of the *Communications Outlook* describes and discusses a number of new and innovative tariff models with charges based on a range of options reflective of a competitive market. These include charging schemes based on attributes such as speed, data or unlimited access, and network capability. Moreover, it is notable that some of the innovative tariffs schemes have emerged in markets where competition forced operators to pioneer the introduction of LTE or become new entrants in a mobile market with initial penetration rates much higher than those faced by incumbents. Alongside the innovation brought to tariffs by MNOs has been that of MVNOs. This highlights a key issue for the future competitiveness of communication markets, which is whether other players in the value chain will become MVNOs. The ecosystems established by Amazon, Apple, Google and Microsoft, to name just a few, could well develop in ways that are difficult to predict.

Finally, the role of backhaul will become increasingly more important as fixed networks are called upon to do the heavy lifting that makes for efficient wireless markets. While there is ongoing debate as regards the schedule for fibre-to-the-residence, all agree that network operators will continue to bring this technology closer to residences and end users. The challenge for regulators is that, regardless of the technology used (e.g. fibre-to-the-residence, DOCSIS-3), many parts of the OECD look likely to face monopolies or duopolies for fixed networks. Wireless can, of course, provide competition for some traditional and some new services, but spectrum availability will always impose limits that are not a constraint for fibre. In this context, it is worth noting that by far the greatest traffic generated by smartphones or tablet has been linked to the use of Wi-Fi associated fixed networks, rather than cellular networks. The 2013 *Communications Outlook* aims to provide information that will help all stakeholders consider these challenges while noting the tremendous opportunities for future economic and social development.

Table 1.1. **Major public telecommunication operators and Internet service providers in the OECD area (fiscal year 2011 unless noted)**

		USD Millions							Units		
	Country	Revenue	Income	Long Term Debt	Capital Expenditures	Mobile revenue	Employment	R&D	Fixed access lines	DSL/cable/ FTTH lines	Mobile subscribers
AT&T	United States	126 723	3 944	61 061	20 110	63 212	256 210		36 734 000	16 427 000	103 247 000
NTT	Japan	111 331	4 956	37 188	19 636	53 126	219 343	3 357	31 672 000	18 886 000	60 129 000
Verizon	United States	110 875	2 404	50 303	16 244	70 154	195 400		24 137 000	8 670 000	107 798 000
Telefonica	Spain	86 300	7 420	76 442	12 477	52 592	291 027		42 842 900	18 915 600	238 748 600
Deutsche Telekom	Germany	80 554	765	52 325	11 545		238 985	169	34 100 000	16 900 000	129 300 000
Vodafone	United Kingdom	71 671	10 742	43 793	12 124	60 005	83 862	490			404 691 000
France Telecom	France	63 087	5 349	46 604	9 217	19 243	165 533	819	44 265 000	14 672 000	167 384 000
Comcast	United States	55 842	4 160	37 942	6 261		102 000		9 342 000	18 147 000	
America Movil	Mexico	48 889	6 088	26 012	8 905	27 457	158 000		29 000 000	15 000 000	242 000 000
Telecom Italia	Italy	41 554	- 6 491	47 459	8 371	30 849	85 126		21 712 000		32 227 000
Vivendi	France	39 572	3 682	17 043	4 624	7 885	51 300			5 042 000	23 894 000
KDDI	Japan	37 848	2 528	9 029	4 183	34 169	18 418	81	2 189 000	3 410 000	35 109 000
Softbank	Japan	33 931	3 324	10 807	4 821	26 875	21 799	11	2 160 000	4 209 000	28 949 000
Sprint Nextel	United States	33 679	- 2 890	20 266	3 388	30 301	40 000				46 607 000
BT	United Kingdom	29 178	3 428	11 733	3 981		92 600	1 344	35 223 000	15 721 000	
VimpelCom	Netherlands	20 262	543	25 724	6 260	18 240	42 025		3 142 000	12 300 000	205 200 000
Time Warner Cable	United States	19 675	1 665	24 620	2 937		47 500			9 954 000	
Telstra	Australia	19 338	2 582	8 996	2 994	8 236	41 183	6	8 234 000	3 319 000	11 744 000
KPN	Netherlands	18 078	2 127	15 988	2 848	10 731	31 084	78	6 197 000	2 714 000	34 143 000
KT	Korea	16 950	1 115	6 849	2 473	6 147	31 980	159	15 900 000	7 823 000	16 563 000
BCE	Canada	16 832	2 020	10 982	2 811	5 284	50 200	231	8 730 431	2 975 121	7 572 651
Telenor	Norway	15 376	1 118	3 483	2 070	23 120	33 000	131	1 714 000	1 665 000	140 422 000
CenturyLink	United States	15 351	571	21 356	2 411		49 250		14 584 000	5 554 000	
TeliaSonera	Sweden	13 476	2 358	8 758	2 255	10 533	28 412	78	4 805 000	2 481 000	54 360 000
Swisscom	Switzerland	12 884	780	9 336	2 354	3 830	19 832		3 426 000	3 437 000	6 049 000
SK Telecom	Korea	12 324	1 243	2 739	2 743	11 822	20 955	263	4 203 567	4 191 892	26 497 267
Rogers Communications	Canada	10 729	1 349	8 663	1 885	7 210	25 900			1 793 000	9 335 000
Telus	Canada	8 976	1 052	4 755	1 595	5 556	33 900		3 590 000	1 290 000	7 340 000
Belgacom	Belgium	8 798	1 040	2 652	1 040	1 533	15 676		4 794 000		5 213 000
Bouygues Telecom	France	8 019	513	807							
Turk Telekomunikasyon	Turkey	7 663	1 328	1 935	1 493	1 845	23 940	17	15 200 000	6 800 000	12 800 000
LG Uplus	Korea	7 135	65	2 001	1 323	3 079	6 190		3 588 267	2 809 690	9 391 000
OTE[1]	Greece	6 920	164	5 685	984	2 885	28 675		5 921 000	2 233 000	20 467 000
NII Holdings	United States	6 735	227	4 245	1 061		13 500				10 712 000
Cablevision Systems	United States	6 701	292	10 759	826		16 350			2 701 000	
Virgin Media	United States	6 164	117	8 922	1 014	892	11 300			4 102 900	2 995 500
Telekom Austria	Austria	6 118	- 346	4 031	1 015				2 608 700	1 465 100	20 266 200
Turkcell Iletisim Hizmetleri	Turkey	5 610	752	1 040	859	5 610	12 834				64 800 000
Hikari Tsushin	Japan	5 290	83	96	28		5 326				
Tele2	Sweden	5 272	630	1 668	716	4 629	7 517		1 894 000	1 183 000	31 109 000
Frontier Communications	United States	5 243	150	8 206	825		15 250		5 290 676	1 764 160	
Telephone and Data Systems	United States	5 180	201	1 530	972	4 343	4 343		1 071 900	267 300	5 891 000
TDC	Denmark	4 898	3 913	392	637	1 353	1 882		1 401 000	928 000	2 891 000
MetroPCS Communications	United States	4 847	301	4 430	964		3 600				9 347 000
Telekomunikacja Polska[2]	Poland	4 693	603	1 309	820		14 715	20	5 445 000	2 715 000	
Freenet	Germany	4 539	197	706	31	4 375	4 057				15 190 000
Level 3 Communications	United States	4 333	- 756	8 385	494		5 500				
Windstream	United States	4 286	172	8 937	724		10 086		1 931 700	1 207 800	
Shaw Communications	Canada	4 093	389	4 538	768				1 233 041	1 877 231	
Portugal Telecom	Portugal	4 017	587	2 916	1 700	5 963	72 347	304	2 648 000	1 105 000	50 708 000

Notes : 1. Deutsche Telekom has a 40% stake in OTE's capital; 2. France Telecom is the parent company of Telekomunikacja Polska.

StatLink http://dx.doi.org/10.1787/888932800900

Chapter 2

Recent communication policy developments*

This chapter notes recent policy developments in fixed and mobile communications. In view of the "smartphone" effect which is shaping communication markets, it devotes special attention to spectrum policy and wireless technology evolution. It also looks at traffic prioritisation and interconnection issues, both in mobile and fixed markets, and at the emergence of new areas relevant to policy makers such as connected televisions. Finally, it summarises trends in key partner countries (Brazil, China, India, Indonesia and South Africa).

* All tables associated with this chapter are available online only. See *www.oecd.org/sti/broadband/communications-outlook.htm*.

All stakeholders now consider communication infrastructures and services to be critical for economic and social development. Governments, in particular, are increasingly aware that widespread availability of broadband networks, including fixed and wireless broadband networks, is crucial for competitive economies and the creation of opportunities across all types of social and civic activity.

Information and communication technologies (ICTs) are a general purpose technology. In other words, they constitute a fundamental infrastructure supporting economic and social activities, similar to the continuing role played by various transportation and energy networks. From a public policy perspective, broadband communication networks, and the services provided over them, not only support existing economic and social activities, but also hold potential for tremendous innovation in these areas and others created in the future.

To take advantage of these opportunities, policy makers need to ensure effective competition, wherever possible, to allow markets to drive these developments. The decision by OECD governments to open markets for telecommunication competition has brought tremendous innovation and benefits across their economies and societies. This was emphasised and reinforced at an OECD high-level meeting (HLM), "The Internet economy: Generating innovation and growth", held in June 2011, where ministers and other stakeholders encouraged the extension of broadband infrastructures to reach maximum national coverage and provide access at affordable prices. They recognised that fostering the use of the Internet in critical areas (health, education, transport, energy) could increase efficiency and enable people to enjoy better lives.

The HLM highlighted two further areas for attention. One was the need for governments to benchmark developments in broadband networks and quantify the Internet's effects on the economy, so as to facilitate evidence-based policies. The second concerned the critical role to be played by all stakeholders in this and other areas. This was subsequently developed into an OECD Council Recommendation on Principles for Internet Policy Making, which encourages countries to follow a number of basic principles to ensure that the Internet remains open and dynamic.

A key message from the HLM and the Council Recommendation, for policy makers in this area, is the need to ensure a multi-stakeholder approach and, where possible, avoid regulation. As a decentralised network of networks, the Internet has achieved global interconnection without the development of any international regulatory regime. The development of a formal regulatory regime, they noted, could risk undermining its growth.

The Internet's openness to new devices, applications and services is a key feature in its success and, consequently, its rapid adoption. In turn, this has enabled service providers and users to develop and adapt new features and capabilities to meet their requirements. There is continuously evolving interaction and independence among the Internet's various technical components, enabling collaboration and innovation, even while they operate independently from one another. This independence permits policy and regulatory changes in some components without requiring changes in others or effecting innovation and collaboration.

A key outcome of the Internet's openness is the long-predicted convergence among networks and services, such as telecommunications (telephony) and cable television (video), through increasing use of the Internet Protocol (IP). Today, the outcomes of this convergence can be witnessed over broadband networks: telephone calls are increasingly provided via data; over-the-top (OTT) video services are gaining customers in many countries; and IPTV is the primary form of television provision in others, such as France. Moreover, convergence over devices, applications and services is occurring in areas that were perhaps less obvious candidates. One example is the success of e-books following the launch of the Amazon Kindle in the United States. Not only have purchases of e-books surpassed physical books, for Amazon, but the company was also able to integrate the communication capabilities of mobile networks and Cloud services to synchronise a user's experience across multiple devices and platforms.

With regard to mobile communications, 3G networks and especially LTE (or 4G) technology provide a platform for the convergence of voice, SMS/MMS and data services over a single data connection. In some ways this can be threatening for service providers, as new providers enter markets in a manner that can be disruptive to traditional business models, but it also creates opportunities. Voice and SMS services, for example, are still responsible for the bulk of revenue for many mobile operators. On the other hand, some operators are exploring new ways to tariff mobile services that treat such services no differently from any other data service over their network. In 2012, Swisscom introduced a tariff scheme that charges users a flat rate for all domestic services provided over their LTE network, but offers that differentiated pricing options for customers depending on the speeds they choose. In addition, most mobile operators are exploring new services. In 2012, Telefonica created a unit to develop opportunities around "big data", while mobile operators in Turkey have introduced mobile payments, money transfers and so on, using SIM cards and Near Field Communications (NFC), in a number of cases leading the world.

The number of active communication service providers in OECD countries remains high (Table 2.1), and the market share of new entrants is increasing continuously (Table 2.2). All communication service providers listed in the tables here are, of course, not equally significant at the national or international level in terms of their competitive footprint. Many cable or telecommunication operators, for example, provide service only on a local or regional basis and some MVNOs may provide service to a relatively small number of users, compared with some of the larger mobile operators which serve hundreds of millions of customers in different OECD countries.

Limited spectrum and the increasing demand for data services mean that mobile networks will strive to offload traffic to fixed networks. At the same time, policy makers and regulators need to ensure enough supply to maintain sufficient backhaul for wireless networks, especially if there is insufficient fixed access network competition. In addition, governments in countries such as the United Kingdom and the United States are exploring innovative ways to make more spectrum available, from greater use of "white spaces" through to incentive auctions. Other countries (e.g. Mexico) are moving towards improving the legal framework for rights of way or making available government buildings for network deployment.

All these challenges mean that OECD countries are exploring different models to encourage the development of competitive choice where possible. One example is the continuing use of unbundling local access networks where there is insufficient facilities-based competition. In other areas, such as in Australia and New Zealand, governments are funding structurally separated national broadband networks. In still others, they are subsidising infrastructure, such as in rural areas, making available government sites for network deployment, or encouraging shared use of facilities (e.g. from ducts and towers to spectrum and cables).

For their part, many network operators are requesting greater flexibility in network management as they explore different business models. In terms of the Internet model for traffic exchange, all the available evidence indicates an efficient and highly competitive market where commercial negotiations are possible and the norm. In other words, the model works well without regulation and most agreements to exchange traffic are made on a "handshake" basis in recognition of the mutual benefits. In such cases, the level of regulatory oversight needs to be concomitant. Issues around traffic prioritisation, sometimes more commonly referred to as "network neutrality", in access networks, come to the fore where there may be insufficient competition for one or more market segments. They are less prevalent where regulators judge there is sufficient competition and where relevant network management practices are transparent for consumers.

Several OECD countries, such as Chile, the Netherlands and Slovenia have introduced "network neutrality" legislation and others have provided non-binding guidelines. Some believe that this issue will draw greater attention in the coming years as policy makers seek to create an environment attractive for investment as well as maintaining effective competition. The two are not, of course, incompatible. In many cases, such as in the field of 4G, competition drives investment and there is no evidence that lack of competition stimulates investment in network infrastructure. Indeed, faced with greater competition in one country, it is likely that operators will invest in those locations prior to those where competition does not push them to do so. On the other hand, there will be cases where, for whatever reason, there will be insufficient competition in some locations or market segments.

Major questions about the level of competition and adequate investment to roll out fixed network infrastructures will remain at the forefront of policy and regulatory concerns. The market-driven separation of "transport with apps and services", witnessed with the arrival of smartphones, will be just as influential for fixed networks. From the perspective of an increasing proportion of consumers, the devices, applications and services they use to connect to fixed networks are the same as those they use to access mobile connectivity. Just as smartphones have taken on the features of computers, so too have computers become more like mobile telephones through the rapid evolution and development of tablets with 3G and 4G connectivity. Fixed networks have, in effect, become the backhaul for mobile and wireless devices with some studies claiming that 80% of data used on mobile devices is received via Wi-Fi connections to fixed networks.

This chapter emphasises the importance of wireless communications, as some of the most significant policy and regulatory development are taking place in this field. Questions surrounding spectrum, new wireless broadband devices and services, competitive backhaul (i.e. fixed networks such as fibre, DSL, cable) and new business models are the key issues facing policy makers and regulators. To these can be added the need to make competitive infrastructures and services available across borders. For this reason, the OECD Council adopted a recommendation, in 2012, aimed at making international mobile communications more competitive. Governments have been increasingly active in the areas of international mobile roaming and mobile termination rates as ineffective competition and its outcomes, such as high prices, act as a barrier to efficient trade and travel. Consequently, all OECD countries are increasingly undertaking action in this area. It could be assessed as being a significant market failure that users switch off devices when they cross borders, as opposed to the increasing use they make of these devices in domestic situations that reflect a functioning competitive market.

Overall trends in competition

Developments in fibre networks

Irrespective of the labels applied to fixed networks, such as fibre-to-the-curb, DSL, cable DOCSIS 3 and so on, a common characteristic they all share is the increasing deployment of fibre optic cables deeper into those networks. Infrastructure with fibre deployed closer to the premises has been an ongoing process in all OECD countries for many years. More recently, the main decision taken by network operators has concerned whether to bring fibre directly to a premise or to a nearby point and use existing or upgraded DSL and cable infrastructure. As of June 2012, the majority of fixed wired broadband connections were provided over DSL (54.8%) and cable modem (30.4%) technologies. In 2011, the share of direct fibre connections in the OECD area was 14.2%, up from 11.1% in December 2009.

Fibre's share of direct connections to premises, while still much smaller than DSL or cable, is increasing at a faster pace than the market's overall growth. Between 2009 and 2011, the number of fibre connections increased by 36.2%, from 31.6 million to 43 million (45.7 million in June 2012), in contrast with a mere 11.1% increase in the number of fixed wired broadband subscriptions. Japan and Korea have by far the largest proportion of fibre connections (Figure 2.1).

Australia and New Zealand have taken decisions to publicly fund national fibre networks. Like neighbouring Singapore, Australia and New Zealand have chosen to structurally separate the provision of access facilities from the provision of retail services. All three countries have indicated that the pace of communication infrastructure deployment prior to taking these initiatives was not sufficient to meet their policy objectives. In Australia, a public owned company, NBN Co., is building a national broadband network which will provide access to high-speed broadband to 100% of Australian premises. It will connect 93% of homes, schools and businesses to a high-speed fibre network, with the remaining 7% of premises served by a combination of next-generation fixed wireless and satellite technologies. In November 2011, NBN Co. published its Wholesale Broadband Agreement which, as its Standard Form of Access Agreement, sets out the terms and conditions for the supply of wholesale products to retail providers. NBN Co.'s terms and conditions are subject to regulation by the competition regulator. According to NBN Co., the construction of the fibre network had commenced or been completed for around 784 592 premises by the end of 2012 and is planned for around 3.5 million homes and businesses by mid-2015.

In New Zealand the government has put in place two initiatives to improve broadband services: the Ultra-Fast Broadband Initiative and the Rural Broadband Initiative. Together,

Figure 2.1. **Percentage of fibre connections in total broadband subscriptions, June 2012**

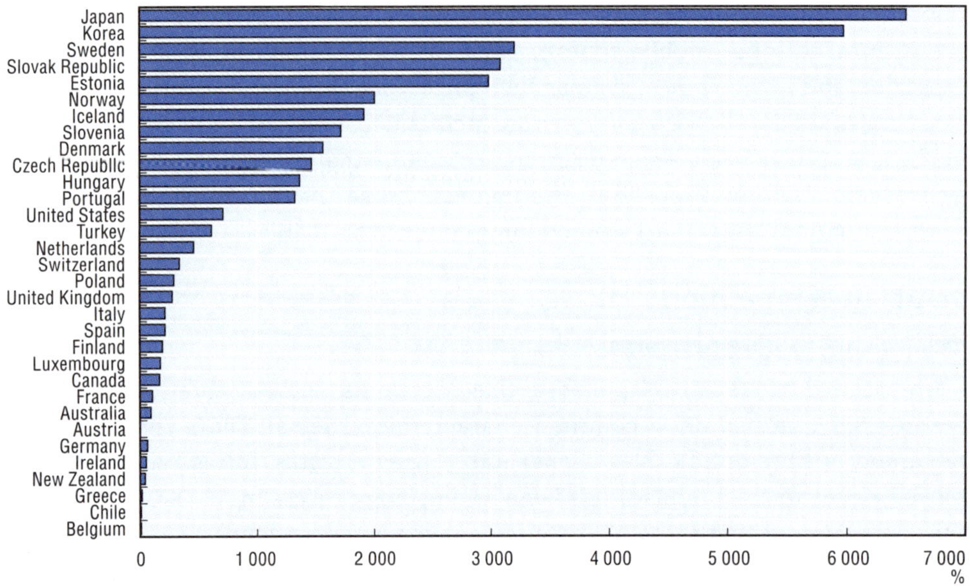

StatLink http://dx.doi.org/10.1787/888932798506

these programmes aim to cover 97.8% of the population. By 2020, the goal is to connect 75% of the population to fibre networks. Schools, hospitals and 90% of businesses will be connected by 2015. Homes and the remaining 10% of businesses will be connected by 2019. Meanwhile, the aim of the Rural Broadband Initiative is to deliver broadband to 252 000 rural households at prices and levels of service comparable with urban areas. In rural areas a combination of copper and fixed wireless infrastructures will be used.

In the Netherlands the incumbent, KPN, has formed a joint venture with Reggefiber, a new entrant focused on deploying fibre networks. The joint venture was cleared, in early 2012, by the Dutch authorities, which incorporated some undertakings in order to maintain sufficient competition, such as transparency and non-discrimination conditions (e.g. the obligation to include a reference offer), as well as access for third parties on regulated terms. More recently, KPN has announced for the first time that it will cease operation of a DSL network in one specific town, which means that DSL customers in that town will have to be migrated to fibre service before the end of 2012.

In some countries, such as Denmark, Norway and Sweden, the deployment of fibre networks has been driven largely by local and regional utilities, which leverage their existing infrastructure and customer relationships to provide broadband services over their networks. The incumbent, TeliaSonera, is deploying fibre to single dwelling units, and is facing competition from these local or regional networks. This demonstrates that alternative models to those of incumbent telecommunication and cable operators deploying nationwide infrastructure may be an option in some areas. Nevertheless, the Nordic experience is based on inherited models and these may not be readily replicated in other countries that lack such local or regional utilities.

In other countries new fibre deployments are underway or operators have expressed a preference to use new technologies for the "local loop". With very-high-speed digital subscriber line (VDSL) fibre is rolled out to a point near the customer's premises. The CEO of Deutsche Telekom announced, in September 2012, that they would extend VDSL coverage to cover 24 million households, twice as much as its current footprint. Vectoring technology potentially enables operators to double the rates achieved on VDSL, up to 100 Mbit/s download and 40 Mbit/s upload speeds, for supported distance. It consists of noise cancellation techniques that reduce interference between copper pairs within the same bundle of lines, and thus increase speeds. One of the factors significantly affecting the performance of DSL lines is "crosstalk". Vectoring seeks to reduce this, thereby enabling clear transmission of IP traffic.

There are different approaches in OECD countries to the regulatory requirements for new fibre networks, or other upgraded high-speed infrastructures. Historically, many countries have used a combination of tools such as functional separation and unbundling of copper local loops to provide more competitive outcomes, where there would otherwise be limited choice for consumers, and continue to do so. Some countries, particularly those investing in new national fibre networks, have adopted structural separation of wholesale and retail business to remove, by structural means, the incentive and ability of integrated operators, generally of monopoly infrastructure, to favour downstream operations over those of competitors that need to use the network. A further group of countries use a combination of tools such as functional separation and unbundling of copper local loops to provide more competitive outcomes, where there would otherwise be limited choice for consumers. The challenge, for these countries, is that the use of unbundling may prove difficult with new fibre networks. Moreover, if technologies such as vectoring are used in copper local loops connected to fibre, unbundling may also be problematic compared to traditional approaches.

While infrastructure competition is preferred this may not be possible in all locations or in all markets. In Korean cities and in Hong Kong, China, there are a substantial number of residential apartments that lend themselves to competing fibre networks connecting to the basements of buildings. As a result these locations have some of the most competitive outcomes for consumers in the world. In countries that have traditional cable television networks there can also be competitive pressure. In Switzerland, Swisscom is actively engaged in partnerships with utilities in different cities and places to deploy fibre networks, mainly as a competitive response against the growing success of cable operators.

Where there is insufficient infrastructure competition to drive new network investment, policy makers face the difficult choice of taking an approach such as in Australia and New Zealand, or finding ways to increase private investment to meet their policy objective or some middle path. Understandably, all operators say that they will not invest in upgraded fixed broadband networks unless they can fully benefit from returns on these investments. They and investors may have different views, however, about the nature of these returns: whether they resemble long-term and stable utility-like returns or the higher rates associated with greater risk. Today, different OECD countries are progressing with a variety of approaches which depend on factors such as the performance or otherwise realignment of current market settings.

For their part, most new entrants, including some mobile operators, contend that new fixed network facilities should also be regulated, as copper networks were, in order to avoid abuse of dominant positions. Nonetheless, in those areas where mandated wholesale access is the rule and a key driver for competition, such as in the European Union area, some say that the current framework does not provide sufficient incentives for investment in fibre-to-the-premises. In that respect, the European Commission has recently presented a draft recommendation on consistent non-discrimination obligations and costing methodologies, in order to promote competition and enhance the broadband investment environment.

Japan is one of the few OECD countries where fibre networks have been widely deployed. It is also noteworthy that fibre unbundling has been in place since 2001, and investment in fibre access has taken place since the late 90s. However, Japan's authorities have expressed concerns about the excessive market power of the incumbent for fibre-based services, as its market share is far higher for fibre than for DSL services (Figure 2.2). This has raised the question of whether change in the regulated access conditions, to these networks, should be undertaken.

Figure 2.2. **Broadband market shares for DSL (left) and FTTH (right) in Japan**

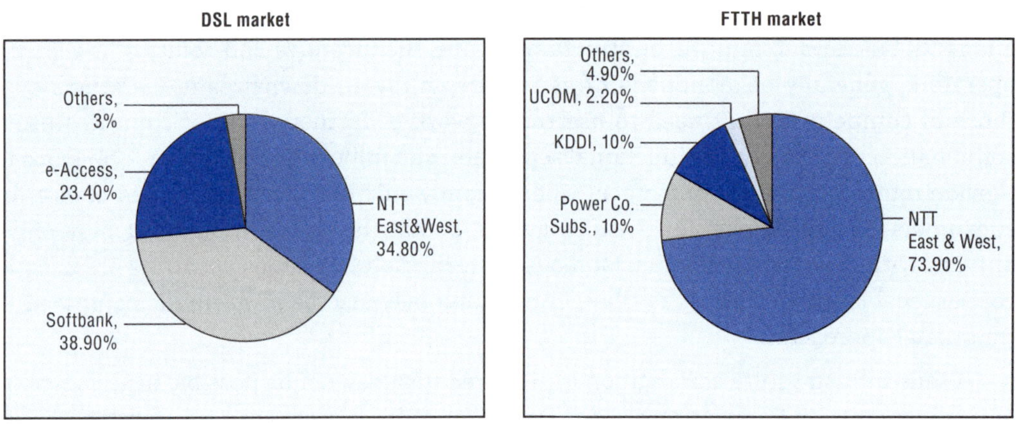

Source: Ministry of Internal Affairs and Communications (MIC), Government of Japan.

StatLink http://dx.doi.org/10.1787/888932798525

Mobile markets

Policy makers and regulators face numerous challenges in maintaining the competition that drives innovation and investment. In mobile markets the scarce nature of spectrum and the need to ensure adequate supply is foremost among these challenges. Availability of spectrum enables authorities to ensure competitive supply and enables operators to pioneer new business models and new services. In France and Israel, for example, new facilities-based mobile entrants entered these markets in 2012, an occurrence less common in OECD countries in recent years particularly after penetration rates rose above 100 mobile telephones per 100 inhabitants and the proliferation of MVNOs. In both countries the market outcomes were more competitive offers for users and greater innovation, just as occurred in New Zealand following the introduction of a new operator in 2009.

In other OECD countries the ability to introduce new operators where there is deemed to be insufficient competition is frequently limited by the availability of spectrum. In the Netherlands a spectrum auction enabled a new entrant (Tele2) to enter the market, adding to the existing three operators. It might be noted that the push to introduce "network neutrality" legislation followed a move by the largest Dutch mobile operator to react to perceived loss of market share to OTT providers of SMS services in that country. Clearly, a key factor for policy makers and regulators is the number of facilities-based providers. At the same time, spectrum for 4G services has enabled incumbents like Swisscom to trial new business models that are potentially beneficial for both infrastructure providers and OTT service providers.

The number of players present in mobile markets in OECD countries has remained relatively stable in recent years (Table 2.3). Existing players show a willingness to consolidate, which brings significant benefits in terms of economies of scale and access to a competitor's spectrum. Mergers among small players may also make them more effective competitors to larger players. Nevertheless, the number of operators is a key factor in ensuring a competitive market and, from the consumers' perspective, increased consolidation may result in lower competition. Few would argue, for example, that the market in the Netherlands is more competitive with three facilities-based operators than in earlier years when it had five. Meanwhile, increased competition is readily evident in countries such as France and Israel that have recently introduced new operators.

Such considerations have been uppermost in the minds of regulatory authorities when considering mergers or the introduction of new entrants. While circumstances vary in each OECD country, it is noteworthy that several large countries arrived at four nationwide facilities-based providers from different starting points. These are France (an increase from three to four), the United Kingdom (a decrease from five to four) and the United States (maintained four).

Starting with the United Kingdom, arguably one of the most competitive mobile markets in the OECD area, authorities allowed Orange (France Télécom) and T-Mobile (Deutsche Telekom) to form the joint venture Everything Everywhere. This represented a merger between the third and fourth largest operators in the United Kingdom by market share. The two operators have subsequently launched the first LTE network in the United Kingdom. In France, a fourth operator was introduced in 2012, with evident benefits for consumers, discussed elsewhere in this *Communications Outlook*.

In the United States, in 2012, regulatory authorities blocked a proposed purchase of the fourth largest operator (T-Mobile) by the second largest operator (AT&T). Regulators determined it would seriously harm competition and would not be in the public interest. One of the main arguments used by AT&T was that the merger would give it access to more spectrum, while critics noted that much of the innovation and choice in this market had in recent years come from smaller players. In October 2012, Japan's Softbank announced that it would bid for Sprint Nextel, the third largest operator in the United States, a move expected to strengthen competition.

There have been a number of changes in smaller markets in OECD countries. In Australia, the third and fourth largest operators merged in order to compete more effectively with the two largest networks. In New Zealand, the previous duopoly was disrupted by the entry of the long-awaited third operator providing immediate benefits for consumers. In 2012, Israel also experienced the same results with the entry of greater competition in its market, following the introduction of a new facilities-based operator (Golan Telecom) and the entry of a cable company which uses the facilities of a previous entrant and leased facilities (Hot Mobile). While both operators offered significantly lower prices, Golan Telecom attracted attention by following a similar pricing structure to Iliad/Free in France.

In Chile and Spain, recent spectrum auctions have resulted in new mobile operators. They have mostly been already active in fixed communications. In Chile, the largest cable operator, VTR, started providing mobile services in 2012. In both countries it is expected that the ability of the fixed operators to gain spectrum will enable them to launch convergent quadruple-play offers. In all countries that have recently introduced new operators, or are about to do so, mobile number portability has proven essential and remains an essential precursor for ensuring competition (Table 2.4).

As discussed in Chapters 3 and 4, the transport of data is now the major source of growth for network operators. While there are significant new opportunities in new services, such as mobile payments, essentially they involve the transport of data in association with partners such as credit companies. Arguably the exploitation of "big data" is one of the largest opportunities for mobile operators, but this is also derived from the transport and location of data. Few expect there to be growth in traditional services such as telephony or SMS as measured by their share of revenue. Revenues corresponding to data services are growing at double-digit rates in most OECD countries, with some exceptions, like Japan, which experienced earlier growth and where data revenues represent the highest share of mobile revenue in the OECD.

Appstores and the emergence of new business models

The ecosystems and business models around broadband networks are extremely dynamic. The smartphone ecosystem is unrecognisable from little more than five years ago. Apple's App Store was launched in just mid-2008 and was quickly followed by many others. Today, an abundance of service providers from Netflix to Spotify and Skype provide OTT services. At the time of the previous *Communications Outlook* the number of mobile operators that permitted Skype over their 3G networks was extremely limited. Hutchison's 3 in Europe and Verizon in the United States were notable exceptions in welcoming Skype. Just two years later an increasing number of operators are adapting tariff options to take into account the availability of VoIP services. They offer flat-rate voice options or charges for data usage metred by speed or downloads and treat VoiP no differently from any other use made by their customers.

App stores have played a critical role in these developments, making OTT services widely available around the world, including in countries where operators have not permitted use of some applications on their 3G networks. Coupled with the wider availability of Wi-Fi, the mere accessibility of these applications and over-the-top use makes the issue of why such services would be blocked a moot one for all stakeholders. The onset of 4G simply increases focus on this question.

Laptop computers increasingly offer 3G or 4G slots for SIM cards. Devices such as Google and Samsung's Chromebook not only include 3G connectivity, but also shift some of the traditional capabilities found in computers to the Cloud and provide these services over broadband networks. Commentators noted that the Chromebook, launched in October 2012, used a chip previously only employed in mobile telephones as opposed to one developed for a laptop. At the same time, they include apps that provide the traditional features of mobile phones. For consumers, services are provided in a seamless manner and may have no direct relationship to the infrastructure provider.

Noteworthy are the different business models being attached to attract consumers. One of the first service providers to bundle connectivity with a device was Amazon for its

Kindle. The so-called Whispersync capability enables users (with Kindles that incorporate a SIM card) to access the Kindle store and download e-books without a direct subscription with a mobile provider. Amazon recoups the cost for this connectivity through the sale of products and services and in turn pays network providers such as AT&T or Vodafone. In turn these providers pay other networks if users roam.

In the United States, Samsung's Chromebook includes two years of access to Verizon's 3G network for up to 100 MB per month. While Chromebook users can purchase additional data from Verizon, for the most part they are expected to connect via Wi-Fi access to fixed networks (some of the devices lack an Ethernet connection). The difference in price between the Wi-Fi-only and 3G-enabled Chromebook is USD 80 (USD 249 versus USD 329). In other words, a vendor such as Samsung is including 100 MB of data per month for USD 3.33 per month over the two-year period.

Taxation issues

During the monopoly era, when public ownership was common, communication services represented a significant source of revenue for some governments. In those countries, this frequently resulted in lower re-investment in network expansion and, as a result, low penetration rates. In countries with privately owned monopolies, penetration rates were much higher, at least in OECD countries, but there was a tendency for over investment (so called gold plating) based on the incentives provided under rate of return regulation. The reforms that separated telecommunication operators and regulation from government departments, as well as those that introduced privatisation and competition, make today's market very different. For one thing, telecommunication revenues, in many cases underpinned by artificially high prices for some services (e.g. international calls), were replaced by overall growth of services, private investment, and more efficient management and operation of networks. Today most OECD countries take the wide availability of services for granted though the requirements of users are ever changing. Revenue for general public expenditure is raised through taxation of communication services. Most countries operate value added taxes (VAT) or goods and services taxes (GST) on the final household consumption of most goods and services, including telecommunication services.

In recent years, a relatively resilient telecommunication industry during the global financial crisis has once again become a potential source for additional revenue in some countries. Spectrum auctions are still a significant source of revenue, though the prices paid in auctions are much lower than during the years of the "dotcom bubble". The most important feature and benefit that auctions brought was the ability to introduce a transparent and explainable tool to promote market entry and, therefore, competition to benefit consumers. Auctions were not or at least should not have been about maximising revenue if that meant artificially distorting the level of competition. If they are about maximising revenue they simply become taxation by other means because virtually everyone uses mobile communications.

In some OECD countries, such as Hungary, concerns have been raised regarding additional taxes on telecommunication services, over and above the general taxes applied across an economy, which may hinder the development of the sector. In Mexico, for example, some stakeholders have requested the withdrawal of a special tax of 3% on telecommunications services (IEPS), as it may harm the development of the sector and impose an additional burden on consumers, including those that do not have service. In other countries, fees have been levied on telecommunication operators for funding public television broadcasters, such as in France and Spain. Policy makers say that the introduction of specific taxes in areas such as mobile or fixed communication networks brings them into line with ownership of televisions and they are therefore technologically neutral. These cases lie now before the European Court of Justice as they may not be in line with European law. Nevertheless, some content, such as e-books or online newspapers, is in principle subject to the standard VAT rate in the European Union whereas their physical equivalents may benefit from preferential rates. Moreover, given the additional features of e-books and newspapers (e.g. text to speech used by some with impaired vision or reading challenges), differences between taxation policies can seem inequitable and contrary to other policy objectives.

Questions over taxation in relation to communication services are, of course, a global phenomenon. In many developing countries there is also a temptation to apply additional taxes on telecommunication services. In these countries, it may be challenging to collect taxes through other means and applying regular taxation is entirely appropriate. Conversely, there is a recent trend in some countries to discriminate between international traffic termination and domestic traffic termination for the purposes of applying taxation. This results in double taxation for international traffic, as taxes in one country get applied on taxes in another country. There is growing evidence that not only are these policies ineffective, but that they also result in an overall welfare loss for consumers in those countries. In the area of VAT/GST, the OECD is developing International VAT/GST Guidelines to minimize risks of double taxation and unintentional non-taxation.

While some special industry levies or fees may be justified for specific purposes, such as funding the sector regulator or contributing to universal service goals, additional tax burdens on the telecommunication sector may harm both consumers and the industry itself, which is a key driver for economic and social development in ways that will result in larger revenue from general taxation and lower and more efficient public expenditures.

Foreign direct investment restrictions and state ownership

Legal limitations on foreign direct investment in telecommunication operators continue to exist in some countries (Table 2.5). In 2012, Canada, Israel, Korea and Mexico maintained certain restrictions on foreign investment in telecommunication operators. Other countries, such as Greece, Japan, New Zealand or Switzerland, maintain limitations on the ownership of the incumbent operator. Some progress has been made in certain countries with a view to removing these barriers. For example, in March 2012, the Canadian government announced that it would lift the current limit for foreign investment in small telecommunication operators, which was previously set at 47.6%. This change affects local telecommunication companies that have less than 10% of market share. In Mexico, the new telecommunication reform will remove previous limits (49%) to foreign ownership of fixed-line and satellite telecommunication operators. In Korea, following the enforcement of a free trade agreement (FTA) with the United States and the European Union, the Korean government will allow a government or a person from those regions to participate in indirect acquisitions of 100% of facilities-based suppliers of public telecommunication services, on the condition that the KCC examines whether the acquisition harms the public interest, as specified by a Presidential decree.

A few OECD governments own a share in domestic incumbent operators (Table 2.6). The governments of Belgium, Luxembourg, Norway and Switzerland own majority stakes in their domestic incumbent operators, while Austria, France, Germany, Greece, Japan, Portugal, the Slovak Republic and Turkey own minority stakes (sometimes as high as 49% as in the Slovak Republic). The governments of Finland and Sweden jointly hold 49% of TeliaSonera, the respective incumbent operators in these two countries. It can be further noted that the creation of government-owned companies to build and manage structurally separated national broadband networks reintroduces public ownership into that segment of the market in these countries. Examples inside the OECD area include Australia and New Zealand. In Australia, legislation provides for the eventual sale of the network company subject to a clear and robust inquiry process following completion of the network build.

Regulatory trends

Open access for fixed broadband networks

Competition in fixed networks continues to be strongly influenced by regulatory oversight in most OECD countries. A recent OECD report, *Broadband Networks and Open Access*, pointed out that the arrangements surrounding wholesale access to fixed networks (e.g. LLU, bitstream, line-sharing) contribute significantly to shaping the industry's dynamics by promoting service-based competition (Tables 2.7 and 2.8).

Among the countries to more recently join the OECD (Chile, Estonia, Israel and Slovenia), local loop unbundling at mandated, regulated prices, is not available in Chile and Israel. The Israeli authorities are in the process of developing framework conditions for mandated access to the incumbent's networks. They are also promoting the entry of the Israeli Electric Company (IEC) into the market, as a carriers' carrier, setting up a subsidiary that will have access to IEC's infrastructure and which aims to roll out a FTTH network within seven years.

Traffic prioritisation (network neutrality)

In the past two years, the debate around traffic prioritisation or network neutrality seems to have expanded from North America to other parts of the world, such as Asia, Europe and South America. There is worldwide interest in the issue and some countries, such as Chile, the Netherlands and Slovenia, have enacted network neutrality rules in legislation. These developments are generally not new but the rapid evolution of the ecosystems, particularly around mobile services, throws up some features worth noting. It is also true that an increasing number of countries are launching public consultations on network neutrality or have developed guidelines, stemming from these processes (Table 2.9).

The main issues at question are not necessarily novel – whether ISPs are permitted by regulators to exclude, slow down or prioritise certain types of traffic for OTT services that compete with their own services, in an absence of sufficient overall market competition. Examples have included restrictions of VoIP services through to apps that enable tethering. As noted, in a growing number of countries competitive forces have encouraged operators to develop tariff options that do not exclude these capabilities and services demanded by customers. All stakeholders, in large part, accept that ISPs should be able to conduct reasonable network traffic management to provide service commensurate with what consumers pay for and expect. While the Internet is a best-effort network these practices should be transparent where they involve rival OTT services.

Following a Europe-wide consultation, the European Commission issued a statement, in November 2010, reporting that a near consensus had been achieved in favour of the importance of preserving the openness of the Internet. The European Commission further noted that the consultation had not suggested a need for additional regulation, over and above a new regulatory framework adopted in 2009. It added the caveat that additional guidance may be needed in the future. It can be noted that the new European rules allow national regulators to set minimum quality of service (QoS) requirements in order to promote network neutrality, as well as additional requirements for information that needs to be provided in contracts, such as traffic management techniques.

The regulatory authority in the United Kingdom, Ofcom, issued a statement in November 2011 following a multi-stakeholder consultation. It raised the concern that "best-effort" Internet access could not coexist with managed services unless sufficient capacity was left available for "best-effort" services. Under these circumstances, Ofcom said it would be willing to act to guarantee a minimum quality of service, should such a case arise. That being said, Ofcom said it believed there is generally sufficient competition in the market in the United Kingdom to discourage discriminatory blocking or unreasonable network management practices. Ofcom's guidance also set out a list of necessary elements in terms of technical information on traffic management practices and transparency, surrounding services blocked or discriminated against.

ARCEP, the regulator in France, conducted a consultation on network neutrality and published its conclusions in September 2010. The regulator set out a list of 10 proposals that it said were aimed at ensuring a dynamic and long-lasting balance in the ecosystem. Among other principles, ARCEP said there was a need to promote increased transparency for end-users, to monitor ISPs' traffic management practices, and to conduct regular evaluations of quality of service. According to these rules, service blocking (e.g. VoIP, P2P) should no longer be undertaken on mobile networks.

In September 2012, following its legal obligations, ARCEP responded to a request of the French Parliament on the status of network neutrality. ARCEP noted a decrease in discriminatory traffic management practices, which it attributes to increasing competition, and especially on mobile networks, where it said they had been more frequent in the past. ARCEP is also working towards the development of a framework of QoS indicators for fixed Internet access. These are common in a number of other OECD countries (Table 2.10). Moreover, ARCEP has recently requested all electronic communications providers registered in France to submit information, on a biannual basis, regarding the 20 most important interconnection agreements they have engaged in. With this decision, ARCEP intends to monitor Internet interconnection markets more closely. This decision took into account remarks from stakeholders, notably those arguing that the number of agreements is so high that information filing can represent an unreasonable burden on them, especially given the overwhelming number of such agreements closed on a handshake basis. Some also said that reporting on some agreements may discourage them from entering into them.

In December 2011, KCC (the Korea Communications Commission) announced its "Guidelines for Network Neutrality and Internet Traffic Management". These contain basic principles on network neutrality and traffic management practices. The guidelines include provisions on transparency in traffic management and a rejection of unreasonable discrimination or blocking, while recognising the need for reasonable network practices.

In France, in January 2013, an incident arose when Free, the second largest broadband network, enabled a default setting in its most recent set-top box that blocked some online advertising associated with OTT services. As in Korea, the authorities in France are examining what further action, if any, may be required.

Other countries have also undertaken specific initiatives surrounding network neutrality, either by enacting them by law or by adopting principles and guidelines. Examples include Canada (2009), Chile (2010), the United States (2010), Norway (2009), Luxembourg (2011), Japan (revised in 2010) and Italy (only public consultation, 2011). Most other OECD countries are monitoring developments. In the case of Chile, one of the rationales for the development of the legislation was to enable the possible development of OTT service providers, in addition to the factors mentioned above. In December 2012, the Slovenian Parliament passed a new electronic communications law, which also enforces network neutrality principles.

Internet interconnection models

Data on the increasing amount of traffic carried over the Internet are presented in Chapter 5. All entities with a connection to the Internet pay for that connection. If they are service providers they build their own networks and either peer or pay transit to other networks to reach the rest of the Internet. Some network access providers believe that other networks should contribute to the cost of carrying or terminating traffic to their customers. Others believe that it is the customers of these networks, including those that request this traffic, that should pay. Such discussions form part of the natural to and fro of commercial negotiations under the Internet's model for traffic exchange.

What everyone accepts is that video traffic has increased and may constitute the largest single type of Internet traffic. Nowadays, video traffic provided by entities such as YouTube, Dailymotion and Netflix represents a significant and growing share of Internet traffic in OECD countries. This has triggered reactions from some network operators who say that entities that provide video content should pay them to terminate that traffic. These companies say that OTT providers of services should contribute to financing their communication network infrastructure, as otherwise the necessary investments will not take place. Operators taking this view say that they should be able to provide higher levels of quality of service to OTT providers willing to pay for those services with others receiving best effort services.

In 2012, this issue came to a head when the association of European Telecommunications Network Operators (ETNO) – whose largest members include Orange, Deutsche Telekom, Telefónica and Telecom Italia – put forward a proposal to amend Article 3 of the International Telecommunications Regulations with language that would entail OTT service providers paying them to terminate traffic. ETNO stated that this would be consistent with the principle "sending party network pays" used in some telecommunication services for the Internet, and would ensure the sustainability of the Internet ecosystem and allow all stakeholders to invest and innovate.

Critics of the ETNO proposal, including other network operators in Europe, as well as North America and Asia, opposed the ETNO proposal stating that the existing Internet model for traffic exchange provides the ability and flexibility for commercial negotiations to take place. Over-the-top service providers were, not surprisingly, also critical of the proposal. They said that customers of network operators request the data under discussion and should therefore be responsible for payment. They also argued that they make significant investment to carry traffic to local exchange points or pay content delivery networks to undertake that task for them. Finally, critics from the Internet technical community said that quality of service is not possible over the best effort Internet, and can be guaranteed only on intranets.

The OECD report *Internet Traffic Exchange: Market Developments and Policy Challenges* argues that the current model for Internet traffic exchange has worked extremely well. It has allowed substantial price reductions since the commercialisation of the Internet, and has underpinned dramatic increases in the number of Internet users. For these reasons a number of policy makers have spoken out against the ETNO proposal, which they say would require an enforcement mechanism overseen by regulatory authorities. They also believe that the ETNO model would not be practical, particularly at the international level, and may conflict with their guidance on network neutrality.

Connected televisions and convergence

Mobile telephones, tablets, laptops and other devices are increasingly using video services over the Internet. For their part, device manufacturers such as LG, Samsung and Sony are making televisions capable of connecting to the Internet. This can enhance users' experience in watching video services such as premium video content, short clips or video on demand (VoD). By way of example, Samsung and LG have launched app stores for connected televisions, which resemble the stores serving users of smartphones. The apps can include well-known OTT services from other providers, depending on the country, such as Netflix, Hulu or the BBC's iPlayer. Digital televisions can also be connected to the Internet by a large number of devices such as Apple TV and Western Digital's Media Player, or may incorporate features such as Google TV.

A preliminary discussion of some of the issues surrounding connected televisions can be found in Chapter 6. They are complex and as a result policy makers and regulatory authorities around the world are closely considering their approach in these areas. Some issues resemble closely those between network operators and OTT service providers. Others relate to broadcasting regulation and the objectives for public policy in this area – from the protection of children online to traditional concerns including cultural diversity and a level playing field for competition if regulation were imposed on some providers and not others.

In this area, distinct regulatory frameworks applied to communication and broadcasting services may result in conflicts arising from, on the one hand, outmoded obligations being imposed on innovative services thus limiting their development and, on the other hand, increased difficulties in applying proportionate regulations to achieve other public policy objectives. The emergence of connected televisions will likely provide further impetus to the trend towards "converged regulators" for all communication services, and an increase in harmonisation of legal frameworks applied for telecommunication and broadcasting, which are in many cases managed by different regulatory agencies.

Role and structure of regulators

Some OECD countries are undertaking organisational changes in regulatory agencies for telecommunication and broadcasting services to reflect convergence. This brings them into line with countries that have regulatory authorities dealing with all communication services. In addition, some countries are merging one or more communication regulators with the general competition authority. In 2013, in the Netherlands, the regulator OPTA will be merged with the competition authority. Meanwhile, in Spain, a draft bill put forward by the government is being discussed in Parliament and proposes to merge all sector regulators together with the competition authority into a single body (the National Commission for Competition and Markets). Reducing public expenditure is one of the primary reasons behind this move towards the consolidation of regulatory bodies.

Merging the roles of *ex-ante* communication regulators and competition authorities was envisaged as a positive outcome of telecommunication and cable liberalisation and of competitive markets, although most OECD countries still acknowledge the need for a sector-specific regulator. New Zealand relied for some time on competition law only, but later recognised the need for a more specialised agency, which would provide closer monitoring of the sector and the required expertise.

Some countries like the United Kingdom or the United States provide their sector regulator with some antitrust and merger review powers, at least for the communications and media sectors, and Germany's Bundesnetzagentur has responsibility over many different network industries (e.g. energy, transport, telecommunications). Nevertheless, the experience of a macro-regulator with *ex-post* and *ex-ante* responsibility over all sectors is rare in OECD countries.

The world goes wireless

Mobile broadband services

Just after the turn of the century a number of OECD countries auctioned spectrum for, so-called, 3G services. Some of the large sums paid reflected an expectation that mobile Internet access would become very popular. It took some years, however, before there was significant demand for mobile Internet access among users. At the time, most users had feature phones with additional limited capabilities on some devices such as email on Blackberry mobile telephones. Some of the first smartphones, such as the original iPhone released in 2007, were launched using 2G or pre-3G devices, such as General Packet Radio Service (GPRS) and Enhanced Data rates for GSM Evolution (EDGE). Arguably, therefore, the initial attraction was the new capabilities of smartphones rather than bandwidth.

The dramatically enhanced capabilities of smartphones were matched by the innovation of the "app" market place. The agreement between AT&T and Apple, around the introduction of the iPhone, separated the sale and management of some applications from provision by the infrastructure provider. Undertaken for commercial rather than regulatory reasons this has stimulated breathtaking innovation to the extent that five years after its introduction more than half the population of the United States had a smartphone. The same model has been adopted by other ecosystems (i.e. mobile operators and providers of other operating systems such as Android, Blackberry and Windows Phone). In turn, the take up of smartphones stimulated ever-increasing use of 3G and has created a driver for the introduction of "4G".

The evolution of mobile communication technology has enabled the seamless provision of wireless broadband services, facilitated by the extended use of smartphones. Adding mobility has substantially expanded the range of possible applications that can be provided over a broadband connection. Services such as e-health applications, mobile instant messaging or geo-location services need an efficient mobile data connection. Smartphone uptake has been very rapid in OECD countries and more affordable smartphones are being sold, with considerable expansion in the potential customer base. The prices for smartphones and tablets are falling rapidly, opening them to a wider market, including in developing countries.

Wireless broadband overtook the number of fixed broadband subscriptions in 2008. In June 2012, wireless broadband accounted for 698.7 million subscriptions and is growing at two-digit rates (Figure 2.3). Fixed broadband subscriptions, while still growing, are not expected to match wireless broadband growth. That being said, the majority of these wireless devices will remain connected by Wi-Fi at work places and home residences.

Figure 2.3. **Wireless and fixed broadband subscriptions in OECD countries**

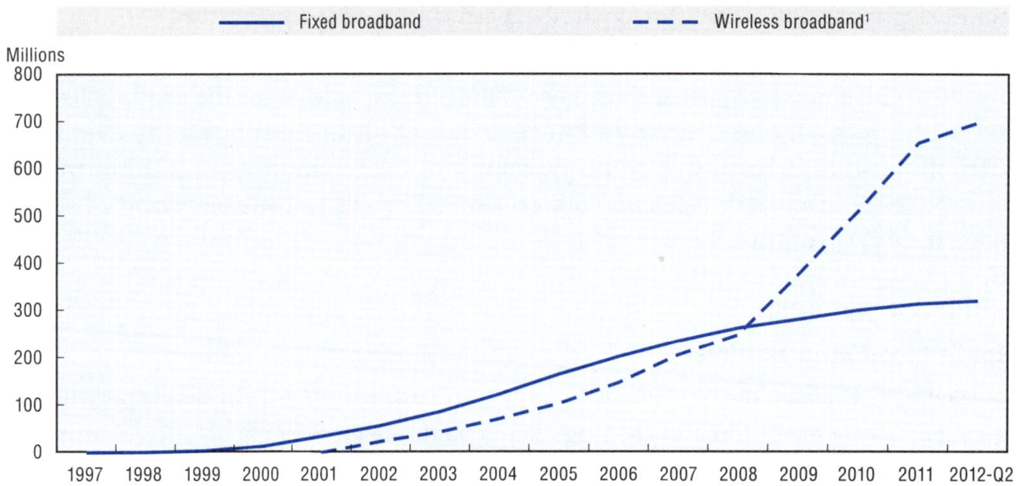

1. Data for Wireless broadband from 2001 to 2007 are estimates.

StatLink http://dx.doi.org/10.1787/888932798544

Mobile termination rates

Reductions in mobile termination rates (MTRs), witnessed in previous years in OECD countries, have continued. Implementation of the 2009 Recommendation of the European Commission, as well increased regulatory attention to MTRs, have contributed to this trend. In that decision the European Commission said that mobile termination rates should be based on the real costs of efficient operators. These costs, they determined, should be calculated on the basis of forward-looking long-run incremental costs (LRIC) whereby only the efficiently incurred costs in providing that service are included. Recent OECD work has also recognised that innovation and greater flexibility in business models is more likely if termination rates are set low or at zero. The current level of MTRs in the OECD area, as of October 2012, and the downward trend since 2004 are evident (Figures 2.4 and 2.5). Examination of the rates for October 2012 shows that there is scope for further reductions in MTRs (Table 2.11) as the current OECD average is just above USD 0.04.

Figure 2.4. **MTRs in OECD countries, USD, 25 October 2012**

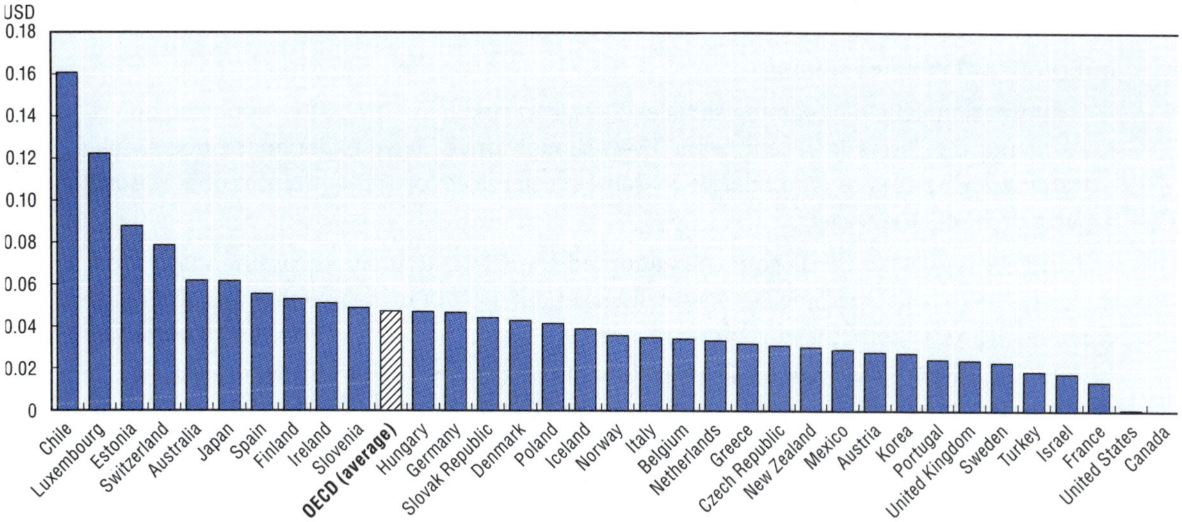

StatLink http://dx.doi.org/10.1787/888932798563

Figure 2.5. **Average (blue) and maximum (red) MTR in OECD countries**

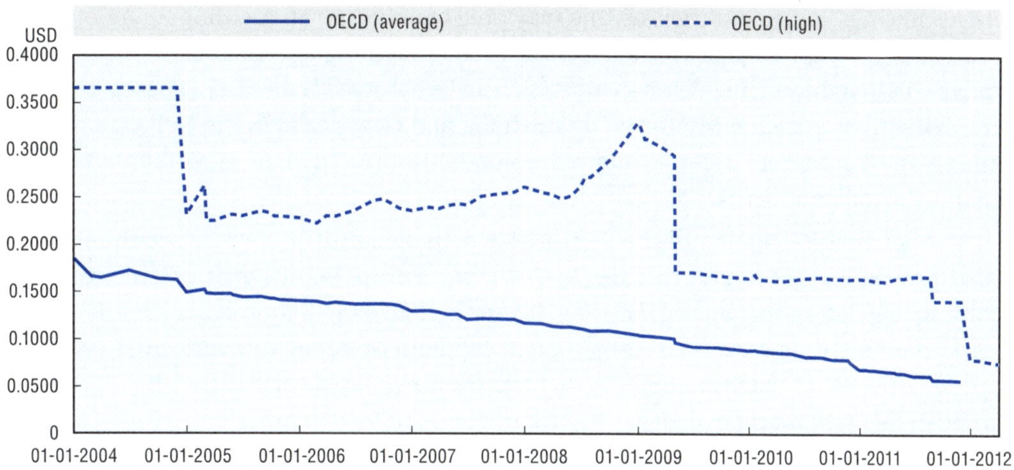

StatLink http://dx.doi.org/10.1787/888932798582

The decline in MTRs has already proven beneficial for competition and innovation in some markets. The entry of Free Mobile (Iliad) has considerably altered the market in France. The company gained 3.6 million customers in the first six months of operation – a much faster pace than new entrants in recent years. One of the components of Free's offer is unlimited calls to fixed and mobile lines in France, plus some international destinations. In Israel, Golan Telecom has adopted the same model. Undoubtedly, such competitive market entry would not have been possible without the substantial reduction in mobile termination rates in recent years.

Policy makers and regulators are becoming increasingly aware of the importance of empowering consumers with tools that can facilitate switching their service provider in a seamless and inexpensive manner, such as number portability. The amendment of a European Union Directive, enacted in 2009, that provides consumers with the right to switch mobile providers within one working day, has proven critical to market competitiveness in that area.

International mobile roaming

International mobile roaming services have attracted increasing attention from policy makers and regulators in recent years. Even though prices have experienced decreases, extremely high prices remain relative to domestic services for some international routes, especially for data services.

In February 2012, OECD countries adopted the OECD Council Recommendation on International Mobile Roaming Services, which set out a list of measures that countries can consider in order to achieve a more reasonable level of roaming prices and improve competition dynamics. Some of the measures proposed in the Recommendation are: to promote awareness of roaming services, their prices and of possible substitute services, for example, on the risk of uncontrolled data roaming connections or on available financial limits; to facilitate trans-national networks and alliances; to improve transparency in wholesale roaming markets; and, in the event that other measure do not work, to implement wholesale and/or retail price regulation.

National regulators lack the jurisdiction to regulate wholesale prices charged by foreign operators and, sometimes, have little motivation to do so regarding wholesale charges in their home countries, as this may only benefit foreign roaming roamers in that country. As a result, a surge of international mobile roaming agreements, or initiatives to explore whether these agreements are possible, can be observed (Table 2.12). These include the joint market investigation conducted by Australia and New Zealand, the Gulf Co-operation Council and South-East Asian Economic Co-operation (ASEAN), or agreements between Finland, Poland and Russia.

The roaming regulation of the European Union constitutes the most comprehensive set of provisions developed to date. In 2012, the European authorities amended this Roaming Regulation for the second time. It now includes regulation on voice and SMS as well as data roaming services, setting a price cap both on retail and wholesale prices. As such, in 2012, retail data roaming prices have been regulated for the first time. The new European regulation acknowledges the lack of competitive dynamics for roaming services and that there were little incentives for market players to launch competitive offers. This is reflected in the results with actual prices being set at, or very close to, price cap levels. In that regard, the new regulation included some structural measures, such as allowing MVNOs to benefit from regulated wholesale prices. In addition, the European Union took the decision to implement unbundling of roaming services from the mobile bundle, to take place in 2014. It also took the decision to allow a higher mark-up, in relative terms, between the regulated retail and wholesale prices.

In the changes made by the European Union it is noteworthy that provisions for separate sale of roaming services include obligations to allow MNOs in the visited countries to offer data services, based on the local break-out (LBO) mechanism included in 3GPP specifications for local data offloading, directly to visiting European customers. This means that roamers will have the possibility to choose a local provider of data services in the visited country. These measures are expected to bring more competition to roaming markets within the European Union area, and the experience will be followed closely by other parts of the OECD area.

Spectrum policy: Main trends

Spectrum resources, previously used for broadcasting, are currently being re-allocated for use in mobile communications and, especially, for wireless broadband services. This "digital dividend" has been examined by the International Telecommunications Union (ITU) in an attempt to harmonise the use of frequencies all over the world. In that respect, in Europe, the European Conference for Postal and Telecommunications Administrations (CEPT) has already attributed the 790-862 MHz band (800 MHz) band to wireless communication services. There are another two alternative frequency arrangements, for the use of mobile communications, corresponding to the APT/CITEL band plan and the United States band plan (Figure 2.6).

Figure 2.6. **Band plans for the digital dividend**

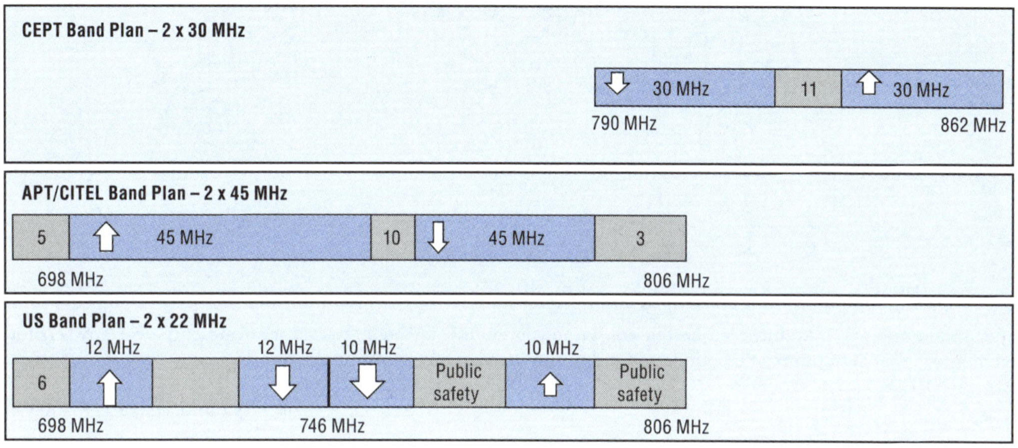

Source: GSMA.

StatLink http://dx.doi.org/10.1787/888932798601

There is significant interest in making more spectrum available for mobile use in the UHF bands used for broadcasting, although this has to be balanced with continued demand for spectrum for broadcasting. Key decisions were taken at the ITU World Radiocommunications Conference (WRC-12), which took place in Geneva in early 2012, in relation to the 700 MHz band and discussions are ongoing in preparation for WRC-15. The availability of more spectrum for mobile in this frequency range would be beneficial for consumers, recognising the need for balance with broadcasting requirements, and increased harmonisation would be beneficial in terms of economies of scale for network devices and terminal equipment.

Notwithstanding this pending harmonisation work, LTE networks are being deployed and, in many cases, commercial services are already operational. According to the Global mobile Suppliers Association (GSA), there are to date 145 commercial networks operating in 66 countries, and 381 operators are investing in LTE networks in 114 countries. The state of deployment of LTE networks, including the starting date of commercial services for OECD countries can be shown (Figure 2.7). As of December 2012, 25 OECD countries have commercial LTE networks in service. Many more are expected to provide these services soon, depending on the availability of spectrum. In the United Kingdom, for example, Everything Everywhere (EE) has recently been allowed to "refarm" its 1 800 MHz spectrum to provide LTE services, given the availability of terminal devices that support LTE technology on that spectrum band.

Figure 2.7. **Launch date of commercial LTE services, OECD countries**

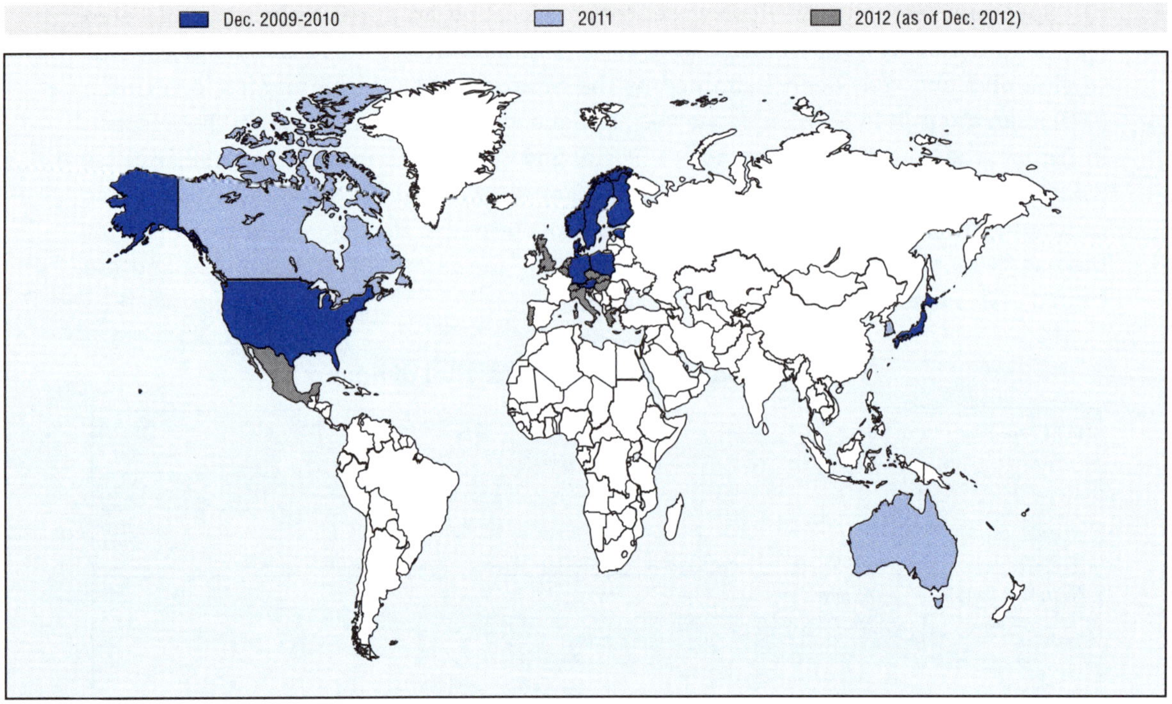

Note: This document and any map included herein are without prejudice to the status of or sovereignty over any territory, to the delimitation of international frontiers and boundaries and to the name of any territory, city or area.
Source: GSMA and OECD.

StatLink http://dx.doi.org/10.1787/888932798620

A high number of spectrum tenders are being conducted in OECD countries. The aim is for the digital dividend frequencies and additional bands to be made available for mobile communications, with wireless broadband services at the forefront. A list of processes to award spectrum that have been initiated, completed or are in the course of being finalised can be shown (Table 2.13). These tenders do not only affect bands within the digital dividend, but also other bands that will be used for wireless communications, such as the 2.6 GHz band in Europe (2.5 GHz band in other regions), the 850 MHz, 1.9 GHz and 1.7/2.1 GHz bands in Australia and North America, or the 900 MHz band. Most countries have also undertaken refarming processes, thereby confirming their support for the technology neutrality principle and liberalising the adoption of different technologies for the bands involved (2G, 3G or 4G).

Between 2010 and 2012, virtually all OECD countries launched spectrum tenders. Austria, Germany and Mexico completed their auctions in 2010, while Belgium, Estonia, France, Greece, Hungary, Italy, Portugal, Poland, Spain and Sweden conducted spectrum auctions in 2011. Chile, Denmark and Japan, Switzerland tendered spectrum in 2012. Finally, Australia, Canada, Czech Republic, Finland, New Zealand, Norway, Slovak Republic and the United Kingdom are in different stages of the process of tendering spectrum resources, either at the assessment phase or at various milestones throughout the design and execution process.

It should be noted, however, that in many of these cases the bands will not be fit for use until their current licensees are migrated to other bands or somehow "refarm" the frequencies currently assigned to them, which may take some time. For example, even though the Spanish spectrum auctions were completed in 2011, the 800 MHz band will not be ready for use for wireless communications services until 2014-15.

Spectrum trading is widely perceived as a useful mechanism that can allow more efficient use of this scarce resource. Although trading is increasingly permitted, market mechanisms have not always been used, as some entities prefer not to sell or lease unused spectrum resources. This is possibly because it is perceived as a strategic asset or because there was previously no incentive to do so. Some recent examples show this may be changing. In August 2012, Verizon completed a USD 3.9 billion purchase from several large cable companies of unused frequencies. The FCC and the Department of Justice (DoJ) cleared the transaction after substantial amendments to the initial proposal, aimed at preserving competition in broadband markets.

The convergence towards LTE and the importance of compatibility can be highlighted. Alongside traditional handset/operating systems manufacturers, new players such as Amazon, Apple, Google and Microsoft want to be able to take advantage of economies of scale and produce products demanded by global markets. Accordingly, if operators cannot support the most popular handsets demanded by customers they will risk losing market share to competitors. There may, of course, be other factors at work in different countries. The fees paid for spectrum licenses for 4G have been much lower than those paid during the years of the "dot-com bubble". In Germany, for example, operators bid just over 10 times less for 4G than 3G spectrum licenses. Significantly, governments have not moved away from holding auctions, as a transparent tool for allocating spectrum. Rather, many believe that the market there has resulted in a more rational assessment of the value of spectrum. Ironically, demand for bandwidth is far more evident now than was the case for 3G auctions.

Whether lower spectrum fees have led to earlier investment, as argued by many industry players, is undoubtedly a consideration as they take decisions based on traditional risk and reward parameters. High license fees are likely most associated with risk, but so too can be the dangers of not proceeding in a competitive market. In addition the potential for rewards are sometimes undervalued.

Developments in selected countries

Brazil

Brazil's telecommunications industry structure is partly a result of the liberalisation policy, which created three regional local telephone providers plus one long-distance operator. As a result, international long-distance services are dominated by Embratel (now owned by America Movil), while most national long-distance services are provided by TIM, Embratel and other carriers like Telefonica and Oi.

Brazil's mobile telephony penetration is in line with OECD countries (119.2% as of end-2011) and the mobile market is dominated by four operators, most of which are owned by foreign companies: Vivo (Telefonica), Tim (Telecom Italia), Claro (América Móvil) and Oi (participated in by Portugal Telecom), all with market shares between 25% and 30%, except for Oi (19%). Cable television services in Brazil had long been underdeveloped, largely due to stringent licensing rules and a ban on telephone providers from providing cable television services, which constrained their development. Anatel, the industry regulator, removed these limitations in 2010 on the grounds of unmet demand for these services, and penetration has recently increased.

In 2012, Anatel imposed a ban on three mobile operators (TIM, Claro and Oi) from selling new mobile services, suggesting that they had underinvested and provided an unacceptable level of quality of service. These operators presented proposals on planned network deployments and upgrades to guarantee quality of service and Anatel lifted the ban. The Brazilian mobile market has been growing steadily in past years. In August 2012, Anatel launched a programme for measuring fixed broadband QoS, defining a set of indicators which have to be reported by operators with more than 50 000 active connections.

Brazil is playing a major role in the region's effort to improve backbone connectivity. It has put forward an initiative to deploy a regional fibre backbone ring, which has been submitted to UNASUR (Union of South-American Nations) and ECLAC (UN Economic Commission for Latin America and the Caribbean) for discussion. The plan could enable a significant improvement in international Internet connectivity. The project will include a 10 000 km-long fibre ring and will be managed by state-owned companies of the member countries of UNASUR. The aim is to overcome the present situation, where 80% of international Internet traffic, between countries in the region, has to be routed through third countries. It is projected that the new network could reduce cost and increase performance. The initiative should also decrease international connectivity prices for other countries in the region, such as Chile. In turn, this could contribute to reducing retail broadband prices and increasing broadband take-up, and to the development of high-speed broadband supply.

An important aspect of the Brazilian telecommunication market is the high level of taxes that are imposed on services. The Minister of Communications has expressed concerns about a tax burden as high as 43% for certain services. Changing the tax system in Brazil can be challenging as there needs to be an agreement between the federal governments and the states. For example, the ICMS (special) state tax may amount up to 35% of total phone bills. The Ministry argues that this burden prevents the industry from growing faster and being available to more users.

China

In terms of size and macroeconomic performance China is playing an increasingly important role in the world's economic landscape. The significance of its telecommunication industry is rising too. China is home to the world's largest mobile operator, China Mobile, with over 700 million customers. Every year, nearly 30 million new subscribers sign up for a fixed broadband subscription. This is about twice as many as the OECD area, though penetration rates are low compared to most OECD countries.

As in many countries, Chinese broadband customers have expressed displeasure regarding speeds and prices for services. In 2008 China passed an antitrust law, which is now being applied to the two largest fixed operators, China Telecom and China Unicom. This was the first antitrust probe for relevant state-owned enterprises (SOE). In November 2011, the National Development and Reform Commission (NDRC) launched an inquiry into these two operators, emphasising margin squeeze practices and underperforming interconnection services. As a result, both operators have committed to address these problems, improve performance and decrease prices. These inquiries were conducted by the antitrust authority. Responsibilities for interconnection agreements also fall under the powers of the Ministry of Industry and Information Technology (MIIT).

Since 2011, competition among international roaming services provided by China's three operators has increased substantially. China Mobile, for example, decreased its international roaming fee six times between 2010 and 2011. China's consumers roaming in many OECD countries can make local calls at prices lower or equivalent to those paid by local users. Prices to call back to China from these countries can be 10 times less expensive than a call in the opposite direction for a consumer from an OECD country. MIIT say that the lower prices are the result of successful negotiation with foreign operators to reduce wholesale rates. This may reflect increasing travel by consumers from China, but raises the question as to why similar rates seem unavailable between OECD countries with large amounts of travel. It also begs the question of why the benefits of these lower rates are not evident for OECD consumers travelling to China, given that the wholesale rates are likely to be reciprocal.

In March 2012, the local telecommunication authority of Zhejiang province directed China Unicom, Zhejiang branch, to refrain from offering and advertising its 2G free call service, which allows unlimited calls within its network. The operator's behaviour was said to be contrary to current regulations that prohibit price discrimination between on-net calls and off-net calls in China. This suggests that larger Chinese operators may be facing difficulties in retaining customers, in part due to regulation designed to assist smaller operators. In this context, international roaming services may be a tool to attract and retain customers if other instruments are not feasible.

India

India's telecommunication market is the second largest in the world in terms of total subscriber numbers. As of 2012, there were some 930 million mobile telephone subscribers and 31.5 million fixed line subscribers. The mobile market is among the most competitive with some 15 operators providing services and among the lowest prices globally. Broadband penetration is still low, however, at 1.2%.

There remains an extremely large divide between telephone densities in urban and rural areas in India. As of December 2011, there were 167.2 telephones per 100 inhabitants in urban areas but only 37.52 in rural areas. The government plans to take various measures under the Universal Service Obligation Fund (USOF) to extend networks to rural areas. The expansion of rural telephones is increasingly based on mobile technology.

By December 2011, 97.11% of Indian villages were covered by the Village Public Telephone (VPT) facility, some replacing the previous Multi Access Radio Relay (MARR), installed before 2002. Under an infrastructure-sharing scheme 7 353 towers have been set up across 500 districts in 27 states.

In November 2011, the government approved the National Optical Fibre Network project, which aims to provide broadband connectivity to 250 000 municipalities over approximately two years at a cost of USD 3.8 billion. This scheme will provide some 900 000 wireline broadband connections to consumers and government institutions and set up 28 672 kiosks by 2014. Wireless broadband infrastructure will provide broadband coverage to about 500 000 communities at 512 Kbit/s. In parallel, the government will provide broadband backhaul connectivity and plans to deploy a network available to third-party service providers on a non-discriminatory basis.

India has a vigorously competitive market for mobile services, based on very low termination rates, which have contributed to boosting usage. In this regard, the regulator

TRAI has proposed to further reduce mobile termination rates from USD 0.037 per minute to USD 0.019 per minute in 2012, down to zero by 2014. In India, the interconnection system is based on Calling Party Pays (CPP). The low rates have enabled high usage in the range of 400 minutes per month per mobile user, largely above the OECD average of 136 minutes in 2011.

One of the latest developments in Indian telecommunications was the revocation of 122 mobile licenses awarded in 2008, including those of Unitech Wireless (participated in by Norway's Telenor), Etisalat, Russia's Sistema, Loop Telecom and Tata Teleservices. These were cancelled on the ground that the awards were arbitrary and unconstitutional. The licenses will be reauctioned by TRAI.

Indonesia

The Indonesian telecommunications industry is characterised by a highly competitive mobile market. Strong price competition, gathering momentum from 2007, has contributed to a steep rise in mobile subscription numbers per 100 inhabitants. Subscriber figures rose from 40.2 per 100 inhabitants in 2007 to 91.7 in 2010, although these numbers may include a substantial proportion of multiple SIM users, as in other countries. The government has favoured price-based competition and has implemented a series of policy initiatives, such as: lowering interconnection fees, issuing a relatively high number of mobile licenses (11 mobile operators in 2009), and promoting infrastructure sharing among operators.

Indonesia has a much lower penetration of fixed broadband access. In 2010, the ITU placed the country at 0.8 broadband subscriptions per 100 inhabitants. The geography of the country with population spread across some 6 000 islands is challenging in this respect. A severe digital divide exists between urban and rural areas.

Taking into consideration the existing challenges for Internet access, WiMAX licenses were awarded to several operators and commercial services started in 2010, with the aim of expanding broadband coverage and decreasing prices. Under the existing universal service framework, operators can receive financial aid from the government-managed universal service fund with their obligation of providing rural areas with basic telecommunications services including low-speed Internet access. The government has also initiated an optical fibre backbone development project called "Palapa Ring", consisting of 35 280 km of undersea optical fibre and 21 708 km of underground fibre optics. By 2012, the project had established seven rings covering 33 provinces and 460 districts across Indonesia.

South Africa

In South Africa, fixed-line telecommunication and broadband penetration rates are low relative to the OECD area. For fixed telecommunication lines the rate stood at 8 per 100 inhabitants in 2011, while the rate for broadband subscriptions per 100 inhabitants was 2%. By way of contrast there were 127 mobile subscriptions per 100 inhabitants. If the use of multiple SIMs is taken into account some estimates put the penetration rate at 80% of the population. GSM mobile coverage reaches 98% of the population.

In 2007 the Marwala Commission recommended the implementation of local loop unbundling (LLU). The regulator ICASA has subsequently endeavoured to develop a policy and regulatory framework that would create the necessary conditions for LLU. Throughout 2010 and 2011, ICASA advanced the process, but has encountered fierce opposition from the incumbent and has been subject to court challenges. This is despite the fact that, at a 2% penetration rate, the copper local loops are underutilised. Some new market entrants have also filed complaints before the Complaints and Compliance Commission (CCC), an independent body related to the regulator ICASA. One complaint resulted in an instruction to ICASA to specify the terms and conditions of the Facility Leasing Regulation, one of the instruments developed as a framework for LLU. ICASA may also conduct a full market investigation on LLU implementation, but the process could take several years.

South Africa also plans to replace analogue broadcasts with digital terrestrial television by 2015. The spectrum resources released as a result of the migration process could be used for communication services such as wireless broadband. The process itself has experienced delays, but these may have certain advantages; for example, the ITU plans for digital dividend bands are becoming clearer, which would enable release of the 800 MHz and 700 MHz shortly after the analogue switch off.

Wireless broadband infrastructure may be especially important for South Africa, given the state of development of fixed communication infrastructure and the low utilisation of available facilities, such as copper local loops. In a positive development, international connectivity in South Africa has increased substantially. This is due to the development of new submarine fibre cables deployed across the East and West African coastline (e.g. EASSy SEACOM, WACS), though this process is ongoing.

The following tables are available at *www.oecd.org/sti/broadband/communications-outlook.htm:*

- Table 2.1. Number of communications providers by country, mid-2012
- Table 2.2. Fixed line subscriber market share of new entrants
- Table 2.3. Market share of mobile network operators in the OECD, end-2011
- Table 2.4. Number portability: number of fixed and mobile numbers ported, 2011
- Table 2.5. National treatment for foreign-controlled enterprises in telecommunications
- Table 2.6. Government ownership of public telecommunication network operators
- Table 2.7. Local loop unbundling
- Table 2.8. Number of unbundled local loops
- Table 2.9. Net neutrality
- Table 2.10. Quality of service
- Table 2.11. Mobile network interconnection
- Table 2.12. Policy developments in international mobile roaming services (since 2010)
- Table 2.13. Spectrum tendering processes

OECD Communications Outlook 2013

Chapter 3

Telecommunication market size

Telecommunication revenues experienced a notable decline in 2009 but stabilised in 2010 and rebounded in 2011. This can be attributed to the strength of mobile communication markets and specifically to the rapid increase in smartphone penetration during this period. The value users associate with smartphones, as opposed to feature phones, is reflected in their willingness to pay higher amounts for monthly subscriptions that comprise voice, text and data bundles as opposed to just voice and text.

The statistical data for Israel are supplied by and under the responsibility of the relevant Israeli authorities. The use of such data by the OECD is without prejudice to the status of the Golan Heights, East Jerusalem and Israeli settlements in the West Bank under the terms of international law.

Telecommunication markets now seem to be rebounding from the global financial crisis (GFC) and are experiencing new growth with increasing revenues. Compared to other sectors, these markets proved relatively resilient during the crisis. Growth in telecommunication revenue was 4.91% in 2011 and 1.24% in 2010, in contrast with a 3.91% decrease in 2009 (Figure 3.1, Table 3.1). The overall turnover of the industry in 2011 (USD 1.363 trillion) was higher than the pre-crisis peak of 2008 (USD 1.343 trillion).

Figure 3.1. **Trends in public telecommunication revenue, investment and access paths, 1980-2011**

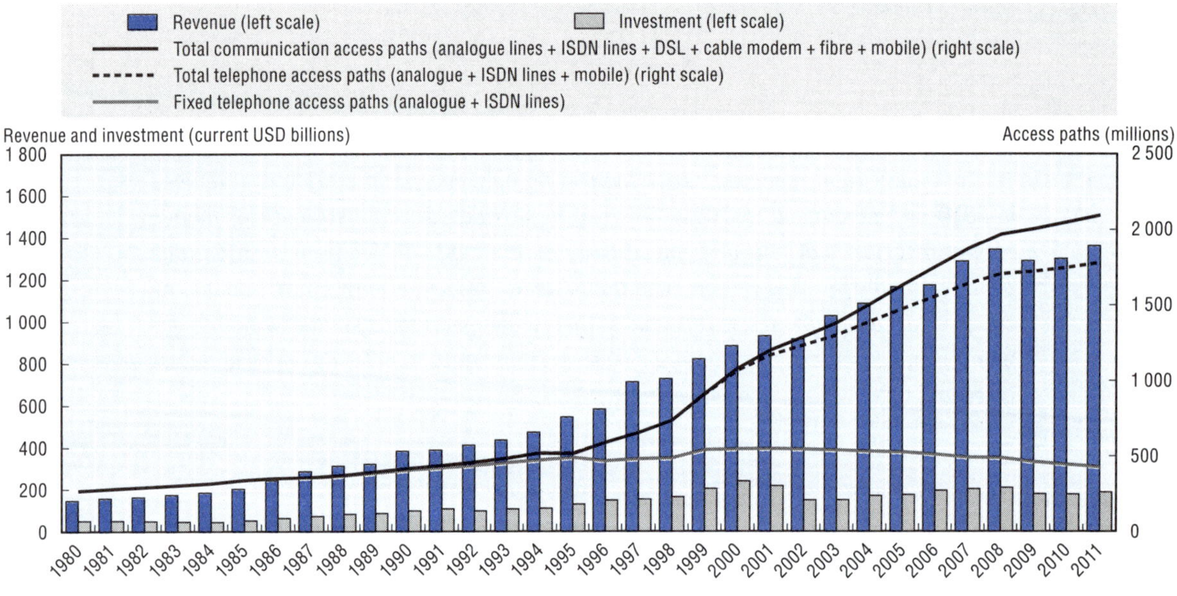

StatLink http://dx.doi.org/10.1787/888932798639

During 2010 and 2011, when measured in PPP, sector growth remained slightly above overall economic growth, and telecommunication revenues as a percentage of GDP decreased slightly. Overall, however, the percentage has remained relatively stable since 2006 at around 3%. The countries where telecommunication revenues were highest as a percentage of GDP were Estonia (4.64%), Korea (4.36%) and Portugal (3.82%). Luxembourg (1.19%), Norway (1.18%), Sweden and Austria (1.51%) had the lowest shares among OECD countries. These numbers take into consideration the growth of the overall economy in current USD PPP exchange rates, which results in higher growth figures being reported than those usually made public by official statistics (using constant USD PPP exchange rates) (Figure 3.2, Table 3.2).

Figure 3.2. **Telecommunication revenue as a percentage of GDP for total OECD, 1985-2011**

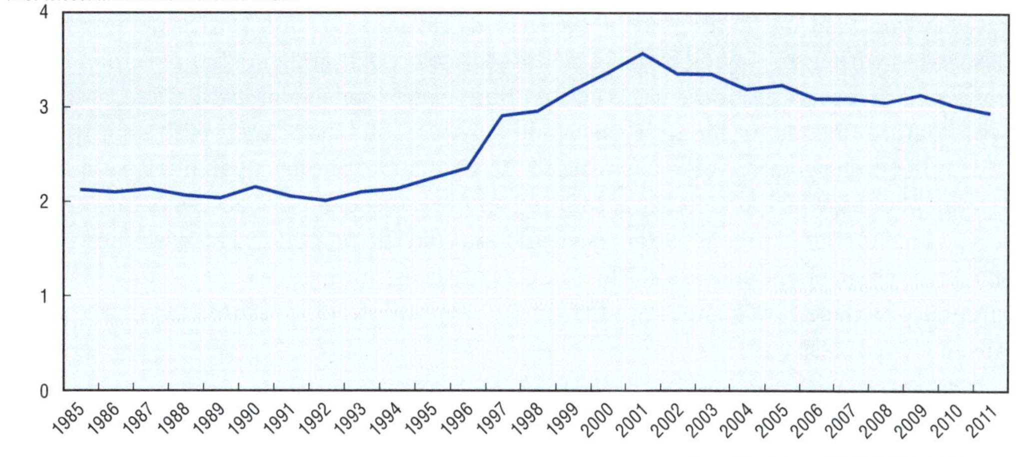

StatLink http://dx.doi.org/10.1787/888932798658

Revenue trends per access path

Following declines over several years, revenue per access path increased between 2009 and 2011 (Figure 3.3, Table 3.3). The average revenue per access path reached USD 655 per annum in 2011, from USD 648 in 2009 and USD 639 in 2010. In 2011 there were just over 2 billion access paths generating USD 1.363 trillion in revenue. The most notable increases in revenue over this period took place in Australia (from USD 763 to USD 1 135), Israel (from USD 447 to USD 596) and Switzerland (from USD 1 027 to USD 1 177). In some cases, such as Australia and Switzerland, this trend may be linked to the relative appreciation of the local currency, as well as the comparative resilience of these economies to the effects of the GFC.

Figure 3.3. **Public telecommunication revenue per communication access path, 2009 and 2011**

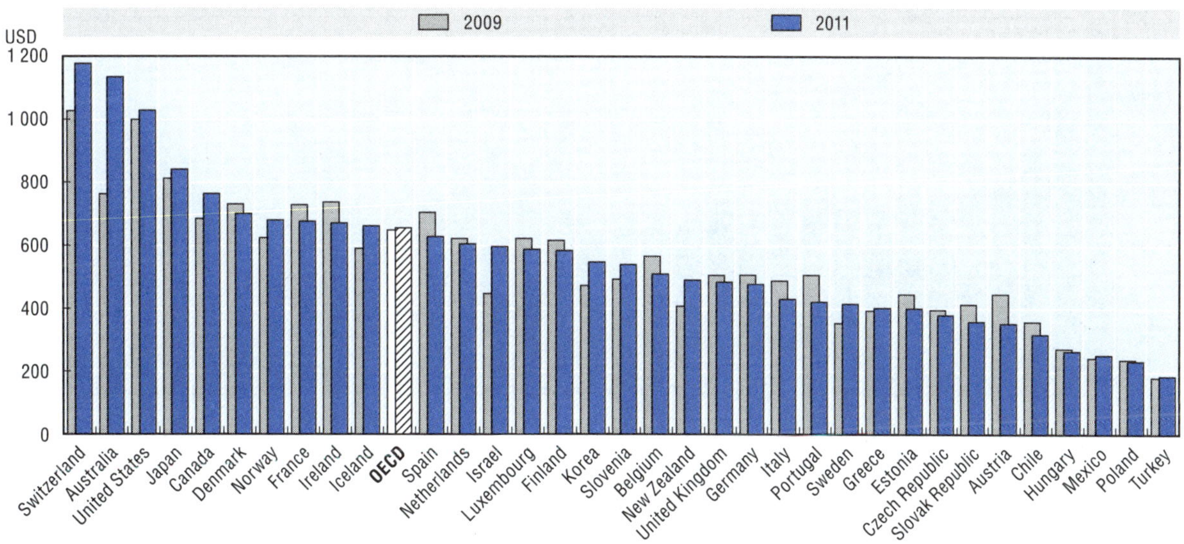

StatLink http://dx.doi.org/10.1787/888932798677

However, the above increases in revenue per access path generally reflect the "smartphone effect". By the end of this period, around half of mobile users in some OECD countries had smartphones. This is in contrast to much lower smartphone penetration rates in the lead up to the end of 2009. The first smartphone tariffs were generally associated with users paying higher amounts for voice, text and data bundles than for just voice and text services. Some users shifted from less expensive prepaid offers, associated with feature phones, to monthly post-paid subscriptions associated with smartphones. This reflects the greater value users associate with smartphone capabilities compared to feature phones, which previously accounted for the largest share of access paths. Recently, a growing number of prepaid offers have also entered the market in line with the decrease in cost of smartphone handsets. The number of access paths does not capture dedicated circuits purchased by business, which may have an effect, presumably small, on the revenue per access path.

The highest revenues per access path in OECD countries were in Switzerland (USD 1 177), Australia (USD 1 135), the United States (USD 1 028) and Japan (USD 842). All these countries have among the highest penetration rates for smartphones. In contrast, Turkey (USD 186), Poland (USD 233) and Mexico (USD 253) had the lowest ratios of revenue per communication access path. It is noteworthy that the latter three countries experienced increases between 2009 and 2011.

Another measure for benchmarking telecommunication revenues is revenue per capita (Figure 3.4). Given that users have increasingly adopted multiple access paths, this provides a different perspective on the increasing value people attach to communication services as reflected in their expenditure. Since the turn of the century, there has been a constant increase in this measure reflecting the increasing adoption of telecommunication services, especially fixed broadband access and mobile communication services. While the

Figure 3.4. **Public telecommunication revenue per capita, 2009 and 2011**

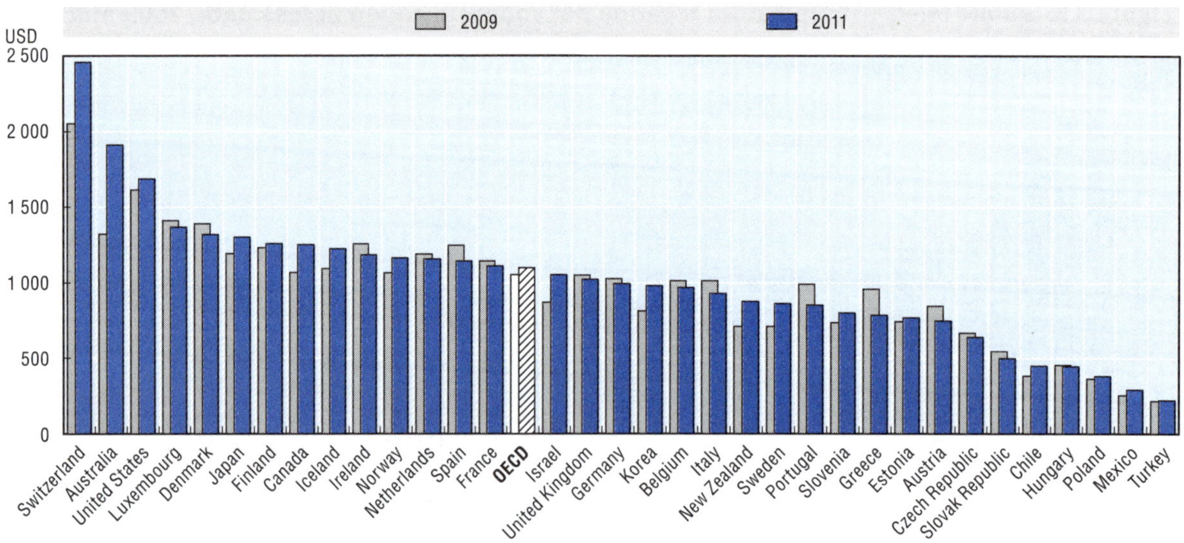

StatLink http://dx.doi.org/10.1787/888932798696

GFC clearly had an impact in 2009 (USD 1 052.6), the smartphone effect had a notable stabilising influence in 2010, as revenue slightly increased per capita in that year (USD 1 053.4). The effect increased further in 2011 with a substantial increase particularly in those countries that better weathered the GFC or began to emerge more quickly from its influence (USD 1 098). While revenue per capita and revenue per access path were around the same level, following the turn of the century, average revenue per capita is now twice that for communication paths (USD 1 098 versus USD 655).

Mobile revenue

The share of revenue generated by mobile communication services continues to increase across the OECD area. In the 13 years between 1998 and 2011 mobile revenues increased sixfold. In 2010 and 2011, mobile revenues increased by 2.9% and 6.7% respectively. Mobile revenues reached USD 651 billion in 2011, up from USD 593 billion in 2009. The share of mobile revenues in total telecommunication revenues rose to 47.8% (46.0% in 2009). Notably, this share was just 28.6% a decade ago (Figure 3.5, Table 3.4).

Figure 3.5. **OECD share of mobile and fixed telecommunication revenues, 1998-2011**

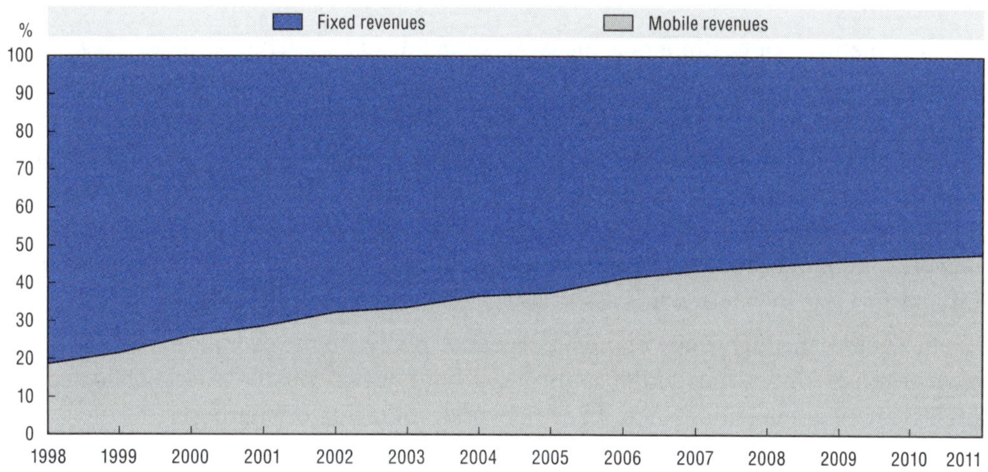

StatLink http://dx.doi.org/10.1787/888932798715

While the trend toward greater shares of mobile revenues is common across the OECD area, there are some differences. Some countries with a lower penetration rate for fixed networks have, as a result, a larger share of mobile revenue in their total telecommunication revenue. This is, however, not always the case. In 2011, 13 out of 34 OECD countries had a mobile revenue share higher than 50% (Figure 3.6).

Some countries have high fixed penetration rates and a relatively high share of mobile revenues: Japan (84.4%), Austria (65.7%), Luxembourg (52.9%), Norway (52.9%) and Italy (52.1%). The drivers of mobile expenditures in those countries are different from those with less-developed fixed infrastructure, such as the Czech Republic, Mexico or Turkey.

Japan is clearly an exceptional country in terms of the distribution of revenues between fixed and mobile communications. The mobile revenue share is as high as 84%, almost 20 percentage points above the next country, Austria. One possible factor that can be used to explain this trend is the pricing patterns in Japan for mobile voice and data services. As shown in Chapter 7, mobile operators in Japan promote flat-rate tariffs that are

Figure 3.6. **Share of mobile revenue in total telecommunication revenue, 2011**

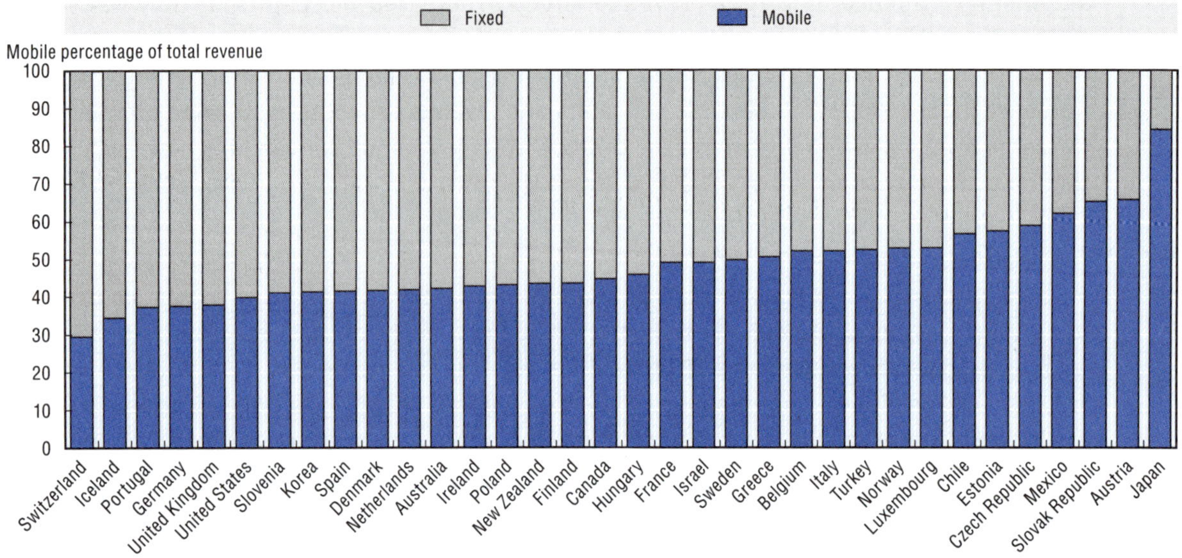

StatLink http://dx.doi.org/10.1787/888932798734

more competitive for customers using larger amounts of data. This promotes fixed mobile substitution, for example, by a larger share of customers making use of mobile devices for Internet access. Historically, Japan was an early adopter of data applications for mobile devices, with i-mode technology in the late 1990s. Mobile operators provide integrated services, including handsets, network services, content applications and platforms (i.e. billing), and this is all reflected in the share of mobile revenue relative to fixed networks. In many countries, this revenue would more likely be found in the accounts of third parties (e.g. in value chains associated with the apps economy).

In Austria, the high share of mobile revenue, relative to the overall market, is also due to substitution. Unlike most other countries, its authorities consider that mobile networks can be viewed as substitutable for fixed networks, for regulatory purposes.

Other differences may be explained by different levels of competition in fixed and mobile markets in a given country. The share of mobile revenues in the United Kingdom, for example, is among the lowest as a proportion of the total market in the OECD. Until 2010, there were five mobile operators competing strongly in this market and this led to some of the lowest prices in the OECD area. Another possible element that can shape this comparison is the importance of pay-television services, and whether these are adopted by a majority of consumers. As fixed telecommunication and cable operators bundle television with voice and broadband services, this can have a significant effect on revenue distribution across services. Countries with well-developed cable infrastructure are Canada, the Netherlands, Portugal and the United States, all of which have a share of mobile revenues below the OECD average of 46.8%.

In the case of the United Kingdom and the United States, a further factor in the lower share of mobile revenues, relative to overall market size, may be the higher presence of international communication hubs for business, with a larger number of international backbone networks providing connectivity or acting as a global hub that covers routes between these countries and others.

The average revenue per subscription remained relatively stable between 2009 and 2011, increasing from USD 39.3 to USD 40.2 (Figure 3.7). There were two factors pulling this figure in different directions. First, revenue per subscriber (i.e. actual people) has increased, particularly with the rapid take up of smartphones. Second, there are more "subscriptions" due to the increasing take up of prepaid cards and the introduction of connected devices (including machine-to-machine connections) that include SIM cards. Between 2009 and 2011, usage of prepaid cards associated with feature phones increased in countries such as the United States. Historically, the United States had mobile penetration rates below 100%. This has been attributed to a market preference for post-paid rather than prepaid services. While this continues to be the case, there has been significant growth in prepaid sales in recent years with lower associated average revenues per user compared to post-paid offers. Meanwhile, the introduction of devices such as e-book readers that include SIM cards (e.g. Amazon's Kindle) all contribute to reducing average revenue per SIM card (i.e. subscriptions). The effect of growth in prepaid cards and devices that include SIM cards has been countered by the increasing revenue per subscription associated with smartphones. The metric "mobile revenue per capita" is included here to provide a different perspective on these developments (Table 3.5).

Figure 3.7. **Mobile revenue per subscription, monthly, 2009 and 2011, USD**

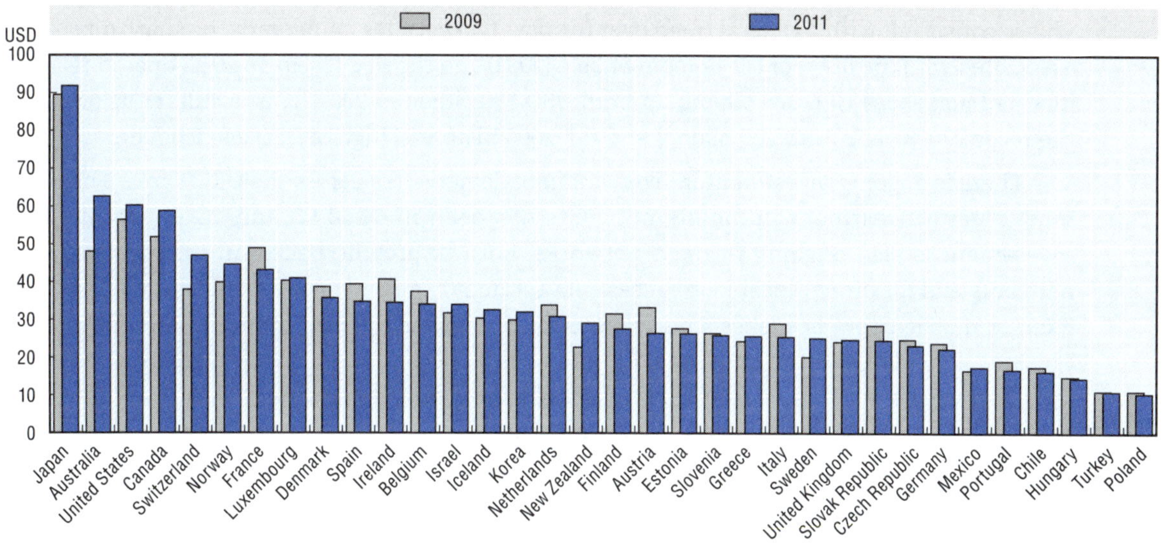

StatLink http://dx.doi.org/10.1787/888932798753

While Australia, Canada, Switzerland and the United Kingdom managed to increase their mobile revenue per subscriber, other countries experienced notable decreases (France, Ireland, the Slovak Republic and Spain). The GFC undoubtedly had a larger effect in some countries than others, particularly for some European countries. It is also worth posing the question as to why average revenue per subscription declined in France but rose in the United Kingdom. Even though some of these relative changes in revenue may also be due to fluctuations in exchange rates, it should be noted that a merger between two operators took effect in the United Kingdom in the second half of 2011, decreasing the number of operators from five to four. Meanwhile, the market in France underwent substantial changes in the lead-up to the introduction of a fourth operator at the beginning of 2012, with incumbents offering more attractive tariffs. Nevertheless, in both markets the main influence should have been felt from 2012, rather than in 2011. In France, for example, prices fell significantly following the introduction of a fourth operator, but there was also significant growth stimulated in the market.

Aside from the GFC, relative strength of currencies and so forth, a notable further factor in the increase in revenues for some countries is likely linked to the launch of 4G networks. Countries where 4G networks have been launched, between 2009 and 2011, include Japan, Norway, Sweden and the United States. This may have allowed some operators to leverage this technology, adding a premium on 3G connections as they were upgraded to 4G.

In the future, average revenue per mobile subscription will likely decline. While operators may be able to increase revenue per subscriber, the number of subscriptions added to networks is likely to outstrip growth from individual customers joining the network. Revenue accruing to operators from connected devices, for example, may be less than for traditional connections. Conversely, the trend towards greater revenue per capita will likely continue as individuals make greater use of these devices in areas such as games, navigation and e-books.

Investment

Investment in telecommunication infrastructure has begun to increase once more. This is consistent with a general trend over the past two decades, punctuated by sharp dips associated with the burst of the dotcom bubble (2001) and the GFC (2009). In 2011, operators invested just above USD 188 billion, up from USD 180 billion in 2009. This is still far below the recent 2008 peak (USD 209 billion) and the 2000 historical high of USD 240.5 billion.

There is ongoing investment in fixed and mobile networks across the OECD area. The most recent investment in mobile networks has been in 3G, 3.5G and 4G networks. Investment is also ongoing in fixed networks, both to provide backhaul for wireless networks and for direct "local loop" access. Investment can sometimes seem "lumpy" if a fixed or mobile operator makes a decision to rapidly upgrade part of their infrastructure. This can sometimes be the case in relation to competition or the need to reach a certain level of coverage related to a regulatory requirement or commercial target.

A decision to take fibre to the home or closer to users can generate a sharp increase in the level of investment. In Australia, the decision to invest in a national broadband network (NBN), which will provide fibre to the home to 93% of Australian premises and a combination of next generation fixed wireless and satellite technologies to the remaining 7%, will involve capital expenditure of approximately USD 37 billion to the end of the rollout in 2012. At the same time, in some countries cable operators are devoting resources to DOCSIS 3.0 upgrades that will allow them to compete with fibre networks deployed by incumbents or new entrants.

A further caveat should be noted concerning investment data: resources devoted to spectrum fees are not included, even though they may represent a non-negligible share of resources used by mobile operators. Spectrum is, of course, a vital asset for their business.

The largest share of public telecommunication investment (i.e. in public telecommunication networks) in the OECD area corresponds to the Americas region (USD 87 billion), followed by Europe (USD 67 billion) and Asia/Pacific (USD 34 billion). The United States accounted for USD 70 billion of public telecommunication investment,

Figure 3.8. **Public telecommunications investment by region, 1997-2011, excluding spectrum fees**

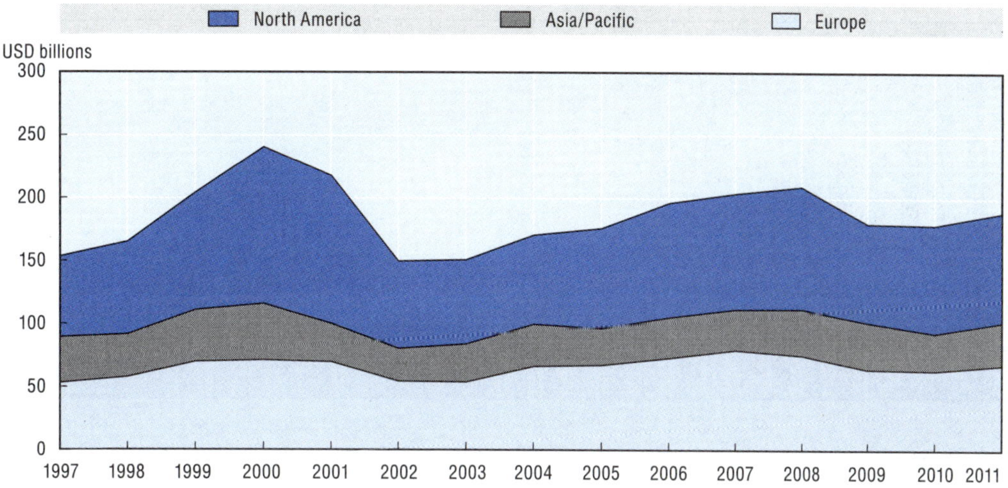

StatLink http://dx.doi.org/10.1787/888932798772

representing over one third (37%) of the global OECD telecommunication investment (Figure 3.8 and Table 3.6 and 3.7). A significant part of this investment was in 4G networks.

The countries with the highest level of investment, when considering investment as a percentage of telecommunication revenue (Table 3.8), were New Zealand (31.7%) in 2010, associated with the introduction of a new mobile operator and the beginning of a publicly funded rollout of a fibre-to-the-home network, Chile (31.1%) and Luxembourg (22.4%). The OECD average remained relatively stable at 14.0% in 2009 and 13.9% in 2011, reflecting a return to higher revenue rather than a decrease in investment. While there have been spikes in investment, such as with the first liberalisation of the market, the trend in recent years has been for these numbers to stabilise at between 14% and 16% of telecommunication revenues.

Some countries are able to report disaggregated investment in mobile and fixed networks (Table 3.9). As expected, those countries with low fixed communication adoption reported high percentages of investment in mobile networks, such as Hungary (63.6%), Turkey (64.5%) or the Slovak Republic (66.6%) and Chile (58.5%). For countries with higher fixed-line penetration, this share remained between 20% and 35%, such as in Denmark (17.7%), Switzerland (18.7%), the United Kingdom (24.2%), Canada (24.5%) and Germany (31.6%).

An additional way to compare investment levels is by telecommunication access paths (Figure 3.9, Table 3.10). This provides an indicator of relative investment levels. In 2011, on average, the investment per total communication access path was USD 90.8 in the OECD area, slightly above the investment in 2009. Australia (USD 195.7), Canada (USD 168.0) and Switzerland (USD 161.4) had the highest level in 2011. The countries with lowest investment per access path were Turkey (USD 26.3), Mexico (USD 39.6) and Austria (USD 39.9).

In the context of converging telecommunication services, both within fixed networks (e.g. voice, data and video services) or between fixed and mobile networks (fixed and mobile convergence), it is challenging to assign investment to specific services. By way of example, many operators are investing in backbone and backhaul fibre networks that can

Figure 3.9. **Public telecommunications investment per access path, USD**

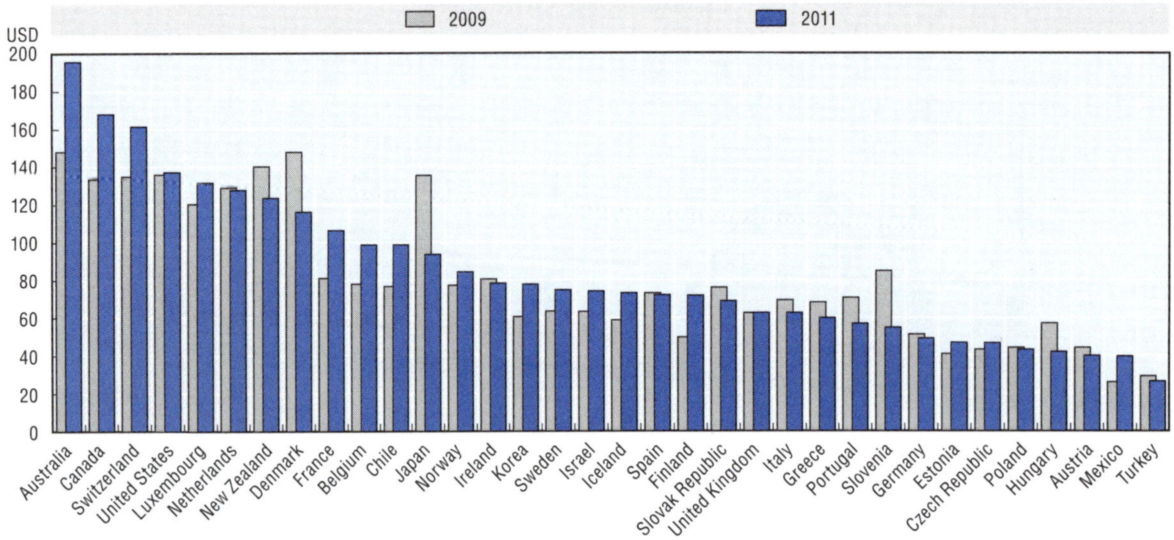

StatLink http://dx.doi.org/10.1787/888932798791

be used both for fixed and mobile communication services to end users. Notwithstanding, there remain a few fixed-only or mobile-only operators in the OECD for which this exercise is somewhat easier. Moreover, fixed-mobile convergence has been limited so far, at least from the point of view of the consumer.

If investment per capita is considered, as opposed to investment per communication path, there are only slight variations in overall investment figures. By way of example, the three countries with the lowest investment also have low ratios for these two metrics. With slight changes in position, the same group of countries led the list using any of the two indicators. The highest investment levels per capita were found in Luxembourg (USD 306.2), Switzerland (USD 337.1) and Australia (USD 329.7). Australia, Canada and Switzerland have experienced high increases in their investment levels since 2009, which may be largely due to the appreciation of their national currencies. This approach is preferred rather than the use of PPPs, as these figures refer to industry data rather than prices faced by consumers. That being said, significant investment has been made in increasing fibre access in Switzerland and DOCSIS 3.0 in Canada. In Australia, the effects of the NBN are likely to increase from 2012 onwards (Figure 3.10, Table 3.11).

Voice traffic

Telephony has become increasingly just another application over networks using the Internet Protocol (IP). Historically, it was the main source of revenue for operators and the main reason for collecting data for billing requirements. That is still the case for many operators including VoIP providers such as Skype, which need to record traffic data for billing purposes when it terminates on some access paths (i.e. Skype-out). However, traffic data for VoIP-to-VoIP calls are rarely recorded in official data for network traffic; and when they are, they are not recorded as telephony.

Today, users can make calls that are, for the most part, recorded as data. This is the case for services such as Skype or Apple's Facetime, as well as any number of ways to chat over social networks and games. For this and other reasons, operators are increasingly

Figure 3.10. **Public telecommunications investment per capita, USD**

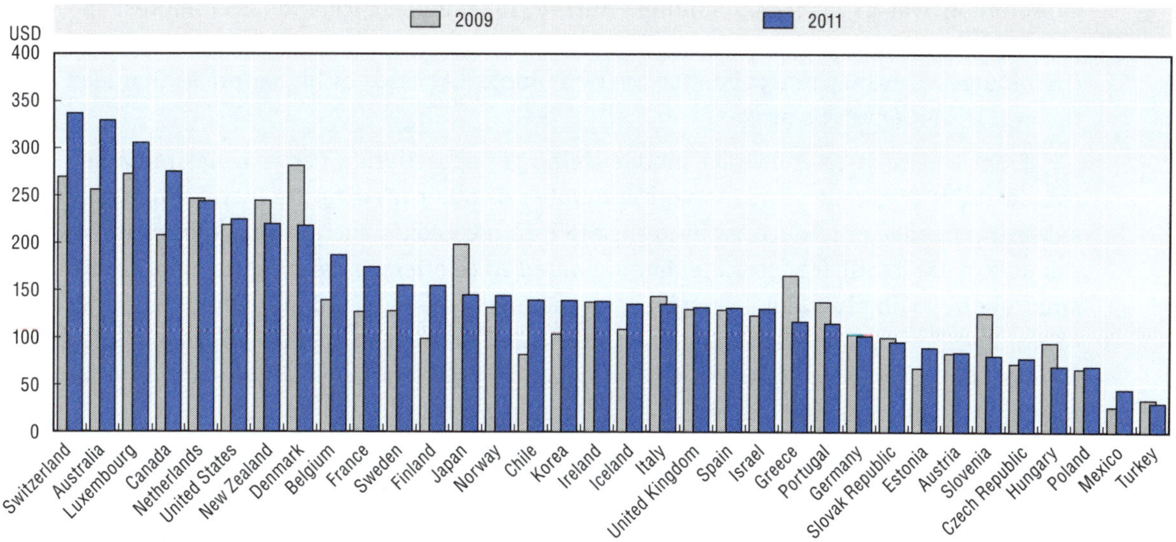

StatLink http://dx.doi.org/10.1787/888932798810

bundling telephony in their general access offers for fixed and mobile networks. Unlimited local calls have been the rule in Canada and the United States for many years and as a result there are less data available for that market segment. Canada, for example, does not report official data for minutes of local calls.

One of the prerequisites for traditional tariff structures in Canada and the United States has been sender-keep-all models or negligible termination rates. This is increasingly being replicated in other OECD countries. As such, following decreases in fixed termination rates and the adoption of IP technologies, operators are increasingly launching offers that include national and international calls as part of a bundle with no disaggregated charge.

While first evident in fixed networks, lower termination charges are leading to the same outcomes in mobile networks with billing shifting from "minutes" or "texts" to the amount of data used or speed made available. It is expected that these developments will continue in a context of decreasing mobile termination rates, which so far have acted as a price floor and a deterrent for operators to launch bundled mobile tariffs in some countries.

One further caveat is notable on traffic data. The United States uses both incoming and outgoing minutes for the production of voice traffic data. To obtain an estimate for mobile traffic in the United States, the data are halved for mobile to mobile and all the wireless to landline minutes. As all other countries reporting data do so on the basis of outgoing minutes, this provides greater comparability with other countries. Even then, the level of voice traffic in the United States has always been significantly higher than all other countries – a combination of much larger bundles or unlimited minutes as well as the multiple use of SIM cards in some other countries that result in lower average usage per SIM card.

Between 2009 and 2011, on average, the number of mobile minutes per subscriber per month in the OECD remained stable at approximately 134 minutes per subscription per month (Tables 3.12 and 3.13). The United States and Canada, as expected, are clear outliers at 356 and 345 minutes per month, even taking into account adjustments made for improved comparability. After the United States and Canada, the highest traffic, per mobile subscription, was in Israel (205 minutes), Turkey (188 minutes), Iceland (185 minutes) and Norway (184 minutes). The lowest traffic was in Poland (74 minutes) and New Zealand (76 minutes). This may reflect greater use of multiple SIM cards or higher prices that lead to lower levels of average usage.

From a peak in 2007, cellular mobile traffic per subscription per year has remained overall stable in recent years. As with a number of other metrics the increasing use of subscriptions does not necessarily translate into increasing average usage in terms of minutes of use (MoU). It is possible that unlimited voice offers, increasingly common with smartphones, will raise MoU, though they generally come with unlimited text and other communication options that can act as substitutes. At the same time, some connected devices, such as eBooks or tablets will not generate voice traffic (i.e. if they are used for telephony it will more likely be recorded as data).

Finally, as noted elsewhere, mobile subscriber data in Estonia include close to 2 million subscriptions for a mobile virtual network operator that uses Estonian numbers but provides international roaming service outside the country. These subscribers have been removed from the figures reported here in order to improve comparability. Monthly data for these subscribers is likely to be low, which may distort average traffic overall.

Recorded international telecommunication traffic has decreased between 2009 and 2011, continuing the downward trend noticeable since 2005 (Table 3.14). The average across the OECD area is now 78.2 outgoing minutes per communication access path, including fixed and mobile lines. Over-the-top (OTT) providers such as Skype, which are not accounted for by most OECD countries, largely drive international telephony traffic today. Where international VoIP is reported effectively, such as in France, it accounts for a large share of international communication traffic.

Employment trends

Direct employment in the telecommunication infrastructures and services sector, in OECD countries for which data are available, slightly decreased in 2011. There has been, on average, a reduction of 1% per year in the previous five years (Table 3.15). Countries like Denmark, Ireland, Italy, Spain and the United States have been gradually reducing their total employment in telecommunication infrastructures and services above the average for the last three years. The inverse trend is visible in Chile, Israel, Mexico, New Zealand and Slovenia. These countries increased their employment level in the three years leading up to 2011. These data do not capture all employment in the sector much of which is outsourced. They also does not record the increases in related employment such as in the production of applications.

Trade in communication equipment

In 2011, trade in telecommunication equipment, for OECD countries, reached its highest level of USD 428 billion (i.e. exports and imports by OECD countries to the world, Tables 3.16 and 3.17). Exports in telecommunication equipment are highly related to global economic conditions. The previous two decades were characterised by high growth followed by a sharp decrease in exports. The first peak was reached in 2000, the period called the "dotcom bubble", and the second one was in 2008, just before the effects of the full financial crisis became evident. In 2011, the data reflect a recovery well underway.

China has a growing share of the world's telecommunication equipment production and exports. China alone exports the equivalent of 55% of the OECD area's exports in telecommunication equipment. In 2004, China overtook the United States as the largest player in this market and it seems that it will soon approach the total level for OECD countries. In 2011, in OECD countries, imports of telecommunication equipment reached their highest level ever at USD 260 billion, which was 1.55 times the value of exports (Table 3.17).

The OECD is currently working on a project to measure trade in value added (TIVA). This aims to show the value of exports and imports of goods by allocating the value added to the countries of origin. This will allow an evaluation of the trade of goods in a different light. Smartphones, for example, may be manufactured in one country and reflected in their current trade data, but this will not necessarily capture the value added in areas such as the operating system and applications developed in other countries.

Table 3.1. **Telecommunication revenue in the OECD area**

USD millions

	1990	1995	1996	1997	1998	1999	2000	2001	2002	2003	2004	2005	2006	2007	2008	2009	2010	2011	CAGR 2009-11	CAGR 2001-11
Australia	8 730	11 130	13 109	16 385	16 385	16 385	..	15 454	..	19 391	..	26 614	..	30 833	30 805	29 244	37 210	43 508	22.0	10.9
Austria	2 809	4 321	4 010	3 721	4 118	4 991	4 423	5 043	5 307	6 662	7 509	7 731	7 557	7 830	7 890	7 058	6 253	6 286	-5.6	2.2
Belgium	2 689	4 317	4 465	4 229	5 100	5 896	7 267	7 088	7 516	9 528	10 855	11 413	10 249	11 057	11 637	10 903	10 402	10 623	-1.3	4.1
Canada	12 123	12 180	13 361	17 080	19 251	19 272	20 578	20 876	21 161	22 854	25 745	28 516	31 917	35 541	37 639	35 980	40 441	43 144	9.5	7.5
Chile	..	..	..	..	..	..	..	..	..	..	..	4 310	4 947	5 811	6 612	6 464	6 677	7 744	9.5	..
Czech Republic	502	995	1 130	1 452	1 833	2 110	2 316	2 558	3 270	4 000	4 439	4 882	5 396	6 849	8 430	7 008	6 665	6 723	-2.1	10.1
Denmark	2 354	3 730	3 641	3 485	3 760	4 430	4 173	4 246	4 384	5 527	6 356	6 574	6 786	8 162	8 125	7 674	7 207	7 341	-2.2	5.6
Estonia	..	..	..	..	..	..	..	340	418	558	667	757	886	1 014	1 058	995	927	1 028	1.6	11.7
Finland	2 233	2 550	2 700	3 081	3 634	4 041	4 004	4 189	4 728	5 169	5 670	5 312	5 638	6 131	6 269	6 569	6 342	6 778	1.6	4.9
France	18 918	30 159	30 612	28 630	29 803	33 781	34 016	36 734	40 123	49 270	56 730	62 156	62 275	70 570	78 985	73 653	70 325	72 325	-0.9	7.0
Germany	25 004	46 296	41 899	43 430	49 111	51 170	51 560	54 018	58 491	71 798	82 469	84 125	82 875	87 534	91 618	83 889	77 895	81 111	-1.7	4.1
Greece	1 277	2 797	3 117	3 291	4 291	4 240	5 089	5 603	6 658	8 553	9 774	9 661	10 456	11 633	12 088	10 788	9 075	8 871	-9.3	4.7
Hungary	359	1 541	1 841	2 138	2 513	3 071	3 210	3 440	3 869	4 686	4 810	5 099	5 009	5 779	5 820	4 537	4 276	4 422	-1.3	2.5
Iceland	85	133	156	151	167	191	253	216	228	319	382	464	471	579	485	349	355	391	5.8	6.1
Ireland	1 290	1 759	1 977	2 126	1 910	1 927	2 249	2 478	3 197	4 022	4 864	4 898	5 357	6 214	6 641	5 612	5 088	5 311	-2.7	7.9
Israel	..	..	..	..	..	..	..	..	..	..	..	..	..	..	..	6 515	7 325	8 153	11.9	..
Italy	16 029	18 513	24 125	24 091	27 778	29 894	29 767	32 346	36 424	46 751	54 528	58 064	58 829	63 208	66 479	60 833	55 468	56 372	-3.7	5.7
Japan	46 333	113 012	118 336	116 505	113 184	143 183	163 253	156 796	129 352	139 225	134 732	132 042	129 868	129 990	139 372	151 953	152 292	167 500	5.0	0.7
Korea	5 167	10 623	14 696	14 616	12 021	17 699	23 655	23 667	27 056	29 115	33 399	39 898	44 899	48 870	44 193	39 927	45 244	48 724	10.5	7.5
Luxembourg	146	301	317	305	341	363	340	372	394	473	528	567	612	676	744	702	653	710	0.6	6.7
Mexico	4 027	6 492	6 755	8 770	9 654	11 298	14 314	16 057	17 076	17 214	19 090	22 297	26 046	29 094	30 230	27 198	30 286	31 824	8.2	7.1
Netherlands	4 986	8 468	8 413	7 890	9 491	10 719	10 150	11 607	12 988	16 604	18 655	18 993	19 202	18 410	21 012	19 636	18 434	19 286	-0.9	5.2
New Zealand	1 448	2 097	2 142	2 249	2 041	2 173	2 224	2 117	2 465	3 095	3 526	3 749	3 195	3 603	3 472	3 075	3 511	3 866	12.1	6.2
Norway	2 336	3 132	3 437	3 609	2 466	2 603	2 620	2 814	3 360	3 991	4 475	4 754	4 851	5 336	5 641	5 132	5 416	5 754	5.9	7.4
Poland	520	2 158	2 535	2 593	3 620	4 592	5 427	6 583	6 905	7 650	9 589	11 443	12 915	15 153	17 884	13 849	14 308	14 598	2.7	8.3
Portugal	1 381	3 048	3 822	3 959	4 215	4 730	5 049	5 995	6 452	7 844	9 030	9 218	9 223	10 633	11 333	10 516	10 000	9 066	-7.2	4.2
Slovak Republic	..	316	417	451	480	444	801	942	1 024	1 346	1 625	1 856	1 960	2 534	3 154	2 949	2 607	2 707	-4.2	11.1
Slovenia	..	..	..	..	..	..	..	..	..	..	..	1 190	1 311	1 508	1 813	1 500	1 380	1 641	4.6	..
Spain	8 715	11 000	11 649	18 002	19 627	22 389	22 737	23 992	31 462	38 812	45 735	51 090	52 850	60 567	64 938	57 261	52 370	52 710	-4.1	8.2
Sweden	5 330	6 993	7 577	6 910	7 393	4 623	4 416	4 826	5 186	6 251	6 805	6 636	6 640	7 371	7 562	6 608	7 219	8 125	10.9	5.3
Switzerland	4 890	8 064	7 687	6 794	7 699	8 729	8 244	8 745	9 516	11 368	12 909	12 917	13 041	14 316	16 092	16 017	16 878	19 343	9.9	8.3
Turkey	2 063	1 672	3 120	4 033	5 031	5 446	6 168	5 867	6 714	10 423	11 441	12 390	12 025	16 253	17 808	15 543	17 053	16 402	2.7	10.8
United Kingdom	25 796	28 552	30 539	34 227	34 227	38 789	43 862	47 146	48 813	58 145	69 301	71 183	75 643	84 061	78 677	64 824	62 361	64 109	-0.6	3.1
United States	146 147	199 147	212 645	326 753	326 753	365 587	402 143	419 514	411 645	415 827	429 430	443 681	459 315	480 030	498 058	495 062	507 533	526 054	3.1	2.3
OECD	353 688	545 497	582 095	710 957	727 896	824 766	884 307	931 669	916 181	1 026 466	1 085 068	1 164 480	1 172 241	1 287 149	1 342 562	1 289 825	1 299 478	1 362 549	2.8	0.0

Notes: Data for Hungary (2011), Japan (2011) and Poland (2010 and 2011) are estimates. The total OECD CAGR (2001-2011) calculation excludes Chile, Israel and Slovenia.

StatLink http://dx.doi.org/10.1787/888932801014

Table 3.2. **Telecommunication revenue as a percentage of GDP**

	1985	1990	1995	1997	1998	1999	2000	2001	2002	2003	2004	2005	2006	2007	2008	2009	2010	2011	GDP per capita 2011 (USD)
Australia	1.92	2.81	2.99	3.75	4.19	3.84	0.00	3.95	0.00	3.47	0.00	3.50	0.00	3.15	2.93	2.89	2.89	2.86	66 827
Austria	1.68	1.75	1.82	1.80	1.93	2.35	2.31	2.64	2.55	2.64	2.59	2.52	2.33	2.09	1.90	1.84	1.66	1.51	49 597
Belgium	1.27	1.37	1.56	1.70	2.00	2.32	3.14	3.06	2.97	3.07	3.02	3.01	2.57	2.40	2.28	2.30	2.22	2.07	46 790
Canada	2.21	2.12	2.09	2.67	3.11	2.92	2.85	2.92	2.88	2.64	2.59	2.51	2.49	2.49	2.51	2.68	2.56	2.49	50 345
Chile	..	..	..	..	..	..	..	..	..	..	..	3.50	3.20	3.36	3.68	3.75	3.09	3.12	14 395
Czech Republic	..	1.69	1.91	2.44	2.87	3.39	3.94	3.97	4.17	4.20	3.89	3.75	3.64	3.79	3.74	3.55	3.35	3.10	20 676
Denmark	1.49	1.77	2.07	2.04	2.17	2.55	2.61	2.64	2.52	2.60	2.60	2.55	2.47	2.62	2.36	2.47	2.31	2.21	59 605
Estonia	..	..	..	..	..	..	..	5.47	5.70	5.69	5.58	5.41	5.29	4.61	4.43	5.20	4.86	4.64	16 530
Finland	1.50	1.62	1.95	2.50	2.80	3.11	3.30	3.37	3.49	3.16	3.02	2.70	2.72	2.49	2.30	2.74	2.70	2.58	48 820
France	1.65	1.55	1.94	2.01	2.03	2.32	2.58	2.75	2.76	2.76	2.78	2.89	2.77	2.73	2.78	2.81	2.76	2.61	42 547
Germany	1.60	2.91	1.87	2.02	2.26	2.40	2.74	2.88	2.91	2.98	3.04	3.03	2.87	2.63	2.52	2.54	2.37	2.25	44 031
Greece	1.33	1.55	2.38	2.44	3.18	3.05	4.03	4.32	4.55	4.45	4.31	4.00	4.00	3.81	3.53	3.35	3.03	2.97	26 413
Hungary	..	..	3.45	4.59	5.24	6.37	6.92	6.52	5.83	5.61	4.72	4.62	4.45	4.25	3.77	3.58	3.34	3.19	13 910
Iceland	1.29	1.35	1.92	2.04	2.01	2.18	2.91	2.72	2.56	2.91	2.88	2.85	2.83	2.83	2.88	2.88	2.82	2.79	43 967
Ireland	2.31	2.15	2.08	2.60	2.14	1.98	2.32	2.35	2.58	2.54	2.62	2.40	2.40	2.39	2.51	2.52	2.48	2.44	48 385
Israel	..	..	..	..	..	..	..	..	..	..	..	..	..	..	..	3.34	3.37	3.35	31 415
Italy	1.48	1.46	1.68	2.01	2.28	2.48	2.71	2.88	2.97	3.10	3.16	3.23	3.15	2.97	2.87	2.88	2.71	2.57	36 115
Japan	1.58	1.52	2.14	2.69	2.89	3.23	3.45	3.77	3.25	3.24	2.89	2.89	2.98	2.98	2.87	3.02	2.77	2.85	45 582
Korea	2.05	2.05	2.17	2.75	3.36	3.83	4.43	4.69	4.70	4.52	4.63	4.72	4.72	4.66	4.74	4.79	4.46	4.36	22 424
Luxembourg	1.03	1.33	1.66	1.65	1.76	1.72	1.68	1.85	1.74	1.63	1.56	1.50	1.44	1.32	1.28	1.35	1.23	1.19	114 507
Mexico	0.52	1.53	2.27	2.00	2.09	2.14	2.25	2.35	2.40	2.46	2.52	2.64	2.74	2.82	2.77	3.09	2.94	2.76	10 543
Netherlands	1.45	3.75	2.05	2.05	2.36	2.61	2.65	2.90	2.96	3.10	3.08	2.96	2.84	2.35	2.40	2.47	2.38	2.31	50 092
New Zealand	2.46	3.33	3.44	3.30	3.64	3.69	4.16	3.99	4.01	3.75	3.50	3.32	2.92	2.69	2.66	2.62	2.47	2.40	36 378
Norway	1.91	2.02	2.14	2.28	1.63	1.64	1.56	1.65	1.75	1.77	1.72	1.56	1.43	1.36	1.24	1.37	1.30	1.18	98 083
Poland	..	0.88	1.69	1.65	2.10	2.74	3.17	3.45	3.48	3.53	3.80	3.77	3.78	3.57	3.38	3.21	3.05	2.84	13 357
Portugal	2.66	1.93	2.83	3.41	3.44	3.75	4.32	4.99	4.87	4.87	4.90	4.78	4.59	4.58	4.48	4.49	4.40	3.82	22 287
Slovak Republic	..	..	1.72	2.12	2.15	2.17	3.93	4.48	4.17	4.04	3.85	3.88	3.53	3.38	3.35	3.38	3.01	2.82	17 628
Slovenia	..	..	..	..	..	..	..	..	..	..	..	3.31	3.38	3.18	3.31	3.04	2.95	3.27	24 473
Spain	1.44	1.69	1.89	3.14	3.27	3.63	3.93	3.95	4.57	4.41	4.40	4.49	4.29	4.20	4.06	3.93	3.79	3.57	32 019
Sweden	1.78	2.24	2.91	2.73	2.90	1.79	1.79	2.12	2.07	1.99	1.88	1.79	1.66	1.59	1.56	1.63	1.56	1.51	57 158
Switzerland	2.15	2.14	2.62	2.49	2.76	3.19	3.22	3.33	3.32	3.41	3.44	3.37	3.21	3.18	3.06	3.15	3.06	2.93	83 781
Turkey	1.03	1.37	1.08	1.56	1.86	2.19	2.33	3.00	2.89	3.44	2.93	2.56	2.27	2.51	2.44	2.53	2.33	2.11	10 511
United Kingdom	2.36	2.59	2.50	2.50	2.33	2.59	2.97	3.19	3.06	3.12	3.18	3.10	3.06	2.98	2.95	2.96	2.76	2.62	38 980
United States	2.67	2.54	2.71	3.95	3.74	3.93	4.06	4.10	3.89	3.75	3.64	3.53	3.45	3.44	3.50	3.56	3.52	3.51	48 043
OECD	2.13	2.23	2.36	2.92	2.97	3.20	3.37	3.58	3.36	3.36	3.20	3.24	3.10	3.10	3.06	3.14	3.02	2.95	37 230

Notes: Calculations make use of estimates in Table 3.1.

StatLink http://dx.doi.crg/10.1787/888932801033

Table 3.3. **Telecommunication revenue ratios**

USD

	2000		2001		2002		2003		2004		2005		2006		2007		2008		2009		2010		2011	
	Per TCAP	Per capita	Per TCAP	Per capita	Per TCAP	Per capita	Per TCAP	Per capita	Per TCAP	Per capita	Per TCAP	Per capita	Per TCAP	Per capita	Per TCAP	Per capita	Per TCAP	Per capita	Per TCAP	Per capita	Per TCAP	Per capita	Per TCAP	Per capita
Australia	..	..	715	791	..	..	762	969	..	..	850	1 296	..	..	860	1 450	837	1 418	763	1 322	1 017	1 658	1 135	1 911
Austria	459	552	497	627	511	657	615	821	629	919	616	940	559	914	552	943	524	946	445	844	369	745	352	747
Belgium	709	709	569	689	566	728	678	919	729	1 042	729	1 090	630	972	633	1 041	626	1 087	566	1 010	523	956	510	968
Canada	694	671	629	673	604	675	614	722	648	806	666	884	695	980	733	1 079	741	1 130	685	1 067	739	1 185	763	1 251
Chile	..	..	..	..	..	..	..	..	..	..	383	265	368	301	381	350	408	394	357	382	309	391	317	448
Czech Republic	281	225	241	250	272	321	307	392	317	435	319	477	336	526	399	664	471	808	395	668	380	634	379	641
Denmark	628	782	576	793	548	815	652	1 025	698	1 176	685	1 213	676	1 248	781	1 495	766	1 479	727	1 390	686	1 300	700	1 318
Estonia			276	249	314	307	371	411	389	494	396	562	415	659	422	755	437	789	444	742	364	692	400	767
Finland	588	774	572	807	611	909	648	991	680	1 084	601	1 013	618	1 071	643	1 159	624	1 180	616	1 230	568	1 182	584	1 258
France	572	560	549	601	579	651	666	794	713	908	727	987	685	982	745	1 106	810	1 232	728	1 142	675	1 085	675	1 110
Germany	585	627	553	656	573	709	660	870	686	1 000	653	1 020	597	1 006	570	1 064	559	1 116	507	1 025	471	953	477	992
Greece	435	466	407	512	441	606	535	776	584	884	533	870	529	938	514	1 039	471	1 076	393	956	407	803	401	784
Hungary	481	314	407	338	377	381	411	463	392	476	394	506	349	497	368	575	341	580	273	453	252	428	265	443
Iceland	669	900	533	756	515	794	675	1 104	773	1 305	870	1 568	846	1 549	1 012	1 858	816	1 517	590	1 093	601	1 116	661	1 225
Ireland	615	591	559	641	661	813	780	1 006	869	1 196	789	1 177	778	1 257	831	1 424	870	1 495	736	1 256	658	1 137	671	1 183
Israel	..	..	..	..	..	..	..	..	..	..	..	..	..	..	..	..	..	..	447	871	523	961	596	1 052
Italy	445	523	424	568	462	637	549	812	589	937	565	991	529	998	514	1 065	536	1 111	488	1 011	430	917	429	928
Japan	1 262	1 286	1 128	1 232	864	1 015	867	1 090	794	1 054	746	1 033	712	1 016	697	1 017	736	1 091	812	1 192	789	1 189	842	1 301
Korea	444	503	391	500	409	568	428	608	468	695	543	829	607	928	640	1 006	542	903	473	812	519	916	547	979
Luxembourg	617	778	544	842	540	883	591	1 046	564	1 151	550	1 217	579	1 295	633	1 405	669	1 520	622	1 410	567	1 287	588	1 368
Mexico	542	146	450	161	415	169	367	169	332	186	323	215	333	249	320	275	293	284	243	253	248	280	253	291
Netherlands	529	637	591	724	650	804	764	1 024	739	1 146	766	1 164	742	1 175	610	1 124	661	1 278	621	1 188	594	1 110	605	1 155
New Zealand	564	575	502	543	560	621	681	765	698	860	651	904	520	761	526	850	472	811	406	710	458	801	491	874
Norway	464	583	469	624	535	740	601	874	611	975	615	1 029	623	1 041	668	1 134	687	1 183	623	1 063	643	1 108	680	1 162
Poland	332	142	322	172	290	181	276	200	282	251	272	300	254	339	276	398	312	469	238	363	239	371	233	379
Portugal	483	494	502	582	482	622	542	751	584	860	555	874	522	871	564	1 002	559	1 067	507	989	471	940	421	851
Slovak Republic	268	148	254	175	237	190	270	250	291	302	316	345	311	364	334	470	440	583	413	544	347	480	359	498
Slovenia	..	..	..	..	..	..	..	..	..	..	463	595	500	653	561	747	604	897	493	735	455	674	540	799
Spain	545	565	504	589	601	762	679	924	762	1 071	761	1 177	732	1 199	790	1 350	817	1 424	704	1 247	635	1 137	628	1 143
Sweden	353	498	358	542	360	581	407	698	440	757	413	735	398	731	429	806	428	820	353	711	375	770	415	860
Switzerland	936	1 144	919	1 200	930	1 296	1 035	1 535	1 120	1 732	1 037	1 722	984	1 726	999	1 879	1 070	2 087	1 027	2 053	1 063	2 168	1 177	2 458
Turkey	184	96	157	90	159	102	222	156	211	169	193	181	162	173	192	231	200	250	181	216	200	234	186	222
United Kingdom	610	745	611	798	594	823	660	976	711	1 158	667	1 182	665	1 249	701	1 378	633	1 281	506	1 049	478	1 002	485	1 022
United States	1 343	1 424	1 288	1 471	1 199	1 430	1 142	1 431	1 094	1 464	1 056	1 499	1 020	1 537	1 020	1 591	1 027	1 635	1 000	1 611	1 001	1 638	1 028	1 686
OECD	823	768	778	803	713	784	745	872	723	916	716	976	671	976	689	1 064	689	1 102	648	1 053	639	1 053	655	1 098

Notes: Total communication access paths (TCAP) = analogue lines + ISDN lines + DSL + cable modem + fibre + mobile subscribers. Revenue calculations rely on estimates derived for Table 3.1.

StatLink http://dx.doi.org/10.1787/888932801052

Table 3.4. **Mobile telecommunication revenue**

USD millions

	1997	1998	1999	2000	2001	2002	2003	2004	2005	2006	2007	2008	2009	2010	2011
Australia	..	..	..	..	..	..	..	..	..	..	12 357	14 097	13 942	16 373	18 398
Austria	763	1 358	1 736	2 126	2 438	2 759	3 574	4 396	4 678	4 648	4 878	5 036	4 564	4 055	4 132
Belgium	..	..	..	..	2 954	3 121	4 086	4 835	5 116	5 422	5 737	6 086	5 558	5 214	5 528
Canada	2 092	2 957	2 955	3 604	3 852	4 593	5 759	7 292	9 105	11 236	13 527	14 987	14 819	17 479	19 317
Chile	..	..	..	..	..	..	..	..	2 033	2 499	3 079	3 591	3 446	3 842	4 381
Czech Republic	368	597	850	1 162	1 414	1 651	2 208	974	1 798	3 385	3 959	5 004	4 243	3 890	3 955
Denmark	762	829	897	983	1 037	1 276	1 768	2 133	2 418	2 652	3 219	3 276	3 169	3 038	3 067
Estonia	..	..	..	..	115	137	188	263	288	333	392	565	520	436	590
Finland	5 299	1 295	1 588	1 666	1 796	2 137	2 528	2 948	2 672	2 825	3 078	2 980	2 917	2 829	2 958
France	4 708	4 385	6 393	7 146	8 954	11 121	14 880	18 356	20 249	30 231	34 337	39 076	36 063	34 486	35 451
Germany	10 092	10 556	13 936	15 963	17 143	18 774	23 708	28 148	28 750	28 875	31 781	33 529	30 833	29 211	30 417
Greece	787	1 127	1 564	1 819	2 096	2 925	4 022	5 043	4 949	5 661	6 293	6 615	5 926	4 712	4 483
Hungary	768	712	764	1 043	1 312	1 574	2 016	2 249	2 656	2 731	3 375	3 522	2 108	1 962	2 029
Iceland	27	36	46	111	104	96	112	159	199	220	254	179	124	122	135
Ireland	291	385	777	1 045	1 252	1 110	1 569	2 230	2 282	2 407	2 810	3 025	2 517	2 229	2 279
Israel	..	..	..	..	..	..	..	..	..	..	..	..	3 641	3 826	3 997
Italy	6 591	9 444	12 128	11 643	14 250	16 427	21 722	26 365	28 699	30 049	32 832	34 474	31 392	28 849	29 358
Japan	43 619	45 697	60 028	74 948	75 383	74 706	74 706	78 942	82 983	87 140	95 804	109 142	120 561	128 514	141 347
Korea	3 534	3 798	7 284	10 705	10 462	11 996	12 684	14 378	17 283	19 464	21 194	18 986	17 217	19 415	20 175
Luxembourg	23	26	81	82	112	123	193	242	284	310	344	371	348	335	376
Mexico	659	1 025	1 772	3 511	4 983	6 226	6 978	8 660	10 958	13 664	16 371	17 735	16 524	18 616	19 746
Netherlands	1 423	2 164	2 580	3 412	4 129	4 434	6 067	5 108	5 136	5 128	8 844	9 015	8 400	7 808	8 086
New Zealand	207	315	481	625	612	660	828	1 121	1 380	1 251	1 452	1 408	1 281	1 489	1 685
Norway	830	622	760	832	999	1 319	1 588	1 850	2 091	2 254	2 574	2 744	2 560	2 782	3 042
Poland	368	668	1 416	1 931	2 621	2 941	3 617	4 704	5 282	6 092	6 071	7 620	5 981	6 179	6 304
Portugal	984	1 155	1 541	1 721	1 791	2 015	2 618	2 817	2 938	3 071	3 555	3 902	3 671	3 338	3 378
Slovak Republic	..	25	13	275	354	416	719	952	1 117	1 274	1 635	2 017	1 880	1 697	1 765
Slovenia	..	..	..	..	..	..	..	..	507	507	643	753	666	607	675
Spain	3 183	4 327	3 638	4 490	5 639	7 025	9 848	12 712	14 977	16 564	20 233	22 158	24 138	21 764	21 921
Sweden	1 104	1 351	1 532	1 538	1 560	1 709	2 052	2 223	2 295	2 339	2 846	3 093	2 832	3 365	4 040
Switzerland	946	1 237	1 670	1 868	2 298	2 703	3 313	3 820	3 843	3 787	4 065	4 671	4 248	4 694	5 699
Turkey	575	336	669	854	756	2 512	3 658	4 750	6 436	6 758	9 841	9 602	8 369	9 226	8 597
United Kingdom	..	6 067	7 760	9 188	10 067	10 352	16 021	21 691	23 642	25 693	29 946	28 635	23 316	22 942	24 291
United States	..	42 584	54 180	65 846	86 336	99 193	112 089	127 602	139 920	156 013	170 583	181 475	185 684	195 525	209 873
OECD	90 004	145 079	189 040	230 137	266 820	296 029	345 119	396 965	436 963	484 483	557 904	599 370	593 457	610 844	651 474

	% of total revenue														
	1997	1998	1999	2000	2001	2002	2003	2004	2005	2006	2007	2008	2009	2010	2011
Australia	..	..	..	..	..	..	..	..	..	..	40.1	45.8	47.7	44.0	42.3
Austria	20.5	33.0	34.8	48.1	48.3	52.0	53.7	58.5	60.5	61.5	62.3	63.8	64.7	64.9	65.7
Belgium	..	..	..	..	41.7	41.5	42.9	44.5	44.8	52.9	51.9	52.3	51.0	50.1	52.0
Canada	12.2	15.4	15.3	17.5	18.5	21.7	25.2	28.3	31.9	35.2	38.1	39.8	41.2	43.2	44.8
Chile	..	..	..	..	..	..	..	..	47.2	50.5	53.0	54.3	53.3	57.5	56.6
Czech Republic	25.3	32.6	40.3	50.2	55.3	50.5	55.2	21.9	36.8	62.7	57.8	59.4	60.5	58.4	58.8
Denmark	21.9	22.1	20.3	23.6	24.4	29.1	32.0	33.6	36.8	39.1	39.4	40.3	41.3	42.2	41.8
Estonia	..	..	..	..	33.7	32.8	33.7	39.5	38.1	37.6	38.6	53.4	52.3	47.0	57.4
Finland	172.0	35.6	39.3	41.6	42.9	45.2	48.9	52.0	50.3	50.1	50.2	47.5	44.4	44.6	43.6
France	16.4	14.7	18.9	21.0	24.4	27.7	30.2	32.4	32.6	48.5	48.7	49.5	49.0	49.0	49.0
Germany	23.2	21.5	27.2	31.0	31.7	32.1	33.0	34.1	34.2	34.8	36.3	36.6	36.8	37.5	37.5
Greece	23.9	26.3	36.9	35.7	37.4	43.9	47.0	51.6	51.2	54.1	54.1	54.7	54.9	51.9	50.5
Hungary	35.9	28.3	24.9	32.5	38.1	40.7	43.0	46.8	52.1	54.5	58.4	60.5	46.5	45.9	45.9
Iceland	17.9	21.4	24.2	43.8	48.2	42.2	35.2	41.8	42.9	46.8	43.8	36.9	35.5	34.3	34.5
Ireland	13.7	20.2	40.3	46.5	50.5	34.7	39.0	45.8	46.6	44.9	45.2	45.5	44.9	43.8	42.9
Israel	..	..	..	..	..	..	..	..	..	..	..	..	55.9	52.2	49.0
Italy	27.4	34.0	40.6	39.1	44.1	45.1	46.5	48.4	49.4	51.1	51.9	51.9	51.6	52.0	52.1
Japan	37.4	40.4	41.9	45.9	48.1	57.8	53.7	58.6	62.8	67.1	73.7	78.3	79.3	84.4	84.4
Korea	24.2	31.6	41.2	45.3	44.2	44.3	43.6	43.1	43.3	43.4	43.4	43.0	43.1	42.9	41.4
Luxembourg	7.4	7.6	22.2	24.1	30.0	31.2	40.9	45.9	50.2	50.6	50.9	49.9	49.6	51.3	52.9
Mexico	7.5	10.6	15.7	24.5	31.0	36.5	40.5	45.4	49.1	52.5	56.3	58.7	60.8	61.5	62.0
Netherlands	18.0	22.8	24.1	33.6	35.6	34.1	36.5	27.4	27.0	26.7	48.0	42.9	42.8	42.4	41.9
New Zealand	9.2	15.4	22.2	28.1	28.9	26.8	26.8	31.8	36.8	39.2	40.3	40.6	41.7	42.4	43.6
Norway	23.0	25.2	29.2	31.8	35.5	39.3	39.8	41.3	44.0	46.5	48.2	48.6	49.9	51.4	52.9
Poland	14.2	18.5	30.8	35.6	39.8	42.6	47.3	49.1	46.2	47.2	40.1	42.6	43.2	43.2	43.2
Portugal	24.9	27.4	32.6	34.1	29.9	31.2	33.4	31.2	31.9	33.3	33.4	34.4	34.9	33.4	37.3
Slovak Republic	..	5.3	2.8	34.3	37.6	40.6	53.4	58.6	60.2	65.0	64.5	64.0	63.8	65.1	65.2
Slovenia	..	..	..	..	..	..	..	..	42.6	38.7	42.7	41.6	44.4	43.9	41.1
Spain	17.7	22.0	16.3	19.7	23.5	22.3	25.4	27.8	29.3	31.3	33.4	34.1	42.2	41.6	41.6
Sweden	16.0	18.3	33.1	34.8	32.3	32.9	32.8	32.7	34.6	35.2	38.6	40.9	42.8	46.6	49.7
Switzerland	13.9	16.1	19.1	22.7	26.3	28.4	29.1	29.6	29.8	29.0	28.4	29.0	26.5	27.8	29.5
Turkey	..	..	..	..	..	..	..	..	..	56.2	60.5	53.9	53.8	54.1	52.4
United Kingdom	..	17.7	20.0	20.9	21.4	21.2	27.6	31.3	33.2	34.0	35.6	36.4	36.0	36.8	37.9
United States	..	13.0	14.8	16.4	20.6	24.1	27.0	29.7	31.5	34.0	35.5	36.4	37.5	38.5	39.9
OECD	12.7	19.9	22.9	26.0	28.6	32.3	33.6	36.6	37.5	41.3	43.3	44.6	46.0	47.0	47.8

Notes: Data for Australia (2007 to 2011) are estimates based on Telstra mobile services revenues (revenues for June 2010 are used for the year 2009 in this report, the same applies to the previous years). Data for Hungary (2011), Israel (2010 and 2011), Japan (2008 to 2011) and Poland (2010 and 2011) are estimates. Calculations make use of estimates in Table 3.1.

StatLink http://dx.doi.org/10.1787/888932801071

Table 3.5. **Cellular mobile telecommunication revenue per cellular mobile subscriptions**

USD

	1996	1997	1998	1999	2000	2001	2002	2003	2004	2005	2006	2007	2008	2009	2010	2011	Monthly 2011	Per capita, 2011
Australia	..	..	..	..	..	..	..	..	..	..	..	581	637	576	726	751	63	808
Austria	..	655	590	404	347	373	410	504	550	559	502	495	475	399	331	317	26	491
Belgium	..	..	..	..	..	384	385	475	529	533	551	534	515	449	410	408	34	504
Canada	610	499	553	428	413	362	383	433	485	535	599	667	678	622	677	705	59	560
Chile	..	..	..	..	..	..	..	..	..	192	201	221	243	209	194	196	16	254
Czech Republic	755	705	618	437	267	204	192	227	90	153	273	299	362	298	270	278	23	377
Denmark	581	528	429	341	292	262	285	371	413	444	455	511	500	464	435	428	36	551
Estonia	..	..	..	..	..	155	156	178	210	199	201	208	298	332	234	317	26	440
Finland	2 765	2 533	455	485	447	430	473	533	590	496	498	506	436	379	337	331	28	549
France	1329	818	391	310	241	242	288	357	412	421	585	621	674	586	530	517	43	544
Germany	1 571	1 234	759	594	331	305	318	366	379	363	337	327	313	285	268	267	22	372
Greece	915	839	548	402	307	263	314	389	456	398	408	388	350	292	318	308	26	396
Hungary	1 284	1 088	687	477	339	264	229	254	258	285	274	306	288	179	163	174	14	203
Iceland	434	413	337	267	515	441	369	402	549	654	683	778	530	364	357	391	33	422
Ireland	698	569	407	486	518	452	355	459	589	542	520	565	599	488	423	414	35	507
Israel	..	..	..	..	..	..	..	..	..	..	..	..	..	381	401	408	34	516
Italy	724	560	465	403	275	279	309	383	417	399	374	364	382	349	308	306	26	483
Japan	1 388	1 140	966	1 056	1 122	1 008	921	862	863	860	857	893	974	1 075	1 075	1 103	92	1098
Korea	1 338	513	272	311	399	360	371	378	393	451	484	487	416	359	382	384	32	405
Luxembourg	465	335	199	387	271	258	260	359	375	395	434	502	525	484	461	491	41	723
Mexico	501	378	306	229	249	229	240	232	225	233	247	246	236	199	204	209	17	181
Netherlands	732	843	647	380	310	359	376	463	321	315	301	479	437	406	379	370	31	484
New Zealand	..	292	251	312	286	253	260	319	370	391	329	342	308	273	316	350	29	381
Norway	572	495	300	285	256	278	348	391	409	440	463	511	527	478	497	534	45	614
Poland	..	453	347	363	286	244	212	208	204	181	166	147	173	133	132	126	10	164
Portugal	1 023	653	376	330	258	225	219	262	267	258	251	264	261	229	203	201	17	317
Slovak Republic	..	..	55	19	212	165	142	195	223	246	260	269	365	342	286	295	25	324
Slovenia	..	..	..	..	..	..	..	..	..	288	279	334	367	317	286	311	26	329
Spain	767	735	614	244	188	190	210	265	329	351	362	418	447	473	424	417	34.7	475
Sweden	444	348	329	299	241	217	215	233	253	252	243	281	284	241	265	302	25	428
Switzerland	1 134	906	728	546	403	436	471	535	609	562	509	495	525	456	487	563	47	724
Turkey	345	357	96	86	57	41	108	131	137	148	128	159	146	133	149	132	11	116
United Kingdom	571	0	467	324	230	225	209	303	361	359	367	406	373	291	283	298	25	387
United States	532	0	615	630	601	672	700	698	691	687	679	684	695	677	686	723	60	673
OECD	810	527	588	518	444	430	430	454	463	461	464	486	493	472	470	482	40	525

Notes: Revenue calculations rely on estimates derived for Table 3.4.

StatLink http://dx.doi.org/10.1787/888932801090

Table 3.6. **Public telecommunication investment in the OECD area**

USD millions (excluding spectrum fees)

	Average 1988-90	Average 1991-93	Average 1994-96	1997	1998	1999	2000	2001	2002	2003	2004	2005	2006	2007	2008	2009	2010	2011
Australia	..	..	..	..	..	..	..	..	..	4 166	4 497	4 440	4 373	6 050	6 101	5 672	6 172	7 505
Austria	965	1 308	1 283	996	1 662	2 002	2 619	1 620	905	411	436	949	937	1 203	749	701	917	711
Belgium	614	779	927	719	670	746	952	1 427	1 203	1 181	1 238	1 328	1 266	1 360	1 584	1 503	1 633	2 056
Canada	3 479	3 353	2 811	4 181	4 357	3 904	4 943	5 138	4 154	3 272	3 978	4 573	6 099	7 633	11 215	7 018	8 155	9 495
Chile				1 077	1 045	1 006	1 118	986	610	561	577	736	1 199	1 393	1 754	1 387	1 920	2 406
Czech Republic	..	226	818	1 421	1 164	854	471	599	455	1 267	512	576	627	783	877	763	793	826
Denmark	490	431	612	890	1 077	986	1 116	1 324	970	851	955	1 137	1 237	1 681	1 891	1 555	1 266	1 216
Estonia				91	94	85	100	86	61	67	63	75	99	130	148	91	91	120
Finland	670	510	632	835	595	572	629	657	475	493	511	453	475	515	556	528	711	833
France	4 548	6 081	6 175	6 423	6 153	6 286	7 194	8 198	5 376	6 109	6 781	7 928	8 769	8 411	9 601	8 194	9 645	11 375
Germany	9 263	15 808	12 717	11 896	8 000	8 298	9 083	10 268	6 698	6 180	7 037	7 250	8 125	9 726	10 588	8 472	7 763	8 333
Greece	291	808	751	843	1 552	1 398	1 346	1 534	1 291	1 263	1 343	901	1 006	1 774	2 016	1 871	1 408	1 324
Hungary	216	456	754	764	662	812	820	750	713	625	653	638	635	669	711	951	676	699
Iceland	12	23	30	29	52	56	69	37	24	44	80	90	78	132	94	35	43	43
Ireland	174	202	260	462	515	460	704	442	575	575	665	674	590	627	759	614	583	620
Israel				..	..	..	..	..	..	..	..	..	..	..	..	922	969	1 012
Italy	7 365	8 657	5 065	5 555	5 959	7 187	6 526	7 208	8 936	8 862	8 746	8 609	8 444	9 812	9 837	8 639	8 045	8 221
Japan	15 389	20 339	33 120	32 815	29 023	33 546	36 516	23 917	19 257	20 422	23 191	18 930	21 037	18 487	23 549	25 322	16 936	18 627
Korea	2 587	3 167	4 615	3 049	4 495	7 038	7 766	5 990	6 396	5 205	5 283	5 199	6 251	6 996	6 244	5 101	5 537	6 923
Luxembourg	39	72	96	79	30	55	15	30	49	44	73	56	88	109	129	136	121	159
Mexico	1 409	2 214	1 862	1 971	3 164	4 028	5 226	5 751	3 130	2 584	3 615	3 513	3 699	3 272	3 648	2 891	5 673	4 985
Netherlands	1 144	1 572	1 511	3 274	5 900	10 418	3 174	2 671	1 564	1 821	3 774	2 702	3 306	3 754	4 171	4 073	3 858	4 073
New Zealand	362	367	340	389	298	352	379	377	320	376	418	515	596	787	835	1 059	1 112	972
Norway	500	483	361	541	477	541	578	597	707	524	550	576	640	683	709	636	662	714
Poland	140	489	896	1 006	1 365	1 862	2 434	1 965	2 326	1 363	1 492	2 086	2 598	3 113	3 058	2 560	2 644	2 698
Portugal	562	973	938	1 078	1 216	1 248	1 179	1 274	967	645	838	916	974	1 667	1 470	1 464	1 419	1 227
Slovak Republic	..	..	287	..	..	..	1 354	1 406	642	345	425	420	443	531	571	543	472	520
Slovenia	..	..	..	..	..	..	..	..	..	..	..	237	311	497	590	258	199	167
Spain	4 517	4 265	3 220	2 654	5 090	6 573	9 346	7 313	5 242	5 104	5 821	6 894	7 107	7 884	7 605	5 940	5 887	6 052
Sweden	1 079	1 164	1 197	1 404	1 159	1 014	1 637	1 714	1 423	1 452	1 577	1 182	1 382	1 583	1 470	1 190	974	1 463
Switzerland	1 597	1 786	1 761	1 637	1 275	2 034	2 245	1 643	1 653	1 580	1 661	1 624	5 190	1 992	2 110	2 103	2 073	2 653
Turkey	548	787	500	553	4 225	3 777	3 541	2 949	2 159	2 204	368	1 389	1 154	1 907	3 212	2 494	2 507	2 315
United Kingdom	4 830	3 738	4 887	9 971	8 987	12 800	14 122	14 159	10 185	10 933	20 869	18 779	17 696	18 934	10 776	8 020	7 897	8 279
United States	23 401	26 064	37 751	56 963	65 079	84 433	113 301	105 607	61 442	60 522	62 715	70 760	79 884	79 717	80 651	67 262	70 149	70 149
OECD	88 514	108 296	129 227	153 564	165 339	204 368	240 500	217 640	149 907	151 051	170 743	176 134	196 314	203 813	209 280	179 968	178 910	188 773

Notes: Data for Australia (2004, 2006, 2008 and 2009), Hungary (2011), Ireland (2005), Israel (2010 and 2011), Japan (2011), Norway (2009 to 2011) and Poland (2010 and 2011), United Kingdom (2010 and 2011) and United States (2011) are estimates.

StatLink http://dx.doi.org/10.1787/888932801109

Table 3.7. **Telecommunication investment by region**

USD millions

	Average 1988-90	Average 1991-93	Average 1994-96	1997	1998	1999	2000	2001	2002	2003	2004	2005	2006	2007	2008	2009	2010	2011	Average 2009-11
Europe	39 603	50 662	60 353	53 120	57 879	70 061	71 251	69 874	54 599	53 942	66 469	67 470	73 176	79 478	75 283	64 254	63 256	67 710	65 074
(%)	*45*	*47*	*35*	*35*	*35*	*34*	*30*	*32*	*36*	*36*	*39*	*38*	*37*	*39*	*36*	*36*	*35*	*36*	*36*
America	28 289	31 631	77 069	64 192	73 644	93 371	124 588	117 482	69 336	66 940	70 885	79 581	90 882	92 015	97 268	78 558	85 897	87 035	83 830
(%)	*32*	*29*	*44*	*42*	*45*	*46*	*52*	*54*	*46*	*44*	*42*	*45*	*46*	*45*	*46*	*44*	*48*	*46*	*45*
Asia/Pacific	20 622	26 003	37 002	36 252	33 816	40 936	44 661	30 285	25 973	30 169	33 388	29 083	32 257	32 320	36 729	37 155	29 757	34 027	33 646
(%)	*23*	*24*	*21*	*24*	*20*	*20*	*19*	*14*	*17*	*20*	*20*	*17*	*16*	*16*	*18*	*21*	*17*	*18*	*18*
OECD	88 514	108 296	174 424	153 564	165 339	204 368	240 500	217 640	149 907	151 051	170 743	176 134	196 314	203 813	209 280	179 968	178 910	188 773	182 550

Notes: Calculations include estimates derived for Table 3.6.

StatLink http://dx.doi.org/10.1787/888932801147

Table 3.8. **Public telecommunication investment as a percentage of telecommunications revenue**

	Average 1988-90	Average 1991-93	Average 1994-96	Average 1997-99	1997	1998	1999	2000	2001	2002	2003	2004	2005	2006	2007	2008	2009	2010	2011
Australia	50.8	24.1	33.4	..	..	..	..	..	..	..	21.5	..	16.7	..	19.6	19.8	19.4	16.6	17.2
Austria	47.9	48.6	37.5	35.7	26.8	40.4	40.1	59.2	32.1	17.0	6.2	5.8	12.3	12.4	15.4	9.5	9.9	14.7	11.3
Belgium	32.9	30.5	28.1	14.3	17.0	13.1	12.6	13.1	20.1	16.0	12.4	11.4	11.6	12.4	12.3	13.6	13.8	15.7	19.4
Canada	38.0	27.6	23.3	22.5	24.5	22.6	20.3	24.0	24.6	19.6	14.3	15.5	16.0	19.1	21.5	29.8	19.5	20.2	22.0
Chile	..	..	..	..	..	..	..	..	..	..	..	..	17.1	24.2	24.0	26.5	21.5	28.8	31.1
Czech Republic	..	68.6	131.5	67.3	97.9	63.5	40.5	20.4	23.4	13.9	31.7	11.5	11.8	11.6	11.4	10.4	10.9	11.9	12.3
Denmark	29.9	19.3	21.6	25.5	25.5	28.7	22.2	26.7	31.2	22.1	15.4	15.0	17.3	18.2	20.6	23.3	20.3	17.6	16.6
Estonia	..	..	..	..	..	..	..	..	25.3	14.5	12.0	9.5	10.0	11.2	12.8	14.0	9.2	9.8	11.7
Finland	47.8	25.1	35.1	19.2	27.1	16.4	14.2	15.7	15.7	10.0	9.5	9.0	8.5	8.4	8.4	8.9	8.0	11.2	12.3
France	30.6	32.7	26.9	20.6	22.4	20.6	18.6	21.1	22.3	13.4	12.4	12.0	12.8	14.1	11.9	12.2	11.1	13.7	15.7
Germany	47.8	48.5	34.6	20.0	27.4	16.3	16.2	17.6	19.0	11.5	8.6	8.5	8.6	9.8	11.1	11.6	10.1	10.0	10.3
Greece	32.7	66.8	38.0	31.6	25.6	36.2	33.0	26.4	27.4	19.4	14.8	13.7	9.3	9.6	15.2	16.7	17.3	15.5	14.9
Hungary	82.9	122.3	71.5	29.5	35.7	26.3	26.4	25.6	21.8	18.4	13.3	13.6	12.5	12.7	11.6	12.2	21.0	15.8	15.8
Iceland	17.6	27.8	28.8	26.4	18.9	31.1	29.2	27.5	17.3	10.6	13.7	20.9	19.5	16.6	22.9	19.5	10.0	12.1	11.1
Ireland	21.7	20.2	24.0	24.2	21.7	26.9	23.9	31.3	17.8	18.0	14.3	13.7	13.8	11.0	10.1	11.4	10.9	11.5	11.7
Israel	..	..	..	..	..	..	..	..	..	..	..	..	..	..	..	..	14.2	13.2	12.4
Italy	64.3	54.0	27.7	22.9	23.1	21.5	24.0	21.9	22.3	24.5	19.0	16.0	14.8	14.4	15.5	14.8	14.2	14.5	14.6
Japan	40.2	43.1	45.3	25.7	28.2	25.6	23.4	22.4	15.3	14.9	14.7	17.2	14.3	16.2	14.2	16.9	16.7	11.1	11.1
Korea	87.5	59.6	61.7	32.7	20.9	37.4	39.8	32.8	25.3	23.6	17.9	15.8	13.0	13.9	14.3	14.1	12.8	12.2	14.2
Luxembourg	49.6	53.5	39.8	16.6	25.8	8.9	15.1	4.5	8.1	12.4	9.3	13.8	9.9	14.3	16.2	17.4	19.3	18.5	22.4
Mexico	112.5	55.9	24.0	30.3	22.5	32.8	35.7	36.5	35.8	18.3	15.0	18.9	15.8	14.2	11.2	12.1	10.6	18.7	15.7
Netherlands	33.2	17.8	23.5	67.0	41.5	62.2	97.2	31.3	23.0	12.0	11.0	20.2	14.2	17.2	20.4	19.9	20.7	20.9	21.1
New Zealand	32.2	25.6	23.4	16.0	17.3	14.6	16.2	17.0	17.8	13.0	12.1	11.8	13.7	18.7	21.8	24.0	34.5	31.7	25.2
Norway	25.5	21.9	14.4	18.4	15.0	19.3	20.8	22.0	21.2	21.1	13.1	12.3	12.1	13.2	12.8	12.6	12.4	12.2	12.4
Poland	29.8	69.8	59.4	39.0	38.8	37.7	40.5	44.8	29.9	33.7	17.8	15.6	18.2	20.1	20.5	17.1	18.5	18.5	18.5
Portugal	62.1	70.2	43.5	27.5	27.2	28.8	26.4	23.4	21.3	15.0	8.2	9.3	9.9	10.6	15.7	13.0	13.9	14.2	13.5
Slovak Republic	..	..	197.3	..	..	..	..	169.0	149.3	62.7	25.7	26.2	22.6	22.6	20.9	18.1	18.4	18.1	19.2
Slovenia	..	..	..	..	..	..	..	..	..	..	..	..	19.9	23.8	33.0	32.6	17.2	14.4	10.1
Spain	109.0	51.5	31.3	23.3	14.7	25.9	29.4	41.1	30.5	16.7	13.1	12.7	13.5	13.4	13.0	11.7	10.4	11.2	11.5
Sweden	34.5	23.2	23.0	19.3	20.3	15.7	21.9	37.1	35.5	27.4	23.2	23.2	17.8	20.8	21.5	19.4	18.0	13.5	18.0
Switzerland	45.1	39.0	28.4	21.3	24.1	16.6	23.3	27.2	18.8	17.4	13.9	12.9	12.6	39.8	13.9	13.1	13.1	12.3	13.7
Turkey	52.6	37.3	20.8	55.7	13.7	84.0	69.4	57.4	50.3	32.2	21.1	3.2	11.2	9.6	11.7	18.0	16.0	14.7	14.1
United Kingdom	28.6	15.3	19.2	19.8	..	26.3	33.0	32.2	30.0	20.9	18.8	30.1	26.4	23.4	22.5	13.7	12.4	12.7	12.9
United States	17.6	17.6	21.9	14.3	..	19.9	23.1	28.2	25.2	14.9	14.6	14.6	15.9	17.4	16.6	16.2	13.6	13.8	13.3
OECD	31.6	29.7	29.4	23.0	21.6	22.7	24.8	27.2	23.4	16.4	14.7	15.7	15.1	16.7	15.8	15.6	14.0	13.8	13.9

Note: Calculations include unofficial estimates derived from Tables 3.1 and 3.6.

StatLink http://dx.doi.org/10.1787/888932801166

Table 3.9. **Investment in cellular mobile infrastructure in the OECD area**

USD millions, excluding spectrum fees

	1997	1998	1999	2000	2001	2002	2003	2004	2005	2006	2007	2008	2009	2010	2011	Percent of total investment 2011
Australia	..	..	..	..	..	..	..	..	..	..	..	..	..	..	..	..
Austria	..	1 211	1 069	1 958	833	502	205	212	483	..	..	..	..	..	..	..
Belgium	..	..	..	..	571	368	302	410	402	513	473	593	473	514	576	28.0
Canada	1 371	988	811	1 346	1 223	1 232	929	846	1 157	1 504	1 776	5 701	1 930	1 748	2 323	24.5
Chile	308	498	616	515	319	224	236	281	361	696	703	906	722	1 082	1 406	58.5
Czech Republic	337	101	317	731	625	355	238	250	368	..	246	279	238	343	442	53.5
Denmark	..	..	..	..	..	..	..	..	..	..	..	..	..	229	215	17.7
Estonia	40	44	34	42	46	33	43	32	31	42	62	76	47	49	58	48.4
Finland	1 352	..		..	..	..	..	..	..	..	..	..	208	263	306	36.7
France	..	..	..	..	..	..	..	..	..	..	..	..	..	..	..	..
Germany	..	2 000	2 872	3 211	2 768	2 264	2 809	3 210	3 125	3 375	2 740	3 235	2 917	2 632	2 632	31.6
Greece	..	..	..	620	533	489	522	730	530	595	666	710	678	589	578	43.7
Hungary	163	..	..	376	422	419	210	265	251	..	..	..	630	445	445	63.6
Iceland	3	6	10	..	..	..	10	10	19	..	..	..	..	..	..	..
Ireland	..	..	..	..	..	..	..	..	..	..	346	353	272	332	354	57.1
Israel	..	..	..	..	..	..	..	..	..	..	..	..	382	382	382	37.8
Italy	1 170	1 745	2 274	3 034	3 318	4 840	4 135	4 605	4 129	..	4 477	4 551	3 399	3 243	3 351	40.8
Japan	..	..	..	..	..	..	..	..	..	..	..	..	..	..	..	..
Korea	1 609	2 088	3 147	3 545	2 045	2 645	2 864	2 640	2 441	3 236	3 379	2 843	2 364	2 755	3 666	52.9
Luxembourg	..	..	..	..	..	..	..	..	..	..	28	37	36	28	36	22.4
Mexico	324	829	1 275	2 134	1 810	1 188	1 151	1 509	1 384	1 077	1 059	1 487	1 181	3 083	2 261	45.4
Netherlands	..	..	..	..	..	..	..	..	..	..	..	..	..	..	..	..
New Zealand	..	..	..	..	..	..	40	45	63	..	..	..	275	112	157	16.2
Norway	..	..	..	..	..	..	..	..	..	..	..	..	..	..	..	..
Poland	..	..	..	..	..	279	355	..	728	902	1 181	1 371	1 105	1 105	1 105	40.9
Portugal	329	674	739	552	484	460	372	501	522	545	818	621	401	462	462	37.6
Slovak Republic	..	..	..	381	..	222	160	148	192	191	319	328	225	172	346	66.6
Slovenia	..	..	..	..	..	..	..	..	..	116	146	139	82	59	59	35.7
Spain	478	..	..	2 642	1 756	..	1 612	2 277	2 753	..	..	..	1 377	1 421	1 218	20.1
Sweden	302	174	192	162	224	591	640	530	392	293	477	506	264	249	789	53.9
Switzerland	171	248	745	616	509	586	627	695	515	389	417	429	421	448	496	18.7
Turkey	..	..	..	..	..	..	..	..	..	..	..	2 003	2 972	1 774	1 493	64.5
United Kingdom	1 866	..	..	..	..	..	..	..	..	..	..	3 069	2 444	2 222	2 000	24.2
United States	..	8 228	14 422	25 482	24 028	23 033	20 989	23 998	27 337	27 969	22 987	25 272	20 651	23 029	23 029	32.8

Notes: Data for Germany (2011), Hungary (2011), Israel (2010 and 2011), Poland (2010 and 2011), Portugal (2011), United Kingdom (2010) and United States (2011) are estimates.

StatLink http://dx.doi.org/10.1787/888932801128

Table 3.10. **Public telecommunication investment per total communication access path**

USD

	Average 1988-90	Average 1991-93	Average 1994-96	Average 1997-99	1997	1998	1999	2000	2001	2002	2003	2004	2005	2006	2007	2008	2009	2010	2011	Average 2009-11	Monthly average 2009-11
Australia	294.8	248.8	328.4	..	..	..	..	..	..	..	163.6	158.4	141.7	130.5	168.8	165.7	148.0	168.7	195.7	170.8	14.2
Austria	310.4	377.6	343.3	251.9	210.5	288.8	256.5	272.0	159.8	87.2	38.0	36.5	75.6	69.3	84.7	49.7	44.2	54.1	39.9	46.1	3.8
Belgium	164.3	183.1	196.8	106.1	119.7	103.2	95.3	92.9	114.6	90.6	84.1	83.1	84.9	77.8	77.9	85.2	78.0	82.2	98.6	86.3	7.2
Canada	238.6	206.1	159.4	167.9	182.4	174.9	146.3	166.6	154.7	118.5	87.9	100.2	106.8	132.9	157.3	220.9	133.7	149.1	168.0	150.3	12.5
Chile	..	..	..	..	..	..	..	..	..	..	..	59.4	65.4	89.1	91.3	108.1	76.7	88.8	98.6	88.0	7.3
Czech Republic	25.2	123.4	333.2	256.9	374.4	247.7	148.5	57.1	56.4	37.9	97.2	36.5	37.6	39.1	45.6	49.0	43.0	45.2	46.6	44.9	3.7
Denmark	171.9	143.4	189.4	190.8	193.1	209.8	169.5	168.0	179.7	121.3	100.4	104.9	118.5	123.2	160.8	178.2	147.3	120.5	116.0	127.9	10.7
Estonia	..	..	..	123.3	148.8	126.2	95.0	91.7	69.6	45.7	44.4	36.9	39.5	46.3	54.0	61.0	40.7	35.7	46.8	41.1	3.4
Finland	260.2	186.1	221.1	120.1	166.6	102.6	91.0	92.3	89.7	61.3	61.8	61.3	51.3	52.1	54.0	55.3	49.5	63.6	71.8	61.6	5.1
France	168.6	199.9	187.3	146.2	169.6	145.6	123.4	121.0	122.6	77.6	82.5	85.2	92.7	96.5	88.9	98.5	81.0	92.6	106.2	93.3	7.8
Germany	312.2	438.3	298.6	173.7	243.5	147.2	130.5	103.1	105.0	65.6	56.8	58.5	56.2	58.5	63.3	64.6	51.2	47.0	49.0	49.1	4.1
Greece	76.8	180.4	145.7	161.1	132.3	204.3	146.7	115.1	111.3	85.6	78.9	80.3	49.7	50.9	78.4	78.5	68.1	63.2	59.9	63.7	5.3
Hungary	233.8	349.5	337.7	166.3	198.0	146.1	154.9	122.9	88.8	69.5	54.8	53.2	49.3	44.3	42.5	41.6	57.1	39.9	41.9	46.3	3.9
Iceland	96.6	166.5	198.5	164.0	129.1	195.7	167.3	183.7	92.2	54.7	92.7	161.9	169.5	140.4	231.6	159.0	58.7	72.9	73.1	68.3	5.7
Ireland	191.4	182.2	197.8	191.3	229.6	203.3	141.0	192.4	99.8	119.1	111.5	118.8	108.5	85.6	83.9	99.4	80.6	75.5	78.3	78.1	6.5
Israel	..	..	..	..	..	..	..	..	..	..	..	..	..	..	..	..	63.3	69.2	74.0	68.8	5.7
Italy	346.8	366.0	202.7	137.2	150.1	131.2	130.5	97.5	94.5	113.3	104.1	94.4	83.8	75.9	79.7	79.3	69.3	62.3	62.6	64.7	5.4
Japan	294.8	350.9	530.4	290.1	324.6	264.0	281.6	282.2	172.1	128.6	127.2	136.6	107.0	115.3	99.1	124.3	135.3	87.7	93.6	105.5	8.8
Korea	194.5	202.8	244.8	130.9	109.8	129.2	153.6	145.7	99.0	96.6	76.5	74.0	70.8	84.6	91.7	76.6	60.5	63.5	77.8	67.2	5.6
Luxembourg	222.5	353.6	409.7	151.5	240.6	85.0	129.0	27.5	44.2	67.2	54.7	77.6	54.2	82.7	102.5	116.2	120.2	105.0	131.6	118.9	9.9
Mexico	289.7	325.6	213.8	211.1	179.2	238.3	215.9	197.8	161.3	76.1	55.2	62.8	50.9	47.3	36.0	35.3	25.8	46.4	39.6	37.3	3.1
Netherlands	170.7	212.4	185.0	509.3	302.7	530.9	694.5	165.5	136.1	78.2	83.9	149.6	109.0	127.8	124.5	131.2	128.8	124.3	127.7	126.9	10.6
New Zealand	254.5	242.8	205.2	121.1	157.8	98.8	106.8	96.1	89.4	72.6	82.6	82.7	89.4	97.0	114.8	113.5	140.0	144.9	123.5	136.1	11.3
Norway	241.1	213.1	145.1	113.6	130.2	104.9	105.8	102.2	99.6	112.7	78.9	75.1	74.5	82.1	85.5	86.4	77.2	78.6	84.4	80.1	6.7
Poland	44.8	123.1	155.6	130.2	120.9	131.1	138.5	148.7	96.2	97.6	49.3	43.9	49.6	51.1	56.6	53.3	44.0	44.1	43.0	43.7	3.6
Portugal	267.6	325.2	257.7	173.6	200.5	174.5	145.7	112.8	106.6	72.2	44.6	54.2	55.2	55.1	88.4	72.5	70.6	66.8	56.9	64.8	5.4
Slovak Republic	..	71.8	256.0	..	..	..	..	452.5	379.5	148.3	69.2	76.2	71.6	70.3	69.9	79.7	76.0	62.9	68.9	69.3	5.8
Slovenia	..	..	..	..	..	..	..	..	..	..	..	..	92.2	118.7	185.2	196.5	84.7	65.5	54.8	68.3	5.7
Spain	383.1	309.4	212.5	183.9	130.0	216.4	205.3	223.9	153.8	100.1	89.2	97.0	102.7	98.4	102.9	95.7	73.1	71.4	72.1	72.2	6.0
Sweden	188.7	196.3	197.6	118.6	151.8	113.7	90.3	130.8	127.0	98.9	94.6	101.9	73.6	82.8	92.2	83.2	63.5	50.6	74.7	62.9	5.2
Switzerland	421.7	425.0	389.3	268.2	307.3	215.3	282.1	254.9	172.6	161.7	143.8	144.1	130.4	391.4	139.1	140.2	134.9	130.6	161.4	142.3	11.9
Turkey	92.9	79.1	35.8	128.1	31.9	206.4	146.1	105.8	79.0	51.1	46.9	6.8	21.7	15.5	22.5	36.1	29.1	29.5	26.3	28.3	2.4
United Kingdom	195.4	141.7	166.5	231.0	260.4	202.2	230.3	196.4	183.5	124.0	124.1	214.1	175.9	155.6	157.8	86.6	62.6	60.6	62.6	61.9	5.2
United States	178.8	182.2	238.3	312.9	304.1	320.3	314.3	378.5	324.4	179.0	166.2	159.8	168.5	177.4	169.3	166.3	135.8	138.4	137.1	137.1	11.4
OECD	227.8	246.2	261.7	229.4	236.4	226.2	225.6	223.9	181.7	116.7	109.6	113.7	108.3	112.3	109.1	107.4	90.5	87.9	90.8	89.7	7.5

Notes: Calculations include estimates derived from Tables 3.6 and 4.2. Total communication access paths = analogue lines + ISDN lines + DSL + cable modem + fibre + mobile subscribers.

StatLink http://dx.doi.org/10.1787/888932801185

Table 3.11. **Public telecommunication investment per capita**

USD

	Average 1988-90	Average 1991-93	Average 1994-96	Average 1997-99	1997	1998	1999	2000	2001	2002	2003	2004	2005	2006	2007	2008	2009	2010	2011	Average 2009-11	Monthly average 2011
Australia	136.0	121.8	168.7	..	..	..	..	..	..	..	208.2	222.1	216.1	209.5	284.6	280.8	256.3	275.0	329.6	287.0	23.9
Austria	126.2	165.6	159.4	194.6	125.0	208.4	250.5	326.9	201.5	111.9	50.7	53.3	115.3	113.3	144.9	89.8	83.8	109.3	84.5	92.5	7.7
Belgium	61.8	77.6	91.4	69.7	70.7	65.7	72.9	92.9	138.8	116.5	113.9	118.8	126.8	120.1	128.0	147.9	139.3	150.0	187.3	158.9	13.2
Canada	127.6	118.2	95.8	137.6	139.8	144.5	128.4	161.1	165.6	132.5	103.4	124.6	141.8	187.2	231.8	336.6	208.1	239.0	275.4	240.8	20.1
Chile	..	..	..	..	..	..	..	..	..	..	..	35.9	45.2	73.0	83.9	104.6	82.0	112.3	139.3	111.2	9.3
Czech Republic	3.8	21.9	79.2	111.4	137.9	113.1	83.1	45.9	58.6	44.6	124.2	50.2	56.2	61.1	75.8	84.1	72.7	75.4	78.7	75.6	6.3
Denmark	95.4	83.4	117.0	185.6	168.3	203.2	185.2	209.0	247.2	180.4	157.9	176.8	209.9	227.4	307.8	344.3	281.5	228.3	218.3	242.7	20.2
Estonia	..	..	..	64.6	64.4	67.4	61.8	72.8	62.9	44.7	49.2	46.7	56.0	73.6	96.5	110.2	67.9	68.0	89.6	75.2	6.3
Finland	134.9	101.1	123.8	129.6	162.4	115.5	110.8	121.5	126.7	91.2	94.5	97.8	86.4	90.2	97.5	104.6	98.9	132.5	154.7	128.7	10.7
France	80.6	106.3	106.8	104.7	107.4	102.5	104.2	118.5	134.0	87.3	98.5	108.5	125.9	138.3	131.9	149.7	127.1	148.8	174.5	150.1	12.5
Germany	148.8	196.2	155.7	114.5	145.0	97.5	101.1	110.5	124.7	81.2	74.9	85.3	87.9	98.6	118.2	128.9	103.5	95.0	101.9	100.1	8.3
Greece	28.9	78.4	71.9	116.6	78.2	143.2	128.5	123.3	140.1	117.5	114.6	121.4	81.2	90.3	158.5	179.4	165.8	124.5	117.0	135.8	11.3
Hungary	20.8	44.2	73.7	72.7	74.2	64.5	79.3	80.3	73.7	70.2	61.7	64.6	63.3	63.1	66.5	70.9	94.9	67.6	70.1	77.6	6.5
Iceland	47.1	89.1	112.5	165.3	105.3	189.4	201.2	247.0	130.7	84.4	151.6	273.2	305.3	257.2	425.2	295.6	108.8	135.4	135.4	126.5	10.5
Ireland	49.5	57.0	72.2	129.0	126.1	138.6	122.4	185.0	114.4	146.4	143.9	163.6	162.0	138.4	143.7	170.7	137.4	130.4	138.1	135.3	11.3
Israel	..	..	..	..	..	..	..	..	..	..	..	..	..	..	..	..	123.2	127.1	130.6	127.0	10.6
Italy	128.7	152.7	89.3	109.5	97.7	104.7	126.3	114.6	126.5	156.3	153.8	150.3	146.9	143.3	165.3	164.4	143.5	133.0	135.3	137.3	11.4
Japan	125.1	163.5	263.9	251.5	260.1	229.5	264.8	287.7	187.9	151.1	159.9	181.5	148.2	164.6	144.7	184.4	198.6	132.2	144.6	158.5	13.2
Korea	60.9	72.4	102.4	104.8	66.3	97.1	151.0	165.2	126.5	134.3	108.8	110.0	108.0	129.2	144.0	127.6	103.7	112.1	139.1	118.3	9.9
Luxembourg	103.1	182.7	234.2	128.8	187.6	71.5	127.3	34.7	68.5	109.8	96.9	158.3	120.0	185.0	227.7	263.9	272.6	238.0	306.2	272.3	22.7
Mexico	17.0	26.1	20.6	32.0	21.0	33.2	41.7	53.2	57.8	31.1	25.4	35.1	33.8	35.3	31.0	34.2	26.9	52.4	45.6	41.6	3.5
Netherlands	77.0	103.6	97.8	414.8	209.8	375.7	659.0	199.4	166.5	96.9	112.3	231.9	165.6	202.3	229.2	253.7	246.4	232.3	244.0	240.9	20.1
New Zealand	108.6	104.5	93.0	90.7	102.5	78.0	91.7	98.0	96.6	80.5	92.8	101.9	124.0	142.0	185.5	194.9	244.5	253.5	220.0	239.3	19.9
Norway	118.3	112.7	82.9	117.2	122.7	107.6	121.2	128.6	132.3	155.8	114.8	119.9	124.6	137.2	145.0	148.7	131.7	135.5	144.2	137.1	11.4
Poland	3.7	12.8	23.2	36.9	26.3	35.6	48.6	63.6	51.4	60.8	35.7	39.1	54.7	68.1	81.7	80.2	67.1	68.7	70.0	68.6	5.7
Portugal	56.7	98.8	95.2	116.5	106.8	120.0	122.6	115.3	123.8	93.2	61.8	79.8	86.8	92.0	157.2	138.4	137.7	133.4	115.2	128.8	10.7
Slovak Republic	..	7.7	53.6	..	..	..	..	250.7	261.3	119.3	64.2	79.0	78.0	82.2	98.3	105.7	100.2	87.0	95.6	94.2	7.9
Slovenia	..	..	..	..	..	..	..	..	..	..	..	..	118.4	155.1	246.4	291.8	126.2	97.0	81.1	101.4	8.5
Spain	116.5	109.3	82.1	119.9	67.0	128.1	164.6	232.1	179.6	126.9	121.5	136.3	158.8	161.3	175.7	166.8	129.3	127.8	131.2	129.4	10.8
Sweden	127.1	134.3	135.7	134.7	158.7	131.0	114.4	184.5	192.7	159.4	162.1	175.4	130.9	152.2	173.0	159.5	127.9	103.9	154.9	128.9	10.7
Switzerland	239.1	260.0	250.0	230.9	230.2	178.8	283.8	311.4	225.5	225.2	213.3	222.9	216.5	686.7	261.5	273.6	269.6	266.3	337.1	291.0	24.2
Turkey	10.0	13.5	8.1	45.4	9.0	67.6	59.6	55.1	45.3	32.7	33.0	5.4	20.3	16.6	27.2	45.2	34.6	34.3	31.3	33.4	2.8
United Kingdom	84.2	64.4	83.4	180.9	171.0	153.7	218.1	239.8	239.5	171.7	183.6	348.7	311.8	292.1	310.5	175.5	129.8	126.8	132.0	129.5	10.8
United States	94.6	102.1	143.5	248.9	208.7	235.7	302.3	401.2	370.3	213.4	208.2	213.9	239.1	267.3	264.2	264.8	218.9	226.5	224.8	223.4	18.6
OECD	86.8	102.2	119.2	153.5	136.2	145.6	178.8	208.8	187.6	128.3	128.4	144.1	147.7	163.5	168.6	171.8	146.9	145.0	152.1	148.0	12.3

Notes: Calculations include unofficial estimates derived for Table 3.6.

StatLink http://dx.doi.org/10.1787/888932801204

Table 3.12. **Cellular mobile voice traffic**

Millions

	1996	1997	1998	1999	2000	2001	2002	2003	2004	2005	2006	2007	2008	2009	2010	2011
Australia	..	..	..	..	..	..	..	..	..	..	..	..	..	..	..	..
Austria	..	..	..	3 674	5 760	7 055	7 902	9 130	10 408	11 590	13 728	16 977	19 596	21 113	21 956	22 185
Belgium	..	..	..	..	..	..	..	7 912	8 904	10 498	12 242	12 951	13 685	14 105	14 581	14 864
Canada	..	..	10 924	12 611	18 270	21 705	29 820	41 166	49 243	64 253	..	..	106 237	111 993	111 255	113 494
Chile	..	..	..	..	1 312	2 253	3 136	3 742	4 392	5 402	6 261	8 992	12 395	14 737	18 324	22 101
Czech Republic	..	..	..	..	1 316	2 442	2 853	3 456	3 691	4 010	9 598	11 501	12 615	13 758	14 886	15 436
Denmark	979	1 301	1 621	2 117	2 600	3 023	3 501	4 165	5 149	6 485	7 569	8 718	9 747	10 367	11 324	11 900
Estonia	..	..	..	..	..	586	736	855	1 145	1 250	1 697	2 065	2 173	2 221	2 434	2 638
Finland	919	1 832	3 198	4 514	5 294	6 520	7 276	8 161	9 643	10 848	12 493	13 546	14 548	15 120	15 919	16 105
France	..	..	9 968	20 571	35 437	44 419	51 844	63 469	74 248	81 711	94 026	99 525	101 779	100 840	102 950	105 300
Germany	..	..	..	17 401	25 004	31 288	33 970	37 089	41 019	43 000	57 110	70 030	86 140	93 610	102 320	107 000
Greece	..	..	..	..	..	..	4 738	6 826	9 053	11 309	13 997	16 854	20 857	23 957	27 193	28 038
Hungary	..	..	..	1 664	2 766	4 055	5 028	6 114	7 453	9 454	11 582	13 610	15 758	16 666	17 462	17 860
Iceland	..	..	..	..	187	220	..	360	410	476	472	547	703	724	743	766
Ireland	..	..	..	..	..	..	..	4 305	4 784	5 699	7 086	8 770	11 191	10 188	10 752	11 022
Israel	..	..	..	..	..	..	..	..	..	..	..	..	..	24 092	24 092	24 092
Italy	..	..	..	..	34 216	42 355	46 253	51 110	61 838	71 027	80 355	93 358	108 667	113 770	124 000	135 900
Japan	19 140	34 146	50 186	68 104	87 204	97 900	105 200	113 000	109 500	112 980	118 020	123 120	133 500	139 620	145 080	139 400
Korea	..	..	..	28 687	41 687	50 883	60 466	66 621	75 940	80 881	86 154	92 775	99 338	102 099	106 971	107 543
Luxembourg	..	..	..	..	..	..	..	383	444	488	535	570	722	792	867	919
Mexico	..	..	..	2 606	4 556	6 410	8 585	12 299	18 980	25 877	33 083	50 826	74 770	94 154	100 494	114 906
Netherlands	..	..	..	..	..	9 700	..	..	..	..	..	..	21 954	22 132	22 734	23 000
New Zealand	..	..	..	..	..	..	..	1 700	1 900	2 200	2 760	3 165	3 660	4 240	4 435	4 400
Norway	..	..	2 235	2 623	2 993	3 595	4 164	4 699	5 605	6 750	7 759	9 108	10 374	11 329	12 003	12 544
Poland	..	..	..	..	..	11 900	8 659	12 577	..	16 352	26 238	34 692	42 529	49 484	42 235	44 750
Portugal	..	..	..	..	6 187	8 691	9 346	10 004	10 649	11 608	12 452	13 646	15 272	17 753	20 196	21 607
Slovak Republic	..	..	..	..	..	1 467	2 205	2 845	3 669	4 874	5 328	6 029	6 637	7 403	8 297	8 464
Slovenia	..	..	..	..	..	..	..	..	..	2 426	2 614	2 875	3 133	3 504	3 691	3 722
Spain	..	..	..	..	15 041	20 210	24 816	30 942	37 120	48 267	57 857	67 981	71 111	70 557	71 421	72 482
Sweden	..	..	..	3 988	5 021	5 529	6 283	6 739	7 619	9 924	12 642	15 631	18 078	19 897	22 194	23 194
Switzerland	..	..	..	2 623	4 148	4 757	4 941	5 151	5 413	5 931	7 111	8 311	9 524	10 914	11 567	12 431
Turkey	..	..	..	..	..	5 859	6 255	11 715	20 319	35 508	48 118	57 664	74 872	108 065	125 800	147 100
United Kingdom	6 306	8 782	12 903	22 154	35 384	44 633	52 004	58 921	64 157	71 433	82 498	99 875	110 861	118 340	124 947	123 561
United States	28 654	47 767	94 280	166 021	295 792	426 733	402 827	535 271	622 230	844 928	998 091	1 398 307	1 178 540	1 262 776	1 279 874	1 240 396

Notes: Data for Israel (2010 and 2011), Japan (2011), Poland (2010 and 2011), Portugal (2011) and United States (2010 and 2011) are estimates.

StatLink http://dx.doi.org/10.1787/888932801223

Table 3.13. **Cellular mobile traffic per mobile subscriber per year**

	1996	1997	1998	1999	2000	2001	2002	2003	2004	2005	2006	2007	2008	2009	2010	2011	2011 (monthly)
Australia	..	..	..	..	..	..	..	..	..	..	..	..	..	..	..	..	..
Austria	..	..	..	855	942	1 079	1 173	1 287	1 302	1 385	1 483	1 723	1 848	1 846	1 794	1 704	142
Belgium	..	..	..	..	..	..	..	919	975	1 093	1 243	1 206	1 158	1 139	1 146	1 098	92
Canada	..	..	2 043	1 825	2 094	2 038	2 486	3 097	3 278	3 776	..	..	4 809	4 703	4 308	4 144	345
Chile	..	..	..	..	386	442	502	515	474	511	503	644	838	896	923	987	82
Czech Republic	..	..	..	..	303	352	331	356	342	341	774	869	914	965	1 034	1 084	90
Denmark	743	901	840	805	772	763	782	874	997	1 190	1 299	1 383	1 486	1 517	1 622	1 662	139
Estonia	..	..	..	..	..	795	836	813	912	865	1 023	1 097	1 146	1 419	1 309	1 416	118
Finland	622	876	1 124	1 379	1 420	1 561	1 611	1 719	1 929	2 015	2 203	2 228	2 130	1 964	1 897	1 801	150
France	..	..	889	998	1 194	1 201	1 343	1 522	1 667	1 699	1 820	1 799	1 755	1 639	1 583	1 535	128
Germany	..	..	..	742	519	557	575	572	552	543	667	721	803	865	940	938	78
Greece	..	..	..	..	..	..	509	661	819	908	1 009	1 039	1 102	1 180	1 835	1 926	160
Hungary	..	..	..	1 040	899	816	730	770	854	1 014	1 162	1 234	1 289	1 413	1 454	1 528	127
Iceland	..	..	..	..	871	933	..	1 289	1 414	1 564	1 461	1 679	2 087	2 132	2 178	2 226	185
Ireland	..	..	..	..	..	..	..	1 258	1 264	1 352	1 530	1 764	2 217	1 977	2 039	2 004	167
Israel	..	..	..	..	..	..	..	..	..	..	..	..	..	2 523	2 523	2 461	205
Italy	..	..	..	..	809	829	871	901	979	989	999	1 036	1 203	1 264	1 324	1 417	118
Japan	711	893	1 061	1 198	1 306	1 308	1 297	1 304	1 197	1 171	1 160	1 147	1 191	1 245	1 214	1 087	91
Korea	..	..	..	1 224	1 555	1 752	1 870	1 983	2 076	2 109	2 143	2 133	2 178	2 130	2 107	2 048	171
Luxembourg	..	..	..	..	..	..	..	711	687	678	749	833	1 021	1 100	1 193	1 201	100
Mexico	..	..	..	337	324	295	331	409	494	549	597	764	993	1 132	1 100	1 215	101
Netherlands	..	..	..	..	..	843	..	..	..	..	..	..	1 064	1 070	1 102	1 053	88
New Zealand	..	..	..	..	..	..	..	654	628	623	726	746	800	903	942	913	76
Norway	..	..	1 079	985	922	1 000	1 099	1 157	1 239	1 420	1 594	1 808	1 991	2 116	2 144	2 203	184
Poland	..	..	..	..	..	1 107	623	723	..	561	714	838	968	1 104	899	892	74
Portugal	..	..	..	..	928	1 089	1 016	1 000	1 007	1 021	1 018	1 012	1 021	1 106	1 226	1 287	107
Slovak Republic	..	..	..	..	..	683	754	773	858	1 074	1 089	994	1 202	1 346	1 400	1 415	118
Slovenia	..	..	..	..	..	..	..	..	..	1 379	1 437	1 491	1 524	1 668	1 739	1 716	143
Spain	..	..	..	..	628	681	740	831	961	1 131	1 266	1 404	1 433	1 382	1 390	1 378	115
Sweden	..	..	..	778	788	770	790	766	867	1 090	1 316	1 545	1 660	1 693	1 749	1 732	144
Switzerland	..	..	..	858	894	902	861	832	863	868	956	1 012	1 071	1 171	1 199	1 228	102
Turkey	..	..	..	..	..	318	268	420	585	814	914	930	1 137	1 721	2 037	2 252	188
United Kingdom	925	1 038	992	925	884	997	1 050	1 114	1 069	1 086	1 177	1 353	1 445	1 475	1 539	1 514	126
United States	651	864	1 362	1 929	2 702	3 321	2 841	3 333	3 369	4 148	4 347	5 608	4 511	4 604	4 489	4 273	356

Notes: Data for Israel (2010 and 2011), Japan (2011), Poland (2010 and 2011) and United States (2010 and 2011) are estimates. For United States, outgoing minutes estimated as all wireless-to-landline minutes plus 50% of wireless-to-wireless minutes. Data on traffic balance among types of calls not available after 2009.

StatLink http://dx.doi.org/10.1787/888932801242

Table 3.14. **International telecommunication traffic**

	Outgoing MiTT per capita													
	1998	1999	2000	2001	2002	2003	2004	2005	2006	2007	2008	2009	2010	2011
Australia	..	..	..	..	..	..	..	..	..	216.3	159.7	182.3	167.2	171.0
Austria	139.5	147.4	158.8	129.8	135.9	148.7	144.6	149.5	187.8	184.3	189.0	171.1	218.3	233.4
Belgium	..	..	94.9	125.8	133.8	0.2	233.4	247.1	235.8	230.6	238.0	247.1	249.8	258.2 ..
Canada	159.3	191.8	171.3	185.6	202.2	..	..	..	..	..	..	..	..	..
Chile	..	..	14.6	16.4	17.4	15.9	16.5	15.5	14.1	13.3	12.7	12.7	12.2	11.5
Czech Republic	33.0	44.2	42.3	47.1	52.3	50.0	50.6	50.5	59.4	47.2	47.6	39.1	23.5	21.8
Denmark	109.8	123.2	164.0	162.2	147.2	149.5	154.4	156.2	156.1	155.0	165.4	167.5	183.7	205.2
Estonia	..	..	..	..	..	83.9	88.8	87.8	110.3	109.5	94.2	81.6	80.9	82.1
Finland	79.8	83.5	90.4	104.2	90.3	..	..	..	..	..	..	..	..	..
France	66.6	97.7	134.2	164.7	182.0	196.2	201.6	210.1	213.5	209.3	207.2	204.8	205.7	309.9
Germany	..	..	..	..	..	..	..	161.1	175.8	185.5	205.6	213.6	211.3	212.5
Greece	63.2	67.1	64.8	65.6	94.2	105.0	120.8	125.9	137.8	164.1	179.9	135.1	150.5	165.2
Hungary	28.9	31.9	32.3	30.5	29.4	30.0	35.9	39.2	42.4	45.2	49.7	46.9	47.0	50.3
Iceland	166.1	181.7	151.4	147.6	147.1	192.6	180.3	208.1	209.1	184.1	193.2	168.1	145.9	132.6
Ireland	238.3	270.3	249.7	254.3	289.2	258.4	281.8	271.8	609.0	577.4	473.7	402.1	369.2	362.8
Israel	..	..	..	..	..	..	..	..	..	..	..	209.2	102.7	151.5
Italy	40.2	44.7	49.0	53.9	64.5	64.0	101.1	103.0	122.0	140.3	161.2	152.9	151.8	152.8
Japan	14.4	14.1	17.2	20.2	20.5	20.9	26.8	30.1	30.9	33.6	29.8	29.6	27.2	28.2
Korea	..	..	..	..	..	..	24.4	26.1	23.6	29.6	41.2	42.2	46.5	50.1
Luxembourg	691.0	739.9	870.8	892.9	867.8	822.5	810.4	770.9	742.2	713.5	695.0	674.4	606.9	575.7
Mexico	13.8	16.2	19.2	20.5	19.8	20.8	22.3	21.4	24.0	26.5	29.8	28.8	28.9	32.0
Netherlands	114.9	136.0	124.2	132.6	127.1	128.8	127.2	127.5	127.0	109.5	108.6	100.4	99.5	83.1
New Zealand	123.8	148.6	162.4	156.2	155.8	138.9	148.6	141.2	191.8	192.2	196.2	198.5	181.3	173.0
Norway	104.2	127.1	146.8	165.1	154.4	153.3	149.1	165.0	201.4	205.8	239.3	229.2	241.6	250.4
Poland	15.7	16.3	17.7	11.2	11.7	9.5	11.7	11.9	10.5	11.3	10.5	10.2	10.3	10.2
Portugal	46.4	40.3	50.0	62.5	94.4	92.3	97.0	106.9	107.0	113.9	117.5	114.5	111.6	115.0
Slovak Republic	29.8	32.9	35.7	37.9	38.9	39.9	44.6	57.2	84.3	93.5	106.5	109.4	69.6	73.3
Slovenia	..	..	..	..	..	..	..	105.5	107.6	111.2	119.7	114.3	111.0	101.9
Spain	34.4	41.4	54.5	65.9	67.1	77.5	86.4	108.4	120.2	122.1	137.4	128.6	119.0	119.1
Sweden	143.0	171.1	123.0	152.8	152.8	153.4	155.6	153.6	155.8	174.3	183.8	177.7	173.6	17.5
Switzerland	285.2	336.8	390.6	416.5	435.4	398.1	426.6	381.4	368.4	398.2	377.9	397.1	408.3	424.7
Turkey	10.3	11.9	11.4	10.4	9.8	9.6	10.6	10.5	9.9	9.1	13.9	14.3	13.0	12.0
United Kingdom	93.6	110.9	114.7	119.0	105.8	115.1	122.6	110.2	111.0	112.8	116.0	113.0	110.4	114.7
United States	87.8	102.1	106.6	116.7	125.0	157.9	217.1	236.7	242.4	231.9	246.1	237.4	201.4	216.8

	Outgoing MiTT per access path (fixed + mobile)													
	1998	1999	2000	2001	2002	2003	2004	2005	2006	2007	2008	2009	2010	2011
Australia	..	..	..	..	..	..	..	..	..	128.3	94.3	105.3	102.6	101.5
Austria	193.4	150.9	132.1	102.9	105.8	111.4	99.0	98.0	114.9	107.8	104.7	90.2	108.1	110.2
Belgium	..	..	94.9	103.8	104.1	0.2	163.2	165.3	152.7	140.2	137.0	138.3	136.8	135.9
Canada	192.9	218.5	177.3	173.4	180.9	..	..	..	..	..	..	..	..	..
Chile	..	..	65.9	49.4	42.7	33.4	27.4	22.4	17.3	14.5	13.2	11.9	9.6	8.2
Czech Republic	72.2	78.9	52.7	45.3	44.4	39.1	36.9	33.8	38.0	28.4	27.8	23.1	14.1	12.9
Denmark	113.4	112.8	131.9	117.9	99.0	95.1	91.6	88.2	84.6	81.0	85.6	87.6	96.9	109.0
Estonia	..	..	..	..	..	75.6	70.0	61.9	69.5	61.2	52.2	48.8	42.5	42.8
Finland	70.9	68.6	68.7	73.8	60.6	..	..	..	..	..	..	..	..	..
France	94.6	115.7	137.0	150.7	161.9	164.4	158.3	154.6	149.0	141.0	136.3	130.6	128.0	188.6
Germany	0.0	0.0	0.0	0.0	0.0	0.0	0.0	103.0	104.2	99.4	102.9	105.7	104.5	102.3
Greece	90.1	76.6	60.5	52.2	68.6	72.4	79.9	77.1	77.7	81.1	78.7	55.5	76.4	84.6
Hungary	65.4	62.3	49.4	36.8	29.1	26.7	29.5	30.5	29.7	28.9	29.2	28.2	27.7	30.0
Iceland	171.6	151.0	112.6	104.1	95.4	117.7	106.9	115.5	114.2	100.3	103.9	90.7	78.6	71.6
Ireland	349.7	311.2	259.7	221.8	235.2	200.2	204.7	182.0	376.8	337.0	275.8	235.7	213.8	205.8
Israel	..	..	..	..	..	..	..	..	..	..	..	107.4	55.9	85.9
Italy	50.3	46.2	41.7	40.3	46.7	43.3	63.5	58.7	64.6	67.7	77.7	73.8	71.1	70.7
Japan	16.5	15.0	16.8	18.5	17.4	16.6	20.2	21.7	21.7	23.0	20.1	20.2	18.0	18.2
Korea	..	..	..	..	..	..	16.4	17.1	15.4	18.8	24.7	24.6	26.3	28.0
Luxembourg	821.6	749.5	690.2	576.6	530.5	464.6	397.4	348.1	331.7	321.4	306.1	297.4	267.6	247.4
Mexico	99.1	83.8	71.3	57.2	48.5	45.2	39.9	32.3	32.2	30.8	30.7	27.6	25.6	27.7
Netherlands	162.4	143.3	103.1	108.4	102.7	96.2	82.1	83.9	80.3	59.5	56.1	52.5	53.3	43.5
New Zealand	156.7	173.0	159.1	144.5	140.4	123.7	120.6	101.7	131.0	118.9	114.2	113.7	103.6	97.1
Norway	101.6	111.0	116.7	124.2	111.7	105.5	93.4	98.6	120.5	121.2	139.0	134.4	140.3	146.5
Poland	57.9	46.5	41.3	21.0	18.8	13.2	13.1	10.8	7.8	7.8	7.0	6.7	6.6	6.3
Portugal	67.4	47.8	48.9	53.8	73.1	66.6	65.9	68.0	64.1	64.1	61.5	58.7	55.9	56.8
Slovak Republic	80.2	76.6	64.4	55.1	48.3	42.9	43.0	52.5	72.0	66.5	80.3	83.1	50.3	52.8
Slovenia	..	..	..	..	..	..	..	82.2	82.4	83.6	80.6	76.7	75.0	68.8
Spain	58.1	51.6	52.6	56.4	52.9	56.9	61.5	70.1	73.3	71.5	78.8	72.6	66.5	65.4
Sweden	124.1	135.1	87.2	100.8	94.7	89.5	90.4	86.4	84.7	92.8	95.9	88.2	84.5	8.4
Switzerland	343.4	334.8	319.7	318.7	312.6	268.4	275.8	229.8	210.0	211.8	193.7	198.7	200.2	203.3
Turkey	31.5	29.1	21.9	18.1	15.4	13.6	13.2	11.2	9.2	7.5	11.1	12.0	11.1	10.0
United Kingdom	123.1	117.1	94.0	91.2	76.4	77.8	75.3	62.1	59.1	57.3	57.3	54.5	52.7	54.4
United States	119.3	106.2	100.6	102.2	104.8	126.0	162.2	166.8	160.9	148.6	154.5	147.3	123.1	132.3

Notes:

MiTT = minutes of international telecommunications traffic. For Germany the MiTT includes VoIP and local traffic.

Data for Australia (2009 to 2011), Germany (2011), Israel (2010 and 2011), Japan (2011), Poland (2010 and 2011) and United States (2011) are estimates. Data for Canada and Finland are not avalable from 2003.

Data for United States are the "World total" of "Traffic billed in the United States." Most international VoIP minutes, including all pure international VoIP minutes, are not included in these totals.

StatLink http://dx.doi.org/10.1787/888932801261

Table 3.15. **Total staff in telecommunications services**

	1996	1997	1998	1999	2000	2001	2002	2003	2004	2005	2006	2007	2008	2009	2010	2011
Australia	101 000	74 600	76 100	89 900	91 900	84 800	88 700	91 400	99 500	100 400	102 300	95 000	90 500	90 300	86 000	90 000
Austria	17 838	17 820	18 720	22 986	23 975	24 431	20 000	19 900	19 800	19 300	17 940	17 695	16 611	14 959	18 257	15 644
Belgium	24 926	23 611	23 886	22 699	22 390	21 762	19 290	23 291	23 921	22 445	21 284	18 651	19 230	19 309	18 947	19 031
Canada	119 481	119 674	122 521	124 379	127 934	124 652	119 933	117 134	118 038	118 285	117 974	116 997	116 449	116 608	115 735	116 302
Chile	..	..	..	20 261	21 277	19 855	19 391	..	14 238	16 291	19 312	30 334	31 059	31 591	36 513	36 897
Czech Republic	26 598	25 821	24 490	23 685	18 810	24 529	24 012	22 770	20 330	19 617	17 374	19 631	20 743	19 415	21 200	20 100
Denmark	16 314	17 268	17 336	18 864	21 330	22 405	21 873	20 471	19 739	18 513	16 597	17 700	17 244	16 549	15 504	14 693
Estonia	..	..	..	..	..	2 319	3 500	3 300	3 200	2 943	2 258	2 699	2 502	2 419	2 174	2 174
Finland	16 856	17 976	19 445	21 601	24 190	25 015	19 426	17 433	17 422	16 385	14 487	13 385	11 969	12 441	12 284	12 336
France	167 817	170 043	155 992	155 297	154 522	151 191	145 487	137 414	142 296	140 623	133 000	130 000	126 000	124 000	127 000	129 000
Germany	206 800	215 624	221 800	221 900	240 700	240 700	231 500	230 600	225 300	224 100	214 700	204 600	188 100	184 200	176 900	176 000
Greece	23 808	22 741	24 026	24 780	23 940	24 890	24 700	26 500	26 300	24 400	22 400	22 800	23 400	22 200	21 800	22 000
Hungary	21 481	21 765	21 350	21 732	21 047	20 870	21 046	19 763	19 722	16 245	14 803	14 683	13 956	14 320	14 138	14 229
Iceland	937	932	1 191	1 458	1 379	1 305	1 598	1 552	1 555	1 514	1 440	1 620	1 337	1 323	1 269	1 251
Ireland	11 918	11 705	11 705	15 000	20 000	17 000	14 900	14 656	..	..	..	14 000	13 200	13 200	12 700	11 400
Israel	..	..	..	..	..	..	..	..	..	..	..	..	..	30 732	33 053	35 374
Italy	93 983	93 782	97 682	100 026	95 809	90 880	86 469	83 436	88 680	87 378	85 029	87 868	85 307	80 247	75 735	72 996
Japan	401 635	397 652	400 797	419 745	437 989	399 139	353 025	330 876	298 495	267 244	293 996	303 149	335 986	324 701	329 864	330 000
Korea	101 100	81 687	76 197	68 379	74 105	76 287	81 422	77 985	87 085	91 009	100 264	96 607	110 741	109 729	113 353	117 549
Luxembourg	816	828	828	1 356	1 478	1 487	1 500	1 500	1 000	1 200	1 100	1 000	950	1 000	900	900
Mexico	57 750	69 138	74 408	86 785	93 001	94 675	90 516	90 002	92 858	100 892	101 501	109 691	117 161	120 072	123 297	127 877
Netherlands	29 690	31 229	31 229	47 500	47 500	52 171	47 953	39 197	..	..	..	..	..	..	..	..
New Zealand	10 110	9 536	9 536	7 047	7 802	7 459	8 100	8 100	8 078	10 398	9 647	10 035	10 500	11 000	11 378	11 940
Norway	19 624	21 268	21 070	22 067	18 487	13 450	13 550	13 200	12 900	..	..	..	..	..	..	..
Poland	73 695	73 100	73 100	77 187	69 013	65 498	55 901	50 533	..	72 726	94 069	83 489	77 229	98 746	87 987	93 367
Portugal	..	..	..	..	..	..	..	..	..	..	..	..	..	..	..	..
Slovak Republic	15 822	15 871	16 264	15 883	15 111	14 653	12 621	11 552	10 878	10 265	10 243	9 651	9 949	10 089	10 212	9 552
Slovenia										4 635	4 699	4 742	4 852	5 144	5 719	5 974
Spain	75 000	73 000	73 136	69 273	71 521	70 669	94 008	88 226	88 005	88 519	88 766	85 018	81 730	80 080	77 839	74 824
Sweden	35 330	34 035	31 467	30 276	31 411	29 443	21 620	20 087	18 914	19 070	19 057	17 534	17 816	19 173	19 451	19 159
Switzerland	20 602	22 145	22 871	24 150	24 158	24 688	23 568	22 446	21 900	22 338	22 478	23 173	22 858	23 105	23 094	23 484
Turkey	2 604	2 442	2 276	2 291	2 129	1 910	1 867	1 865	1 771	1 815	..	..	..	..	..	..
United Kingdom	11 859	14 600	..	24 103	..	37 766	..	37 795	..	..	..	..	..	..	..	..
United States	..	1108 000	1167 400	1270 800	1396 600	1423 900	1280 900	1166 800	1115 100	1071 300	1047 600	1030 600	1019 400	965 700	902 900	865 300
OECD *(above data)*	1 705 394	2 787 893	2 836 823	3 051 410	3 199 508	3 209 799	2 948 376	2 789 784	2 597 025	2 589 850	2 594 318	2 582 352	2 586 778	2 562 350	2 495 203	2 469 353

Notes: Data for Estonia (2011), Finland (2009 to 2011), Hungary (2009 to 2011), Japan (2011) and Poland 2012 and 20110 are estimates. OECD total is for available data.

StatLink http://dx.doi.org/10.1787/888932801318

Table 3.16. **Communication equipment exports, USD millions, 1996-2011**

	1996	1997	1998	1999	2000	2001	2002	2003	2004	2005	2006	2007	2008	2009	2010	2011	CAGR 2009-11
Australia	356	423	310	378	549	468	194	318	402	430	436	468	575	514	584	620	9.9
Austria	209	475	271	295	412	402	728	856	1 081	1 882	1 930	2 214	2 050	1 302	1 256	1 384	3.1
Belgium	1 069	1 062	1 505	1 464	2 272	2 795	1 445	1 378	1 402	1 701	1 320	1 818	2 102	1 504	1 174	976	-19.5
Canada	3 360	3 893	4 059	5 615	10 497	4 806	3 751	3 500	4 419	5 776	6 898	6 955	5 573	4 171	4 090	3 883	-3.5
Chile	2	4	4	11	10	8	11	12	8	14	17	37	51	44	52	62	18.8
Czech Republic	65	54	96	64	176	467	535	813	905	632	605	1 901	2 583	1 837	2 208	3 904	45.8
Denmark	558	877	1 095	1 211	1 296	1 250	2 173	1 561	1 524	2 513	1 685	1 142	817	623	648	759	10.4
Estonia	9	78	157	179	689	474	253	325	440	491	527	494	506	330	810	1 821	134.9
Finland	3 389	4 052	5 504	5 931	8 259	6 808	7 185	8 130	7 737	10 606	10 572	12 052	12 835	5 697	3 478	2 867	-29.1
France	3 480	4 166	5 795	7 180	9 751	7 253	6 445	5 679	6 814	6 661	9 505	5 476	4 985	4 473	4 621	5 242	8.3
Germany	7 017	8 663	8 508	10 386	12 297	12 753	12 847	12 102	17 463	19 891	18 899	17 165	11 514	8 778	10 661	11 999	16.9
Greece	60	96	132	150	303	218	203	226	312	267	353	283	307	200	255	340	30.4
Hungary	18	39	58	49	756	1 608	2 828	4 027	6 843	6 077	6 249	9 536	11 091	9 845	11 257	12 126	11.0
Iceland	0.002	0.02	0.06	0.07	0.35	0.39	0.41	0.35	0.8	1.0	2.5	1.6	3.0	1.1	0.8	0.5	-29.7
Ireland	795	1 208	1 747	3 337	2 828	2 986	2 168	1 184	1 215	1 133	961	1 258	1 493	898	671	610	-17.6
Israel	1 609	1 991	2 354	2 745	3 741	3 219	2 367	2 224	2 690	2 146	2 505	207	3 453	2 566	2 520	2 496	-1.4
Italy	1 792	2 175	2 520	2 672	2 841	3 395	2 401	2 349	3 116	3 644	3 778	3 966	3 691	2 826	3 337	3 129	5.2
Japan	4 570	5 016	4 891	5 600	7 719	5 726	4 052	4 506	4 338	3 458	2 995	7 147	7 312	6 037	5 194	4 429	-14.3
Korea	1 325	1 833	2 302	4 527	6 559	8 325	10 823	14 650	20 357	20 493	18 336	28 928	34 488	29 574	25 933	25 853	-6.5
Luxembourg	..	..	..	217	450	721	533	263	222	228	170	174	144	125	134	134	3.4
Mexico	1 767	2 537	3 483	4 946	8 595	8 805	7 222	5 816	7 563	8 855	10 367	0	17 752	16 600	18 669	16 413	-0.6
Netherlands	1 296	1 300	1 558	2 692	4 416	4 342	1 945	3 011	4 213	4 397	4 572	14 049	13 500	11 181	13 363	15 237	16.7
New Zealand	77	103	93	75	71	58	69	95	98	95	96	117	140	119	116	151	12.7
Norway	412	484	491	425	412	414	361	418	578	611	634	671	803	699	771	722	1.7
Poland	42	69	75	69	87	105	152	153	183	476	613	805	1 260	1 005	911	887	-6.1
Portugal	48	37	30	49	53	67	64	79	88	145	119	154	165	172	223	220	13.0
Slovak Republic	..	65	51	32	32	44	23	22	46	123	372	248	352	449	1 026	1 833	102.0
Slovenia	106	85	85	50	73	123	120	150	171	117	115	161	164	107	146	192	33.8
Spain	820	929	956	1 204	1 131	1 166	1 049	1 459	1 346	1 292	1 087	671	607	446	445	571	13.1
Sweden	5 426	6 726	7 748	9 479	10 220	4 753	5 344	5 805	7 817	7 900	7 166	8 873	9 814	6 606	9 446	10 940	28.7
Switzerland	619	653	672	632	700	671	524	529	693	1 269	842	648	654	579	605	594	1.3
Turkey	66	54	89	63	83	96	64	55	48	42	53	91	119	82	91	117	19.7
United Kingdom	6 284	5 167	10 569	10 720	14 145	14 870	15 558	11 033	8 787	10 657	13 124	6 081	6 504	6 529	6 872	7 475	7.0
United States	12 373	15 156	15 293	17 307	20 904	17 710	13 973	12 660	15 648	17 025	18 881	21 937	25 202	21 863	26 006	30 131	17.4
OECD	59 019	69 469	82 503	99 753	132 328	116 906	107 406	105 389	128 567	141 048	145 784	155 778	182 608	147 783	157 574	168 117	6.7
Brazil	..	197	227	367	1 065	1 260	1 320	1 294	1 079	2 721	2 990	2 212	2 412	1 677	1 286	1 029	-21.7
China	1 329	1 537	1 874	2 373	4 431	5 860	7 292	9 952	18 141	26 451	37 785	58 885	66 091	63 155	75 559	91 894	20.6
India	40	46	30	32	34	46	50	67	77	107	186	262	315	3 915	2 002	4 565	8.0
Indonesia	257	195	234	142	288	115	125	209	226	363	288	301	410	479	182	361	-13.1
Russian Federation	44	53	32	49	51	31	52	57	89	73	315	233	124	133	98	145	4.5
South Africa	68	103	182	159	183	180	157	161	197	157	165	225	196	188	167	190	0.3
World	71 100	82 700	96 600	115 400	156 000	142 700	136 000	135 200	178 100	220 300	266 700	272 300	320 300	287 400	332 600	393 700	17.0

Notes: Luxembourg is included in Belgium prior to 1999. Trade data for China are estimates corrected for re-exports/re-imports from Hong Kong SAR of China. Data for Austria (2011), the Netherlands (2011), Norway (2011), Spain (2011) and for the World are estimates.

Source: OECD, ITCS database.

StatLink http://dx.doi.org/10.1787/888932801337

Table 3.17. **Communication equipment imports, USD millions, 1996-2011**

	1996	1997	1998	1999	2000	2001	2002	2003	2004	2005	2006	2007	2008	2009	2010	2011	CAGR 2009-11
Australia	1 332	1 331	1 268	2 327	2 999	2 071	1 710	2 174	2 956	3 164	3 764	4 068	4 228	4 271	5 306	6 925	27.3
Austria	518	591	1 105	1 465	1 549	1 165	1 288	1 624	1 872	2 576	2 077	2 421	2 569	2 069	2 167	2 383	7.3
Belgium	953	1 141	1 423	1 825	2 049	2 649	1 758	1 655	1 788	2 571	1 936	2 761	2 911	2 187	2 244	2 333	3.3
Canada	2 310	2 778	2 992	3 630	5 469	4 220	3 605	3 533	4 095	4 255	5 111	6 074	6 682	6 142	8 065	9 992	27.6
Chile	337	453	613	555	573	506	475	413	615	785	1 089	1 215	1 357	1 115	1 591	2 015	34.4
Czech Republic	558	485	442	466	730	585	606	794	959	683	886	1 785	2 186	1 517	2 204	3 403	49.8
Denmark	787	903	1 068	1 122	1 430	1 440	2 110	1 664	2 104	3 882	2 646	1 808	1 539	1 443	1 571	2 033	18.7
Estonia	60	96	96	113	115	128	135	275	181	188	193	306	291	173	572	1 192	162.8
Finland	441	457	564	616	1 180	1 072	710	869	1 138	2 185	2 450	4 411	5 424	2 778	2 259	2 447	-6.1
France	2 151	2 781	3 267	3 959	4 862	5 145	3 910	4 371	5 432	6 724	10 163	8 276	9 144	8 746	10 070	10 679	10.5
Germany	3 234	3 859	4 970	5 956	8 045	9 225	8 432	7 847	13 285	16 530	17 364	14 739	12 932	11 612	13 684	16 954	20.8
Greece	254	457	817	903	820	690	555	906	1 068	933	1 125	1 492	1 655	1 266	922	773	-21.8
Hungary	351	333	363	392	604	665	1 006	1 746	2 402	1 814	1 674	4 524	4 904	4 163	5 120	5 874	18.8
Iceland	31	32	47	47	64	39	36	45	46	70	62	96	75	44	43	61	17.1
Ireland	345	565	890	1 551	1 846	2 386	1 535	991	1 247	1 386	1 486	1 494	1 537	1 139	827	906	-10.8
Israel	759	621	673	869	995	805	736	588	785	865	897	1 149	1 236	1 217	1 253	1 857	23.6
Italy	2 074	3 080	3 816	4 387	5 046	4 224	3 966	4 430	7 294	7 083	6 853	6 870	7 128	5 726	7 657	7 546	14.8
Japan	3 553	3 117	3 177	3 543	4 870	3 854	2 886	2 557	2 807	3 086	3 737	9 266	10 482	10 380	13 574	18 871	34.8
Korea	1 467	1 448	698	1 453	3 005	1 773	1 531	1 423	1 424	1 852	2 610	4 540	5 342	4 610	5 990	8 578	36.4
Luxembourg	..	..	..	299	512	738	499	358	348	459	354	304	293	274	281	300	4.7
Mexico	1 144	1 768	2 359	2 962	4 496	4 092	2 646	2 669	3 528	3 731	5 695	0	11 781	10 149	12 817	13 943	17.2
Netherlands	1 502	1 785	2 323	4 272	5 771	6 062	3 165	3 740	5 755	6 254	5 669	16 159	15 462	12 376	14 891	16 103	14.1
New Zealand	336	327	306	414	450	320	248	327	458	542	488	584	647	572	600	828	20.3
Norway	641	672	756	780	809	711	634	775	1 059	1 001	1 101	1 397	1 464	1 271	1 521	1 758	17.6
Poland	565	848	994	1 204	1 359	1 314	1 212	1 307	1 360	1 808	2 131	2 989	3 578	2 580	2 633	2 709	2.5
Portugal	345	472	639	735	682	718	687	740	877	959	911	1 261	1 368	1 055	1 091	973	-3.9
Slovak Republic	..	268	231	136	136	188	228	275	364	485	757	808	770	670	840	1 277	38.1
Slovenia	75	106	112	180	181	143	146	157	234	167	197	290	336	237	261	335	18.8
Spain	2 189	1 743	2 224	3 716	4 076	3 233	2 708	3 413	4 720	5 587	5 787	6 398	6 245	4 848	5 238	6 290	13.9
Sweden	1 017	1 259	1 688	1 793	2 255	1 560	1 357	1 717	2 737	2 623	2 549	4 174	4 506	3 384	4 430	6 507	38.7
Switzerland	905	1 091	1 234	1 289	1 498	1 198	1 109	1 245	1 541	2 108	1 811	2 195	2 527	2 270	2 687	3 155	17.9
Turkey	459	698	1 104	1 868	2 354	847	659	840	1 441	1 739	2 023	2 725	2 524	2 402	2 719	3 115	13.9
United Kingdom	5 962	4 923	7 742	9 416	12 721	9 622	8 016	9 674	13 332	14 788	15 795	15 087	14 679	12 567	15 407	16 470	14.5
United States	11 128	12 261	14 970	21 258	34 891	29 595	29 270	31 797	39 202	48 531	52 050	58 631	65 906	62 898	75 974	81 723	14.0
OECD	47 785	52 749	64 971	85 500	118 443	102 981	89 572	96 939	128 455	151 412	163 442	190 298	213 710	188 150	226 511	260 307	17.6
Brazil	1 105	1 807	1 645	1 463	1 700	1 847	502	463	782	1 021	1 091	3 090	4 603	2 904	3 413	4 646	26.5
China	1 995	1 695	3 145	3 430	4 347	5 180	4 724	5 464	4 314	3 925	4 950	13 672	15 213	15 449	17 788	17 566	6.6
India	121	225	250	306	406	487	1 187	2 441	3 293	4 576	6 000	7 785	5 390	9 157	10 838	11 112	10.2
Indonesia	1 215	1 401	386	106	225	235	347	455	917	1 085	981	1 430	2 940	2 517	3 821	4 606	35.3
Russian Federatio	903	1 347	1 029	629	697	1 037	1 251	1 310	2 064	3 627	6 074	6 919	8 126	4 296	6 677	7 818	34.9
South Africa	651	1 112	1 880	1 223	1 352	1 100	1 157	1 148	1 662	2 200	2 478	2 678	2 831	2 284	3 134	3 495	23.7
World	69 200	80 900	93 270	111 800	153 000	141 000	125 000	131 000	174 100	214 000	253 000	268 000	314 000	286 500	354 000	421 000	21.2

Notes: Luxembourg is included in Belgium prior to 1999. Trade data for China are estimates corrected for re-exports/re-imports from Hong Kong SAR of China. Data for Austria (2011), the Netherlands (2011), Norway (2011), Spain (2011) and for the World are estimates.

Source: OECD, ITCS database.

StatLink http://dx.doi.org/10.1787/888932801356

OECD Communications Outlook 2013

Chapter 4

Network dimensions and development

In 2011, the total number of OECD communication access paths was 2 080 million, equating to 168 subscriptions per 100 inhabitants. Mobile subscriptions represented 65% of communication access paths, not much changed from 64% in 2009, though traditional fixed telephony subscriptions continue to decline. Fibre broadband subscriptions grew rapidly at 16.61% year on year between 2009 and 2011. Greater use of mobile broadband access has been stimulated by the popularity of smartphones. The average subscription rate of mobile Internet access, in OECD countries as a whole, rose to 56.6% in June 2012, up from just 30.7% in 2009. To meet the rapidly increasing demand for mobile data, network operators are undertaking several initiatives including off-loading traffic onto fixed networks by Wi-Fi. Their mobile broadband services are advertised with the OECD median speed of 12Mbit/s in 2012, which is not far from the fixed broadband median speed observed two years before.

The statistical data for Israel are supplied by and under the responsibility of the relevant Israeli authorities. The use of such data by the OECD is without prejudice to the status of the Golan Heights, East Jerusalem and Israeli settlements in the West Bank under the terms of international law.

At the close of 2011, the total number of communication access paths (analogue + ISDN, DSL, cable modem, fibre and mobile subscriptions) was just over 2 billion in the OECD area. Mobile subscriptions represented 64.92% of communication access paths, up from 63% in 2009, whereas fixed subscriptions continue to decline. Analogue connections accounted for 18.81% of the total number, down from 21% in 2009. It should be noted, however, that mobile subscriptions include around 656 million mobile broadband subscriptions that fall within the broader mobile subscription category (Figure 4.1). Some mobile broadband connections are used together with a voice connection (standard mobile broadband subscriptions) or are dedicated to mobile broadband services exclusively (dedicated mobile broadband subscriptions), but are counted as a mobile subscription (Table 4.1).

Figure 4.1. **Total fixed, mobile and broadband access paths**

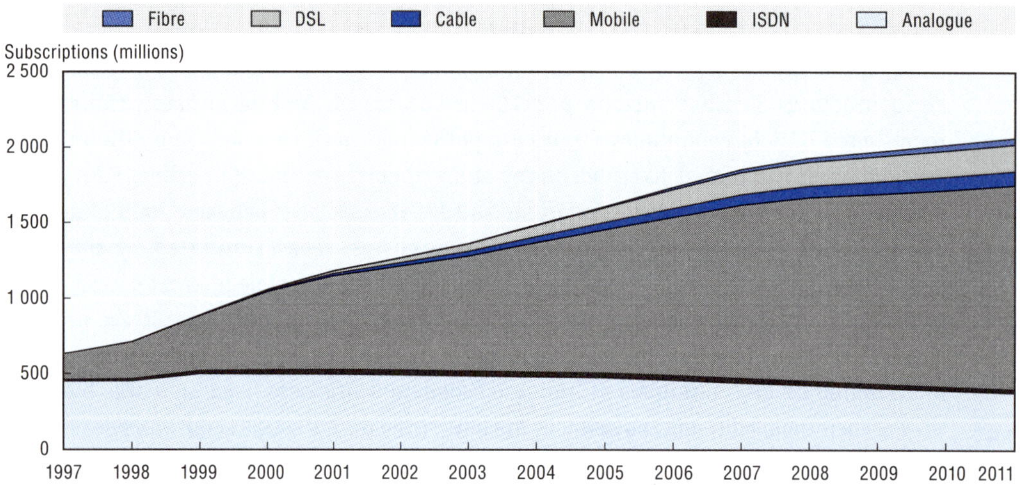

StatLink http://dx.doi.org/10.1787/888932798829

General trends in access to communication networks over the past decade persisted between 2009 and 2011. The decline in fixed telephone access paths, which began in 2001, continued across the OECD area. Until 2008, the high growth rate of mobile subscriptions led to a year-on-year increase of 7% to 8% in the total number of communication paths, but following this period, overall growth slowed to around 2% with annual growth at 2.24% in 2011 (Tables 4.2 and 4.3).

Between 2009 and 2011, overall growth in communication access paths reflected growth in its major component, mobile subscriptions, which grew by 3.59% year on year. The fastest-growing component was fixed broadband subscriptions, especially fibre. Fibre broadband subscriptions grew at 16.61% year on year between 2009 and 2011, but had a limited effect on the overall growth rate, as fibre-to-the-home (FTTH) represents only 2.07% of the total number of communication paths (Figure 4.2). Cable and DSL broadband achieved growth rates of CAGR 7.46% and 2.48% respectively. ISDN lines and standard

Figure 4.2. **Percentage growth in communication access paths, by technology, 2009-11**

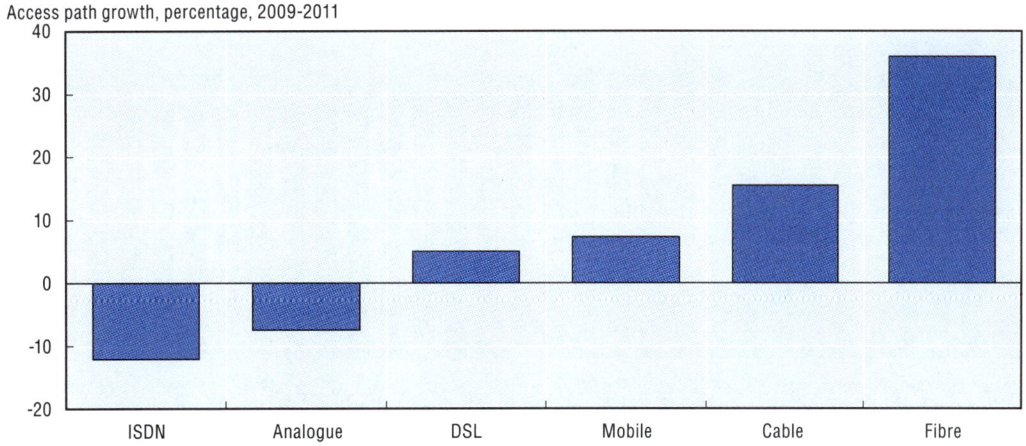

StatLink http://dx.doi.org/10.1787/888932798848

analogue access lines decreased by CAGR 6.23% and 3.73% over the same period. It is expected that the number of mobile subscriptions will keep rising in the coming years, stimulated by user adoption of multiple devices (e.g. smartphones and tablets). Mobile broadband subscriptions are growing at double-digit rates. In 2011, the number of wireless broadband subscriptions increased from 508 to 656 million in the OECD area, a 30% increase.

In 2011, the total number of OECD communication access paths was 2 080 million, equating to 168 subscriptions per 100 inhabitants (Figure 4.3). The OECD average is well over one subscription per capita reflecting the fact that many users have more than one communication path available to them. For example, a household of three members may have one fixed-line subscription, one broadband connection and three to five mobile subscriptions, as one or more members may have private and professional mobile contracts. In addition, they may have different devices with associated SIM cards such as a tablet or ebook reader. Moreover, some users may have more than one SIM card to benefit from varying tariffs at different times of the day or from on-net tariffs, or for other specific services.

Estonia provides one example of data reflecting multiple SIMs: a mobile virtual network operator (Top Connect and World Mobile OÜ), providing international mobile roaming services in some 190 countries, has some 2.1 million subscriptions for a country with a population of 1.3 million. Since the operator is using Estonian numbers, the inclusion of subscription numbers would have strongly affected the figures presented in this chapter; its subscriptions have therefore been excluded. This operator provides worldwide callback services (e.g. TravelSIM). Austria, Finland, Italy and Luxembourg top the list of countries with over 200 subscriptions per 100 inhabitants, whereas Mexico and Turkey are below 120 communication access paths per 100 inhabitants.

Fixed line developments

The number of fixed telephone access paths was 428 million in 2011, including analogue and ISDN fixed lines. This represents a decrease of 7% between 2009 and 2011. Some countries have experienced steep falls in the number of fixed telephone access

Figure 4.3. **Total communication access paths per 100 inhabitants, 2011**

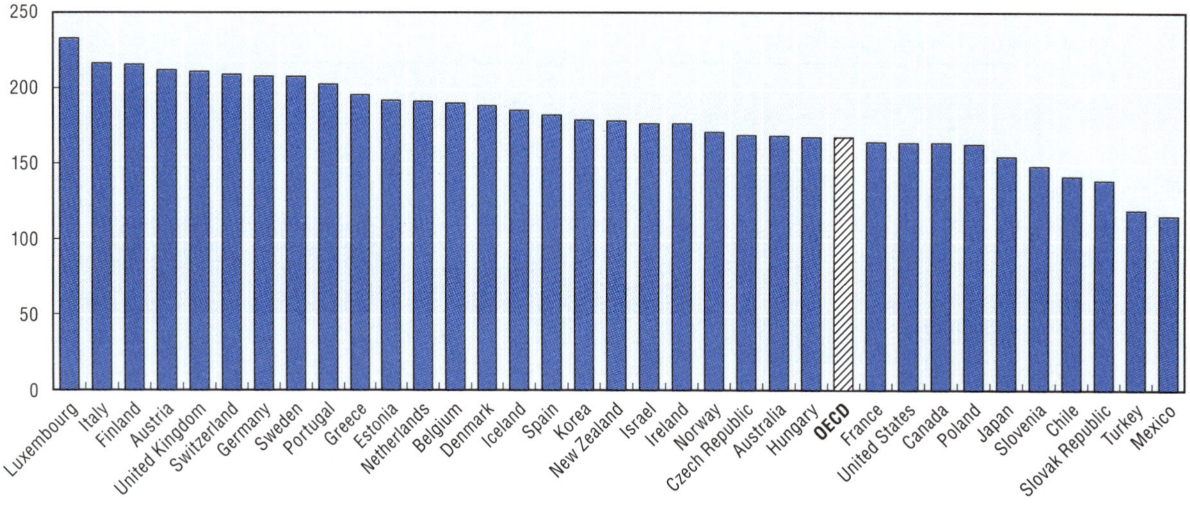

StatLink http://dx.doi.org/10.1787/888932798867

paths, particularly Israel (38.6%), Denmark (27%), the Netherlands (26%) and Finland (24.5%). This is most likely due to increasing fixed-mobile substitution and the popularity of stand-alone broadband, which may play a crucial role if households decide to discontinue their fixed telephone subscription and use VoIP or other over-the-top (OTT) services, including those provided by their fixed broadband access provider. By way of contrast, Luxembourg (5.9%), Austria (4%) and Mexico (0.9%), among others, have experienced some growth, although small in overall terms (Figure 4.4).

Figure 4.4. **Net additions of fixed telephone access paths (analogue + ISDN lines), 2009-11**

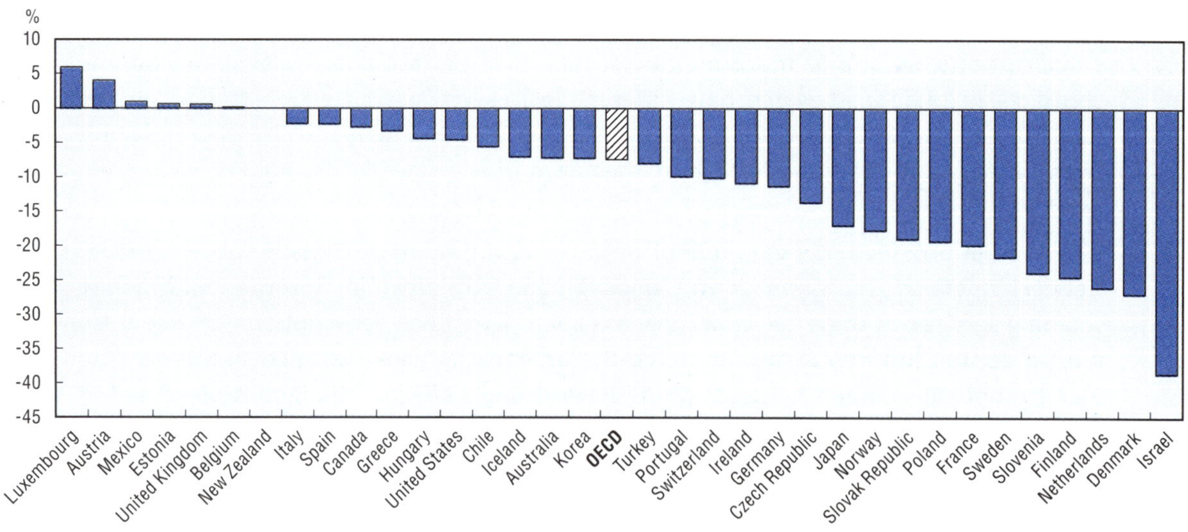

StatLink http://dx.doi.org/10.1787/888932798886

In 2011, Luxembourg (53.9), Canada (52.8) and the United Kingdom (48.2) topped the list of OECD countries ranked by number of fixed telephone access paths per 100 inhabitants. The OECD average was 34.5 for the same date (Table 4.4). Slovak Republic (15.1), Czech Republic (17.4) and Mexico (18.0) had the lowest penetration rate per 100 inhabitants. Households or businesses subscribe to fixed telephone lines, but not generally single users. In consequence, using household-based statistics to complement these data can provide a better understanding of fixed telephone penetration in OECD countries.

The number of ISDN lines continued to decline in the OECD area between 2009 and 2011. Those countries still reporting ISDN data have recorded an overall figure for the OECD of 23.6 million in 2011, down from 26.9 million in 2009, representing a decline of 6.23% per annum. ISDN technology was a digital evolution of the traditional telephone network. Its number peaked in 2003 with some 33 million subscriptions in the OECD area and has since declined, mostly due to the emergence of inexpensive, reliable and faster broadband services. The ISDN communication standards permitted voice, data and video communication services over the public switched telephone network.

Mobile developments

Mobile subscriptions account for 65% of the total number of communication paths in the OECD area. Notwithstanding slower growth rates, mobile subscription grew by 3.80% in 2011 and by 3.39% in 2010 (Figure 4.5). This is far from the growth rates experienced towards the end of the previous century of up to 50% per annum. This reflects maturity in mobile markets in the OECD, though a significant increase may emerge in the future depending on how machine-to-machine subscriptions are recorded.

Figure 4.5. **Cellular mobile subscriptions in OECD countries**

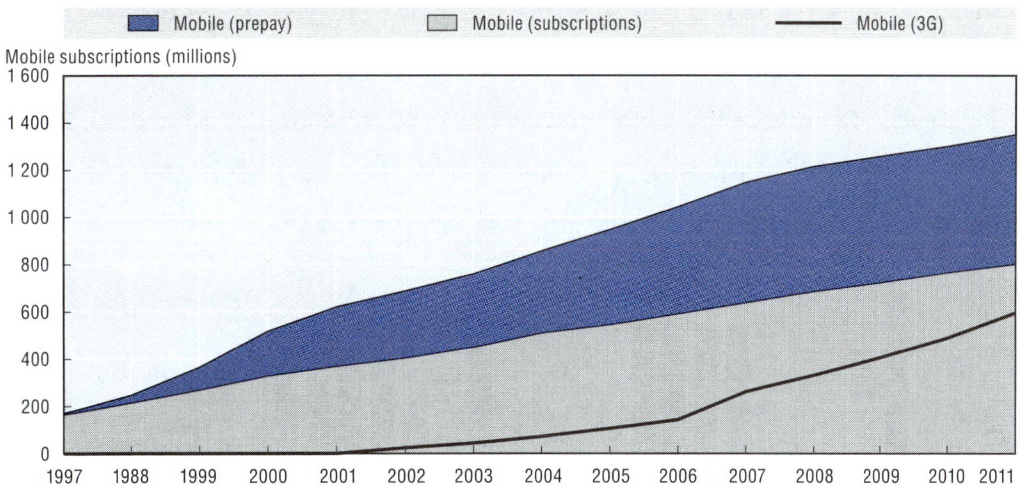

StatLink http://dx.doi.org/10.1787/888932798905

Today, the bulk of growth in the wireless sector corresponds to mobile broadband services enabled by 3G and, in the future, by 4G technologies. The uptake of 3G technologies together with quickly growing smartphone penetration has contributed to the current increase in mobile broadband subscriptions (Table 4.6). The OECD accounts for a decreasing share of the number of mobile subscriptions around the world. This reflects the success of the prepaid model for mobile services and feature phones, as well as earlier mobile service uptake in the 90s in the OECD area. The share was 85% in 1993 but accounted for just 23% in 2011. This also reflects the high growth rates of many emerging economies such as Brazil, China, India, Indonesia and South Africa. These countries are now looking to replicate the success of the prepaid model for mobile devices capable of Internet access.

The growth rate in mobile subscriptions is uneven across OECD countries, although the trend overall is slowing. For example, Chile's growth rate was CAGR 16.69% between 2009 and 2011, whereas Finland's was 7.75% and Canada's 7.24%. This is in contrast to a negative growth rate of 15% in Greece, and almost non-existent growth in Australia, the Czech Republic and Hungary. The decrease of mobile subscriptions in Greece is likely associated with the deep economic crisis in that country. However, users tend to be reluctant to give up their mobile telephones and prefer to limit their expenditure through the use of prepaid cards. With this in mind, the reduction in Greece may be associated with the obligation introduced in 2009 for prepaid customers to register their mobile subscriptions; users may prefer to register only one subscription rather than retaining multiple SIM cards (Tables 4.7 and 4.8).

In 2011, Austria, Finland, Italy and Portugal had over 150 mobile subscriptions per 100 inhabitants (Figure 4.6). By way of contrast, Canada (79.4), Mexico (86.6), Turkey (88.3) and the United States (93.0) had lower penetration rates, below the OECD average of 109 mobile subscriptions per 100 inhabitants. High subscription numbers are usually present in countries with a high share of prepaid users, such as Italy or Portugal (Table 4.8). This includes countries where multiple SIM cards may permit customers discounted call rates on different networks (e.g. a high price differential between on-net/off-net calls and/or at different times of day). While the prepaid model has been less used in countries such as Canada, Japan, Korea and the United States, it is starting to gain greater market share in the latter, particularly for feature phones.

Figure 4.6. **Cellular mobile subscriptions per 100 inhabitants, 2G and 3G, 2011**

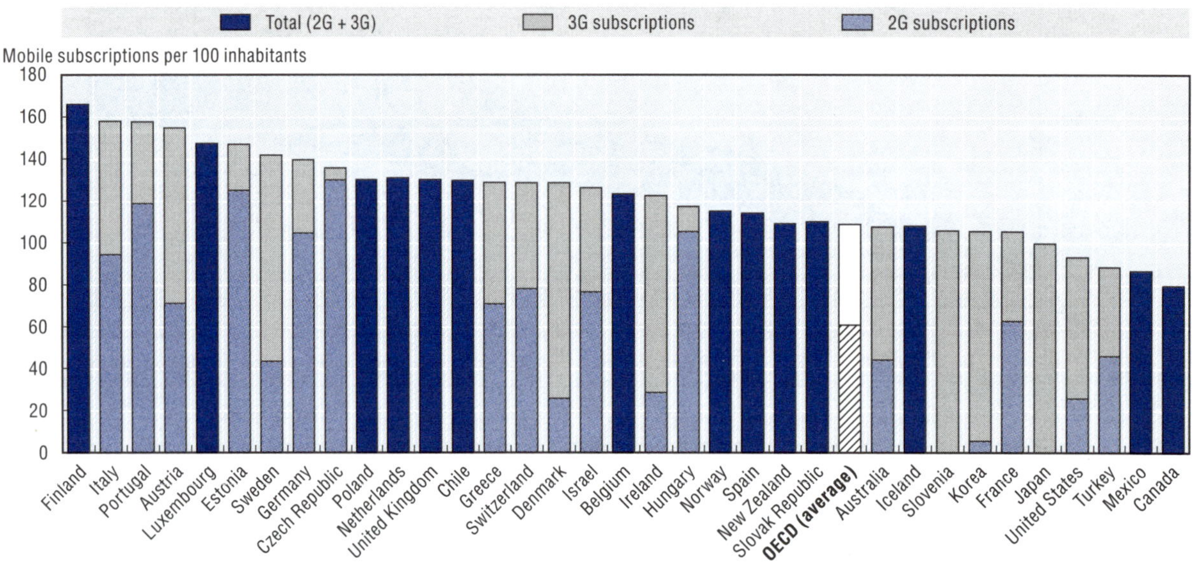

StatLink http://dx.doi.org/10.1787/888932798924

The share of 3G-enabled mobile subscriptions has grown from 32.22% in 2009 to 44.1% in 2011, with a total number of just below 596 million (Table 4.9). Not all countries have been able to report data on the share of 3G technology in mobile subscriptions. More recently, given the rising uptake of wireless broadband services, an increasing number of countries have been able to report data for this indicator. In 2011, for selected countries that were able to report data, 3G adoption reached 100% of mobile subscribership in Slovenia and Japan, while in Korea the share of 3G subscriptions is 94.9% of total mobile subscriptions (Figure 4.7). Other countries that have recently started reporting data, such as the United States, also showed a high proportion of 3G subscriptions (72.3%).

Figure 4.7. **3G cellular mobile adoption (3G subscriptions as a percentage of total subscriptions), 2011**

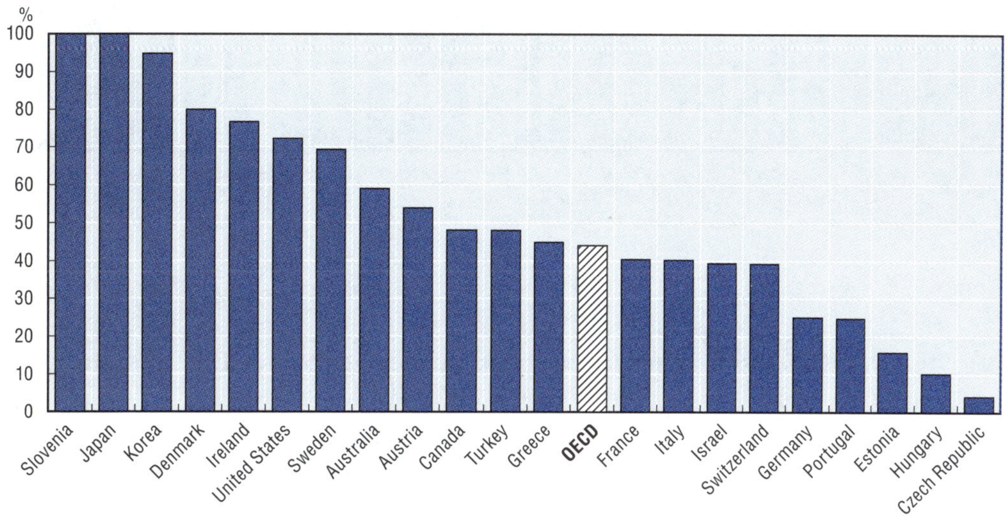

StatLink http://dx.doi.org/10.1787/888932798943

Broadband developments

Shift from dial-up to broadband

The remaining transition from dial-up technology to broadband, as a means to access the Internet, has continued between 2009 and 2011. Even though not all countries reported separate data for dial-up connections, less than 10% of fixed Internet connections in 2009 were using dial-up technology and this share declined further in 2011 to around 3.3%. The overall number of dial-up Internet access subscriptions was 10.5 million in 2011. Dial-up only accounts for a share higher than 10% of fixed Internet connections in Australia (10.4%) and New Zealand (19.9%), while in 21 out of 30 OECD countries the reported data indicates shares of 3% or less (Figure 4.8).

In OECD countries, the average number of fixed (wired) broadband subscriptions per 100 inhabitants was 25.88 in June 2012 (Figure 4.9). For the first time since the OECD began collecting broadband subscription numbers, Switzerland led the ranking with 41.64 subscriptions per 100 inhabitants, just above the Netherlands (39.38) and Denmark (38.27). The growth rate in fixed network broadband subscriptions has diminished in recent years and was at 4.3% (for 2011). Growth has been below 8% since 2009, far from the higher rates achieved in the early 2000s (more than 300% in 2000, 50.1% in 2003 and 25% in 2006) when overall penetration rates were also low (Tables 4.10 and 4.11). Nonetheless, it is worth noting that broadband uptake has significantly exceeded general economic growth in this period for many countries.

Figure 4.8. **Dial-up and broadband shares of total fixed Internet subscriptions, December 2011**

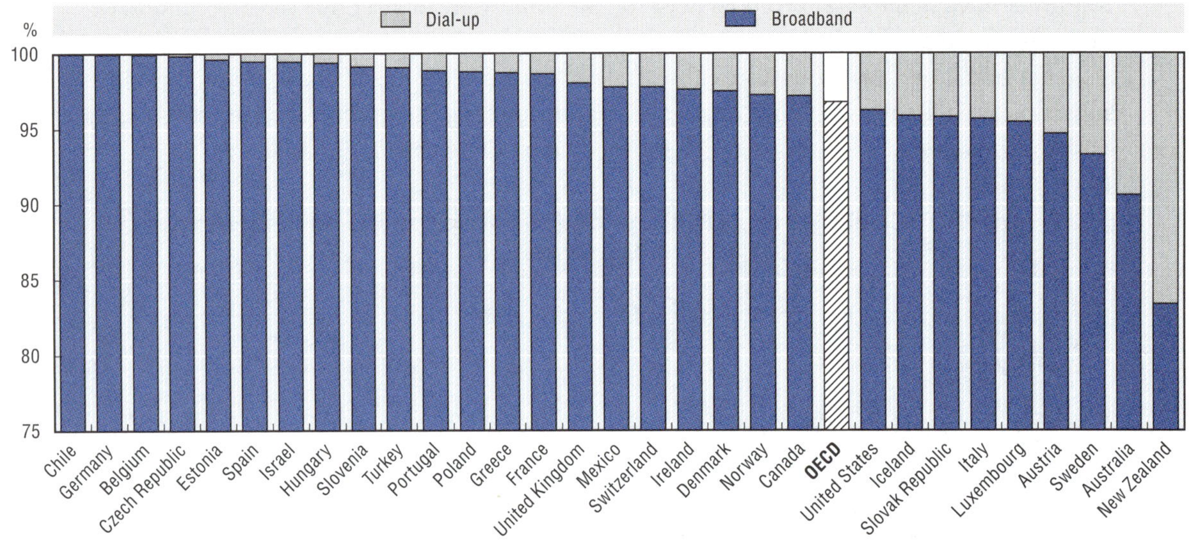

StatLink http://dx.doi.org/10.1787/888932798962

Figure 4.9. **OECD fixed (wired) broadband subscriptions per 100 inhabitants, by technology, June 2012**

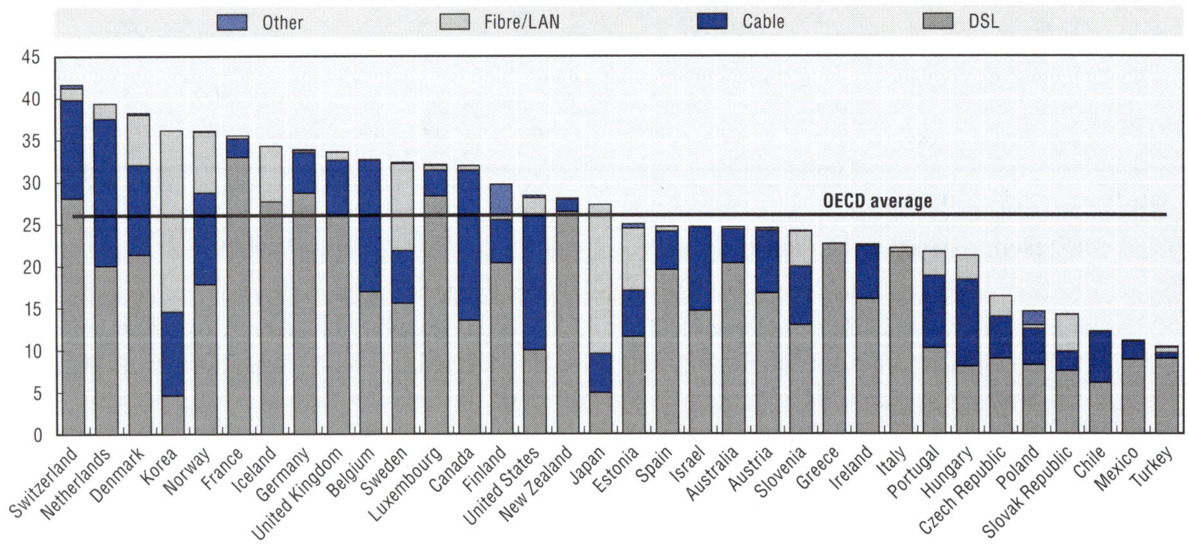

StatLink http://dx.doi.org/10.1787/888932798981

In June 2012, the overall share of DSL subscriptions continued to decrease, down to a share of 54.8%. While there was a slight increase for cable's share of the overall market (30%), this was attributable mostly to the increase in FTTH subscriptions, which represented 14.2% of the total number of fixed broadband subscriptions. The countries with the highest ratio of fibre connections in broadband subscriptions were Japan (65.0%), Korea (59.7%) and Sweden (32%). Only nine OECD countries have more than 15% of fibre subscriptions (Table 4.12).

Another important measure of broadband availability is network coverage. Broadband coverage indicators raise a number of issues. These indicators are usually broken down into coverage figures by technology, for example, DSL or cable broadband coverage, or coverage indicators for various wireless technologies (e.g. 3G, 3.5G, LTE). DSL coverage reached between 98% and 100% in 16 out of 34 OECD countries. However, many were not able to report coverage data or did so on the basis of the percentage of lines served by DSL-enabled local exchanges (Table 4.13). Accordingly, not all lines covered by these exchanges may be able to offer services as the length of local loops may exceed that sufficient for DSL service. There are less data available for coverage of cable networks but it is significant for some countries: Canada, Hungary, the Netherlands and the United States report over 80% coverage.

Upgrading broadband

Although growth of fibre broadband subscriptions was relatively modest between 2009 and 2011, some trends can be noted with regard to the competitive positioning of operators when launching broadband offers. In some countries operators have already deployed extensive fibre networks, such as in Japan and Korea, the two OECD countries with the most extensive fibre infrastructure, or Verizon in the United States. In some OECD countries with extensive cable coverage, infrastructure upgrades from cable operators have accelerated investment in fibre. An example is KPN in the Netherlands, which formed a joint venture with the fibre operator Reggefiber to deploy fibre networks. As of September 2012, they had provided FTTH infrastructure to 1.379 million homes out of some 7.2 million households and 436 000 had subscribed. This has been in response to increasing competitive pressure from cable operators.

In Spain, the number of FTTH subscriptions has increased by 101.6% year-on-year from September 2011 to September 2012. As of September 2012, there were 270 621 customers, the majority of which belong to Telefónica. However, Telefónica has gradually been losing broadband market share for DSL. It lost 3% of its customer base over the same period, mainly in the wealthier areas of large cities where competitive pressure from cable operators may be more intense.

Cable operators have been using DOCSIS 3.0 to increase advertised speeds. Shaw in Canada offers up to 250 Mbit/s, DNA Welho in Finland up to 350 Mbit/s, and VTR in Chile up to 120 Mbit/s, the highest speed for broadband services in South America. Infrastructure competition, mainly in large cities, is likely to further increase uptake of higher broadband speeds and promote fibre infrastructure deployment.

Mobile broadband

Mobile broadband growth

The popularity of smartphones has stimulated greater use of mobile broadband access. The average subscription rate of mobile Internet access in OECD countries as a whole rose to 56.6% in June 2012, up from just 30.7% in 2009. In June 2012, countries such as Korea (104.3), Sweden (101.8), Australia (97.4), Finland (95.8), Denmark (92.0), Japan (83.5) and Norway (80.5%) had mobile broadband penetration rates higher than 80% of their overall mobile penetration (Figure 4.10). The average mobile broadband growth rate in 2011 was 29.10%. Some countries had grown more than 100%, such as Chile (114.0%), the Czech Republic (324.8%), Estonia (121.2%), Mexico (156.8%), Spain (135.9%) and Turkey (343.3%), while only a few had grown below 10%. Some countries may encounter challenges in recording actual usage levels consistent with other countries. For example, the number of

Figure 4.10. **OECD wireless broadband subscriptions per 100 inhabitants, by technology, June 2012**

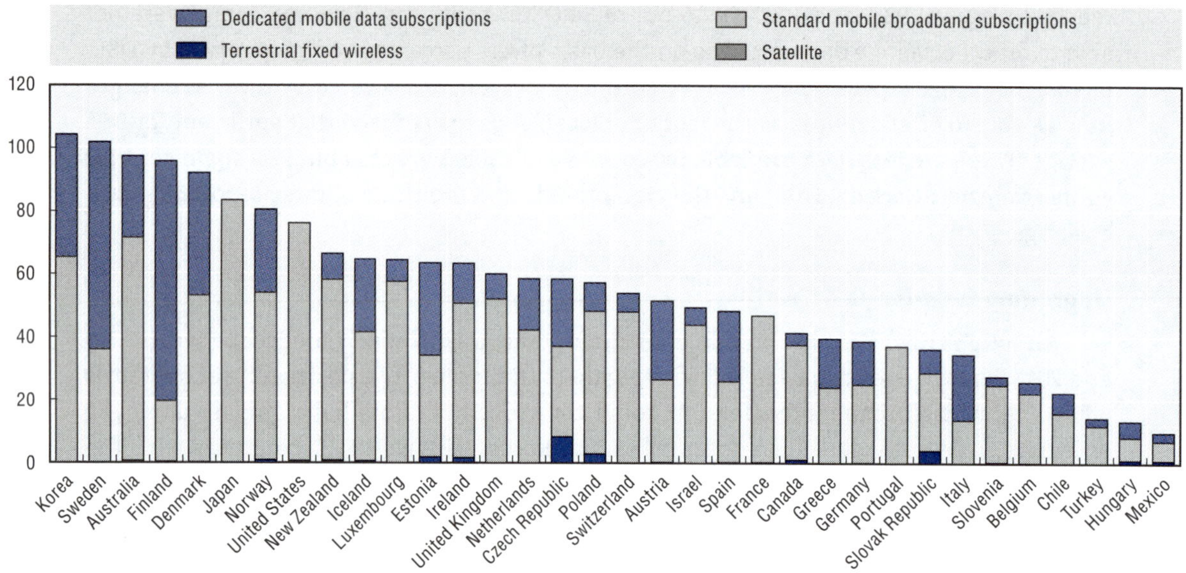

Note: Standard mobile broadband subscriptions may include dedicated mobile data subscriptions when breakdowns are not available.

StatLink http://dx.doi.org/10.1787/888932799000

standard mobile subscriptions should only include those used over the previous three months. Nonetheless, this indicator has provided reasonably uniform data since 2010, covering both standard and dedicated mobile broadband subscriptions (Table 4.14 and 4.15).

In 2012, smartphones, which by some estimations consume 35 times more data than feature phones, are used by more than half of the population in Australia, Norway, Sweden, United Kingdom and the United States. Globally, more than 500 million smartphones are now said to be purchased annually. There has also been extraordinary growth in usage of Wi-Fi, 3G and now 4G-enabled tablets, following Apple's release of the iPad in 2010. Along with traditional manufacturers, such as LG, Samsung, Sony and others, convergence has occurred among firms such as Amazon, Google and Microsoft, now marketing tablets under their own brands. As tablets are said to use large data volumes (3.5 times greater than smartphones) consumers are expected to adopt 4G more quickly than the initial launch of 3G.

Laptops and netbooks are among the highest data-using devices for wireless services, using on average four times that of tablets. However, reports show their sales volumes are not rising. Overall, the combination of all mobile devices using SIM cards has generated increasing demand for wireless services. Cisco's Visual Networking Index (VNI) reports that smartphone data usage rose from 189 MB in 2011 to 342 MB a month in 2012. Over the same period, they reported that average tablet data usage rose from 416 MB to 821 MB. Meanwhile, average laptop or netbook data consumption grew by 0.2 GB from 2.3 GB to 2.5 GB. According to Cisco, these factors combined contributed to 1.7-fold growth in global mobile data traffic between 2011 and 2012. All indicators point towards this rapid growth continuing in the foreseeable future. Cisco estimates that growth in the use of all mobile devices will result in a thirteenfold increase in global mobile data traffic between 2012 and 2017.

Network operators are undertaking several initiatives to meet the rapidly increasing demand for mobile data. One approach is to offload traffic from wireless networks onto fixed networks by using WiFi access points, femtocells, picocells and other microcells. The selection of appropriate "small cells" depends on availability, reliability and costs of the different technologies. The most commonly used technique to offload mobile traffic onto fixed networks is Wi-Fi. Most major mobile network operators across OECD countries have strategies in place to expand their use of Wi-Fi. Those with their own fixed networks on their own local access infrastructure or as resellers of these facilities are striving to leverage the expanded coverage and backhaul capabilities they provide.

In France, at the beginning of 2012, Iliad launched "Free Mobile" as a new market entrant with a reported 5 million Wi-Fi hotspots across the country (some 25% market share). Free's fixed network subscribers share their bandwidth via Wi-Fi-enabled routers connected to DSL or fibre loops (i.e. the so-called Freebox), thus creating the hotspots. Shortly thereafter, the company enabled Extensible Authentication Protocol Method for GSM Subscriber Identity Module (EAP-SIM) capabilities for smartphone connectivity to the Freebox. This enables their mobile customers to use those five million hotspots without individually authenticating access as they enter the area served by Wi-Fi. From the customer's perspective this is seamless. Data uploaded and downloaded does not count towards fair use caps and, depending on mobile network conditions, may provide better quality of service, thus providing incentives to seek out hotspots. From the perspective of Free, the company can offer its customers better service as well as increased flexibility in pricing.

The widespread use of EAP-SIM occurred at a commercial level in two other OECD countries, Sweden and Switzerland, prior to its launch by Free Mobile. Swisscom, the first company to make use of EAP-SIM in its public Wi-Fi hotspots, also became one of the first network operators to tariff its mobile services via speed. In Korea, Korea Telecom's mobile network offloads over 10% of its mobile data traffic equivalent to over 2 000 TB per month to Wi-Fi access points. On a global basis, Cisco forecasts that 46% of mobile data traffic, some 5.1 Exabytes per month, will be offloaded from smartphones and tablets by 2017. Many operators such as AT&T (United States), Network Norway, Orange Telecom, Softbank (Japan) and T-Mobile use femtocells in addition to Wi-Fi.

According to Ericsson, Enhanced Data rates for GSM Evolution (EDGE) subscribers will still account for a considerable portion of the global market in coming years, particularly with the addition of more users outside the OECD area, but the total number of GSM/EDGE subscriptions is expected by Ericsson and other market observers to decline from 2012. Uptake of High Speed Packet Access (HSPA) will coincide with the further development of Long Term Evolution (LTE) networks. By September 2012, 64 LTE networks had been deployed in 19 OECD countries, with 11 in the United States. The earliest adoption of LTE took place in the Nordic countries (TeliaSonera), traditionally leaders in the introduction of new wireless standards, Korea (LG UPlus and SK Telecom), Japan (NTT Docomo) and the United States (Verizon).

In Korea, LG UPlus was the first player in the mobile market to introduce LTE. It launched in July 2011 and completed the first national LTE network in March 2012. These efforts by Korea's smallest operator (by number of subscriptions) saw the company gain around 5 million LTE subscriptions as of December 2011. SK Telecom, the largest mobile operator in Korea, launched LTE at the same time and has acquired 6 million LTE customers over the same period, while KT had 2 million LTE customers in August 2012. In Japan, NTT Docomo has led the market since launching its LTE service "Xi" in December 2010, gaining 4 million users over the first 20 months. The Verizon Wireless 4G LTE network, also launched in 2010, offers 4G LTE coverage to more than 245 million people in the United States – approximately four out of five people. The company gained more than 10 million subscriptions in the first two years.

Some commentators have noted that networks that use Code Division Multiple Access (CDMA) such as Verizon Wireless and LG UPlus, have a stronger incentive to develop LTE to enhance their ability to compete and develop new services. In 2008, as LTE was strongly emerging as a global standard, Qualcomm took the decision not to develop CDMA's fourth-generation technology, UMB (Ultra Mobile Broadband). As a result, CDMA operators faced a technology gap which impeded their ability to provide users with increased downlink speeds, compared to other advanced 3G technologies such as HSPA, unless they adopted LTE. To ensure competitivity and enhance compatibility with other networks, CDMA operators therefore had more incentive to be among the first to shift to LTE networks.

There are few comprehensive or official data on fixed and mobile broadband traffic. Some private sources, such as Cisco, provide a useful overview of the evolution of the industry. In addition, some official sources in OECD countries have published broadband traffic data (Table 4.16), although they sometimes correspond to different periods of time or subscriber profiles (e.g. with or without active use requirements).

Notwithstanding these differences, official statistics on broadband traffic provide indications of market trends. For example, fixed broadband traffic per access has experienced considerable increases in Australia, Portugal and New Zealand between 2010 and 2012, but has remained relatively stable in Germany and Italy. The former may be due to the significant increase in data caps in Australia and New Zealand during this period, previously constrained by relatively low caps for fixed broadband access.

In 2012, Australia had the highest traffic per mobile access of the selected countries (1.41 GB). This includes, however, some "fixed wireless" accounts. In areas of the country that lack access to DSL, the network with the current greatest coverage, users can opt for packages on mobile networks that are tariffed differently from regular mobile subscriptions. These options have higher data allowances but at a fixed location. These data are therefore not comparable to those for other countries.

A further issue is that usage patterns for smartphones, tablets and other devices are changing rapidly, and data more than one or two years old may not reflect this evolution. Nonetheless, official data reveal that growth rates in mobile traffic have been relatively uneven across OECD countries. Between 2010 and 2011, mobile traffic per access only increased by 5.39% in Germany and 0.81% in Portugal. In contrast, it grew by 80.58% in Japan and by 59% in France over the same period. In France, this rapid growth was from a relatively low base among the sampled countries at 146 MB in 2011, and is likely to continue given the increase in competitivity of that country's market in 2012.

Broadband speeds

Advertised broadband speeds

In most countries, advertisements for fixed broadband services are based on download speeds. While some may argue that subscribers should consider other parameters (e.g. latency, jitter), speed remains a factor easily understood by consumers. One exception to this approach is broadband plans where data capacity is limited. In such cases, the maximum data allowance plays an important role in advertising, even though data-capped offers predominate in only a few OECD countries such as Australia, Canada, New Zealand and Turkey. In addition, differences in advertising practices may exist across countries; for example, operators in some countries advertise speeds close to the theoretical maximum for DSL, while others advertise speeds close to those actually achieved either by regulation or market equilibrium. Nevertheless, advertised speeds are a useful indicator to benchmark broadband markets across the OECD area.

The results presented here are based on data drawn from 691 standalone broadband offers across the 34 OECD countries. Bundled offers (e.g. those including pay television and telephony in addition to broadband service) were initially included in the sample, but were removed to avoid repetition and bias in the calculation of average and median advertised speeds. Only when no stand-alone broadband package was sold by a particular operator was the bundled offer included in the sample. If both stand-alone and bundled broadband were available to customers, only the stand-alone offer was retained and considered for the calculation of speeds. The average advertised download speed increased from 37.5 Mbit/s in September 2010 to 44.44 Mbit/s in September 2012 (a 18.5% increase in two years). This is in line with the higher presence of fibre-based offers and uptake of DSL (ADSL2+) and cable (DOCSIS 3.0) technologies with enhanced capabilities. The dataset was composed of 326 DSL-based, 160 cable-based and 205 fibre broadband offers.

Average advertised download speeds for a given country can greatly differ from median speeds. The reason for this divergence is the strong impact of offers with higher speeds (e.g. 1 Gbit/s, 400 Mbit/s) on country averages (Figure 4.11). For example, in September 2012, the average download speed in Sweden was 136.28 Mbit/s, but the median speed was only 30.72 Mbit/s. A similar observation is valid for Japan (with an average speed of 95.00 Mbit/s versus 19.46 median speed) or Slovenia (with 63.93 versus 15.00).

Across all OECD countries, the median broadband download speed was 20.48 Mbit/s in September 2012, up from 15.4 Mbit/s in September 2010. The dataset also allows for the possibility of breaking average and median speeds by technology (Figure 4.12). The average advertised download speed for DSL connection was 16.54 Mbit/s, significantly lower than cable (44.14 Mbit/s) and fibre (89.03 Mbit/s). These average advertised speeds have experienced notable increases for DSL (16.5%) and cable connections (48.6%) and have remained stable for fibre.

Advertised upload speeds have decreased for fibre connections (37.62 Mbit/s on average in September 2012 versus 60.0 Mbit/s two years earlier) and DSL subscription (2.13 Mbit/s vs. 2.5 Mbit/s). This generally reflects the arrival on the market of new fibre offers with lower upload speeds and reduced average advertised speeds, but also the withdrawal of some advertised fibre offers. Upload speeds for cable broadband subscription experienced a notable increase from 2.7 Mbit/s in September 2010 to 3.63 Mbit/s in September 2012.

Figure 4.11. **Average and median advertised download speeds, September 2012**

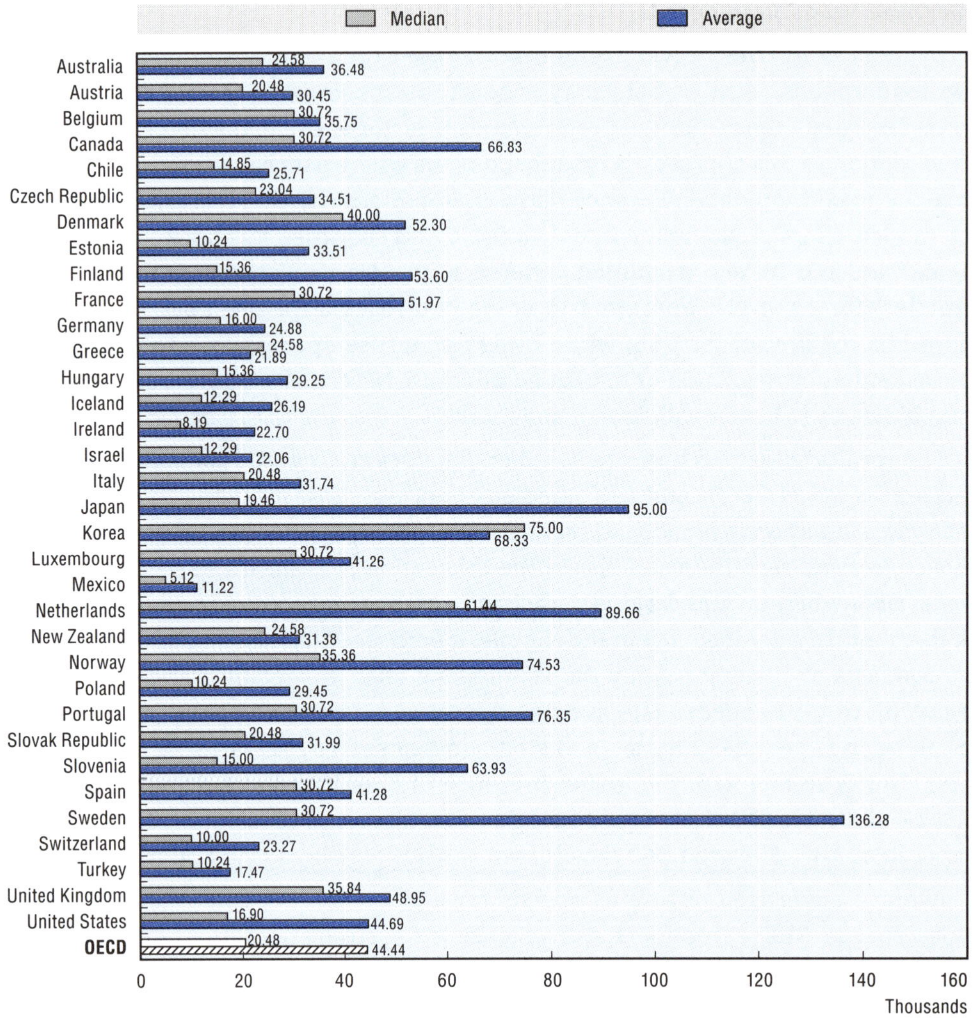

Source: OECD and Teligen.

StatLink http://dx.doi.org/10.1787/888932803978

High download speeds can be compared for incumbent and non-incumbent broadband providers in the OECD area (Figure 4.13). In some cases, non-incumbents launch broadband offers with higher speeds to challenge established players and attempt to gain market share. Incumbents then respond by offering high download speeds. This trend has been reported in some OECD countries. In 2010, non-incumbents offered higher speeds than incumbents in 27 countries. This number has since decreased to 19 countries. In nine countries, incumbents and non-incumbents offered similar maximum download speeds, while incumbents offered faster speeds in six countries.

The range of advertised download speeds offered by broadband providers in the OECD area is extremely wide (Figure 4.14). Entry-level speeds usually start at 1 or 2 Mbit/s, although in some countries download speeds of 256 kbit/s (New Zealand, Poland) or around 512 bit/s (Mexico, Switzerland) are still being advertised. At the other end of the scale, three countries (Japan, Slovenia and Sweden) are offering 1 Gbit/s. Surprisingly, some operators in Portugal and the Slovak Republic which used to offer 1 Gbit/s have ceased to do so. This may be the result of limited take-up of what were costly services (around USD 300 and

Figure 4.12. **Average advertised download and upload speeds, by technology, September 2012**

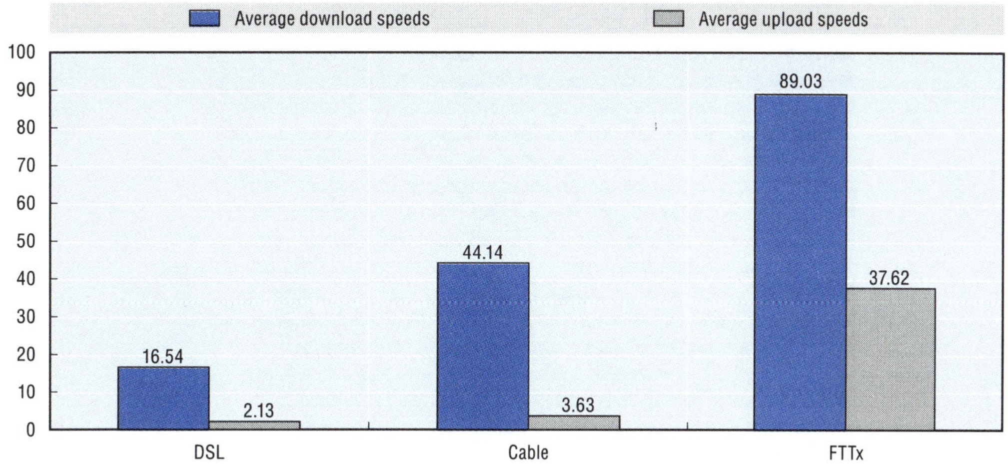

Source: OECD and Teligen.

StatLink http://dx.doi.org/10.1787/888932803997

USD 500 per month), and the limited availability of applications that require such download speeds.

The countries with the narrowest range of broadband speeds were Greece (from 2 Mbit/s to 50 Mbit/s) and Iceland (from 6 Mbit/s to 50 Mbit/s). In contrast, Poland has offers from 256 Kbit/s to 150 Mbit/s, Slovenia from 1 Mbit/s to 1 Gbit/s, and Japan from 960 Kbit/s to 1 Gbit/s. Some countries offer a significant number of broadband offers, allowing customers to choose among different data allowances or various combinations of download and upload speeds (e.g. Slovenia). For this data collection 67 offers were retained for Turkey, 61 for New Zealand and 52 for Slovenia, whereas only five different offers in terms of speed and data allowance were identified in Italy, six in Korea and seven in France. More recently some countries have launched faster speeds, in some cases with limited availability. In late 2012, Axtel began offering a FTTH service (up to 150 Mbps) in Mexico's three largest cities, while in Germany Deutsche Telekom has deployed a FTTH network in some 30 small and middle-sized cities.

Actual broadband speeds

In recent years, discussions of broadband performance have paid increasing attention to the sometimes significant gaps between advertised speeds and the "actual" speeds consumers may experience. These issues have become a matter for concern for policy makers and regulators in some OECD countries. For example, in the United Kingdom, data reported by Ofcom indicated an increase in the average gap between advertised and actual speeds from 7.6 Mbit/s in November/December 2010 to 8.7 Mbit/s in November 2011. In this case, consumers experienced only 7.6 Mbit/s in 2011 – on average around half of the advertised speed of 16.3 Mbit/s. In other cases, however, such analysis has revealed that providers are generally meeting advertised speeds or improving efforts in this area.

In the United States, the FCC has reported that actual performance delivered by most ISPs was more consistent with advertised figures in 2012 than in 2011. Reporting in July 2012, the FCC found that ISP performance had improved overall, with service providers

Figure 4.13. **Fastest average connection offered by incumbent and non-incumbent operators, logarithmic scale, September 2012**

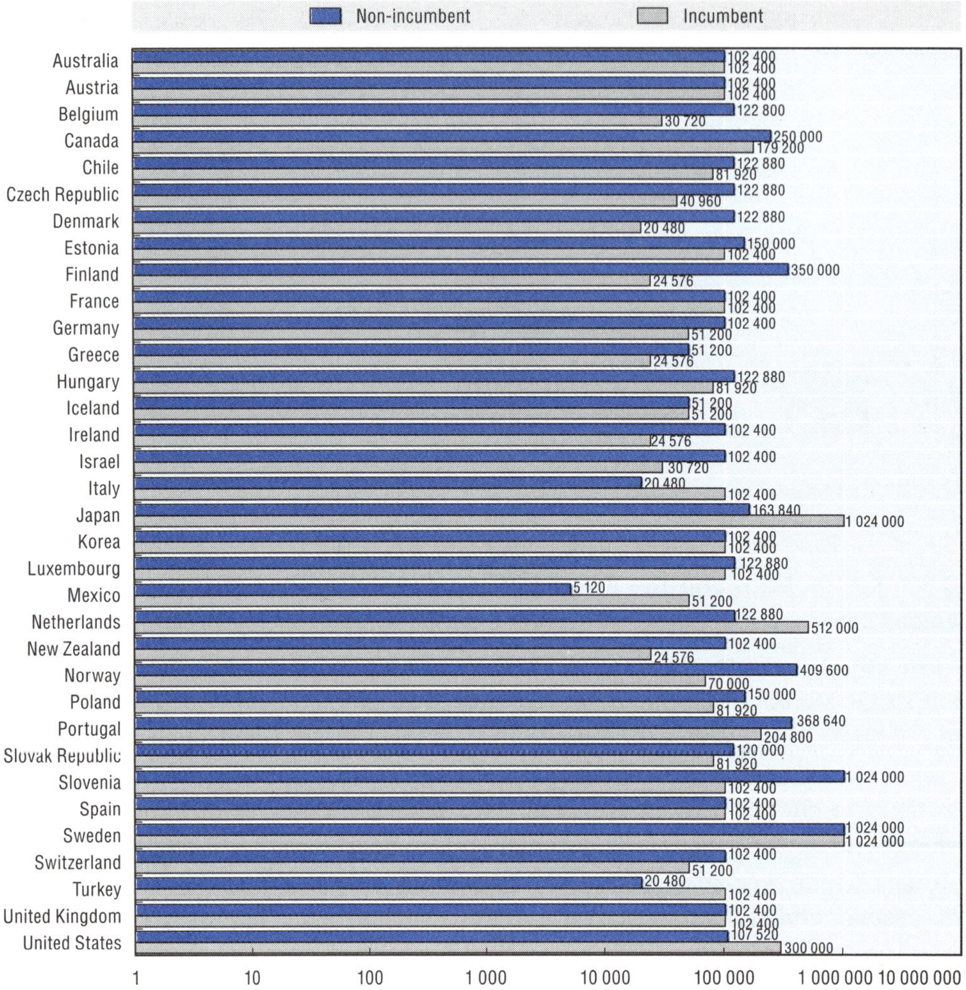

Source: OECD and Teligen.

StatLink http://dx.doi.org/10.1787/888932804016

delivering on average 96% of advertised speeds during peak intervals, and with five ISPs routinely meeting or exceeding advertised rates.

In 2011 and 2012, the OECD organised two workshops on broadband metrics, which considered measurement of advertised and actual speeds. Participants noted the emergence of several tools for measuring actual download and upload speeds, among other quality-of-service parameters, such as latency or jitter. They also noted the existence of significant barriers to the creation of a unified methodology, arising from the high number of technical choices required. The results from various performance-measuring tools have been compiled using data from Akamai, M-Lab and Ookla, and are presented here to allow consideration of the different approaches (Figure 4.15, Table 4.17).

The data from these three sources strongly correlate (average correlation coefficient equals 0.85), except in the case of Japan. It can also be observed that Ookla delivers systematically higher download speed measurement than the other two tools. Even though M-Lab does not report any data for Korea for this *Communications Outlook*, it seems

Figure 4.14. **Broadband advertised speed ranges, all technologies, logarithmic scale, September 2012**

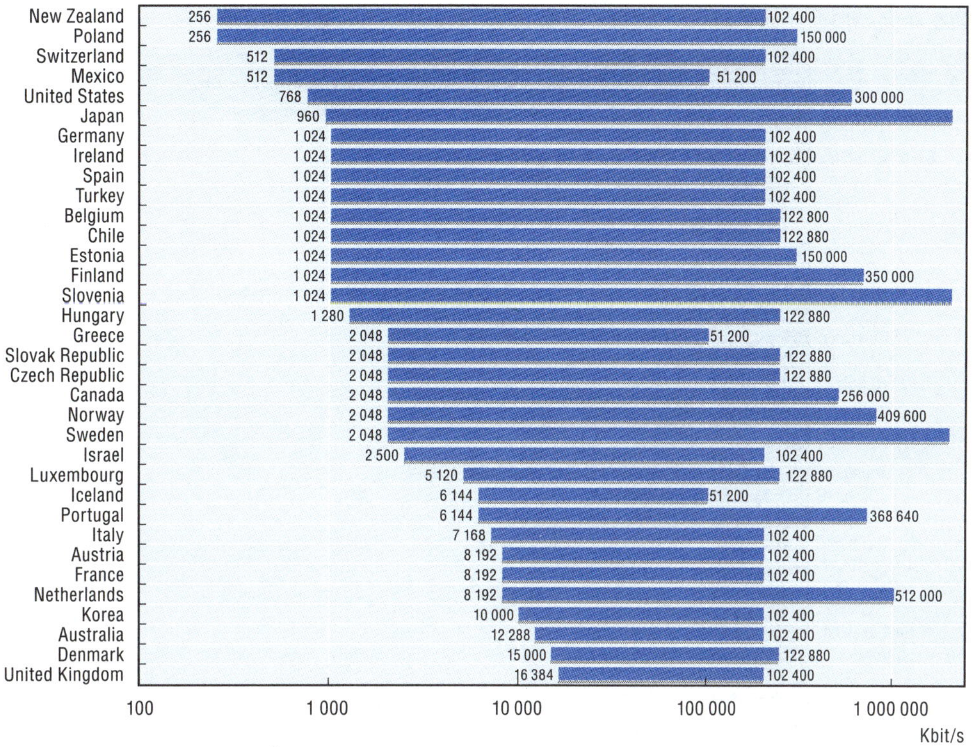

Source: OECD and Teligen.

StatLink http://dx.doi.org/10.1787/888932804035

very likely that Korean broadband subscribers enjoy the highest actual download speeds in OECD countries. Like Japan, this is no doubt attributable to greater use of fibre for final connections to business and residential premises. Akamai ranks Japan second only to Korea while Ookla accords Sweden this position. If M-Lab is used as the reference point, Japan delivers the best results, followed by the Netherlands and Belgium. Among these countries, Japan, Korea and the Netherlands were ranked highest in the previous edition of the *Communications Outlook*, based on Akamai data from 2010.

It is worth noting some of the characteristics of the different tools when drawing conclusions from these data. Akamai is a company offering global content distribution services, which gathers quarterly speed data through its server network located around the world. It does not, therefore, rely on speed tests initiated by consumers, but rather on recording the results of delivering content based on this requests. By way of contrast, M-Lab and Ookla compile results from speed tests conducted by users who actively measure their actual speed to access the Internet.

M-Lab enables visualisation of the results of the Network Diagnostic Tool (NDT) test, which measures the maximum amount of data that can be transferred from an M-Lab server to the user's device within a defined period of time ("download throughput"). Ookla employs a method for their speed tests that aims at "filling the pipe" of a testing user, to assess the capability of single computers to perform multiple downloads of one type or another simultaneously.

Figure 4.15. **Actual download speeds, comparison between Akamai, M-Lab and Ookla, Mbit/s**

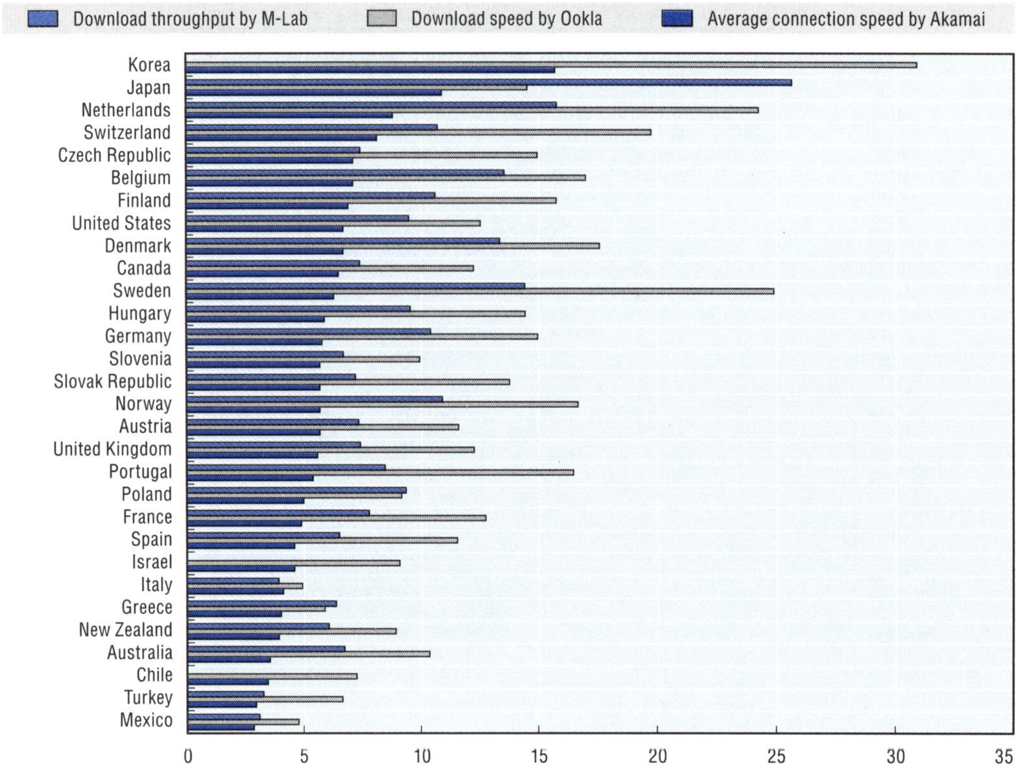

Note: The indicator based on M-Lab data is an average of all tests undertaken in the quarter using the M-Lab sites (servers), located in the corresponding region of country. See the notes of Table 4.17 for details.
Source: Akamai, M-Lab, Ookla.

StatLink http://dx.doi.org/10.1787/888932804054

A number of factors can be considered when comparing speed measurement based on voluntary tests conducted by broadband customers. First, although speed tests are valuable tools to inform consumers of actual performance of broadband services, consumers may show different degrees of willingness or incentives across different countries to perform those tests. Users more interested in online services and seeking faster speeds are more likely to conduct tests. One academic report suggests that users in the Netherlands of self-selected speed measurement tools are more likely to subscribe to higher speed services. Second, speed tests are more or less popular depending on country, so the available sample may be significantly larger or smaller. In Ookla's dataset collected for 1Q2012 around 654 tests per 100 inhabitants were undertaken in Hungary and 519 tests in the United Kingdom, whereas only 8 tests/100 inhabitants were observed in Japan and 13 in Korea. It is not difficult to draw the conclusion that speed may not be an issue for users in Japan and Korea. In addition, factors such as the overall broadband adoption rate, the extent to which ISPs promote the tool, and the languages spoken may affect the number of tests. Ookla, for example, lists several languages in its speed test settings, but not Japanese or Korean.

M-Lab data, like that for Ookla, show similar differences for the number of speed tests. In this case the number of NDT tests per 100 inhabitants in the same quarter also varies widely, from 1.18 in Hungary and 1.06 in the Netherlands to 0.01 in Japan and 0.04 in

Mexico. If a conclusion is drawn that speed is not an issue in Japan the reverse may be true for Mexico. In other words, users of offers with low advertised speeds may also have little incentive to test their speeds. One of the strengths of M-Lab's approach is to provide a high degree of openness, which allows consideration and discussion of these differences. More generally, a range of observations from a variety of providers gives all stakeholders a view of where services are likely placed across different countries. If a country is consistently high or low in different independent comparisons this is likely an accurate indication of its performance.

A further difference regarding Akamai's dataset is its inclusion of mobile broadband services for some mobile networks. Even though data are not available for some countries, Akamai data revealed that a German wireless operator delivered the highest average speed with 6.0 Mbit/s, followed by one from Greece (5.0 Mbit/s) and a Czech operator (4.7 Mbit/s). Other datasets used here do not currently report speed indicators specific to mobile networks. Considering the ongoing rapid uptake of mobile broadband services in the OECD area, performance measurement of mobile networks (e.g. Akamai, speed tests made via mobile devices) will become increasingly important in the future. Regulators, academics and consumers who use those data to inform their decisions, will also need to assess the implications of different methodologies and variations across tools.

Faster speeds reported for a country do not necessarily imply wider deployment or adoption of broadband in that country. The following figure provides an overview of the relationship between broadband adoption and speed performance (Figure 4.16). It seems that there is no clear consistency between the number of subscriptions per 100 inhabitants and the ratios of three actual speed tiers for more than 10 Mbps, 4-10 Mbps and less than 4 Mbps.

Figure 4.16. **Broadband penetration rates by speed tiers, 2012**

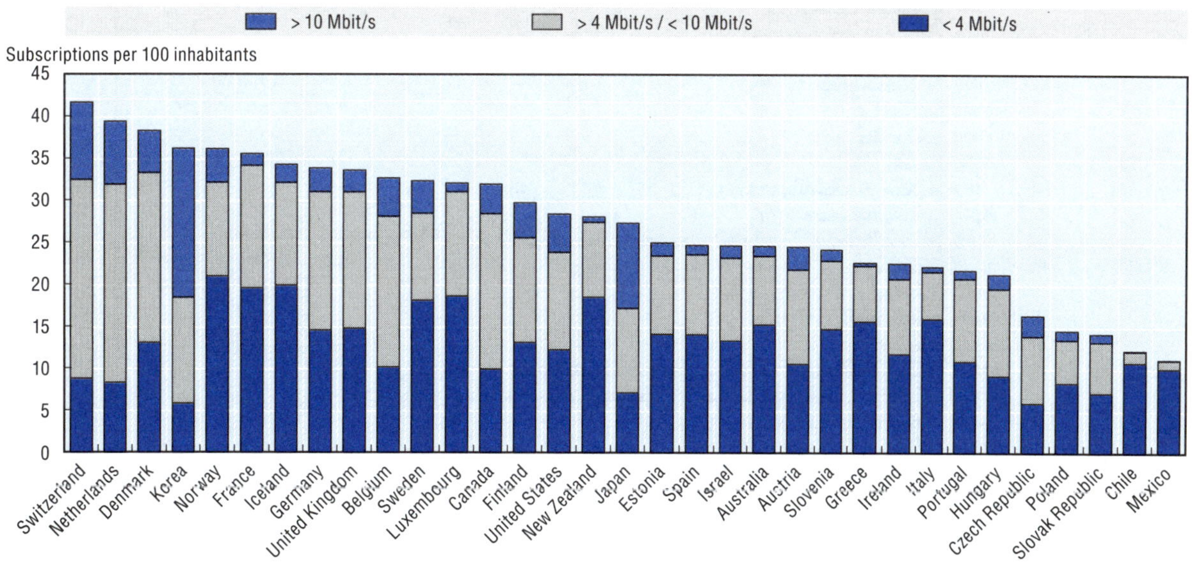

Note: Based on OECD subscription data (June 2012) merged with Akamai's actual speed data (2nd quarter, 2012).
Source: OECD and Akamai.

StatLink http://dx.doi.org/10.1787/888932799019

Mobile broadband speeds and data consumption

Mobile broadband providers are increasingly advertising broadband speeds at levels much closer to those of fixed broadband offers in some countries. For example, the OECD median speed for September 2012 for broadband mobile (12 Mbit/s) is not far from the fixed broadband median speed in 2010 (15.4 Mbit/s). In September 2012, Denmark had the fastest advertised mobile broadband speed among OECD countries (Figure 4.17), both in terms of median and average download advertised speeds. It is also worth noting that its median advertised mobile download speed of 80 Mbps is higher than most fixed speeds advertised in Denmark over that period (the median fixed advertised speed was 40 Mbps). Undoubtedly, the early adoption of LTE technology in Denmark has contributed to a sharp increase in advertised speeds. According to this ranking, the second-fastest country is Korea (75 Mbit/s median speed), followed by Greece (42.2 Mbit/s), Canada and Poland (both 42.0 Mbit/s). In addition, there may be differences in advertising practices across countries; for instance, operators in some countries advertise faster speeds closer to the theoretical maximum which are rarely achieved in real usage, as noted above for fixed broadband services.

Figure 4.17. **Average and median advertised mobile download speeds, September 2012**

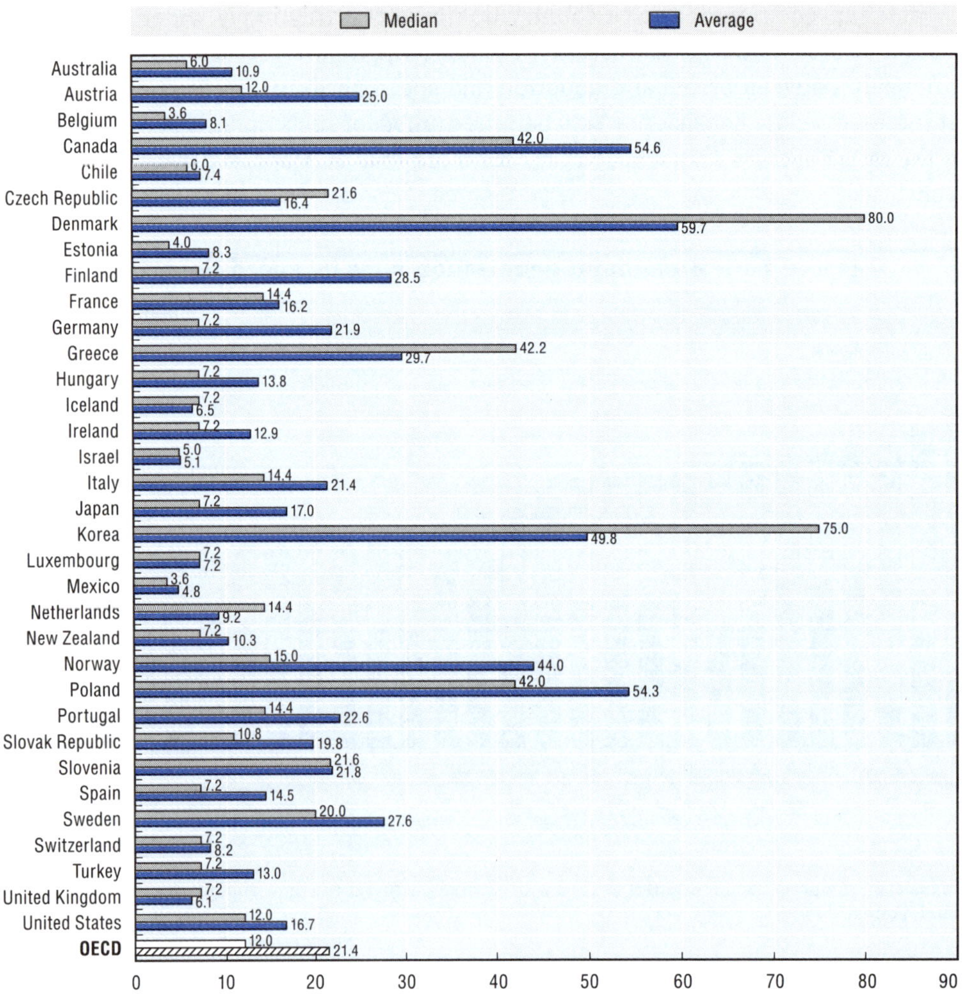

Source: OECD and Teligen.

StatLink http://dx.doi.org/10.1787/888932804073

Denmark is not the only country where mobile broadband is advertised at speeds close to, or even higher than, those of fixed broadband services; this is also the case for Canada, Germany Greece and Korea. Notwithstanding these developments, the technical constraints of wireless technologies should also be taken into account, as speeds may greatly reduce with an increasing number of users, and the difference between advertised and actual speeds is typically larger for wireless broadband services. These developments may be considered in discussions of possible substitutions between fixed and mobile broadband services. However, in most countries and for most users, wireless broadband services are complementary only for fixed broadband connections. This is well reflected in the differing tariff plans for these technologies with mobile broadband services typified by low caps relative to fixed networks.

The range of advertised mobile broadband speeds can also be examined across countries. Some eight countries have at least one advertised offer with a speed of 100 Mbit/s, the fastest among OECD countries (Figure 4.18). Of those countries, Finland has the broadest range of advertised speeds, from 400 Kbit/s up to 100 Mbit/s. Inexpensive mobile broadband offers in Finland usually deliver slow speeds, while higher speeds come with a higher price. This is not, at present, a general trend in OECD countries, as operators usually focus their marketing on data allowances, but it may become more common in the coming years with the diffusion of LTE technology. In Switzerland, as noted above, Swisscom has already launched wireless broadband offers based on advertised speeds rather than data allowances.

Figure 4.18. **Mobile broadband advertised speed ranges, logarithmic scale, September 2012**

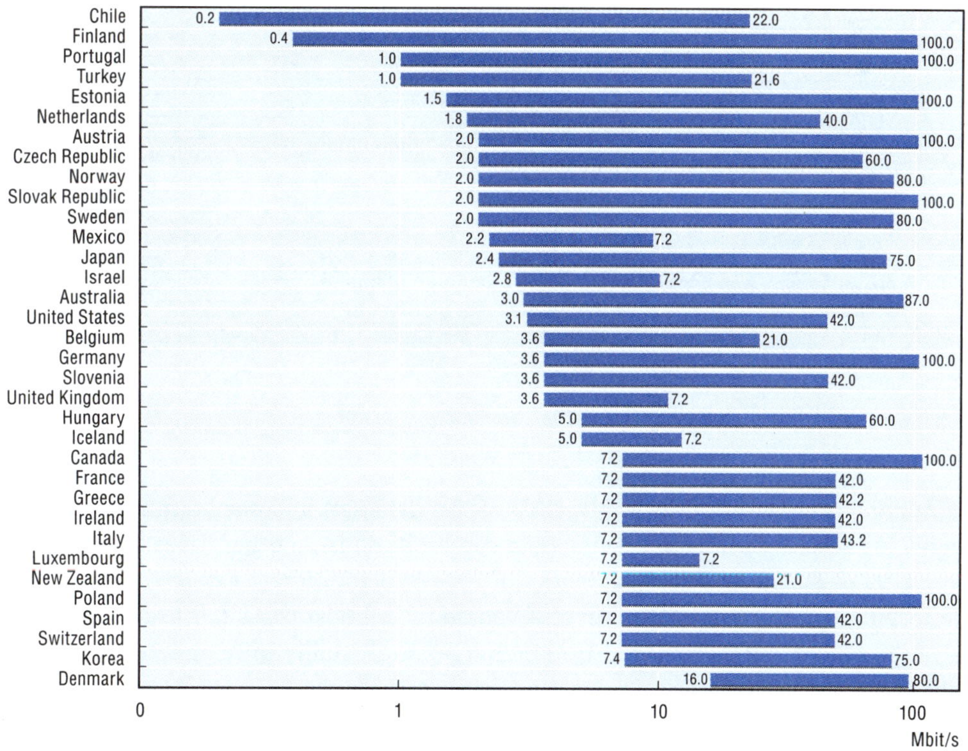

Source: OECD and Teligen.

StatLink http://dx.doi.org/10.1787/888932804092

Data caps for fixed networks

Data caps for fixed broadband offers are only relevant in a minority of OECD countries, even though in some they cover virtually the entire broadband market. Overall, only 32% of fixed broadband offers were subject to bit caps in September 2012 (219 out of 691 offers), up from 29% in 2010. Operators in Australia, Iceland and New Zealand only advertise broadband offers with a bit cap limitation. In five of the countries where data caps exist (Austria, Chile, Hungary, Slovak Republic and Spain) only a small minority of offers are affected (below 10%), with the remaining offers being unlimited (Figure 4.19).

Figure 4.19. **Prevalence of explicit bit/data caps among surveyed offers, by country, September 2012**

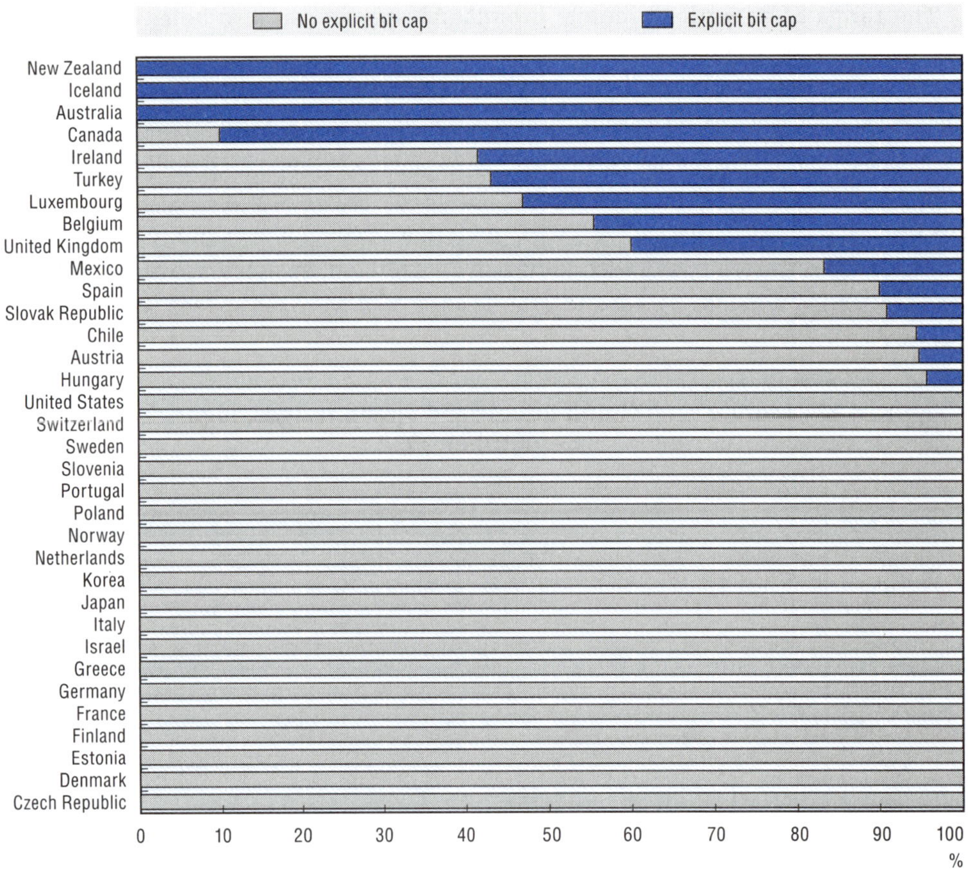

StatLink http://dx.doi.org/10.1787/888932804111

A small number of offers (around 5% of the data set) in some countries have a fair usage limit despite having no specific data cap limitations. In other cases, for example in Iceland, there is a general fair usage limit of 200 GB in addition to explicit data caps (from 1 GB to 140 GB in the case of Siminn and Vodafone). In most cases fair usage limits are in the range of 200-300 GB and additional charges are usually only enacted when users continue to exceed prescribed amounts on an on-going basis.

Average data caps for countries, where they exist, have increased since 2010 up to an OECD average (not considering fair usage limits) of 126.0 GB – or 69.86 GB when calculating across all countries with data caps (Figure 4.20). This is significantly more than in 2010. For

Figure 4.20. **Average data caps by country (GB), logarithmic scale, September 2012**

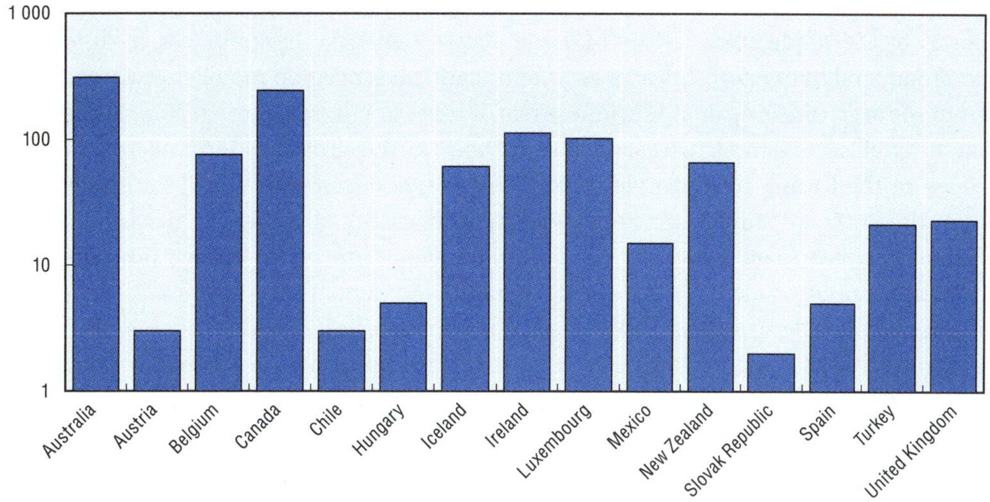

StatLink http://dx.doi.org/10.1787/888932804130

example, the average data cap in Australia went from 75 GB in 2010 to 310 GB in 2012, just as the average Canadian data cap increased from approximately 100 GB in 2010 to 243.7 GB in 2012.

Data caps for mobile networks

Selling mobile broadband services with data allowances is a common practice among mobile providers in OECD countries. When the data cap is reached, users may either be given the option of purchasing extra data for a given price (overage charge) or see their speeds "throttled", reducing it to levels such as 64 Kbit/s or 128 Kbit/s. In some cases, the consumer's price plan may be automatically upgraded to the next higher layer, which involves a higher data allowance and a higher monthly price. This may be understood as equivalent to an overage charge. In very rare cases, the service could be disrupted when reaching the maximum data allowance for a billing period.

The findings summarised here refer only to mobile broadband services for tablets and laptops in 2012. No services for handsets are included as they are commonly purchased as a part of a mobile bundle. In 2012, the median size of data caps among the dataset of 2 249 mobile broadband offers for laptop and tablets was 5 GB, for an average of 8.6 GB. The highest caps were found in the Slovak Republic and Sweden with 100 GB for the surveyed offers. Belgium and New Zealand had the lowest maximum data caps in the OECD area with only 4 GB, which explains why these two countries have the highest prices for 5 GB and 10 GB mobile broadband baskets.

Notwithstanding this, there were also unlimited flat-rate offers without data caps sold by some operators in a few countries. In 2012, KDDI au in Japan advertised an unmetered service with a maximum download speed of 40 Mbit/s, which utilised both 3G and WiMAX technologies. The Finnish mobile operator Elisa offers 4G unlimited data with a maximum advertised speed of 80 Mbit/s. Generous data caps such as these are extremely rare in the OECD, however, as 2 154 of the total 2 249 advertised offers for residential users included some limitation on data consumption.

Mobile communication services are heavily constrained by the availability of spectrum, which is shared among all users accessing a particular geographic location, as opposed to the "dedicated" nature of some access network elements (e.g. local loop) in fixed broadband networks. This acts as a significant constraint on mobile networks, which need to manage traffic against efficient levels of investment and competition. If operators make available sufficient infrastructure to handle the growing demand for wireless services in the future, for example through extensive deployment of LTE technology or substantial Wi-Fi offloading, some operators may choose to relax these limits and promote unlimited mobile broadband offers. Maximum data caps are in fact moderately correlated with maximum download speeds of mobile networks, which may suggest that data cap limitations could be to some extent relaxed if transmission speeds increase due to the deployment of LTE technology. LTE operators charging by speed rather than data allowances may be one indicator of what could happen, but this will be driven as much by consumer demand in competitive markets as by other factors.

Developments in selected Key Partners and Partners

The trend in declining trend in fixed telephony access paths per 100 inhabitants, found in OECD countries, is being increasingly mirrored elsewhere in the world. A notable exception is Brazil, where the number of fixed telephony access paths grew slightly from 21.5 to 21.9 between 2009 and 2011 (Figure 4.21). Nevertheless, penetration rates of both fixed telephony access and total communication access in these non-member countries are still lower than the OECD averages, except for the Russian Federation (over two communication paths per capita in 2011).

Figure 4.21. **Fixed and total communication access paths per 100 inhabitants, for selected Key Partners and Partners**

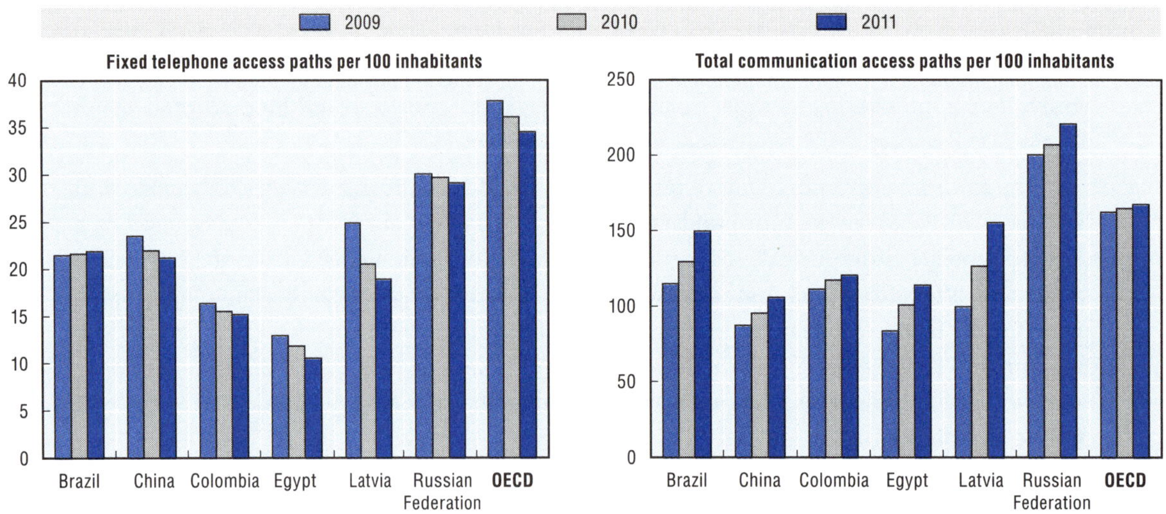

StatLink http://dx.doi.org/10.1787/888932799038

All these Key Partners and Partners are above the OECD average of 108.8 mobile subscriptions per 100 inhabitants in 2011, except for China (Figure 4.22). In addition, their subscription rates are generally growing faster than in the OECD area. Such rapid growth

Figure 4.22. **Cellular mobile subscriptions per 100 inhabitants, for selected Key Partners and Partners**

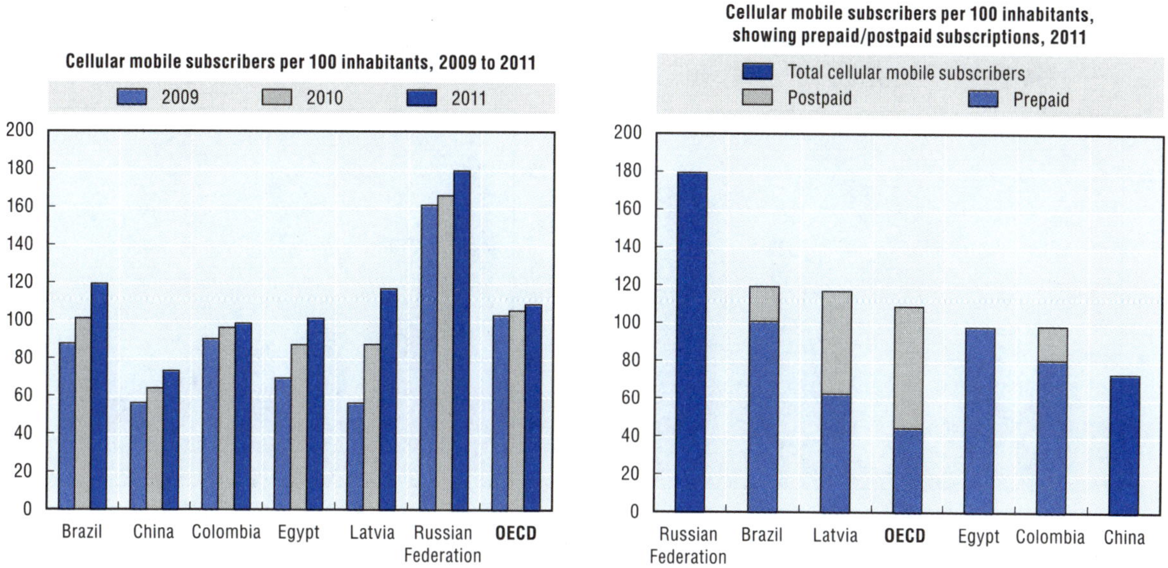

StatLink http://dx.doi.org/10.1787/888932799057

and high subscription rates for mobile communications in these economies are in contrast with the declines in fixed communications paths. In these economies, more markedly than in the OECD area, mobile communication services provide the predominant access to voice and data communication services including for Internet access. Moreover, all selected economies reporting data show a higher proportion of prepaid subscriptions than the OECD average.

In contrast with the remarkable take-up of mobile communication services, in the economies selected, fixed broadband markets have a much larger scope for future development (Figure 4.23). Fixed broadband subscription rates are uneven across these economies. Latvia has the highest subscription rate among these countries and has adopted fibre technology to a larger extent than the OECD average (4.7 as opposed to 3.5 subscriptions per 100 inhabitants).

Figure 4.23. **Broadband subscriptions per 100 inhabitants, for selected Key Partners and Partners**

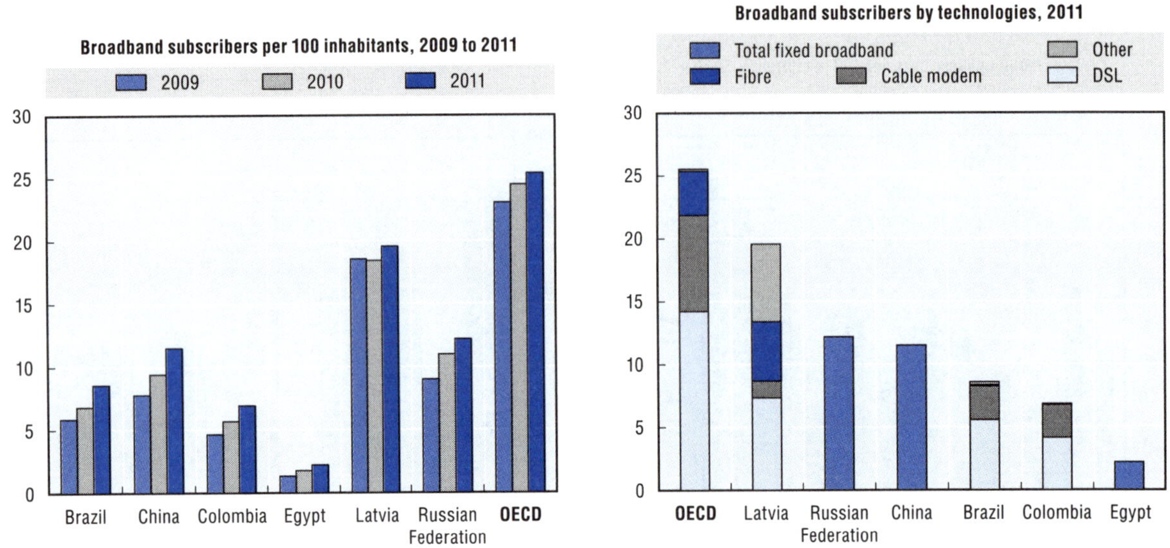

StatLink http://dx.doi.org/10.1787/888932799076

Table 4.1. **Access trends in the OECD area**

	1997	1998	1999	2000	2001	2002	2003	2004	2005	2006	2007	2008	2009	2010	2011	CAGR 2000-11	CAGR 2009-11
Lines/subscribers																	
Standard analogue access lines	468 505 815	469 782 422	516 883 787	514 578 175	513 479 230	509 023 719	502 748 748	495 988 155	490 665 867	474 845 325	456 467 672	441 002 461	422 028 898	406 489 904	391 157 940	-2.46	-3.73
ISDN lines	10 071 488	13 746 080	21 003 739	27 896 863	30 962 650	32 298 130	32 743 892	31 799 678	32 258 250	32 411 137	31 160 184	28 279 978	26 870 121	25 226 857	23 625 190	-1.50	-6.23
ISDN channels	32 154 248	42 392 920	61 385 180	79 471 106	83 509 734	86 602 512	88 080 583	86 762 800	95 930 488	100 439 700	98 704 027	82 003 411	79 664 254	79 372 686	75 443 854	-0.47	-2.68
Mobile subscriptions	170 909 682	246 751 278	364 988 936	518 317 143	620 799 295	688 549 104	759 684 404	856 712 920	947 032 598	1 045 079 326	1 148 282 025	1 215 693 512	1 258 291 605	1 300 914 025	1 350 386 775	9.10	3.59
Including wireless broadband subscriptions													*371 863 857*	*508 149 076*	*655 999 145*		*32.82*
Dedicated Mobile data													115 887 480	185 171 530	257 864 532		49.17
DSL lines		27 531	583 019	5 896 051	17 080 170	30 515 181	48 975 846	73 152 237	99 143 853	125 204 351	146 351 600	158 188 652	167 032 989	173 012 192	175 430 638	36.13	2.48
Cable modem subscriptions	96 000	679 921	2 761 073	7 619 618	15 054 014	22 891 690	31 664 278	40 058 009	48 849 969	60 710 443	69 063 374	76 857 534	81 691 420	88 954 353	94 334 461	25.70	7.46
Fibre to the home/building subscriptions			312 204	523 402	1 106 904	2 035 699	2 376 574	4 392 972	7 917 301	14 267 681	20 152 407	25 936 064	31 660 966	37 802 456	43 052 126	49.32	16.61
Other subscriptions			109 939	341 846	1 133 372	1 925 029	2 275 050	2 620 782	4 708 076	2 980 455	3 877 030	4 628 410	1 764 984	1 945 351	1 962 884		
Telephone access																	
Fixed telephone access paths (analogue + ISDN lines)	481 270 589	486 575 200	540 955 843	545 777 536	547 920 372	544 788 863	538 744 703	531 132 935	526 384 762	510 640 050	490 952 116	489 136 109	462 554 986	445 230 069	428 203 828	-2.18	-3.78
Total telephone access paths (analogue + ISDN lines + mobile)	652 180 271	733 326 478	905 944 779	1 064 094 679	1 168 719 668	1 233 337 967	1 298 429 107	1 387 845 855	1 473 417 360	1 555 719 376	1 639 234 141	1 704 829 621	1 720 846 591	1 746 144 094	1 778 590 602	4.78	1.66
Communication access																	
Fixed communication access paths (analogue lines + ISDN lines + DSL + cable modem + other + fibre)	481 366 589	487 282 652	544 612 139	559 816 607	581 161 460	600 231 433	621 761 402	648 736 153	682 295 885	710 822 525	726 519 497	750 118 359	742 940 361	744 999 070	741 021 052	2.58	-0.13
Total communication access paths (analogue lines + ISDN lines + DSL + cable modem + fibre + other + mobile)	649 486 985	730 986 832	905 928 833	1 074 330 845	1 197 996 324	1 284 767 125	1 378 184 258	1 501 783 692	1 626 732 855	1 748 308 013	1 867 529 194	1 948 405 437	1 989 363 334	2 034 345 138	2 079 950 014	6.19	2.25
Broadband																	
DSL lines as percentage of fixed communication access paths		0.1	0.1	1.1	2.9	5.1	7.9	11.3	14.5	17.6	20.1	21.1	22.5	23.2	23.7	32.70	2.62
Cable subscribers as percentage of fixed communication access paths	0.02	0.1	0.5	1.4	2.6	3.8	5.1	6.2	7.2	8.5	9.5	10.2	11.0	11.9	12.7	22.54	7.60
Fibre subscribers as percentage of fixed communication access paths			0.1	0.1	0.2	0.3	0.4	0.7	1.2	2.0	2.8	3.5	4.3	5.1	5.8	45.43	16.76

StatLink http://dx.doi.org/10.1787/888932801375

Table 4.2. **Total communication access paths in the OECD area**

In thousands

	1993	1996	1997	1998	1999	2000	2001	2002	2003	2004	2005	2006	2007	2008	2009	2010	2011	CAGR 2009-11	CAGR 2000-11
Australia	9 590	13 430	14 288	15 343	16 565	18 450	21 616	23 434	25 459	28 394	31 325	33 516	35 850	36 816	38 332	36 585	38 350	0.02	6.88
Austria	3 924	4 297	4 732	5 755	7 806	9 629	10 141	10 375	10 833	11 930	12 552	13 515	14 196	15 055	15 857	16 934	17 835	6.05	5.76
Belgium	4 701	5 258	6 012	6 490	7 819	10 250	12 453	13 276	14 045	14 899	15 652	16 280	17 469	18 602	19 278	19 872	20 849	3.99	6.67
Canada	18 902	21 471	22 917	24 914	26 680	29 661	33 209	35 054	37 215	39 721	42 835	45 895	48 509	50 762	52 500	54 697	56 520	3.76	6.04
Chile	..	319	410	964	2 261	3 402	5 163	6 411	7 590	9 712	11 253	13 459	15 253	16 219	18 096	21 631	24 403	16.13	19.62
Czech Republic	2 409	3 018	3 795	4 700	5 752	8 254	10 628	12 016	13 036	14 007	15 295	16 056	17 168	17 886	17 733	17 551	17 726	-0.02	7.20
Denmark	3 563	4 571	4 608	5 134	5 815	6 640	7 370	7 996	8 472	9 105	9 597	10 033	10 452	10 610	10 552	10 513	10 481	-0.34	4.24
Estonia	..	499	609	745	898	1 089	1 234	1 333	1 503	1 713	1 911	2 135	2 401	2 421	2 239	2 548	2 569	7.12	8.12
Finland	3 270	4 346	5 011	5 801	6 288	6 815	7 326	7 743	7 977	8 339	8 835	9 125	9 538	10 045	10 664	11 173	11 609	4.34	4.96
France	33 170	34 431	37 883	42 273	50 922	59 468	66 866	69 265	74 032	79 576	85 523	90 857	94 664	97 492	101 139	104 164	107 074	2.89	5.49
Germany	41 199	46 746	48 863	54 350	63 561	88 073	97 756	102 032	108 801	120 302	128 902	138 883	153 559	164 022	165 457	165 339	169 940	1.35	6.16
Greece	5 191	5 861	6 370	7 595	9 534	11 693	13 776	15 085	15 996	16 723	18 124	19 779	22 639	25 681	27 462	22 270	22 096	-10.30	5.96
Hungary	2 282	3 154	3 859	4 530	5 241	6 674	8 449	10 253	11 388	12 285	12 939	14 350	15 723	17 084	16 646	16 945	16 710	0.19	8.70
Iceland	166	201	221	265	333	378	404	443	473	494	533	558	572	594	591	591	591	-0.06	4.14
Ireland	1 370	1 680	2 011	2 531	3 261	3 658	4 431	4 834	5 158	5 599	6 211	6 888	7 480	7 630	7 622	7 730	7 917	1.92	7.27
Israel	..	1 984	..	..	5 848	7 295	8 497	9 261	9 496	10 086	10 233	12 438	11 442	12 118	14 571	14 003	13 675	-3.12	5.88
Italy	26 065	31 436	37 023	45 434	55 065	66 899	76 264	78 875	85 113	92 655	102 743	111 246	123 081	124 121	124 617	129 126	131 361	2.67	6.33
Japan	63 453	89 539	101 103	109 934	119 128	129 376	138 981	149 715	160 533	169 749	176 922	182 385	186 498	189 482	187 164	193 120	199 024	3.12	3.99
Korea	19 397	23 131	27 762	34 787	45 832	53 308	60 514	66 233	68 078	71 378	73 453	73 930	76 323	81 476	84 382	87 234	88 998	2.70	4.77
Luxembourg	234	295	327	358	426	551	684	731	801	935	1 031	1 058	1 067	1 111	1 130	1 151	1 209	3.43	7.40
Mexico	9 187	9 848	10 995	13 276	18 659	26 419	35 643	41 151	46 856	57 580	68 948	78 235	91 015	103 324	111 981	122 303	125 973	6.06	15.26
Netherlands	8 237	9 168	10 818	11 114	15 001	19 184	19 630	19 992	21 721	25 233	24 783	25 863	30 167	31 804	31 616	31 036	31 902	0.45	4.73
New Zealand	1 846	2 195	2 463	3 018	3 301	3 946	4 215	4 404	4 544	5 052	5 756	6 144	6 856	7 353	7 566	7 672	7 876	2.03	6.48
Norway	2 801	3 746	4 152	4 547	5 109	5 651	5 998	6 276	6 636	7 328	7 735	7 789	7 987	8 208	8 234	8 420	8 466	1.40	3.74
Poland	5 744	6 749	8 322	10 413	13 437	16 362	20 429	23 822	27 677	33 949	42 078	50 875	54 985	57 374	58 195	59 970	62 718	3.81	12.99
Portugal	3 687	4 407	5 374	6 969	8 564	10 456	11 950	13 394	14 477	15 462	16 597	17 669	18 864	20 285	20 749	21 252	21 549	1.91	6.79
Slovak Republic	1 122	1 275	1 592	2 005	2 319	2 992	3 704	4 327	4 992	5 579	5 872	6 304	7 591	7 165	7 139	7 509	7 551	2.84	8.78
Slovenia	..	..	..	..	..	..	..	2 410	2 473	2 635	2 569	2 622	2 686	3 003	3 044	3 032	3 040	-0.06	..
Spain	15 353	18 507	20 415	23 519	32 020	41 745	47 557	52 382	57 186	59 999	67 149	72 219	76 626	79 492	81 308	82 427	83 972	1.62	6.56
Sweden	6 863	8 557	9 244	10 198	11 223	12 518	13 493	14 389	15 343	15 483	16 054	16 692	17 173	17 677	18 728	19 260	19 594	2.28	4.16
Switzerland	4 677	4 834	5 328	5 923	7 210	8 808	9 519	10 228	10 986	11 530	12 452	13 258	14 327	15 042	15 594	15 879	16 436	2.66	5.84
Turkey	14 268	15 092	17 354	20 466	25 856	33 465	37 336	42 254	46 973	54 316	64 101	74 255	84 588	89 080	85 777	85 092	88 125	1.36	9.20
United Kingdom	30 745	36 646	38 291	44 443	55 589	71 889	77 171	82 131	88 128	97 482	106 760	113 723	119 992	124 361	128 122	130 354	132 277	1.61	5.70
United States	171 687	171 991	187 340	203 193	268 604	299 334	325 590	343 246	364 192	392 554	420 012	450 275	470 789	485 011	495 278	506 963	511 537	1.63	4.99
OECD	515 102	588 683	649 487	730 987	905 929	1 074 331	1 197 996	1 284 767	1 378 184	1 501 784	1 626 733	1 748 308	1 867 529	1 948 405	1 989 363	2 034 345	2 079 950	2.25	6.19

Notes: Total communication access paths = (analogue lines + ISDN lines + DSL + cable modem + fibre + other broadband + mobile subscribers). Data for Chili (2010 and 2011), France (2011), Israel (2010), the Netherlands (2010 and 2011), Poland (2000 to 2004) and United States (2008 to 2011) are estimates.

StatLink http://dx.doi.org/10.1787/888932801394

Table 4.3. **Total communication access paths per 100 inhabitants in the OECD area**

	1990	1995	1996	1997	1998	1999	2000	2001	2002	2003	2004	2005	2006	2007	2008	2009	2010	2011	Rank
Australia	47.1	51.0	72.9	76.8	81.6	87.0	95.7	110.7	118.5	127.2	140.2	152.5	160.6	168.6	169.4	173.2	163.0	168.4	23
Austria	41.8	47.2	54.0	59.4	72.2	97.7	120.2	126.1	128.4	133.4	146.0	152.6	163.5	171.0	180.6	189.6	201.9	211.8	4
Belgium	39.3	46.5	51.8	59.1	63.6	76.5	100.0	121.1	128.5	135.4	143.0	149.4	154.4	164.5	173.7	178.7	182.6	189.9	13
Canada	55.2	60.0	72.5	76.6	82.6	87.8	96.7	107.1	111.8	117.6	124.4	132.8	140.9	147.3	152.4	155.6	160.3	163.9	27
Chile	..	..	2.2	2.8	6.4	14.9	22.1	33.2	40.7	47.7	60.3	69.2	81.9	91.9	96.8	106.9	126.5	141.3	31
Czech Republic	15.7	23.2	29.3	36.8	45.7	55.9	80.3	104.0	117.8	127.8	137.2	149.4	156.4	166.3	171.5	169.0	166.9	168.9	22
Denmark	56.6	62.0	86.9	87.2	96.8	109.3	124.4	137.6	148.7	157.2	168.5	177.1	184.5	191.4	193.2	191.1	189.6	188.2	14
Estonia	..	..	35.0	43.3	53.4	65.1	79.4	90.2	97.9	110.8	126.8	141.8	158.7	178.9	180.6	167.0	190.1	191.7	11
Finland	53.5	55.5	84.8	97.5	112.6	121.7	131.7	141.2	148.9	153.0	159.5	168.4	173.3	180.3	189.1	199.7	208.3	215.5	3
France	49.6	57.3	57.8	63.4	70.4	84.4	97.9	109.3	112.4	119.3	127.3	135.8	143.3	148.4	152.0	156.9	160.7	164.3	25
Germany	50.7	51.4	57.1	59.6	66.3	77.4	107.2	118.7	123.7	131.8	145.8	156.3	168.6	186.7	199.7	202.1	202.2	207.8	7
Greece	39.1	48.5	54.7	59.1	70.1	87.6	107.1	125.8	137.3	145.1	151.2	163.2	177.4	202.3	228.5	243.4	196.9	195.4	10
Hungary	9.6	21.5	30.6	37.5	44.1	51.2	65.4	82.9	100.9	112.4	121.5	128.3	142.5	156.4	170.2	166.1	169.5	167.6	24
Iceland	51.4	55.6	74.6	81.6	96.8	120.3	134.4	141.8	154.2	163.6	168.7	180.1	183.2	183.6	185.9	185.2	185.7	185.2	15
Ireland	28.1	36.5	46.3	54.9	68.1	86.9	96.1	114.7	122.9	129.0	137.7	149.3	161.6	171.3	171.7	170.6	172.7	176.3	20
Israel	..	..	34.9	0.0	0.0	95.3	115.8	131.7	140.6	141.5	147.7	147.1	175.6	158.7	165.0	194.7	183.7	176.4	19
Italy	39.4	43.7	55.3	65.1	79.8	96.7	117.5	133.8	138.0	147.8	159.3	175.3	188.7	207.3	207.4	207.0	213.5	216.2	2
Japan	44.2	49.7	71.1	80.1	86.9	94.0	101.9	109.2	117.4	125.7	132.8	138.5	142.7	146.0	148.4	146.8	150.8	154.5	29
Korea	35.7	42.0	50.8	60.4	75.2	98.3	113.4	127.8	139.1	142.2	148.6	152.6	152.8	157.0	166.5	171.6	176.6	178.8	17
Luxembourg	47.8	56.4	71.1	78.0	84.1	98.7	126.2	154.9	163.6	177.0	203.9	221.4	223.8	222.0	227.1	226.8	226.8	232.7	1
Mexico	6.6	9.8	10.6	11.7	13.9	19.3	26.9	35.8	40.8	46.0	56.0	66.4	74.7	86.1	97.0	104.2	112.9	115.3	34
Netherlands	46.4	52.5	59.1	69.3	70.8	94.9	120.5	122.4	123.8	133.9	155.0	151.9	158.3	184.2	193.5	191.3	186.8	191.1	12
New Zealand	43.8	44.8	58.6	65.0	79.0	85.9	102.0	108.1	110.9	112.3	123.2	138.8	146.4	161.7	171.8	174.7	175.0	178.1	18
Norway	50.3	56.8	85.5	94.3	102.6	114.5	125.8	132.9	138.3	145.4	159.6	167.4	167.1	169.7	172.1	170.6	172.2	170.9	21
Poland	8.6	14.8	17.6	21.7	27.2	35.1	42.8	53.4	62.3	72.5	88.9	110.3	133.4	144.3	150.5	152.5	155.7	162.8	28
Portugal	24.1	36.1	43.8	53.3	68.8	84.2	102.3	116.1	129.2	138.7	147.2	157.3	166.9	177.8	191.0	195.2	199.8	202.3	9
Slovak Republic	..	20.9	23.7	29.6	37.2	43.0	55.4	68.8	80.4	92.8	103.7	109.0	116.9	140.7	132.5	131.8	138.3	138.8	32
Slovenia	..	..	..	..	..	..	..	..	120.8	123.9	132.0	128.4	130.6	133.1	148.5	149.1	148.0	148.1	30
Spain	32.4	38.6	46.9	51.6	59.2	80.2	103.7	116.8	126.8	136.1	140.5	154.7	163.9	170.8	174.3	177.0	178.9	182.1	16
Sweden	68.3	68.6	96.8	104.5	115.2	126.7	141.1	151.7	161.2	171.3	172.1	177.8	183.8	187.7	191.7	201.4	205.4	207.5	8
Switzerland	58.7	65.6	68.0	74.9	83.0	100.6	122.2	130.7	139.3	148.4	154.7	166.0	175.4	188.1	195.1	199.9	203.9	208.9	6
Turkey	12.3	23.0	24.9	28.2	32.8	40.8	52.1	57.3	64.0	70.2	80.2	93.5	107.0	120.5	125.3	119.1	116.6	119.2	33
United Kingdom	44.1	50.3	63.0	65.7	76.0	94.7	122.1	130.5	138.5	148.0	162.9	177.2	187.7	196.8	202.5	207.3	209.4	210.9	5
United States	53.9	64.4	63.8	68.6	73.6	96.2	106.0	114.2	119.2	125.3	133.9	141.9	150.7	156.0	159.3	161.2	163.7	163.9	26
OECD	39.7	45.4	52.6	57.6	64.4	79.2	93.3	103.3	110.0	117.1	126.8	136.4	145.6	154.4	160.0	162.3	164.9	167.5	

Notes: Total communication access paths = analogue lines + ISDN lines + DSL + cable modem + fibre + mobile subscribers. Data for Chili (2010 and 2011), France (2011), Israel (2010), the Netherlands (2010 and 2011), Poland (2000 to 2004) and United States (2008 to 2011) are estimates.

StatLink http://dx.doi.org/10.1787/888932801413

Table 4.4. **Fixed telephone access paths in the OECD area**

In thousands

	1996	1997	1998	1999	2000	2001	2002	2003	2004	2005	2006	2007	2008	2009	2010	2011	Per 100 inhabitants (2011)	CAGR 2009-11	CAGR 2000-11
Australia	9 440	9 710	10 001	10 211	10 440	10 511	10 400	10 460	10 370	10 120	9 940	9 760	9 360	9 020	8 660	8 370	36.8	-3.67	-1.99
Austria	3 698	3 567	3 455	3 455	3 374	3 307	3 187	3 144	3 069	3 305	2 877	2 742	2 680	2 662	2 778	2 769	32.9	1.99	-1.78
Belgium	4 780	5 037	4 734	4 609	4 475	4 315	4 279	4 226	4 148	4 144	4 077	4 015	3 818	3 782	3 792	3 787	34.5	0.07	-1.51
Canada	18 051	18 722	19 384	19 187	19 527	19 810	19 252	19 161	19 069	19 124	19 218	19 257	19 264	18 708	18 394	18 201	52.8	-1.36	-0.64
Chile	2 264	2 693	3 047	3 068	3 302	3 478	3 467	3 252	3 345	3 461	3 384	3 460	3 530	3 564	3 458	3 366	19.5	-2.82	0.17
Czech Republic	2 817	3 273	3 735	3 806	3 898	3 669	3 389	3 279	3 059	2 869	2 548	2 493	2 380	2 120	1 630	1 830	17.4	-7.09	-6.64
Denmark	3 255	3 164	3 203	3 175	3 202	3 172	3 074	2 998	2 914	2 797	2 615	2 354	2 061	1 679	1 439	1 227	22.0	-14.53	-8.35
Estonia	439	469	499	511	514	492	440	423	404	387	376	398	383	372	376	375	28.0	0.29	-2.83
Finland	2 869	2 919	2 955	3 007	3 057	3 082	2 943	2 736	2 560	2 276	2 026	1 841	1 733	1 430	1 250	1 080	20.0	-13.10	-9.02
France	31 991	32 128	31 050	30 253	29 597	29 248	28 980	28 673	28 502	27 969	26 477	23 804	21 773	19 839	17 900	15 900	24.4	-10.48	-5.49
Germany	40 964	40 687	40 437	40 110	39 666	39 696	39 650	39 380	39 081	38 995	38 248	36 867	34 230	32 266	30 403	28 629	35.0	-5.80	-2.92
Greece	5 330	5 432	5 539	5 640	5 760	5 813	5 769	5 656	5 612	5 519	5 395	5 328	5 254	5 249	5 204	5 076	44.9	-1.66	-1.14
Hungary	2 681	3 153	3 494	3 639	3 592	3 454	3 301	3 255	3 197	3 001	3 419	3 299	3 142	3 069	2 977	2 933	29.4	-2.23	-1.82
Iceland	154	155	159	161	161	158	158	152	150	151	147	148	153	147	142	137	42.8	-3.58	-1.48
Ireland	1 390	1 500	1 585	1 661	1 637	1 660	1 701	1 703	1 679	1 727	1 739	1 741	1 686	1 597	1 515	1 424	31.7	-5.59	-1.26
Israel	1 984	..	..	2 808	2 921	2 996	2 961	2 877	2 865	2 476	2 614	2 459	2 852	3 265	3 265	2 005	25.9	-21.63	-3.36
Italy	25 022	25 263	25 134	24 996	24 494	24 753	24 799	26 011	24 800	24 008	22 666	22 850	22 557	22 528	22 434	22 023	36.3	-1.13	-0.96
Japan	62 633	62 849	62 626	62 129	61 957	61 324	60 772	60 218	59 608	58 053	55 165	51 232	47 321	43 339	39 568	35 953	27.9	-8.92	-4.83
Korea	19 950	20 866	20 795	22 118	22 426	22 725	23 490	22 877	22 871	22 920	23 119	23 130	22 132	20 090	19 273	18 633	37.4	-3.70	-1.67
Luxembourg	250	260	228	217	248	251	251	246	245	244	245	254	260	264	268	280	53.9	2.91	1.11
Mexico	8 826	9 254	9 927	10 927	12 333	13 774	14 975	16 330	18 073	19 512	19 861	19 998	20 491	19 504	19 891	19 684	18.0	0.46	4.34
Netherlands	8 152	9 129	7 767	8 211	8 174	7 985	7 852	7 677	7 434	5 942	5 777	6 166	5 422	4 804	4 080	3 556	21.3	-13.96	-7.29
New Zealand	1 719	1 753	1 763	1 759	1 749	1 765	1 801	1 847	1 843	1 847	1 851	1 854	1 861	1 880	1 876	1 880	42.5	0.0	0.66
Norway	2 484	2 475	2 475	2 446	2 388	2 317	2 295	2 202	2 106	1 931	1 669	1 510	1 384	1 247	1 132	1 026	20.7	-9.30	-7.40
Poland	6 532	7 510	8 485	9 533	9 615	9 657	9 809	9 978	10 034	11 741	11 394	10 300	9 454	8 493	7 672	6 853	17.8	-10.18	-3.03
Portugal	3 744	3 867	3 894	3 892	3 766	3 779	3 726	3 657	3 647	3 573	3 436	3 231	2 970	2 787	2 652	2 511	23.6	-5.08	-3.62
Slovak Republic	1 246	1 392	1 539	1 655	1 698	1 556	1 403	1 295	1 250	1 197	1 167	1 151	1 098	1 013	890	821	15.1	-9.99	-6.39
Slovenia	..	..	..	..	..	..	701	683	747	744	721	663	573	494	432	376	18.3	-12.74	..
Spain	15 510	16 085	16 467	17 134	17 748	17 427	17 641	17 759	17 934	19 461	19 865	20 193	20 576	20 334	20 210	19 867	43.1	-1.16	1.03
Sweden	6 065	6 075	6 089	6 093	6 053	5 951	5 846	5 739	5 601	5 377	5 077	4 785	4 471	4 037	3 574	3 167	33.5	-11.43	-5.72
Switzerland	4 171	4 284	4 224	4 153	4 108	4 101	4 077	4 016	3 941	3 831	3 760	3 698	3 635	3 494	3 261	3 142	39.9	-5.17	-2.41
Turkey	14 286	15 744	16 960	18 060	18 402	18 913	18 928	18 933	19 139	18 993	18 846	18 216	17 519	16 551	16 218	15 226	20.6	-4.09	-1.71
United Kingdom	29 829	29 828	31 442	31 646	31 823	32 070	31 213	32 053	31 336	31 092	30 650	30 580	30 350	30 055	30 043	30 225	48.2	0.28	-0.47
United States	127 948	132 027	133 484	180 683	183 671	184 709	182 261	175 848	170 502	167 398	160 270	151 175	162 763	152 873	148 572	145 875	46.7	-2.32	-2.07
OECD	470 475	481 271	486 575	540 956	545 778	547 920	544 789	538 745	531 133	526 385	510 640	490 952	489 136	462 555	445 230	428 204	34.5	-3.78	-2.18

Notes: Fixed telephone access paths: analogue + ISDN lines. Data for Chili (2010 and 2011), France (2011), Israel (2010), the Netherlands (2010 and 2011), Poland (2000 to 2004) and United States (2008 to 2011) are estimates.

StatLink http://dx.doi.org/10.1787/888932801432

Table 4.5. **Standard analogue telecommunication access lines in the OECD area**

In thousands

	1993	1996	1997	1998	1999	2000	2001	2002	2003	2004	2005	2006	2007	2008	2009	2010	2011	CAGR 2009-11	CAGR 2000-11	Per 100 inhabitants 2011
Australia	8 900	9 170	9 350	9 540	9 760	10 050	10 060	10 400	10 460	10 370	10 120	9 940	9 760	9 360	9 020	8 660	8 370	-3.67	-1.65	36.76
Austria	3 701	3 656	3 482	3 299	3 202	3 034	2 900	2 754	2 687	2 609	2 562	2 468	2 351	2 317	2 319	2 437	2 439	2.55	-1.97	28.96
Belgium	4 632	4 725	4 939	4 549	4 353	4 042	3 884	3 854	3 805	3 733	3 737	3 678	3 624	3 440	3 425	3 454	3 472	0.69	-1.37	31.63
Canada	17 567	18 051	18 660	19 294	19 082	19 409	19 689	19 153	19 062	18 973	19 035	19 134	19 176	19 184	18 633	18 322	18 132	-1.35	-0.62	52.58
Chile	..	2 151	2 693	3 047	3 109	3 303	3 467	3 467	3 252	3 318	3 436	3 384	3 460	3 526	3 575	..	..	..	..	..
Czech Republic	2 398	2 817	3 273	3 732	3 795	3 872	3 585	3 243	3 094	2 867	2 695	2 388	2 333	2 222	1 976	1 490	1 710	-6.97	-7.16	16.29
Denmark	3 203	3 225	3 104	3 086	2 928	2 827	2 767	2 680	2 621	2 557	2 476	2 332	2 105	1 847	1 504	1 296	1 111	-14.07	-8.14	19.94
Estonia	..	439	469	499	511	512	491	436	419	400	383	372	394	380	369	374	372	0.42	-2.85	27.78
Finland	2 810	2 842	2 861	2 855	2 850	2 849	2 806	2 726	2 500	2 390	2 140	1 920	1 740	1 650	1 430	1 250	1 080	-13.10	-8.44	20.05
France	32 600	31 600	31 572	31 050	30 253	29 597	29 248	28 980	28 673	28 502	27 969	26 477	23 804	21 773	19 839	17 900	15 900	-10.48	-5.49	24.40
Germany	39 200	39 000	37 800	36 200	34 500	32 200	30 500	29 100	27 837	26 986	26 340	25 440	24 050	21 720	20 010	18 670	17 530	-6.40	-5.38	21.44
Greece	5 163	5 329	5 431	5 536	5 611	5 659	5 608	5 413	5 200	5 079	4 933	4 790	4 742	4 699	4 725	4 725	4 642	-0.89	-1.79	41.04
Hungary	2 219	2 675	3 133	3 457	3 614	3 492	3 294	3 092	3 038	2 980	2 792	3 216	3 102	2 952	2 896	2 815	2 798	-1.70	-1.99	28.06
Iceland	149	154	152	151	148	144	140	140	135	134	134	132	135	141	136	132	127	-3.37	-1.14	39.80
Ireland	1 313	1 390	1 500	1 536	1 585	1 590	1 590	1 600	1 610	1 590	1 605	1 631	1 634	1 582	1 502	1 427	1 340	-5.53	-1.54	29.85
Israel	..	..	..	..	2 808	2 921	2 996	2 961	2 877	2 865	2 476	2 595	2 429	2 836	3 243	2 613	1 984	-21.78	-3.46	25.59
Italy	24 854	24 918	24 801	24 251	23 453	22 569	22 244	21 943	23 000	22 400	21 725	20 540	20 883	20 818	20 965	20 970	20 671	-0.70	-0.80	34.03
Japan	61 106	61 526	60 451	58 559	55 446	52 258	50 997	51 162	51 592	51 626	50 563	48 169	44 779	41 392	37 918	34 539	31 319	-9.12	-4.55	24.32
Korea	18 925	19 942	20 845	20 756	21 944	22 326	22 667	23 385	22 773	22 806	22 785	23 092	23 103	22 132	20 090	19 273	18 633	-3.70	-1.63	37.43
Luxembourg	229	248	255	219	189	206	191	191	171	166	165	166	171	180	185	190	204	4.88	-0.12	39.18
Mexico	8 801	8 826	9 254	9 927	10 927	12 318	13 747	14 956	16 315	18 059	19 500	19 850	19 986	20 480	19 493	19 882	19 675	0.47	4.35	18.01
Netherlands	8 020	8 110	8 850	7 767	7 330	6 915	6 569	6 316	6 120	5 922	4 518	4 459	4 994	4 376	3 855	3 391	2 919	-12.98	-7.54	17.49
New Zealand	1 660	1 719	1 753	1 763	1 759	1 749	1 765	1 801	1 847	1 843	1 847	1 851	1 854	1 861	1 880	1 876	1 880	0.00	0.66	42.52
Norway	2 431	2 440	2 325	2 166	1 914	1 685	1 548	1 484	1 417	1 377	1 293	1 159	1 068	1 000	916	849	784	-7.52	-6.72	15.82
Poland	5 728	6 532	7 510	8 479	9 483	10 814	11 225	11 534	11 323	11 174	10 364	9 951	8 942	8 267	7 237	6 441	5 712	-11.16	-5.64	14.83
Portugal	3 586	3 724	3 819	3 803	3 752	3 571	3 528	3 447	3 376	3 366	3 295	3 164	2 968	2 719	2 559	2 444	2 325	-4.68	-3.82	21.83
Slovak Republic	1 118	1 246	1 392	1 539	1 651	1 686	1 525	1 348	1 219	1 155	1 106	1 082	1 071	1 026	947	832	768	-9.95	-6.90	14.11
Slovenia	..	..	..	..	..	..	..	601	562	598	575	561	524	462	405	358	315	-11.90		15.33
Spain	15 095	15 413	15 854	16 285	16 770	17 102	17 427	17 641	17 759	17 934	19 461	18 736	19 198	19 582	19 340	19 146	18 955	-1.00	0.94	41.09
Sweden	6 013	6 032	6 010	5 965	5 890	5 783	5 665	5 581	5 494	5 397	5 198	4 922	4 644	4 338	3 922	3 469	3 066	-11.58	-5.61	32.47
Switzerland	4 410	4 045	4 076	3 883	3 622	3 382	3 240	3 163	3 089	3 012	2 924	2 897	2 876	2 851	2 753	2 566	2 490	-4.91	-2.75	31.64
Turkey	14 184	14 286	15 744	16 960	18 060	18 395	18 904	18 915	18 917	19 125	18 978	18 832	18 201	17 502	16 534	16 201	15 211	-4.09	-1.71	20.57
United Kingdom	28 479	29 668	29 569	31 051	31 045	30 940	31 060	30 135	30 987	30 362	30 197	29 807	29 770	29 611	29 357	29 418	29 624	0.45	-0.39	47.22
United States	156 973	126 379	130 273	131 628	178 650	182 013	183 360	180 941	174 609	169 325	166 779	159 094	150 055	146 304	142 646	139 080	135 603	-2.50	-2.64	43.46
OECD	485 469	462 112	468 506	469 782	516 884	514 578	513 479	509 024	502 749	495 988	490 666	474 845	456 468	441 002	422 029	406 490	391 158	-3.73	-2.46	31.51

Notes: Data for Chili (2010 and 2011) are not available. Data for France (2011), Israel (2010), the Netherlands (2010 and 2011), Poland (2000 to 2004) and United States (2008 to 2011) are estimates.

StatLink http://dx.doi.org/10.1787/888932801451

Table 4.6. **Cellular mobile subscriptions in the OECD area**

	1993	1996	1997	1998	1999	2000	2001	2002	2003	2004	2005	2006	2007	2008	2009	2010	2011	CAGR 2009-11	CAGR 2000-11
Australia	682 000	3 990 000	4 578 000	5 342 000	6 340 000	8 010 000	11 100 000	12 670 000	14 300 000	16 476 000	18 420 000	19 760 000	21 260 000	22 120 000	24 220 000	22 540 000	24 490 000	0.56	10.69
Austria	221 450	598 804	1 164 270	2 300 000	4 300 000	6 117 243	6 541 386	6 736 368	7 094 502	7 991 170	8 369 251	9 254 265	9 855 352	10 605 967	11 434 330	12 241 085	13 022 578	6.72	7.11
Belgium	67 771	478 172	974 494	1 756 287	3 186 602	5 629 000	7 690 000	8 101 778	8 605 834	9 131 705	9 604 695	9 847 375	10 738 121	11 822 190	12 381 229	12 723 355	13 533 190	4.55	8.30
Canada	1 332 982	3 420 318	4 194 761	5 346 026	6 911 038	8 726 636	10 648 824	11 997 000	13 291 000	15 020 000	17 016 600	18 749 100	20 277 400	22 092 500	23 811 900	25 825 400	27 387 200	7.24	10.96
Chile	..	319 474	409 740	964 248	2 260 687	3 401 525	5 100 783	6 244 310	7 268 281	9 261 385	10 569 572	12 450 801	13 955 202	14 796 593	16 450 223	19 852 242	22 399 969	16.69	18.69
Czech Republic	11 151	200 315	521 469	965 476	1 944 553	4 346 009	6 947 151	8 610 177	9 708 683	10 782 567	11 775 878	12 406 199	13 228 631	13 805 466	14 258 404	14 390 000	14 240 000	-0.06	11.39
Denmark	357 589	1 316 592	1 444 000	1 931 101	2 628 585	3 370 020	3 960 165	4 477 845	4 767 100	5 166 912	5 449 206	5 828 157	6 304 000	6 557 000	6 834 000	6 981 000	7 159 000	2.35	7.09
Estonia	..	60 000	140 000	246 000	387 000	574 650	737 100	880 550	1 051 940	1 255 730	1 445 300	1 658 700	1 882 776	1 897 365	1 565 409	1 860 038	1 862 674	9.08	11.28
Finland	459 074	1 476 976	2 091 791	2 845 985	3 273 433	3 728 625	4 175 587	4 516 772	4 747 000	4 999 060	5 384 572	5 670 000	6 080 000	6 830 000	7 700 000	8 390 000	8 940 000	7.75	8.27
France	467 000	2 440 139	5 754 539	11 210 100	20 619 000	29 681 300	36 997 400	38 593 000	41 702 000	44 544 000	48 088 000	51 662 000	55 337 000	57 994 000	61 540 000	65 030 000	68 590 000	5.57	7.91
Germany	1 768 000	5 782 200	8 175 500	13 913 000	23 446 000	48 202 000	56 126 000	59 128 000	64 800 000	74 316 000	79 200 000	85 652 000	97 151 000	107 245 000	108 255 000	108 846 000	114 125 000	2.68	8.15
Greece	28 000	531 488	938 038	2 056 084	3 894 312	5 932 403	7 963 742	9 314 260	10 330 323	11 059 920	12 448 473	13 874 674	16 226 675	18 918 092	20 298 102	14 815 705	14 557 672	-15.31	8.50
Hungary	63 000	473 000	706 000	1 036 000	1 601 000	3 076 000	4 967 430	6 886 111	7 944 586	8 727 188	9 320 169	9 965 720	11 029 930	12 224 163	11 792 475	12 011 823	11 689 937	-0.44	12.90
Iceland	17 409	46 302	65 746	106 000	172 600	215 000	235 400	260 900	279 670	290 068	304 001	322 840	326 098	336 922	339 715	341 077	344 085	0.64	4.37
Ireland	57 065	290 000	510 747	946 000	1 600 000	2 020 000	2 770 000	3 122 148	3 421 261	3 785 052	4 213 436	4 631 197	4 970 719	5 048 127	5 154 289	5 273 313	5 499 790	3.30	9.53
Israel	..	..	..	..	3 039 500	4 373 610	5 500 621	6 300 008	6 618 367	7 221 955	7 757 000	8 403 765	8 982 731	9 266 448	9 550 164	9 550 164	9 791 022	1.25	7.60
Italy	1 206 975	6 413 412	11 760 000	20 300 000	30 068 000	42 290 000	51 096 000	53 100 000	56 700 000	63 153 000	71 838 000	80 416 000	90 135 922	90 341 160	90 032 886	93 634 699	95 912 476	3.21	7.73
Japan	2 131 367	26 906 511	38 253 893	47 307 592	56 845 594	66 784 374	74 819 158	81 118 324	86 654 962	91 473 960	96 483 732	101 698 165	107 338 974	112 050 077	112 182 922	119 535 344	128 204 758	6.90	6.11
Korea	471 784	3 180 989	6 895 477	13 982 919	23 442 724	26 816 398	29 045 596	32 342 493	33 591 758	36 586 052	38 342 323	40 197 115	43 497 541	45 606 984	47 944 222	50 767 241	52 506 793	4.65	6.30
Luxembourg	5 082	45 000	67 208	130 000	208 364	303 274	432 400	473 000	539 000	646 000	719 500	714 000	684 000	707 000	720 000	727 000	765 000	3.08	8.78
Mexico	386 100	1 021 900	1 740 814	3 349 475	7 731 635	14 077 880	21 757 559	25 928 266	30 097 700	38 451 135	47 128 746	55 395 461	66 559 462	75 303 469	83 193 574	91 362 753	94 565 305	6.62	18.90
Netherlands	216 000	1 016 000	1 688 550	3 347 000	6 790 000	11 000 000	11 500 000	11 800 000	13 100 000	15 913 000	16 289 000	17 058 000	18 453 000	20 627 000	20 682 000	20 627 000	21 848 000	2.78	6.44
New Zealand	186 000	476 200	710 000	1 254 900	1 542 000	2 187 000	2 422 000	2 539 000	2 599 000	3 027 000	3 530 000	3 803 000	4 245 000	4 577 000	4 697 000	4 707 448	4 821 164	1.31	7.45
Norway	369 271	1 261 445	1 676 763	2 071 672	2 663 552	3 244 646	3 593 251	3 790 086	4 060 829	4 524 750	4 754 453	4 868 916	5 037 650	5 211 207	5 354 554	5 599 286	5 693 957	3.12	5.25
Poland	15 699	216 900	812 000	1 928 000	3 904 000	6 747 000	10 750 000	13 898 471	17 401 222	23 096 065	29 166 391	36 745 454	41 388 774	43 926 365	44 806 632	46 976 572	50 146 101	5.79	20.00
Portugal	101 231	663 651	1 506 958	3 074 633	4 671 458	6 664 951	7 977 537	9 202 232	10 002 705	10 571 100	11 368 494	12 236 104	13 477 414	14 953 227	16 051 044	16 473 690	16 794 499	2.29	8.76
Slovak Republic	3 125	28 658	200 141	465 364	664 072	1 293 736	2 147 331	2 923 383	3 678 774	4 275 164	4 540 374	4 893 232	6 068 063	5 520 043	5 497 719	5 925 012	5 983 059	4.32	14.94
Slovenia	..	..	..	..	..	..	..	1 708 742	1 761 901	1 848 637	1 759 232	1 819 572	1 928 412	2 054 889	2 100 435	2 121 950	2 168 548	1.61	..
Spain	257 261	2 997 212	4 330 282	7 051 441	14 884 207	23 938 970	29 655 729	33 530 997	37 219 839	38 622 582	42 693 832	45 695 061	48 422 470	49 623 339	51 052 693	51 389 417	52 597 287	1.50	7.42
Sweden	850 000	2 492 000	3 169 000	4 108 793	5 126 100	6 372 367	7 177 813	7 948 518	8 801 266	8 784 536	9 103 505	9 606 661	10 116 852	10 891 660	11 750 105	12 691 619	13 387 910	6.74	6.98
Switzerland	259 200	662 700	1 044 400	1 698 565	3 057 509	4 638 519	5 275 791	5 736 303	6 188 793	6 274 763	6 834 233	7 436 157	8 208 884	8 896 706	9 322 580	9 644 157	10 122 056	4.20	7.35
Turkey	84 187	806 339	1 609 808	3 506 100	7 796 000	15 062 744	18 420 000	23 323 118	27 887 535	34 707 549	43 608 965	52 662 709	61 975 807	65 824 110	62 779 554	61 769 635	65 321 745	2.00	14.27
United Kingdom	2 216 000	6 817 000	8 463 000	13 001 196	23 942 411	40 013 263	44 767 541	49 546 944	52 868 573	60 028 915	65 805 665	70 077 926	73 806 165	76 735 443	80 255 445	81 165 000	81 612 000	0.84	6.69
United States	14 712 000	44 043 000	55 312 293	69 209 321	86 047 000	109 478 000	128 500 000	141 800 000	160 600 000	184 700 000	203 700 000	229 619 000	249 332 000	261 284 000	274 283 000	285 125 000	290 304 000	2.88	9.27
OECD	29 003 773	120 472 697	170 909 682	246 751 278	364 988 936	518 317 143	620 799 295	688 549 104	759 684 404	856 712 920	947 032 598	1 045 079 326	1 148 282 025	1 215 693 512	1 258 291 605	1 300 914 025	1 350 386 775	3.59	9.10
World	34 171 809	145 085 841	215 020 435	318 235 884	491 176 119	738 000 000	961 000 000	1 157 000 000	1 416 000 000	1 763 000 000	2 207 000 000	2 747 000 000	3 369 000 000	4 031 000 000	4 647 000 000	5 311 000 000	5 972 000 000	13.36	20.93
OECD % share of world total	85	83	79	78	74	70	65	60	54	49	43	38	34	30	27	24	23	-8.62	-9.79

Note: Data for Estonia (2008, 2009 and 2010) and for Israel (2010) are estimated.

Source: ITU (for world mobile subscriptions)

StatLink http://dx.doi.org/10.1787/888932801470

Table 4.7. **Cellular mobile penetration, subscriptions per 100 inhabitants**

	1996	1997	1998	1999	2000	2001	2002	2003	2004	2005	2006	2007	2008	2009	2010	2011	CAGR 2009-11	CAGR 2000-11
Australia	21.7	24.6	28.4	33.3	41.6	56.8	64.1	71.5	81.4	89.7	94.7	100.0	101.8	109.5	100.4	107.6	-0.87	9.03
Austria	7.5	14.6	28.8	53.8	76.4	81.3	83.3	87.4	97.8	101.8	111.9	118.7	127.2	136.7	145.9	154.6	6.35	6.63
Belgium	4.7	9.6	17.2	31.2	54.9	74.8	78.4	83.0	87.7	91.7	93.4	101.1	110.4	114.7	116.9	123.3	3.65	7.62
Canada	11.6	14.0	17.7	22.7	28.4	34.3	38.3	42.0	47.0	52.8	57.6	61.6	66.3	70.6	75.7	79.4	6.07	9.79
Chile	2.2	2.8	6.4	14.9	22.1	32.8	39.7	45.7	57.5	65.0	75.8	84.1	88.3	97.2	116.1	129.7	15.54	17.46
Czech Republic	1.9	5.1	9.4	18.9	42.3	67.9	84.4	95.2	105.6	115.1	120.8	128.2	132.4	135.9	136.8	135.7	-0.09	11.17
Denmark	25.0	27.3	36.4	49.4	63.1	73.9	83.3	88.4	95.6	100.6	107.2	115.5	119.4	123.8	125.9	128.6	1.92	6.68
Estonia	4.2	10.0	17.7	28.1	41.9	53.9	64.7	77.6	92.9	107.3	123.4	140.3	141.5	116.8	138.8	139.0	9.09	11.52
Finland	28.8	40.7	55.2	63.4	72.0	80.5	86.9	91.1	95.6	102.6	107.7	115.0	128.5	144.2	156.4	165.9	7.27	7.88
France	4.1	9.6	18.7	34.2	48.9	60.5	62.6	67.2	71.3	76.4	81.5	86.8	90.4	95.4	100.3	105.2	5.00	7.22
Germany	7.1	10.0	17.0	28.6	58.6	68.2	71.7	78.5	90.1	96.0	104.0	118.1	130.6	132.2	133.1	139.6	2.74	8.20
Greece	5.0	8.7	19.0	35.8	54.3	72.7	84.8	93.7	100.0	112.1	124.5	145.0	168.4	179.9	131.0	128.7	-15.41	8.16
Hungary	4.6	6.9	10.1	15.6	30.1	48.8	67.8	78.4	86.3	92.4	99.0	109.7	121.8	117.7	120.1	117.2	-0.18	13.15
Iceland	17.2	24.3	38.7	62.3	76.5	82.6	90.7	96.7	99.1	102.8	106.1	104.7	105.5	106.4	107.3	107.9	0.68	3.18
Ireland	8.0	14.0	25.5	42.6	53.1	71.7	79.4	85.6	93.1	101.3	108.7	113.9	113.6	115.4	117.8	122.5	3.04	7.89
Israel	0.0	0.0	0.0	49.5	69.4	85.2	95.6	98.6	105.7	111.5	118.7	124.6	126.2	127.6	125.3	126.3	-0.52	5.59
Italy	11.3	20.7	35.7	52.8	74.3	89.7	92.9	98.4	108.6	122.6	136.4	151.8	151.0	149.6	154.8	157.9	2.74	7.10
Japan	21.4	30.3	37.4	44.9	52.6	58.8	63.6	67.9	71.6	75.5	79.6	84.0	87.8	88.0	93.3	99.6	6.37	5.97
Korea	7.0	15.0	30.2	50.3	57.0	61.3	67.9	70.2	76.2	79.7	83.1	89.5	93.2	97.5	102.7	105.5	4.02	5.75
Luxembourg	10.9	16.0	30.6	48.3	69.4	97.9	105.9	119.2	140.9	154.5	151.0	142.3	144.5	144.5	143.3	147.3	0.95	7.08
Mexico	1.1	1.9	3.5	8.0	14.3	21.8	25.7	29.5	37.4	45.4	52.9	63.0	70.7	77.4	84.4	86.6	5.75	17.77
Netherlands	6.5	10.8	21.3	43.0	69.1	71.7	73.1	80.8	97.8	99.8	104.4	112.7	125.5	125.1	124.2	130.9	2.27	5.98
New Zealand	12.7	18.7	32.8	40.1	56.5	62.1	64.0	64.3	73.8	85.1	90.6	100.1	106.9	108.4	107.4	109.1	0.29	6.15
Norway	28.8	38.1	46.7	59.7	72.2	79.6	83.5	89.0	98.6	102.9	104.5	107.0	109.3	110.9	114.5	115.0	1.80	4.31
Poland	0.6	2.1	5.0	10.2	17.6	28.1	36.4	45.6	60.5	76.4	96.4	108.6	115.2	117.4	122.0	130.2	5.28	19.93
Portugal	6.6	14.9	30.4	45.9	65.2	77.5	88.8	95.8	100.7	107.8	115.6	127.0	140.8	151.0	154.9	157.7	2.20	8.36
Slovak Republic	0.5	3.7	8.6	12.3	24.0	39.9	54.4	68.4	79.4	84.3	90.8	112.4	102.1	101.5	109.1	110.0	4.10	14.86
Slovenia	0.0	0.0	0.0	0.0	0.0	0.0	85.7	88.3	92.6	87.9	90.6	95.5	101.6	102.9	103.6	105.6	1.33	
Spain	7.6	10.9	17.8	37.3	59.5	72.8	81.2	88.6	90.5	98.4	103.7	107.9	108.8	111.2	111.5	114.0	1.29	6.10
Sweden	28.2	35.8	46.4	57.9	71.8	80.7	89.1	98.3	97.7	100.8	105.8	110.6	118.1	126.4	135.3	141.8	5.93	6.38
Switzerland	9.3	14.7	23.8	42.7	64.3	72.4	78.1	83.6	84.2	91.1	98.4	107.7	115.4	119.5	123.9	128.6	3.75	6.50
Turkey	1.3	2.6	5.6	12.3	23.4	28.3	35.3	41.7	51.2	63.6	75.9	88.3	92.6	87.1	84.6	88.3	0.69	12.82
United Kingdom	11.7	14.5	22.2	40.8	68.0	75.7	83.5	88.8	100.3	109.2	115.7	121.0	125.0	129.9	130.4	130.1	0.08	6.08
United States	16.3	20.3	25.1	30.8	38.8	45.1	49.2	55.3	63.0	68.8	76.8	82.6	85.8	89.3	92.0	93.0	2.08	8.28
OECD	10.8	15.2	21.7	31.9	45.0	53.5	58.9	64.6	72.3	79.4	87.0	95.0	99.8	102.7	105.5	108.8	2.92	8.35

Notes: Data for Estonia (2008, 2009 and 2010) and for Israel (2010) are estimated.

StatLink http://dx.doi.org/10.1787/888932801489

Table 4.8. **Mobile pre-paid subscriptions**

In thousands

	1996	% of total	1997	% of total	1998	% of total	1999	% of total	2000	% of total	2001	% of total	2002	% of total	2003	% of total
Australia	..	..	..	..	..	..	409	6	1 350	17	3 300	30	4 120	33	5 400	38
Austria	..	..	..	..	..	..	2 044	48	3 185	52	3 331	51	3 259	48	3 338	47
Belgium	..	..	..	..	..	..	1 275	40	3 377	60	4 901	64	5 331	66	5 716	66
Canada	..	..	..	..	341	6	1 132	16	1 879	22	2 736	26	2 937	24	3 146	24
Chile	..	..	..	..	..	..	1 285	..	2 333	..	3 810	..	4 861	..	5 795	80
Czech Republic	..	..	..	..	..	..	..	..	..	..	3 016	43	6 732	78	7 268	75
Denmark	..	..	..	..	..	..	980	37	1 238	37	1 474	37	1 354	30	1 118	23
Estonia					20		67		153		204		262		364	35
Finland	..	..	..	..	..	..	30	1	75	2	84	2	90	2	94	2
France	..	..	..	..	..	..	7 279	35	13 806	47	18 022	49	17 108	44	17 149	39
Germany	..	..	..	..	2 087	15	5 533	24	26 318	55	31 374	56	31 338	53	33 307	51
Greece	..	..	..	..	716	35	2 052	..	3 469	58	5 029	63	6 066	65	6 750	65
Hungary	..	..	..	..	..	..	474	30	1 749	57	3 585	72	5 378	78	6 158	78
Iceland	..	..	..	..	6	5	40	23	63	29	88	37	88	34	113	40
Ireland	..	..	..	..	..	..	640	40	1 266	63	1 967	71	2 210	71	2 510	73
Israel	..	..	..	..	..	..	..	..	..	..	..	..	..	..	..	..
Italy	577	9	5 527	47	15 022	74	25 257	84	37 290	88	45 792	90	47 732	90	51 706	91
Japan	..	..	..	..	..	..	1 907	3	1 414	2	1 847	..	2 084	3	2 610	3
Korea	..	..	..	..	..	..	..	..	..	..	..	..	607	2	591	2
Luxembourg	..	..	..	..	..	..	47	22	120	39	179	41	179	38	318	59
Mexico	423	41	982	56	2 282	68	6 327	82	12 450	88	19 974	92	23 922	92	28 070	93
Netherlands	..	..	..	..	1 573	47	3 938	58	7 370	67	7 500	65	7 400	63	8 100	62
New Zealand	..	..	..	..	577	46	879	57	1 487	68	1 661	69	1 737	68	1 798	69
Norway	..	..	..	..	474	23	1 113	42	1 385	43	1 514	42	1 654	44	1 666	41
Poland	..	..	..	..	463	24	942	24	2 606	39	5 120	48	7 375	53	9 467	54
Portugal	..	..	..	..	2 429	79	3 706	79	5 305	80	6 329	79	7 293	79	7 929	79
Slovak Republic	..	..	..	..	..	..	127	19	483	37	1 536	72	1 961	67	2 284	62
Slovenia	..	..	..	..	..	..	..	..	..	..	..	..	902	53	867	49
Spain	..	..	..	..	2 609	37	9 240	62	15 737	66	19 271	65	22 087	66	21 627	58
Sweden	..	..	235	7	1 016	25	1 983	39	2 773	45	3 536	50	4 309	54	5 003	57
Switzerland	36	5	209	20	590	35	1 053	34	1 707	37	2 155	41	2 315	40	2 601	42
Turkey	..	..	..	..	..	..	780	10	6 628	44	11 500	62	17 125	73	20 851	75
United Kingdom	..	..	..	..	2 910	22	12 059	50	27 365	77	30 885	70	33 976	69	35 582	67
United States	..	..	..	..	..	..	4 302	5	6 570	6	11 565	9	11 565	8	11 565	7
OECD	1 037	1	6 953	4	33 115	13	96 900	26	188 499	37	253 284	41	285 358	41	310 860	40

	2004	% of total	2005	% of total	2006	% of total	2007	% of total	2008	% of total	2009	% of total	2010	% of total	2011	% of total
Australia	7 080	52	8 504	46	9 700	49	10 150	46	9 990	45	10 580	44	10 710	48	11 230	46
Austria	3 529	47	3 774	45	3 880	42	3 695	35	3 552	33	3 642	32	3 897	32	4 168	32
Belgium	6 036	66	6 042	63	5 942	60	6 147	52	6 654	56	6 739	54	6 710	53	7 304	54
Canada	3 330	25	3 820	22	4 203	22	4 463	20	4 817	22	5 028	21	5 416	21	5 806	21
Chile	7 645	93	8 638	82	9 807	79	10 432	71	10 763	73	11 933	73	14 066	71	15 886	71
Czech Republic	7 733	73	7 834	67	7 452	60	7 381	53	7 294	53	7 207	51	7 100	49	6 620	46
Denmark	1 013	19	998	18	1 023	18	990	15	1 078	16	1 148	17	1 013	15	987	14
Estonia	481	45	562	39	719	43	700	37	637	34	525	34	624	34	625	34
Finland	350	7	369	7	454	8	550	8	710	10	770	10	790	9	910	10
France	17 124	37	17 561	35	18 102	33	19 028	33	18 736	32	18 805	31	18 626	29	19 573	29
Germany	31 374	54	40 200	51	39 947	47	53 433	50	60 680	57	60 483	56	60 249	55	63 786	56
Greece	7 286	75	8 339	67	9 599	69	11 471	61	13 811	73	15 715	77	10 306	70	10 182	70
Hungary	6 383	73	6 338	68	6 442	65	6 887	56	7 484	61	6 681	57	6 465	54	5 798	50
Iceland	125	46	133	44	144	45	135	40	141	42	145	43	158	46	159	46
Ireland	2 845	85	3 202	76	3 540	75	3 708	73	3 747	74	3 432	67	3 397	64	3 451	63
Israel	..	..	..	..	2 290	..	2 274	25	2 445	26	2 616	27	2 604	27	2 591	26
Italy	57 659	104	65 732	92	72 696	90	80 604	89	80 264	89	77 046	86	78 983	84	79 511	83
Japan	2 858	3	2 726	3	2 494	2	2 109	2	1 541	1	1 099	1	1 258	1	1 370	1
Korea	527	2	662	2	538	1	872	2	909	2	980	2	1 001	2	1 017	2
Luxembourg	381	65	419	58	372	52	310	44	279	39	260	36	230	32	236	31
Mexico	35 943	114	43 861	93	51 092	92	61 361	81	69 053	92	72 971	88	78 803	86	80 469	85
Netherlands	10 064	76	12 028	74	9 382	55	8 171	40	8 393	41	7 628	37	6 357	31	6 252	29
New Zealand	2 115	81	2 461	70	2 595	68	2 878	63	3 096	68	3 104	66	3 162	67	3 168	66
Norway	1 754	38	1 736	37	1 615	33	1 425	27	1 422	27	1 432	27	1 492	27	1 478	26
Poland	13 498	81	18 813	65	24 319	66	26 684	61	24 466	56	23 419	52	24 022	51	26 400	53
Portugal	8 424	87	9 212	81	9 761	80	10 341	69	11 080	74	11 702	73	11 880	72	12 066	72
Slovak Republic	2 445	56	2 393	53	2 382	49	3 097	56	2 307	42	1 979	36	2 140	36	1 930	32
Slovenia	878	42	773	44	735	40	748	36	729	35	669	32	636	30	627	29
Spain	20 067	54	20 714	49	20 881	46	20 765	42	20 313	41	20 865	41	19 969	39	20 377	39
Sweden	4 629	53	4 638	51	4 693	49	4 496	41	4 407	40	4 532	39	4 651	37	4 545	34
Switzerland	2 485	40	2 808	41	3 103	42	3 559	40	3 896	44	4 058	44	4 222	44	4 275	42
Turkey	26 355	76	30 601	70	42 695	81	50 237	76	52 640	80	46 910	75	42 712	69	42 279	65
United Kingdom	39 794	67	43 197	66	45 778	65	47 036	61	46 830	61	47 285	59	44 039	54	41 446	51
United States	15 000	8	23 430	12	36 270	16	44 710	17	46 509	18	54 582	20	58 736	21	62 706	22
OECD	347 207	40	402 515	41	454 643	43	510 847	42	530 672	44	535 972	43	536 421	41	549 229	41

Notes: Data for Estonia (2007, 2008, 2009 and 2010) and Israel (2010) are estimates.

StatLink http://dx.doi.org/10.1787/888932801508

Table 4.9. **3G cellular mobile subscriptions in the OECD area**

	2001	2002	2003	2004	2005	2006	2007	2008	2009	2010	2011	CAGR 2009-11
Australia	..	..	20 000	238 070	532 000	1 560 000	4 560 000	8 550 000	12 280 000		14 470 000	8.55
Austria	..	..	..	180 240	901 812	1 671 000	2 464 715	3 560 744	4 489 620	5 874 145	7 026 508	25.10
Belgium	..	..	..	..	..	..	..	..	..	..	..	..
Canada	..	..	..	..	..	..	..	..	..	12 352 618	13 188 757	..
Chile	..	..	..	..	..	..	..	..	2 217 260	3 210 966	..	..
Czech Republic	..	..	..	..	65 000	119 405	182 495	276 485	354 166	510 000	600 000	30.16
Denmark	..	..	3 425	124 674	..	326 927	666 178	1 025 000	1 408 683	5 670 165	5 725 952	101.61
Estonia	..	..	..	..	3 000	15 000	44 600	99 895	136 600	203 678	293 723	46.64
Finland	..	..	..	13 000	45 000	..	1 040 000	..	..	..	..	..
France	..	..	..	..	..	..	..	..	1 770 000	22 910 000	27 750 000	295.95
Germany	..	..	..	..	..	..	12 400 000	17 900 000	19 000 000	21 200 000	28 600 000	22.69
Greece	..	..	..	18 800	229 537	419 553	1 126 039	7 331 678	7 823 857	6 369 625	6 543 077	-8.55
Hungary								365 393	614 421	777 461	1 188 086	39.06
Iceland	..	..	..	..	..	..	..	..	..	..	..	..
Ireland							994 144	1 262 032	2 220 510	3 198 740	4 221 755	37.89
Israel	..	..	..	..	..	640 606	1 801 235	2 797 542	3 793 848	3 825 755	3 857 663	0.84
Italy			400 000	2 813 000	10 477 700	17 091 000	23 127 000	26 500 000	29 900 000	34 500 000	38 600 000	13.62
Japan	89 400	7 161 100	16 692 000	30 352 700	48 329 400	69 909 200	88 097 400	99 631 300	109 056 900	118 151 100	128 153 700	8.40
Korea	..	16 537 747	24 826 749	32 538 532	36 089 425	38 529 752	42 488 783	44 777 814	47 351 418	49 889 740	49 804 021	2.56
Luxembourg	..	..	..	..	..	..	..	..	..	..	..	..
Mexico	..	..	..	..	..	..	..	..	..	..	..	..
Netherlands	..	..	..	..	..	..	1 311 000	3 755 000	..	..	..	
New Zealand	..	..	..	..	..	..	..	..	..	..	..	..
Norway	..	..	..	..	..	..	..	..	..	..	..	..
Poland	..	..	..	..	..	..						..
Portugal	..	..	..	..	..	..	3 074 319	4 319 850	5 984 053	4 078 377	4 153 778	-16.68
Slovak Republic	..	..	..	..	..	174 999	473 110	827 603	1 100 003			
Slovenia	..		1 276 226	1 451 905	1 367 090	1 376 044	1 868 789	1 935 216	2 100 435	2 121 950	2 168 548	1.61
Spain	..	..	..	..	..	..	3 074 319	..	..	..	..	..
Sweden	..	..	18 000	322 000	..	1 214 000	..	..	5 483 463	7 355 469	9 287 708	30.14
Switzerland	..	..	..		114 806	360 690	1 447 095	1 813 700	2 739 731	3 442 013	3 976 412	20.47
Turkey	..		18 000	322 000		1 214 000	2 258 000	..	7 064 842	19 407 264	31 375 507	110.74
United Kingdom	..	..	230 000	2 567 000	4 611 000	7 714 808	12 514 000	18 444 216	25 503 490	..	..	..
United States	..	13 900	30 700	49 200	257 431	484 277	54 961 800	84 914 180	119 030 000	163 055 000	209 999 000	32.83
OECD	89 400	23 712 747	43 515 100	70 991 121	103 023 201	142 821 261	259 975 021	330 087 647	405 439 247	488 104 067	595 211 079	21.16

StatLink http://dx.doi.org/10.1787/888932801527

Table 4.10. **Total fixed broadband subscriptions in the OECD area**

	1997	1998	1999	2000	2001	2002	2003	2004	2005	2006	2007	2008	2009	2010	2011	June 2012	CAGR 2009-11	CAGR 2000-11
Australia	1 000	5 000	13 800	0	5 000	363 500	698 700	1 548 300	2 785 000	3 816 172	4 830 200	5 336 000	5 092 000	5 385 000	5 490 000	5 601 000	3.83	..
Austria	..	..	50 900	137 400	292 600	451 571	594 429	869 987	1 177 637	1 383 798	1 597 949	1 769 745	1 760 433	1 915 093	2 043 004	2 063 612	7.73	27.81
Belgium	..	..	22 841	145 823	448 349	895 671	1 213 304	1 618 944	1 902 739	2 355 589	2 715 737	2 961 728	3 114 695	3 356 599	3 528 839	3 586 116	6.44	33.60
Canada	21 000	161 000	582 000	1 407 790	2 750 186	3 805 030	4 763 166	5 631 714	6 694 474	7 928 009	8 974 830	9 405 318	9 980 000	10 477 886	10 931 877	11 013 711	4.66	20.48
Chile	..	..	..	..	62 020	166 384	322 071	450 249	683 346	1 008 322	1 298 085	1 422 591	1 645 471	1 779 136	2 002 573	2 101 524	10.32	..
Czech Republic		..	1 500	10 000	12 100	16 900	48 498	166 000	650 000	1 101 758	1 446 420	1 699 697	1 354 986	1 530 800	1 656 000	1 715 700	10.55	59.12
Denmark	..	..	11 800	67 399	237 673	443 297	706 281	1 024 160	1 350 415	1 589 749	1 794 142	1 992 022	2 039 131	2 092 366	2 095 474	2 131 167	1.37	36.68
Estonia	..	400	0	700	4 200	12 700	28 300	53 727	78 900	100 066	120 113	140 690	301 092	311 961	331 890	334 655	4.99	75.09
Finland	..		7 500	30 000	68 000	283 500	494 300	779 929	1 174 200	1 429 200	1 617 100	1 482 000	1 533 800	1 532 700	1 588 700	1 600 400	1.77	43.46
France		13 464	50 000	189 443	620 322	1 691 992	3 656 654	6 529 997	9 465 600	12 718 313	15 522 732	17 725 000	19 760 000	21 234 000	22 584 000	23 167 000	6.91	54.44
Germany	..	..	5 000	205 000	1 934 000	3 254 000	4 621 286	6 904 983	10 706 600	14 982 600	19 541 400	22 532 000	24 935 500	26 089 800	27 185 816	27 677 770	4.41	55.94
Greece	..	..		72	72	1 932	10 476	51 471	156 560	509 081	1 084 115	1 508 630	1 915 006	2 250 410	2 462 661	2 558 830	13.40	158.34
Hungary	..	..	1 453	2 304	27 879	65 704	189 370	360 741	617 966	965 384	1 394 285	1 717 695	1 785 226	1 956 218	2 086 292	2 114 832	8.10	85.70
Iceland	..	..	..	2 035	10 478	24 285	41 406	53 264	78 221	87 738	97 937	103 697	104 770	107 011	110 010	109 359	2.47	43.73
Ireland	..	..	..	300	400	10 600	33 050	135 751	270 645	517 249	767 736	896 346	870 562	941 405	993 667	1 011 730	6.84	108.93
Israel	..	..	..	..	..	..	..	..	..	1 420 357	..	..	1 756 300	1 818 000	1 879 000	1 910 000	3.43	..
Italy	..	..	615	114 900	415 000	976 019	2 401 939	4 701 252	6 896 696	8 164 000	10 095 441	11 222 848	12 056 129	13 057 096	13 425 601	13 444 129	5.53	54.16
Japan	..	..	154 019	634 732	2 837 348	7 618 228	12 765 552	18 667 599	22 384 838	25 522 092	27 927 018	30 110 423	31 642 029	34 016 470	34 865 854	35 142 132	4.97	43.93
Korea	..	..	270 987	4 065 648	8 744 204	10 400 097	11 608 901	11 921 439	12 190 711	10 613 262	9 694 872	13 737 564	16 347 716	17 193 570	17 859 003	18 010 256	4.52	14.40
Luxembourg	..	..	..	..	1 230	6 861	15 571	44 145	67 357	99 030	128 960	143 446	145 400	155 892	163 617	166 630	6.08	..
Mexico	..	..	..	8 622	111 070	247 016	428 377	1 055 115	2 306 971	2 978 359	4 457 247	7 528 969	9 283 529	11 048 951	11 723 336	12 108 478	12.37	92.69
Netherlands	..	..	0	10 000	145 200	340 200	944 200	1 885 561	2 552 052	3 028 000	5 547 902	5 755 000	6 130 000	6 329 000	6 498 000	6 573 000	2.96	80.18
New Zealand	..	..	..	10 334	28 079	64 100	98 042	181 572	379 000	490 067	757 132	914 961	988 993	1 088 730	1 174 790	1 239 569	8.99	53.77
Norway	..	..	200	17 329	87 656	190 731	364 251	697 190	1 049 000	1 250 899	1 439 575	1 611 901	1 633 017	1 688 907	1 746 840	1 789 279	3.43	51.71
Poland	..	..	..	..	21 696	114 000	297 291	818 575	1 170 825	2 735 728	3 296 242	3 994 000	4 895 551	5 320 978	5 719 789	5 602 280	8.09	..
Portugal	..	..	0	25 154	99 466	261 107	502 546	1 244 373	1 656 282	1 996 589	2 154 823	2 361 104	1 911 826	2 126 689	2 243 277	2 309 702	8.32	50.42
Slovak Republic	..	..	..	..	0	420	18 677	53 288	134 594	243 257	372 055	547 315	627 722	694 414	746 727	771 638	9.07	..
Slovenia	..	..	..	..	..	..	28 698	40 252	65 542	81 446	94 973	375 363	449 138	477 850	495 375	495 953	5.02	..
Spain	..	..	1 848	58 415	474 282	1 209 969	2 207 008	3 441 630	4 994 274	6 658 907	7 875 379	9 156 969	9 786 578	10 762 142	11 289 980	11 384 588	7.41	61.37
Sweden	..	..	3 800	92 455	364 585	594 762	803 516	1 097 338	1 573 699	2 008 276	2 271 000	2 315 000	2 941 373	2 994 000	3 039 000	3 053 000	1.65	37.37
Switzerland	..	..	0	60 891	141 688	414 742	778 501	1 315 166	1 786 785	2 061 938	2 419 965	2 510 841	2 777 154	2 973 609	3 172 005	3 276 700	6.87	43.24
Turkey	..	..	..	292	2 818	2 967	153 026	469 048	1 500 000	2 745 881	4 395 800	5 736 619	6 446 374	7 103 911	7 576 798	7 666 609	8.41	151.93
United Kingdom	..	..	..	52 889	332 963	1 371 085	3 205 594	6 117 659	9 861 843	12 995 140	15 606 100	17 275 660	17 812 208	19 145 758	20 440 314	21 066 624	7.12	71.87
United States	74 000	500 000	1 874 108	6 184 800	12 380 936	19 185 138	27 744 352	37 352 520	48 413 345	60 385 969	70 282 049	77 423 044	78 349 000	82 758 000	85 630 000	88 520 000	4.54	26.99
OECD	96 000	707 052	3 052 371	13 538 664	32 661 428	54 484 508	81 787 334	114 421 146	156 776 140	195 972 225	231 619 314	263 429 486	282 172 709	301 714 352	314 780 108	321 317 973	5.62	33.11

Note: The broadband data from 1997 to 2009 are calculated according to the 'old' OECD methodology including DSL, cable, fibre, other, satellite and fixed wireless connections. For 2010, 2011 and June 2012 data, the current or 'new' methodology separates all fixed and wireless connections.

StatLink http://dx.doi.org/10.1787/888932801546

Table 4.11. **Total fixed broadband subscriptions per 100 inhabitants in the OECD area**

	1997	1998	1999	2000	2001	2002	2003	2004	2005	2006	2007	2008	2009	2010	2011	June 2012	Rank 2011	Rank 2009
Australia	0.01	0.03	0.15	0.38	0.84	1.84	3.49	7.65	13.56	18.28	22.75	24.66	23.69	24.08	24.11	24.60	21	17
Austria	..	..	0.64	1.72	3.64	5.59	7.62	10.62	14.37	16.74	19.25	21.22	22.45	23.57	24.26	24.51	18	20
Belgium	..	..	0.22	1.42	4.36	8.67	11.70	15.54	18.17	22.34	25.57	27.67	29.04	31.55	32.14	32.67	12	13
Canada	0.07	0.53	1.84	4.59	8.87	12.14	15.06	17.63	20.76	24.34	27.26	28.22	30.50	31.22	31.70	31.94	13	10
Chile	..	..	..	..	0.40	1.06	2.02	2.80	4.20	6.14	7.82	8.49	12.02	11.15	11.60	12.17	32	31
Czech Republic	..	..	0.01	0.10	0.12	0.17	0.48	2.50	6.46	11.07	14.54	16.97	19.37	15.14	15.78	16.35	29	25
Denmark	..	..	0.22	1.26	4.44	8.25	13.10	18.96	24.92	31.79	35.64	36.81	37.43	37.14	37.63	38.27	3	1
Estonia	..	0.03	0.11	0.36	1.49	3.09	6.04	10.28	13.30	16.97	19.79	23.58	25.24	24.07	24.76	24.97	19	21
Finland	..	..	0.15	0.58	1.31	5.45	9.48	14.92	22.39	27.14	30.58	30.43	27.33	28.91	29.49	29.71	14	14
France	..	0.02	0.08	0.31	1.01	2.75	5.89	10.45	15.03	20.06	24.38	27.63	30.36	33.80	34.65	35.55	4	11
Germany	..	..	0.01	0.25	2.35	3.95	5.59	8.37	12.98	18.19	23.74	27.44	30.51	32.55	33.24	33.84	9	9
Greece	..	..	..	..	..	0.02	0.10	0.47	1.41	4.57	9.69	13.41	17.04	20.78	21.77	22.62	26	28
Hungary	..	..	..	0.02	0.26	0.65	1.99	3.57	6.34	9.59	13.88	16.90	18.76	20.32	20.92	21.21	28	26
Iceland	..	..	..	0.72	3.68	8.45	14.31	18.20	26.37	28.83	31.45	32.47	33.54	33.61	34.48	34.28	7	5
Ireland	..	..	..	0.01	0.01	0.27	0.83	3.32	6.59	12.18	17.59	20.17	21.53	21.50	22.13	22.53	25	22
Israel	..	..	..	..	..	..	..	..	..	20.14	21.99	22.80	23.61	24.23	24.24	24.64	20	18
Italy	..	..	..	0.20	0.73	1.71	4.17	8.08	11.77	14.24	17.06	18.86	20.38	22.34	22.10	22.13	24	24
Japan	..	..	0.12	0.50	2.25	6.36	11.58	17.21	21.89	20.69	22.50	23.61	24.84	26.91	27.07	27.29	16	16
Korea	..	..	0.58	8.65	19.70	24.32	26.16	27.03	28.69	29.01	30.36	31.84	33.54	35.63	35.88	36.18	6	6
Luxembourg	..	..	..	..	0.28	1.54	3.45	9.64	14.48	21.01	26.93	29.42	31.86	31.65	31.50	32.08	10	7
Mexico	..	..	..	0.01	0.11	0.25	0.42	1.01	2.22	2.84	4.22	7.06	8.83	10.60	10.73	11.09	33	34
Netherlands	..	..	0.96	1.63	3.82	7.04	11.79	18.96	25.22	31.00	34.30	35.61	37.09	38.48	38.93	39.38	2	2
New Zealand	..	..	..	0.27	0.72	1.64	2.61	4.74	9.12	11.81	18.04	21.57	23.10	26.09	26.57	28.04	17	19
Norway	..	..	0.11	0.40	1.87	4.20	8.18	15.20	22.62	26.84	30.52	33.72	33.83	34.87	35.27	36.13	5	4
Poland	..	..	..	..	0.06	0.30	0.78	2.14	2.41	7.18	8.65	10.48	12.27	14.08	14.85	14.54	30	30
Portugal	..	..	..	0.25	0.96	2.51	4.81	7.89	11.05	13.45	14.27	15.93	17.89	20.21	21.06	21.69	27	27
Slovak Republic	..	..	..	..	0.01	0.01	0.35	0.96	2.49	5.08	7.66	11.45	11.59	13.31	13.72	14.18	31	32
Slovenia	..	..	..	..	..	..	3.29	5.71	9.74	13.74	16.48	21.10	22.54	23.62	24.13	24.16	23	29
Spain	..	..	0.09	0.15	1.16	2.93	5.25	8.06	11.51	15.11	17.60	20.08	21.31	23.73	24.48	24.68	22	23
Sweden	..	..	0.12	2.16	6.32	9.76	13.24	17.68	24.16	26.41	30.39	31.39	31.49	32.02	32.19	32.33	11	8
Switzerland	..	..	..	0.85	1.97	5.65	10.55	17.67	23.85	27.31	32.00	32.73	35.82	39.46	40.31	41.64	1	3
Turkey	..	..	..	0.01	0.02	0.04	0.29	0.75	2.23	4.00	6.26	8.07	8.97	10.02	10.25	10.37	34	33
United Kingdom	..	..	..	0.10	0.59	2.31	5.37	10.35	16.31	21.45	25.59	28.14	29.49	31.78	32.58	33.58	8	12
United States	0.03	0.18	0.75	2.21	4.37	6.70	9.58	12.78	16.36	20.28	23.20	25.46	25.80	27.35	27.44	28.37	15	15
OECD	0.01	0.06	0.30	1.21	2.93	4.87	7.28	10.34	13.78	16.79	19.65	21.81	23.10	25.03	25.36	25.88		

Notes: See notes of Table 4.10.

StatLink http://dx.doi.org/10.1787/888932801565

Table 4.12. **Total fixed broadband subscriptions by access technology**

	2006				2007				2008			
	DSL	Cable	Fibre	Other	DSL	Cable	Fibre	Other	DSL	Cable	Fibre	Other
Australia	2 995 000	624 300	..	196 872	3 815 000	861 000	..	154 200	4 176 000	916 000	..	244 000
Austria	832 107	521 626	3 662	26 403	985 163	559 225	4 042	49 561	1 150 275	569 758	4 806	44 102
Belgium	1 469 668	878 360	14	7 561	1 620 577	1 071 107	56	24 053	1 734 866	1 200 231	722	26 631
Canada	3 714 335	4 180 751	1 072	32 923	4 096 932	4 747 898	1 072	130 000	4 300 694	4 973 552	1 072	130 000
Chile	584 098	424 224	..	..	760 482	537 603	..	..	816 702	605 889	..	..
Czech Republic	493 402	230 306	35 000	378 050	613 220	309 000	55 000	524 200	697 172	382 512	70 000	620 000
Denmark	1 062 040	506 734	138 588	20 975	1 206 282	541 708	151 700	46 152	1 226 290	538 846	194 382	61 886
Estonia	101 363	59 183	40 883	17 459	115 226	68 592	51 521	16 830	125 074	74 573	66 117	15 383
Finland	1 234 000	181 100	..	14 100	1 348 000	209 600	..	59 500	1 366 200	214 800	..	35 900
France	12 019 313	690 000	..	9 000	14 804 715	718 017	27 268	..	16 825 000	860 000	40 000	..
Germany	14 400 000	490 000	..	92 600	18 500 000	985 000	..	46 000	20 900 000	1 600 000	..	32 000
Greece	484 321	0	760	24 000	1 083 521	0	594	..	1 506 614	0	..	..
Hungary	614 894	335 490	..	15 000	751 860	574 707	1 327	67 718	794 986	763 567	51 816	86 345
Iceland	85 280	..	668	1 790	94 630	..	1 218	2 089	98 762	..	2 615	2 320
Ireland	379 124	55 925	1 780	82 200	549 594	82 477	4 165	131 500	660 025	104 133	5 612	126 576
Israel	879 509	540 848	..	..	963 000	616 165	..	..	999 000	667 583	..	..
Italy	8 156 000	0	229 000	8 000	9 754 680	0	277 000	99 862	10 903 000	0	308 000	72 000
Japan	14 013 248	3 609 625	8 803 898	11 580	12 710 678	3 873 547	12 152 715	12 585	11 594 082	4 083 072	14 417 207	12 966
Korea	5 458 861	5 152 986	3 399 659	1 415	4 603 425	5 091 066	5 015 126	381	3 718 135	5 085 348	6 670 596	852
Luxembourg	90 100	8 710	250	220	116 900	11 500	300	560	124 246	18 972	320	228
Mexico	1 960 557	987 802	..	30 000	3 148 349	1 236 238	..	164 251	5 436 668	1 980 319	..	111 982
Netherlands	3 028 000	1 972 000	65 000	..	3 300 000	2 210 000	70 000	37 902	3 563 000	2 192 000	100 000	..
New Zealand	435 000	27 000	..	28 067	674 000	48 087	..	35 045	815 488	55 858	380	43 235
Norway	975 150	177 800	70 303	27 646	1 085 000	236 675	94 580	20 000	1 109 000	321 750	145 000	32 000
Poland	1 882 045	813 683	1 195	40 000	2 352 100	904 142	1 458	40 000	2 744 000	1 200 000	1 458	50 000
Portugal	881 512	537 552	..	4 623	892 859	605 799	..	14 656	996 561	669 087	2 281	24 377
Slovak Republic	182 391	36 701	46 338	8 678	277 838	52 666	66 649	16 091	354 423	63 806	114 858	85 784
Slovenia	194 339	81 446	..	..	247 404	85 231	..	..	285 853	95 076	44 564	1 154
Spain	5 262 617	1 350 101	..	46 189	6 230 952	1 633 489	23 057	10 938	7 282 928	1 775 842	26 201	71 998
Sweden	1 531 000	454 000	390 000	5 700	1 716 000	536 000	509 000	5 014	1 737 000	563 000	590 000	15 000
Switzerland	1 391 521	598 663	3 934	70 000	1 684 266	710 000	21 462	22 400	1 756 000	730 000	17 006	20 643
Turkey	2 723 547	27 804	..	22 334	4 346 054	35 651	..	14 095	5 660 000	67 408	..	9 211
United Kingdom	9 928 140	3 058 500	..	8 500	12 157 200	3 413 900	..	35 000	13 556 860	3 682 800	..	36 000
United States	25 761 869	32 097 223	1 035 677	1 748 100	29 745 693	36 497 284	1 623 097	2 190 072	30 439 000	41 468 000	3 061 051	2 632 044
OECD	125 204 351	60 710 443	14 267 681	2 979 985	146 351 600	69 063 374	20 152 407	3 970 655	159 453 904	77 523 782	25 936 064	4 644 617

	2009				2011				June 2012			
	DSL	Cable	Fibre	Other	DSL	Cable	Fibre	Other	DSL	Cable	Fibre	Other
Australia	4 178 000	914 000	15 000	..	4 553 000	900 000	37 000	..	4 632 000	917 000	52 000	..
Austria	1 263 038	569 903	6 917	5 033	1 370 275	655 500	9 066	8 163	1 407 398	633 294	18 337	4 583
Belgium	1 775 409	1 331 526	1 407	6 353	1 859 316	1 665 813	2 829	881	1 857 523	1 723 796	3 271	1 526
Canada	4 440 000	5 540 000	..	365 617	4 699 366	6 091 285	135 731	5 495	4 663 027	6 158 815	185 569	6 300
Chile	882 864	758 943	..	..	1 017 225	977 859	1 914	5 575	1 051 376	1 042 805	2 077	5 266
Czech Republic	778 286	441 700	135 000	..	918 000	515 000	223 000	..	939 000	523 700	253 000	..
Denmark	1 238 000	557 000	232 000	19 000	1 196 589	575 738	311 095	12 052	1 190 954	592 816	335 908	11 489
Estonia	135 170	76 756	67 846	20 771	158 587	73 163	90 981	9 159	155 301	73 034	99 759	6 561
Finland	1 185 900	222 700	12 600	6 000	1 125 400	263 600	26 100	173 600	1 097 700	276 100	30 700	195 900
France	18 493 000	1 020 000	69 000	..	20 984 000	1 403 000	197 000	..	21 481 000	1 440 000	246 000	..
Germany	22 424 800	2 300 000	133 700	77 000	23 415 642	3 537 039	159 535	73 600	23 476 720	3 952 000	178 800	70 250
Greece	1 916 630	0	2 000	..	2 458 629	0	3 511	521	2 554 689	..	3 902	239
Hungary	823 275	840 589	120 291	1 071	832 414	988 864	265 014	..	799 728	1 025 010	290 094	..
Iceland	97 862	..	6 908	..	92 745	..	17 265	..	88 331	..	21 028	..
Ireland	714 016	150 910	5 636	..	728 902	259 773	4 992	..	720 099	286 418	5 213	..
Israel	1 035 000	719 000	2 300	2 300	1 111 000	768 000	..	..	1 136 000	774 000	..	..
Italy	11 995 019	0	254 174	3 936	13 134 031	0	278 646	12 924	13 145 926	..	290 039	8 165
Japan	10 134 491	4 300 594	17 195 696	..	7 058 021	5 908 632	21 899 201	..	6 344 476	5 958 367	22 839 289	..
Korea	3 222 419	5 147 994	7 977 303	..	2 428 317	5 030 710	10 399 976	..	2 295 061	4 960 203	10 754 992	..
Luxembourg	131 879	25 896	353	420	146 274	15 609	1 311	423	147 119	15 981	2 903	627
Mexico	7 308 791	2 097 872	..	82 117	9 343 108	2 272 792		107 436	9 645 949	2 331 587	..	130 942
Netherlands	3 645 000	2 351 000	134 000	..	3 433 000	2 793 000	272 000	..	3 347 000	2 924 000	302 000	..
New Zealand	927 427	60 058	1 508	..	1 108 080	63 010	3 700	..	1 169 014	65 155	5 400	..
Norway	1 007 587	415 214	207 589	3 202	898 916	525 883	317 662	4 379	882 562	541 356	359 548	5 813
Poland	2 877 286	1 389 943	67 694	..	3 177 840	1 794 793	135 647	611 509	3 158 944	1 653 339	162 953	627 044
Portugal	1 108 680	760 637	30 745	2 211	1 102 474	902 646	236 563	1 511	1 086 528	914 683	306 789	1 702
Slovak Republic	367 723	80 251	177 574	2 174	404 823	115 796	223 980	2 128	407 429	124 785	237 573	1 851
Slovenia	286 960	104 939	67 452	..	278 341	135 880	79 903	1 251	266 126	142 935	85 627	1 265
Spain	7 885 940	1 866 101	32 322	2 215	9 010 689	2 092 981	175 283	11 027	9 023 365	2 103 880	245 062	12 281
Sweden	1 666 134	579 141	687 403	8 970	1 515 000	612 000	901 000	11 000	1 470 000	593 000	977 000	13 000
Switzerland	1 943 487	795 500	26 281	28 455	2 148 000	890 080	91 000	30 000	2 210 000	925 000	111 000	30 700
Turkey	6 216 028	146 622	41 000	42 724	6 776 036	460 451	267 144	73 167	6 632 661	485 531	469 668	78 749
United Kingdom	14 370 000	3 840 000	3 290	..	15 904 673	4 119 564	398 077	18 000	16 297 666	4 168 129	582 829	18 000
United States	31 190 000	43 392 360	3 945 977	803 000	31 029 000	47 926 000	5 886 000	789 000	31 161 000	50 277 000	6 286 000	796 000
OECD	137 666 101	82 797 149	31 660 966	1 482 569	175 417 713	94 334 461	43 052 126	1 962 801	175 941 671	97 603 719	45 744 330	2 028 253

Note: The broadband data from 1997 to 2009 are calculated according to the 'old' OECD methodology including DSL, cable, fibre, other, satellite and fixed wireless connections. For 2010, 2011 and June 2012 data, the current or 'new' methodology separates all fixed and wireless connections.

StatLink http://dx.doi.org/10.1787/888932801584

Table 4.13. **Availability of digital subscriber lines (DSL) in the OECD area**

	Commercial service launch	Actual coverage by year end (%)												Indicator used to express coverage
		2000	2001	2002	2003	2004	2005	2006	2007	2008	2009	2010	2011	
Australia	August 2000	50.0	72.0	75.0	75.0	81.0	81.0	88.0	91.0	91.0	92.0	92.0	92.0	Population
Austria	November 1999	72.0	77.0	80.0	80.0	87.0	90.0	95.0	95.0	99.0	99.0	99.0	99.0	Lines
Belgium	October 1999	75.0	93.0	98.0	98.0	100.0	100.0	100.0	100.0	100.0	100.0	100.0	100.0	Lines
Canada	1996	69.0	70.0	75.0	75.4	75.4	75.4	89.0	89.0	84.0	84.0	85.0	88.0	Population
Chile		..	..	..	..	..	..	..	..	..	..	..	..	
Czech Republic	March 2003	..	..	..	44.0	84.0	90.0	90.0	90.0	90.0	90.0	97.0	98.0	Lines
Denmark	July 1999	65.0	90.0	95.0	95.0	96.0	98.0	98.0	99.0	99.0	99.0	99.0	98.0	Lines
Estonia		5.0	20.0	30.0	45.0	65.0	70.0	75.0	86.4	93.4	94.0	90.0	90.0	Lines
Finland	May 2000	50.0	60.0	75.0	81.5	94.1	95.6	96.0	96.0	..	..	..	..	Lines
France	November 1999	32.0	66.0	71.0	79.0	90.0	97.0	..	..	..	..	..	..	Population
Germany	August 1999	60.0	70.0	80.0	85.0	90.0	90.0	96.0	98.0	97.0	97.0	98.0	98.0	Households
Greece	June 2003	..	..	..	..	6.0	9.0	..	94.3	95.0	..	..	95.0	
Hungary	September 2000	..	..	..	58.0	70.0	85.0	87.0	89.0	96.9	97.5	98.0	98.5	Population
Iceland	April 2000	33.0	51.0	78.0	90.0	92.0	92.0	..	..	..	..	..	..	Population
Ireland	May 2002	..	..	25.0	50.0	74.0	90.0	..		90.0	90.0	90.0	90.0	Lines
Israel		..	..	..	..	..	..	99.0	99.0	99.5	100.0	100.0	100.0	
Italy	December 1999	45.0	67.5	70.0	80.0	85.0	90.0	89.0	94.0	95.0	96.0	96.0	97.0	Lines
Japan	September 2000	..	73.5	80.0	90.0	93.0	94.0	95.2	98.0	..	..	..	100.0	Households
Korea	April 1999	..	70.0	89.0	93.0	100.0	100.0	..	..	100.0	100.0	100.0	100.0	Lines
Luxembourg	2001	..	65.0	90.0	90.0	100.0	100.0	96.0	98.0	99.0	99.0	99.0	99.0	Population
Mexico	September 2001	..	..	..	58.9	75.5	92.0	..	..	..	..	..	..	Lines
Netherlands	June 2000	40.0	67.0	85.0	85.0	100.0	100.0	100.0	100.0	99.0	99.0	99.0	99.0	Lines
New Zealand	June 1999	60.0	69.0	83.0	84.8	92.0	93.0	92.0	93.0	93.0	93.0	95.0	97.0	Population (customers)
Norway	December 2000	20.0	50.0	58.0	67.0	77.0	91.0	..	..	94.3	94.4	94.4	93.8	Lines
Poland (TPSA)	2001	..						..	..	..	64.0	..	..	Population
Portugal	December 2000	..	..	..	..	..		94.0	95.0	95.0	98.0	99.0	99.0	Lines
Slovak Republic	2003	..	..	..	14.5	..	..	9.6	14.6	17.7	19.4	21.0	21.3	Population
Slovenia		..	..	..	..	..	..	..	..	..	84.0	36.3	35.8	Population
Spain	1999	62.2	81.3	89.3	92.0	92.0	92.0	..	..	..	99.0	99.0	99.0	Lines
Sweden	October 2000	..	70.0	75.0	78.0	90.0	96.0	..	97.8	97.9	98.0	98.0	98.1	Lines
Switzerland	October 2000	..			98.0	98.0	98.0	98.0	98.0	98.0	98.0	98.0	98.0	Lines
Turkey	February 2001	..	..	2.5	5.0	10.0	10.0	..	..	..	38.0	38.0	38.0	Lines
United Kingdom	July 2000	50.0	60.0	64.0	98.0	98.0	99.9	100.0	100.0	100.0	100.0	100.0	100.0	Lines
United States	1997	36.0	50.0	68.0	75.0	77.0	78.0	79.0	82.0	84.0	85.0	84.0	87.8	Lines

Notes: Data for Australia (2010 and 2011), Denmark (2010), the Netherlands (2010 and 2011), New Zealand (2010), Norway (2010), Slovenia (2010), Spain (2010 and 2011), Sweden (2010) and Turkey (2010 and 2011) are estimates.

StatLink http://dx.doi.org/10.1787/888932801603

Table 4.14. **Total wireless broadband subscriptions in the OECD area**

	2009	2010	2011	June 2012	CAGR 2010-11
Australia	2 967 000	12 451 000	20 816 000	22 178 000	67.18
Austria	1 545 929	2 480 217	3 600 271	4 334 468	45.16
Belgium	634 017	1 057 090	2 096 243	2 844 187	98.30
Canada	310 000	10 364 520	13 531 360	14 279 850	30.55
Chile	610 153	1 453 825	3 110 939	3 866 878	113.98
Czech Republic	1 051 762	1 273 579	5 409 519	6 134 301	324.75
Denmark	1 660 053	3 591 784	4 720 310	5 125 855	31.42
Estonia	60 099	268 700	594 329	850 601	121.19
Finland	3 932 800	4 549 000	4 707 300	5 160 000	3.48
France	17 580 000	23 203 438	27 898 000	30 562 000	20.23
Germany	41 900	21 272 150	28 671 280	31 666 400	34.78
Greece	1 376 248	2 788 778	4 131 352	4 480 265	48.14
Hungary	709 421	879 061	1 330 413	1 342 196	51.34
Iceland	99 091	147 242	186 414	206 359	26.60
Ireland	1 814 051	2 105 739	2 764 858	2 845 357	31.30
Israel	4 024 848	3 706 000	4 150 000	3 831 000	11.98
Italy	10 308 718	17 856 686	20 229 396	20 991 956	13.29
Japan	94 713 089	97 722 144	104 748 262	107 508 636	7.19
Korea	42 656 758	47 324 006	50 836 715	51 894 381	7.42
Luxembourg	112 248	253 713	331 800	334 800	30.78
Mexico	429 429	3 250 193	8 347 025	10 725 623	156.82
Netherlands	533 000	6 315 000	8 767 000	9 760 000	38.83
New Zealand	219 563	1 726 007	2 946 260	2 942 301	70.70
Norway	3 344 541	3 670 839	3 807 489	3 987 776	3.72
Poland	16 970 615	19 328 066	20 431 910	22 049 733	5.71
Portugal	996 402	4 100 789	4 153 861	3 960 947	1.29
Slovak Republic	978 523	1 311 425	1 952 945	1 972 551	48.92
Slovenia	1 817	431 151	502 594	566 793	16.57
Spain	16 533 264	12 826 088	30 254 282	22 277 983	135.88
Sweden	6 512 000	7 863 000	9 194 000	9 616 000	16.93
Switzerland	2 741 781	3 444 376	3 978 412	4 245 550	15.50
Turkey	403 437	1 458 436	6 465 606	10 661 171	343.32
United Kingdom	15 318 300	22 642 034	32 822 000	37 629 000	44.96
United States	120 673 000	165 033 000	218 511 000	237 824 000	32.40
OECD	371 863 857	508 149 076	655 999 145	698 656 918	29.10

StatLink http://dx.doi.org/10.1787/888932801622

Table 4.15. **Total wireless broadband subscriptions by access technology**

	2009				2010				2011				2012			
	Satellite	Fixed wireless	Standard mobile	Dedicated mobile data	Satellite	Fixed wireless	Standard mobile	Dedicated mobile data	Satellite	Fixed wireless	Standard mobile	Dedicated mobile data	Satellite	Fixed wireless	Standard mobile	Dedicated mobile data
Australia	107 000	22 000	..	2 838 000	..	24 000	8 197 000	4 230 000	100 000	35 000	15 190 000	5 491 000	94 000	30 000	16 192 000	5 862 000
Austria	..	32 924	213 172	1 299 833	..	24 117	823 000	1 633 100	..	19 671	1 793 300	1 787 300	..	19 565	2 215 933	2 098 970
Belgium	..	19 186	432 217	182 614	..	16 544	792 559	247 987	..	14 920	1 717 918	363 405	..	13 549	2 423 209	407 429
Canada	..	310 000	..	..	..	339 217	9 072 971	952 332	..	345 603	12 000 800	1 184 957	..	348 796	12 614 358	1 316 696
Chile	..	10 647	379 274	220 232	..	10 242	763 213	680 370	..	8 695	1 792 550	1 309 694	..	7 535	2 733 006	1 126 337
Czech Republic	1 124	680 000	..	370 638	1 179	730 000	..	542 400	1 300	839 000	2 637 911	1 931 308	1 000	870 000	3 026 532	2 236 769
Denmark	..	24 129	1 048 607	587 317	..	46 089	1 978 360	1 567 335	0	48 389	2 646 753	2 025 168	..	8 495	2 953 436	2 163 924
Estonia	..	37 753	..	22 346	..	35 922	162 851	69 927	..	31 763	268 843	293 723	..	23 729	434 421	392 451
Finland	..	31 800	2 993 000	908 000	..	26 700	2 886 000	1 636 300	..	17 300	1 140 000	3 550 000	..	14 200	1 042 300	4 103 500
France	..	73 000	17 507 000	..	..	108 438	23 095 000	..	..	153 000	27 745 000	..	..	150 000	30 412 000	..
Germany	40 800	1 100	..	..	42 290	29 860	13 500 000	7 700 000	38 700	32 580	18 100 000	10 500 000	34 400	32 000	20 400 000	11 200 000
Greece	565	1 059	1 149 299	225 325	1 129	1 114	2 497 224	289 311	945	676	2 679 887	1 449 844	920	664	2 719 580	1 759 101
Hungary	..	95 000	214 421	400 000	..	101 600	277 461	500 000	..	105 000	692 289	533 124	..	107 020	723 120	512 056
Iceland	64	2 238	77 034	19 755	50	2 151	99 208	45 833	87	1 487	123 779	61 061	79	1 628	131 023	73 629
Ireland	2 925	88 261	1 273 073	449 792	4 266	74 293	1 455 341	571 839	5 651	71 458	2 094 311	593 438	6 974	65 519	2 202 930	569 934
Israel	..	..	3 793 848	231 000	..	..	3 310 000	396 000	..	..	3 721 000	429 000	..	..	3 402 000	429 000
Italy	27 389	911	5 998 018	4 282 400	731	2 529	10 620 566	7 232 860	720	5 006	8 554 488	11 669 182	699	4 870	8 313 399	12 672 988
Japan	..	12 846	94 700 243	..	..	11 081	97 711 063	..	..	9 363	104 738 899	0	..	9 279	107 499 357	..
Korea	855	..	6 071 912	36 583 991	702	..	11 727 035	35 596 269	519	..	24 021 738	26 814 458	490	..	32 574 572	19 319 319
Luxembourg	..	..	112 248	..	..	..	230 400	23 313	..	..	299 300	32 500	..	..	299 300	35 500
Mexico	45 620	383 809	..	..	46 794	473 921	1 914 267	815 211	68 442	794 692	5 022 760	2 461 131	64 554	958 446	6 683 953	3 018 670
Netherlands	..	..	..	533 000	..	..	4 582 000	1 733 000	..	..	6 351 000	2 416 000	..	..	7 047 000	2 713 000
New Zealand	10 899	28 664	..	180 000	13 723	25 180	1 451 843	235 261	14 325	21 399	2 346 689	563 847	14 300	18 260	2 540 609	369 132
Norway	..	35 214	2 771 605	537 722	..	34 445	2 689 225	947 169	..	39 310	2 546 969	1 221 210	..	40 906	2 637 630	1 309 240
Poland	29	666 332	14 197 701	2 106 553	138	840 530	15 667 575	2 819 823	163	1 294 889	15 833 074	3 303 784	230	1 119 079	17 462 192	3 468 232
Portugal	24	23 563	972 816	..	11	22 401	4 078 377	..	67	16	4 153 778	0	103	18	3 960 826	..
Slovak Republic	184	147 150	552 635	278 554	373	182 166	771 499	357 387	252	206 119	1 358 324	388 250	392	217 591	1 350 595	403 973
Slovenia	8	1 809	..	..	6	3 125	381 973	46 047	5	6 720	444 066	51 803	6	6 990	504 472	55 325
Spain	1 616	57 875	14 719 443	1 754 330	11 237	11 914	3 765 502	9 037 435	9 877	85 701	18 417 999	11 740 705	9 374	101 841	11 907 262	10 259 506
Sweden	..	3 000	4 782 000	1 727 000	..	3 000	4 763 000	3 097 000	..	3 000	3 984 000	5 207 000	..	3 000	3 404 000	6 209 000
Switzerland	2 050	..	2 495 021	244 710	2 363	..	3 081 796	360 217	2 000	..	3 546 509	429 903	2 050	..	3 778 750	464 750
Turkey	7 074	..	396 363	..	10 416	..	1 448 020	..	10 805	..	4 907 380	1 547 421	11 223	..	8 790 698	1 859 250
United Kingdom	..	18 300	11 200 000	4 100 000	..	18 000	17 837 000	4 787 034	..	18 000	27 748 000	5 056 000	..	..	32 640 000	4 989 000
United States	1 116 000	527 000	119 030 000	..	1 176 000	587 000	163 270 000	..	1 236 000	698 000	216 577 000	..	1 197 000	761 000	235 866 000	..
OECD	1 364 226	3 335 570	307 080 950	60 083 112	1 311 408	3 785 579	414 901 329	88 150 760	1 489 858	4 906 757	545 196 314	104 406 216	1 437 794	4 933 980	590 886 463	101 398 681

StatLink http://dx.doi.org/10.1787/888932801641

Table 4.16. **Fixed and mobile broadband traffic per access (GB)**

Monthly fixed broadband traffic per access (GB)

	2010	2011	2012	Growth rate 2010-11	Growth rate 2011-12	Period of data collection	Source
Australia	9.322	15.703	23.117	68.46%	47.22%	2nd quarter	Australian Bureau of Statistics (www.abs.gov.au)
Canada	18.5	21.7		17.30%		Year	CRTC (www.crtc.gc.ca)
Germany	10.2	11.6		13.73%		Year	Bundesnetzagentur (www.bundesnetzagentur.de)
Italy	18.187	20.013		10.04%		Year	AGCOM (www.agcom.it)
Japan	19.278	20.833	22.129	8.07%	6.22%	May	MIC (www.soumu.go.jp)
New Zealand		9	16		77.78%	June	Statistics New Zealand (www.stats.govt.nz)
Portugal	22.041	26.776	33.959	21.48%	26.83%	2nd quarter	ANACOM (www.anacom.pt)
Spain		26.2				Year	CMT (www.cmt.es)
United Kingdom		17	23		35.29%	March	OFCOM (stakeholders.ofcom.org.uk)
United States			50.986			April	FCC (www.fcc.gov) (1)

Monthly mobile broadband traffic per access (GB)

	2010	2011	2012	Growth rate 2010-11	Growth rate 2011-12	Period of data collection	Source
Australia	1.236	1.370	1.409	10.85%	2.86%	2nd quarter	Australian Bureau of Statistics (www.abs.gov.au) (2)
		0.092	0.136		47.19%	2nd quarter	Australian Bureau of Statistics (www.abs.gov.au) (3)
France	0.092	0.146		59.25%		Year	ARCEP (www.arcep.fr) (4)
Germany	0.257	0.271		5.39%		Year	Bundesnetzagentur (www.bundesnetzagentur.de) (5)
Italy	0.152	0.184		20.95%		Year	AGCOM (www.agcom.it)
Japan	0.183	0.330	0.668	80.58%	102.59%	June	MIC (www.soumu.go.jp)
Netherlands	0.121	0.146	0.183	20.41%		1st half of the year (2012:1st quarter)	OPTA (www.opta.nl)
Portugal	1.012	1.020	1.069	0.81%	4.80%	2nd quarter	ANACOM (www.anacom.pt) (6)
		0.253	0.279		10.58%	2nd quarter	ANACOM (www.anacom.pt) (7)
Spain		0.390				Year	CMT (www.cmt.es)
Sweden	1.429	1.630		14.05%		Year	PTS (www.pts.se) (8)
Switzerland	0.158					Year	OFCOM (www.bakom.admin.ch)
United Kingdom		0.11	0.246		90.39%	March for 2011, June for 2012	OFCOM (stakeholders.ofcom.org.uk) (9)

Notes: 1. Data is collected by automated, direct measurement of broadband performance at the homes of voluntary consumers; 2. Average of data volume downloaded by mobile subscribers excluding mobile handset users. 3. Average of data volume downloaded by mobile handset users with Internet access connections (denominator is revised number of subscriptions). 4. Denominator is number of multimedia service customers who used multimedia service such as wap, i-mode, MMS or e-mail at least once in December. 5. Denominator is number of regular 3G users who used the service at least once within the last three months (October – December). 6. Average of mobile broadband customers with actual use, i.e. customers who established at least one IP session to access the Internet using broadband in the period being reported. 7. Average of all mobile broadband subscriptions. 8. Average of stand-alone mobile broadband subscriptions and other MBB subscriptions of at least 1 GB per month. 9. Average of 2G and 3G SIMs. The growth rate is CAGR.

StatLink http://dx.doi.org/10.1787/888932801660

Table 4.17. **Actual download speeds, comparison between Akamai, M-Lab and Ookla, Mbit/s, 1Q2012**

	Average Connection Speed by Akamai	Download Throughput by M-Lab (3)	Download Speed by Ookla (4)	Unique IP addresses per 100 inhabitants, Akamai	Number of NDT tests per 100 inhabitants, M-Lab	Number of tests per 100 inhabitants, Ookla
Australia (2)	3.5	6.7	10.4	61.74	0.37	359.77
Austria	5.7	7.3	11.6	36.72	0.25	311.20
Belgium	7.1	13.6	17.0	39.35	0.56	207.81
Canada	6.5	7.4	12.2	39.28	0.65	463.87
Chile	3.4		7.2	18.80		146.28
Czech Republic	7.1	7.4	15.0	21.25	0.57	314.12
Denmark	6.7	13.4	17.6	56.50	0.90	267.97
Estonia (1)	5.2	8.4	32.0		0.88	787.19
Finland	6.9	10.6	15.8	57.22	0.52	516.63
France	4.9	7.8	12.8	39.17	0.46	77.49
Germany	5.8	10.4	15.0	44.07	0.11	78.74
Greece	4.0	6.4	5.9	24.72	0.85	315.63
Hungary	5.9	9.6	14.5	26.07	1.18	654.15
Iceland (1)	5.4	6.1	20.4	47.89	1.69	416.55
Ireland (1)(2)	7.3	8.4	8.1	35.76	0.45	510.60
Israel (2)	4.6		9.1	37.20		407.35
Italy	4.1	3.9	4.9	27.85	0.69	317.85
Japan (2)	10.9	25.6	14.5	31.63	0.01	8.02
Korea	15.7	30.9		39.79		12.96
Luxembourg (1)	4.7	10.7	21.2	33.96	0.85	515.71
Mexico (2)	2.8	3.0	4.8	10.01	0.04	96.51
Netherlands (2)	8.8	15.8	24.2	49.03	1.06	319.57
New Zealand (2)	3.9	6.1	8.9	46.13	0.20	236.41
Norway	5.7	10.9	16.7	74.88	0.64	472.15
Poland (2)	5.0	9.4	9.2	20.99	0.36	312.22
Portugal	5.4	8.4	16.5	28.38	0.71	305.02
Slovak Republic	5.7	7.2	13.8	16.61	0.65	288.64
Slovenia	5.7	6.7	10.0		0.76	490.44
Spain	4.6	6.5	11.5	30.33	0.53	90.17
Sweden	6.3	14.4	24.9	68.39	0.85	116.62
Switzerland	8.1	10.7	19.7	40.77	0.29	126.62
Turkey	2.9	3.2	6.7		0.16	53.03
United Kingdom	5.6	7.4	12.3	41.00	0.66	519.27
United States	6.7	9.5	12.5	46.94	0.28	294.70

Notes: 1. Excluded from Figure 4.14 due to large difference between indicators; 2. Population uses numbers of 2010. The others use numbers of 2011. 3. These speed indicators are averages of all tests done in the quarter using the M-Lab sites (servers) located in corresponding regions of countries. The regions are specified as Australia, Europe including Turkey, Japan, New Zealand, and North and South America. Number of NDT tests listed on the right corresponds to the same tests done in each region. 4. These speed indicators are averages of all tests done in the quarter and calculated by daily data.

Sources: Akamai, "The State of the Internet", www.akamai.com; Measurement Lab (M-Lab), www.measurementlab.net; Ookla, "NET INDEX",

StatLink http://dx.doi.org/10.1787/888932801679

Table 4.18. **Communications data for key partners and partners countries**

	2008	2009	2010	2011	2008	2009	2010	2011
	Fixed telephone access paths, in thousands				**Fixed telephone access paths per 100 inhabitants**			
Brazil	41 235	41 497	42 141	43 026	21.5	21.5	21.6	21.9
China	340 804	313 688	294 341	285 115	25.7	23.5	21.9	21.2
Colombia	7 907	7 474	7 186	7 127	17.6	16.4	15.5	15.2
Egypt	11 853	10 313	9 618	8 714	15.1	12.9	11.9	10.6
Latvia	586	563	462	424	25.8	24.9	20.5	18.9
Russian Federation	43 306	42 989	42 427	41 576	30.2	30.0	29.7	29.1
OECD	489 136	462 555	445 230	428 204	40.2	37.7	36.1	34.5
	Total communication access paths, in thousands				**Total communication access paths per 100 inhabitants**			
Brazil	199 506	222 178	252 338	294 267	104.2	115.0	129.4	149.6
China	1 064 913	1 165 050	1 279 563	1 426 368	80.2	87.3	95.4	105.8
Colombia	51 044	50 746	54 286	56 564	113.4	111.2	117.3	120.5
Egypt	53 859	66 711	81 705	93 960	68.8	83.7	100.7	113.8
Latvia	2 223	2 251	2 846	3 484	97.9	99.6	126.4	155.3
Russian Federation	252 109	286 389	295 816	315 094	176.1	200.2	206.9	220.6
OECD	1 948 405	1 989 363	2 034 345	2 079 950	160.0	162.3	164.9	167.5
	Cellular mobile subscribers, in thousands				**Cellular mobile subscribers per 100 inhabitants**			
Brazil	148 591	169 386	196 930	234 358	77.6	87.7	101.0	119.2
China	641 230	747 384	859 002	986 253	48.3	56.0	64.0	73.2
Colombia	41 365	41 155	44 478	46 200	91.9	90.1	96.1	98.5
Egypt	41 287	55 352	70 661	83 425	52.7	69.4	87.1	101.1
Latvia	1 257	1 269	1 969	2 621	55.4	56.1	87.4	116.8
Russian Federation	199 522	230 500	237 689	256 117	139.4	161.1	166.3	179.3
OECD	1 215 694	1 258 292	1 300 914	1 350 387	99.8	102.7	105.5	108.8
	Total fixed broadband subscribers, in thousands				**Total broadband subscribers per 100 inhabitants**			
Brazil	9 680	11 295	13 266	16 884	5.1	5.8	6.8	8.6
China	82 879	103 978	126 220	155 000	6.2	7.8	9.4	11.5
Colombia	1 772	2 118	2 622	3 236	3.9	4.6	5.7	6.9
Egypt	720	1 046	1 426	1 820	0.9	1.3	1.8	2.2
Latvia	379	419	415	439	16.7	18.5	18.4	19.6
Russian Federation	9 280	12 900	15 700	17 401	6.5	9.0	11.0	12.2
OECD	263 430	282 173	301 714	314 780	21.8	23.1	25.0	25.4
	Fixed broadband subscribers, in thousands, 2011				**Fixed broadband subscribers, per 100 inhabitants, 2011**			
	DSL	Cable modem	Fibre	Other	DSL	Cable modem	Fibre	Other
Brazil	10 951	5 261	219	453	5.6	2.7	0.1	0.2
China	120 854		500	33 646	9.0	0.0	0.0	2.5
Colombia	1 951	1 214	35	36	4.2	2.6	0.1	0.1
Egypt	1 798			22	2.2	0.0	0.0	0.0
Latvia	165	30	106	139	7.3	1.3	4.7	6.2
Russian Federation	8 006			9 395	5.6	0.0	0.0	6.6
OECD	175 418	94 334	43 052	1 963	14.1	7.6	3.5	0.2

StatLink http://dx.doi.org/10.1787/888932801698

OECD Communications Outlook 2013

Chapter 5

Internet infrastructure

This chapter examines developments in Internet infrastructure. Attempts to measure the Internet can prove challenging, as unlike most communication technologies it relies on different actors and participants, and as a network of networks has no single point at which metrics are gathered. Nevertheless, data from surveys and databases are available for some indicators. These provide country level information on Internet infrastructure in areas such as Internet hosts, domain name registrations, address space, secure servers and network traffic, among others. The chapter also provides guidance on the interpretation of the collected data. The Internet is still growing strongly, but relative growth has decreased compared to previous periods in some categories, as might be expected given widespread adoption of this technology

The statistical data for Israel are supplied by and under the responsibility of the relevant Israeli authorities. The use of such data by the OECD is without prejudice to the status of the Golan Heights, East Jerusalem and Israeli settlements in the West Bank under the terms of international law.

In the two decades since its commercialisation the Internet and its protocols have become ubiquitous. The Internet Protocol (IP) is now an enabling technology for networks and the devices that communicate over them, underpinning modern economies and societies. The adage "IP over everything and everything over IP" has become the central defining characteristic of communication. Most data transport has already moved to an IP-only world.

These developments are most evident in the increasing range of applications and choices available to users for communication purposes. SMS is being replaced by services such as Blackberry Messenger, KakaoTalk and WhatsApp. The Internet has also transformed television, in particular by allowing apps on Smart-TVs. The increasing shift towards IP has facilitated these services, absorbing and extending the capabilities of "traditional" communication services. It enabled the roll-out of Long Term Evolution (LTE) as the first fully IP-based mobile network, and allows cable television networks to offer both traditional broadcast TV and IPTV for specific services.

Future developments around the use of IP will aim to bring billions of "smart" devices online, establishing machine-to-machine (M2M) connections to create an "Internet of Things". Manufacturers are predicting that by 2020, for example, every light bulb sold will be "Internet enabled". This could lead to significant savings with light-emitting diode (LED)-powered and network-enabled streetlights decreasing costs per light/year by USD 50-100, a significant saving in cities that have tens of thousands of streetlights. The use of M2M may be just as significant in private spaces. If households (e.g. families of two adults and two children) have as many as 25 Internet-connected devices in 2017 and 50 in 2022 this could amount to 14 billion M2M devices connected in and around the home in a decade.

This chapter presents and discusses data regarding the foregoing developments in services underpinning Internet infrastructure. It is clear that networks have grown at a remarkable rate over the previous two decades. Examining these data in terms of technological change alone, however, would neglect the underlying story of innovation, competition and service evolution. These developments have been combined with technological advances to create greater choices for communication users spurred along by market liberalisation. Today's Internet ecology consists of a wide variety of roles performed by an even wider array of actors.

Networks on the Internet

The Internet is composed of millions of individual networks, from residential consumer networks through to those that span the globe. These individual networks route traffic from a network, such as a home or business premise, to an Internet service provider (ISP). How the ISP routes traffic to and from these networks and the rest of the Internet is not predetermined. There are no pre-set routes on the Internet, and no standard hierarchy of how traffic should flow from one network to another. For this reason alone it makes little sense to apply concepts of national borders to exchange of traffic. The lack of pre-set routes, however, does mean that every network needs to know all the routes between it and all other networks on the Internet.

There are borders between networks, of course, but these are no longer national borders. The Border Gateway Protocol (BGP) is the method by which routing between networks is undertaken on the Internet. The essence of BGP is that the owner of a network compiles lists of IP-address blocks for which it is responsible. Currently, the smallest routable block on the Internet is 256 IPv4 addresses. Theoretically, BGP could be made to work on an individual IP address basis, but this would lead to an unmanageable list.

The aggregate of IP blocks for which the network is responsible is known as an autonomous system (AS). Such a network is called autonomous because it can determine the routing of its traffic independently from any other network. Every AS is assigned a unique AS number (ASN) by a regional Internet registry (RIR). The network owner then uses BGP to announce or advertise that, for example, ASN 12345 represents IP blocks a.b.c.d and d.e.f.g, and so on. However, simply announcing an ASN with its associated blocks does not allow traffic to traverse; to achieve this, an Autonomous Systems must announce the ASNs for which it can carry traffic.

If a network wants to route traffic to IP address x.y.z.a, for example, it will locate the address block in its routing table and note to which ASN the IP address belongs. It will then ascertain whether it is directly connected to that network or which of the transit providers announce a route to the receiving ASN. Based on criteria such as cost, link saturation and speed, it will select one of the possible transit providers but cannot determine how this network routes the traffic. This system allows end points to quickly announce changes, such as in which IP blocks they advertise. These changes can then be adopted by other networks in seconds and may take less than an hour to propagate to all networks globally. At the same time, networks can change transit providers quickly and signal changes for their preference of transit providers. The outcome of this model for traffic exchange on the Internet has been a rapid and scalable system that allows for a near frictionless and highly competitive ability to interconnect with so-called transit providers.

Data is transported across networks through commercial arrangements known as peering and transit. **Transit** is a purchase of capacity on a network, measured in megabits per second (Mbit/s) irrespective of direction, which will be routed to the whole Internet. An ASN that carries traffic for other ASNs is known as a transit AS. **Peering** occurs when two networks bypass a transit network and exchange traffic directly. This saves both networks the costs of transiting, and networks often agree to peer without exchange of payment.

An Internet routing table is a database that lists all ASNs. The routing table announces transit relations between networks to every other network, but does not announce peering relations, as these are bilateral relations between two networks. Peering relations are therefore much less visible than transit relations. Many network operators treat their peering relationships as commercially sensitive information and only careful analysis of the routes traffic takes in their network would show direct interconnections; however, no information on the commercial relations can be derived.

Use of an ASN is not confined to ISPs and telecommunications companies. Organisations like the European Commission (AS42848), the BBC (AS2818), the Swedish National Police (AS2793-AS2817), Football club Werder Bremen (AS197803), Ahoy event Accommodation (AS48392) and high schools in the Swiss canton of Zug (AS34288) all use an ASN. By using BGP, these organisations can enter the wholesale market as well as buying retail services. In other words, they route their own traffic, connect to Internet Exchange Points (IXPs), buy transit from multiple parties and, as a result, can act independently from telecommunication network operators or ISPs. From the perspective of the Internet, such organisations are indistinguishable from telecommunication network providers, ISPs or content providers. Use of an ASN allows these organisations to determine how their networks function and to scale those networks according to their wishes, without being fully dependent upon a third party. This increases competition in the market.

As of mid-August 2012, 59 934 ASNs had been assigned by RIRs. Of these, 43 051 are visible in the global routing table, up from 39 678 at the end of 2011, 36217 in 2010 and 26 964 at the end of 2007 This represented an increase of 18.9% between 2010 and 2012 from 34% between 2007 and 2010 (Table 5.1). While growth rates have slowed, this rise was still extraordinary, especially in the context of the global financial crisis.

The difference between assigned and visible entries in the routing table indicates that 30% of ASNs have been assigned but are currently not advertised. There may be several reasons for this: it may take time to activate a new entry; networks may have gone out of business, been sold or consolidated behind one AS and the assignment not returned to the RIR; or the network may be using the ASN internally (for which a private ASN should have been used). The rate of allocated but unadvertised ASNs has increased in recent years.

The number of routed AS a country has may be one proxy for the amount of competition in a market. It indicates the ease with which a company may take control over routing its traffic and exchange this traffic with other networks. Iceland has the most routed ASNs per capita with 12.22 per 100 000 inhabitants (Figure 5.1). At the end of 2012 the OECD average was 2.17 with 19 countries above this level. Most countries saw an increase in the number of AS per capita between 2010 and 2012, except for Korea and the Czech Republic. In the case of the Czech Republic, there was an unexpected reverse in routed AS, increasing from 328 to 550 between 2009 and 2010, then decreasing by 20% to 437 in 2012. According to European IP Networks (RIPE) this was the outcome of Russian and Ukrainian organisations registering via the Czech Republic, thus inflating the number of Czech registrations. Japan, Korea and Mexico stand out because of their low number of AS numbers and low growth in this metric. This indicates either lower competition in these markets (Mexico) or a preference to use larger network providers rather than create independent ASNs (possibly the case in Japan and Korea).

While only one of many factors to effect growth rates, some analysis by the OECD has shown that a country with one more AS per 100 000 inhabitants may have a growth rate 0.01 percentage point higher two to three years later than a similar country with less ASNs. In addition, an OECD-UNESCO report on local content highlighted a strong correlation between growth in AS numbers and local content development.

Autonomous Systems reach ASNs with which they do not have a peering relationship through transit. Most of these will be geographically distant from their network and will require third parties to route traffic. Often traffic traverses several networks before it reaches the end destination. Studies by RIPE Labs have shown that traffic has to traverse an average of 4.3 other networks before reaching the destination AS. This number has remained stable despite the growth of the Internet in the last decade. This suggests that most networks are able to locate competitive offers for transit. The number of interconnections of these transit providers has also risen and created a more

Figure 5.1. **Routed AS numbers per 100 000 inhabitants, 2012**

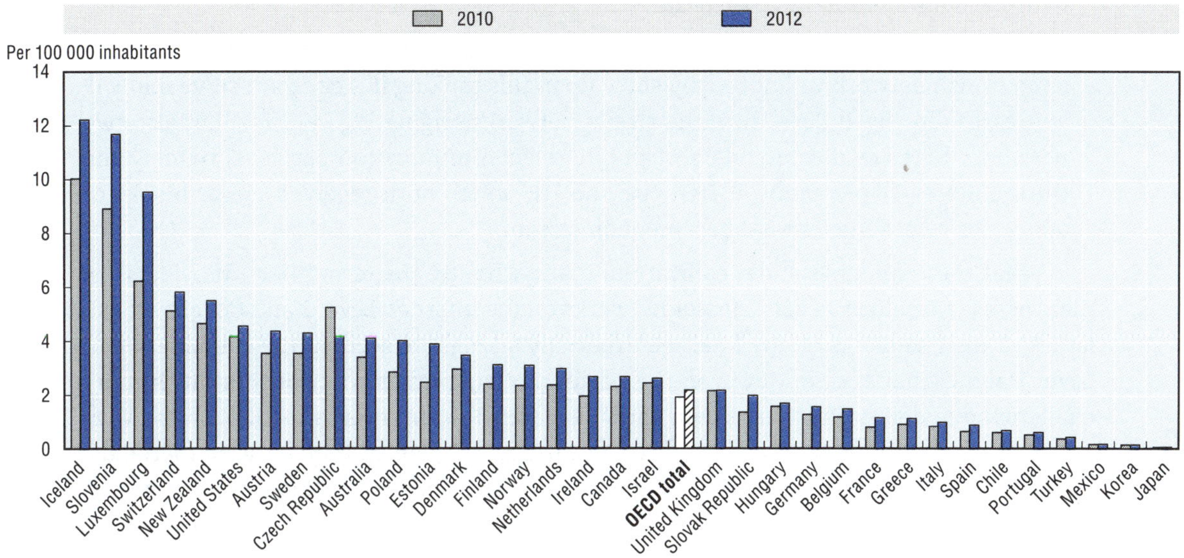

Sources: OECD, Potaroo.

StatLink http://dx.doi.org/10.1787/888932799095

interconnected core Internet. A failure of one network in the core is therefore less likely to create any significant long-term impact, because most networks have many alternative transit providers available.

Autonomous Systems differ significantly and vary considerably in size. Different measures are available, most of which calculate how much of the Internet the network can reach directly. Another approach examines the number of IP addresses behind an AS.

Four different perspectives are presented (Table 5.2). The first set concerns data generated by the Cooperative Association for Internet Data Analysis (CAIDA). The other three present the Internet as seen by University of Oregon Route Views Project. These data only show information from routing tables and not information on number of customers, revenues, geographic size and so on. The four ways of measuring the size of the largest networks around the world are as follows:

- **Customer cone.** This is the set of ASNs that can be reached from a given AS following only customer links. This means that the network Level 3, for example, can reach 83% of Autonomous Systems through its customers, or that 83% of networks are customers of Level 3 or its customers. Level 3 is mentioned twice because it recently merged with Global Crossing and the two networks had not, at that stage, been integrated.
- **IPv4 adjacencies.** These refer to the number of Autonomous Systems for which the network is the preferred route, as seen from the network of Route Views. Noteworthy here are ReTN.net and Anders Telecom, both Russian networks. In their service area many smaller networks use AS numbers for greater control and flexibility of routing. This increases their rankings.
- **IPv4 prefixes announced.** This is the number of IPv4 blocks announced by a network. This provides a sense of the number of networks without an ASN behind the AS. However, this is also dependent upon the level of aggregation of these blocks of IP

addresses. For example, a 1/24-block of 256 addresses is counted equal to a /16 block, which is equivalent to 256/24-blocks.

- **IPv4 addresses originated.** This measures the number of addresses unique to that AS. A large transit network will not easily show up in this ranking if it does not serve many IP addresses directly. It does show the size of networks originating in China, but also two networks that are not providers of public telecommunication services, namely the United States Department of Defense and HP, which both received larger blocks of addresses before the Internet became mainstream.

What these rankings show, contrary to many who use the term "Tier 1 providers", is that there is no definitive list of major networks on the Internet. Level 3 and Cogent are the largest transit networks on the Internet, especially taking into account the combination of Level 3 and Global Access. However, the position of subsequent networks is much more challenging to determine. These data show that transit is an international business with networks from countries like China, Germany, India, Italy, Japan, Korea, the Netherlands, Russia and Sweden all appearing in the rankings. It also shows that there is no dominant party that controls routing on a global scale, though some networks may have a significant presence in their region.

An alternative to purchasing transit is to agree to directly interconnect a network with another network via a peering agreement. Both parties, so-called peers, pay for transit, regardless of the direction of traffic. If a network sends 1 000 Mbit/s peak traffic it will need to buy an equal amount of transit. The receiving network will also need to buy 1 000 Mbit/s of transit. This is irrespective of whether there is 50 Mbit/s or 999 Mbit/s flowing in the opposite direction. Smaller networks (measured in total Mbit/s) frequently benefit most from a peering exchange, as they may not be able to negotiate as good a price for transit as a larger operator. This reflects the fact that larger networks have more to offer in terms of network reach and so on, reflecting greater investment. Some large content providers buy more transit than almost any traditional telecommunication network operator, and therefore benefit less from peering than telecommunication operators.

In principle, a peering relationship can be established at any location where two networks meet. However, most peers meet at data centres at the conflux of multiple networks. Direct interconnections between networks are known as **private peering**. Another way to interconnect is over a central switch managed by a third party, often an independent organisation or association. These organisations are located at Internet exchange points (IXPs). Peering at an IXP is known as **public peering**. Even when networks engage in public peering, they may reserve private peering for peers with larger traffic volumes. As of 2012, IXPs exist in all OECD countries after Mexico established one in November 2012. IXPs benefit peers by allowing each one to connect to a large number of other peers using just one fibre and one interface, whereas direct interconnections require an interface per peer. Public peering is more prevalent in Europe, whereas private peering is used more in North America, largely as a result of historical factors in the development of the Internet.

Clear network effects are visible at peering locations: the presence of more networks facilitates peering with a larger volume of traffic, which in turn makes the peering location more attractive. In a 2011 survey by Packet Clearing House, 142 210 peering relationships were identified among 4 331 ISP networks in 96 countries. Of these peering relationships, 141 512 (99.51%) were "handshake" agreements in which the parties agreed to commonly understood terms without creating a written document. This shows that peering is a well-understood relationship on the Internet. Many IXPs employ a route server on which networks can announce for which routes they always accept peering, allowing automatic configuration of peering relationships without need for human interaction. Though the survey found that on average networks peer with 32.8 networks, over 60% peer with less than 10 networks. Some networks like Google and Akamai peer with over 1 000 networks. The high schools of the Canton of Zug in Switzerland, mentioned above, peer with 920 IPv4 and IPv6 peers at six IXPs with 633 peers configured explicitly and 287 received through route servers. This allows them to peer approximately 70% of their traffic. By buying on the wholesale market instead of the retail market, and peering with networks, the schools have received more bandwidth and higher availability for the same cost.

While some of the engineering necessary to allow the Swiss schools to operate on the wholesale market was sponsored, the schools themselves meet the costs for hardware, transit costs, fibre and memberships of IXPs. The staff is competent in managing BGP and associated technical requirements and the work is done after hours. Overall, the schools demonstrate the potentially diverse background of an ASN relative to traditional network operators.

The principle benefit of peering, apart from the monetary savings, is that it allows traffic to take the shortest logical path between two networks. Generally, this is also the shortest physical path. This benefits the customers of both networks by giving them better network and service performance. As a result, more and more local exchange points are developing. In the past, IXPs focused on capital cities, but now regional peering has become prevalent and there is increasingly greater geographical peering by larger content delivery networks (CDNs).

Large providers of content, such as Akamai, Google and Netflix, are deploying cache servers deeper into the network of their peers. This can save these peers from paying for traffic over interregional backbones. A Google cache deployed in Kenya boosted traffic over the Kenyan IXP (KIXP) from 100-150 Mbit/s to around 1 000 Mbit/s. By peering locally, the Kenyan ISPs did not have to buy transit for this traffic, which then ranged from USD 150 to USD 600 per Mbit/s/month. This saved Kenyan ISPs a considerable amount of money each month. It is difficult to calculate the exact monetary savings as some capacity was unrealised demand and other traffic may have replaced some of the traffic that is now peered locally.

While data do not exist on yearly aggregate traffic growth over all IXPs across the OECD area, and many IXPs only report annual data, some indications can be provided. Data from the AMS-IX in Amsterdam, which is the IXP with the most members (513) and the second-highest peak traffic load (after DE-CIX in Frankfurt and ahead of LINX in London), show that its traffic grew from 90 050 Terabytes per month in August 2008 to 314 000 Terabytes in July 2012, a growth of 350%. If exchanged evenly over a month, 314 000 Terabytes would require over 1 000 000 Mbit/s. The peak 5-minute average is however 60% higher at 1.6T bit/s. This saves the members of AMS-IX between USD 500 000 and USD 1.6 million every month. Individual AMS-IX members are now installing 100 Gbit/s ports to handle expected future growth.

Address space: IPv4 and IPv6

The Internet Protocol (IP) is a communications protocol responsible for transporting data from a host to its destination across the Internet. IP uses a numeric addressing system and routes messages based on IP addresses, which specify the locations of the source and destination nodes. Every device connected to the Internet, whether a desktop computer, tablet or mobile device and so on, needs an IP address in order to communicate with other devices. There are two versions of IP addresses in active use: IP version 4 (IPv4) and IP version 6 (IPv6).

Introduced in 1983, the prevalent communications protocol today is IPv4. It has a 4.3 billion-address space, which seemed ample addresses at the time, as the Internet was used by a relatively small number of academic networks. Given the extraordinary growth of the Internet over the past two decades, this has proven to be insufficient. There are said to be more than 2 billion people using the Internet in 2013, and this is likely to double in a relatively short period. Not only will the number of users exceed the number of IPv4 addresses, many will have several devices connected. The Internet technical community has long anticipated this development and a newer version of the Internet Protocol, IPv6, has been designed to accommodate this tremendous growth. The deployment of the IPv6 protocol began in 1999.

Recent years have seen positive developments in the deployment of IPv6. Though the number of websites and the number of hosts that are IPv6 enabled in 2012 is still limited, the amount of content that can be delivered over IPv6 has greatly increased. Practical experience reported by universities and theoretical models developed by Cisco, a major equipment manufacturer, show that over 40% of content could be delivered over IPv6 if it was activated by networks around the world. This is because major content and service sites, such as those of Apple, Google (including YouTube), Microsoft and Yahoo!, have all activated IPv6 in line with the deployment of IPv6-compatible networks and devices.

The Internet Assigned Numbers Authority (IANA) delegates blocks of IP addresses and Autonomous Systems (AS) numbers to each Regional Internet Registry (RIR) to meet the needs of that region. RIRs follow regional policies to allocate resources to Local Internet Registries (LIRs) or to National Internet Registries (NIRs). LIRs either assign address space to end-users or allocate address space to ISPs who, in turn, assign IP addresses to enterprises and end-users. The IANA assigned the last five unallocated IPv4 address blocks to the regional registries (RIRs) in February 2011. The Asia Pacific Registry (APNIC) assigned all unallocated IPv4 blocks by 2011 and the RIPE NCC (serving members in Europe, the Middle East and parts of Central Asia) reached the same stage in September 2012. The North American Regional registry (ARIN) is expected to assign all unallocated addresses by 2014, while AFRINIC and Lacnic have a small number of blocks available for a longer period (Figure 5.2).

Although IPv4 address space is reaching exhaustion, the number of routed IPv4 addresses per inhabitant still provide one indication of the relative development of Internet infrastructure. This can be illustrated for the OECD area, as well as some selected countries (Figure 5.3). The United States led with 3.2 IPv4 addresses per capita along with the Scandinavian countries, which have also experienced high Internet take up. By way of contrast, countries like Mexico and Turkey in 2012 had one fifth of an IPv4 address per inhabitant (equivalent to every five people in the population sharing one IP address).

Figure 5.2. **IPv4 depletion per RIR, 2012**

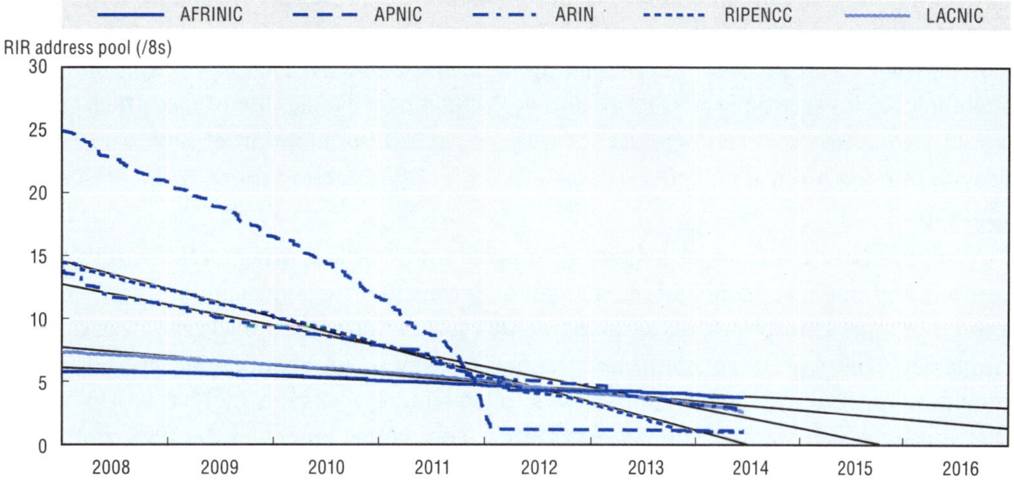

Source: Potaroo, RIR.

StatLink http://dx.doi.org/10.1787/888932799114

Figure 5.3. **Routed IPv4 addresses per capita, 2012**

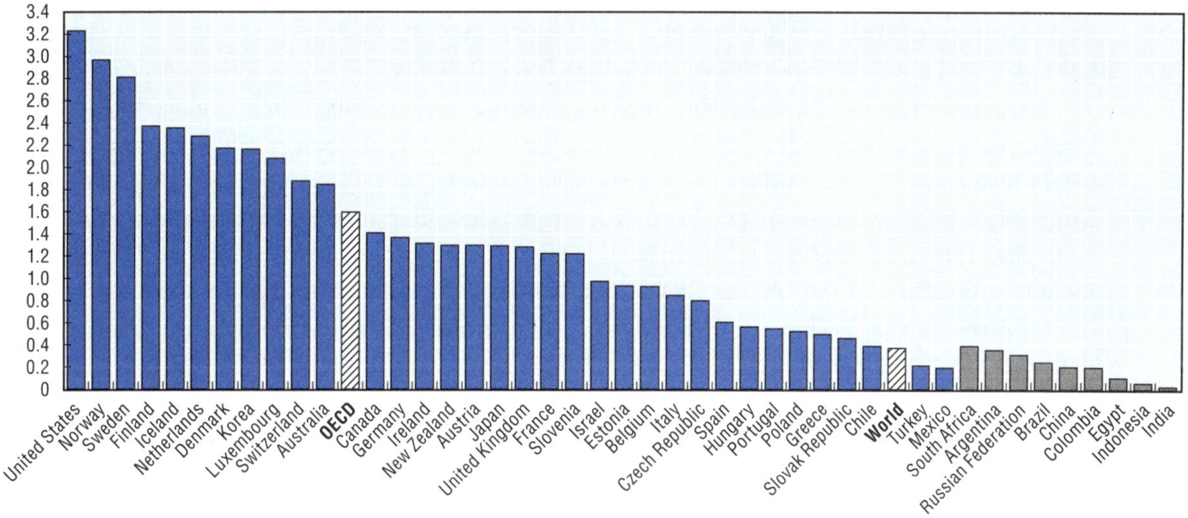

Sources: Potaroo, RIR.

StatLink http://dx.doi.org/10.1787/888932799133

In order to cope with the exhaustion of IPv4 addresses, companies have adopted a pragmatic approach employing complex and expensive layers of network address translation (NAT) to share scarce IPv4 addresses among multiple users. As a NAT uses a common IP address for all terminal devices behind it, this implies that the devices connected to it become invisible to the routing tables. Therefore, it requires the ISP to sustain simultaneous port sessions between the user and the NAT so that information packets reach the user, given that these cannot be directly routed. Technical limitations may restrict the number of port sessions that can be simultaneously opened per user connected to the NAT, therefore, one caveat of using this partial solution to cope with the exhaustion of IPv4 addresses is that the end-user's experience may be affected. Another disadvantage is that the Internet could become a collection of fragmented local networks.

Even though the IPv6 protocol has been available for more than a decade, the main issue hindering full implementation is lack of backwards compatibility. Network operators, service providers and equipment vendors therefore all need to co-ordinate configuration of IPv6 as the default protocol in their equipment and services. This co-ordination among stakeholders is essentially a "chicken and egg" challenge: availability of more IPv6 content would give access providers greater incentive to speed up adoption of IPv6, and content providers need more IPv6-active customers to justify enabling more IPv6 services and content.

In June 2011 the Internet Society sponsored an IPv6 "test drive". The world's largest content and application providers, as well as government agencies, equipment vendors and other large enterprises, all participated in a joint effort to test IPv6 readiness during a single day. This experiment confirmed the readiness of a number of networks and content providers for IPv6 implementation. It was followed one year later by IPv6 World Launch Day, during which some of the largest Internet companies launched the IPv6 protocol by default. As a result, communications or part of communications can now take place automatically over IPv6 if connectivity, content and equipment are available.

Major content providers including Akamai, Bing, Facebook, Google, Netflix, Wikipedia and Yahoo! turned on the IPv6 protocol by default. Several major ISPs such as AT&T, Comcast, Free, Iliad, KDDI, RCS, TWC and Verizon Mobile, among others, committed to enabling IPv6 for their new users. Some equipment vendors, such as Cisco and D-Link, have also configured IPv6 as a default feature in their products.

Data informing the status of IPv6 adoption are imperative to follow this process. Some leading sources regarding IPv6 deployment include the Internet Society statistics of the World IPv6 Launch (*www.worldipv6launch.org/measurements*), APNIC (*www.potaroo.net*), RIPE Labs (*https://labs.ripe.net/statistics/?tags=ipv6*) and Cisco 6lab (*http://6lab.cisco.com/stats/index.php*). Data from these sources illustrate different layers of deployment of IPv6. These include: address space (i.e. IPv6 prefix allocations); Transit Autonomous Systems and total routed Autonomous Systems in IPv6; available web content in this protocol; and, finally, IPv6 user penetration ratio.

Figure 5.4. **IPv6 allocations by RIR, 2012**

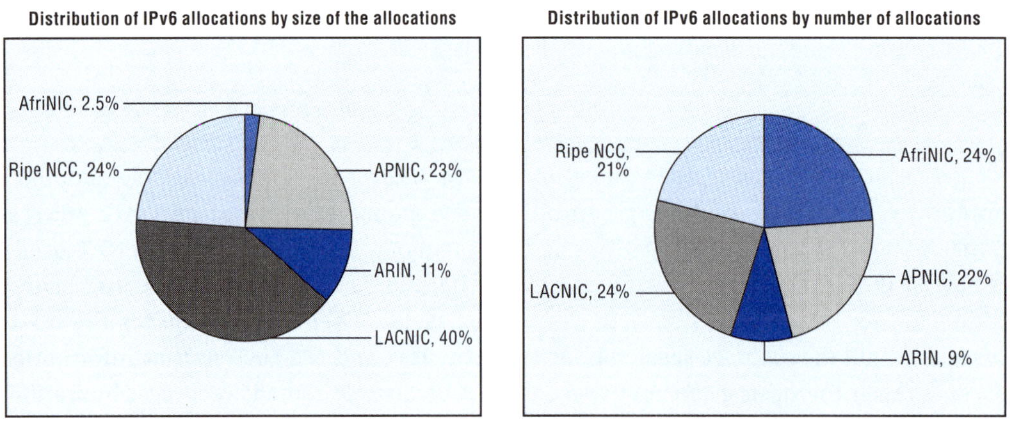

Source: Potaroo, OECD.

StatLink http://dx.doi.org/10.1787/888932799152

IPv6 address space

As with IPv4, IANA delegates IPv6 address space to RIRs, which then allocate it to interested entities based on need. At the end of 2012, RIRs had made 17 729 allocations (Table 5.3 and 5.4). OECD countries accounted for 69.5% of total IPv6 allocations worldwide. The United States (23.89%) was the country with the largest share of IPv6 allocations in total, followed by Germany (6.83%), the United Kingdom (4.33%) and the Netherlands (3.49%).

The size of IPv6 prefix allocation can provide one indication of the scale of planned deployments; however, extremely large allocations (given the magnitude of the IPv6 space) were provided in the past to some operators and large users skewing the "by size" results. Perhaps a more reasonable measure is the number of IPv6 allocations. At the end of 2012 the leader in IPv6 allocations in 2012 was the United States (549 allocations), followed by the United Kingdom (199 allocations) and Germany (187 allocations) (Figure 5.5).

Figure 5.5. **Numbers of IPv6 allocations per year, top eight OECD countries, 1999-2012 (year-end)**

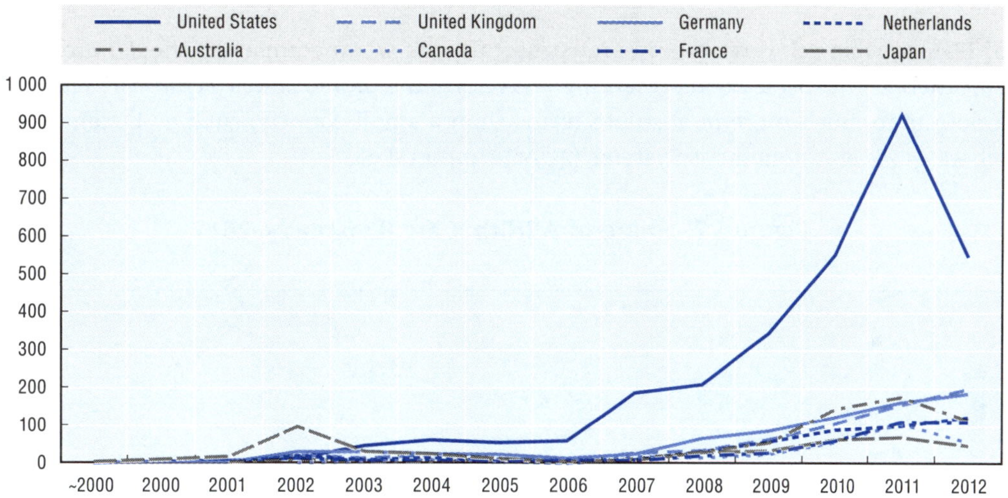

Source: Potaroo, OECD.

StatLink http://dx.doi.org/10.1787/888932799171

IPv6 transit autonomous systems and routed autonomous systems

Data from the Cisco 6lab Transit AS indicator reveal that, globally, the majority of the core of the Internet is now IPv6 enabled (Figure 5.6). The top 20 Autonomous Systems of the world (measured in terms of connections to networks) already provide transit services in IPv6. Of the top 100, 79% provide transit in IPv6 and an additional 10% are IPv6 enabled. "Enabled" here means that even if the transit AS is not currently routing through IPv6, it may very well start providing transit services in IPv6 in the near future. In this way, this ratio may give an idea of potential IPv6 routing over IPv6 of transit autonomous systems. It should be noted that not every IPv4 Transit AS may move to IPv6, as some transit networks use the move to IPv6 to consolidate some of their IPv4 Transit Autonomous Systems.

One way to assess the relative difference in IPv6 capability by country is to examine the percentage of ASNs that are IPv6 capable. Of the whole OECD area, around 17% of announced routed Autonomous Systems were IPv6 capable at the end of 2012. Some countries are leading the way, such as Norway with 55.0%, or the Netherlands with 46.3%

Figure 5.6. **Transit autonomous systems ready for IPv6, 2012**

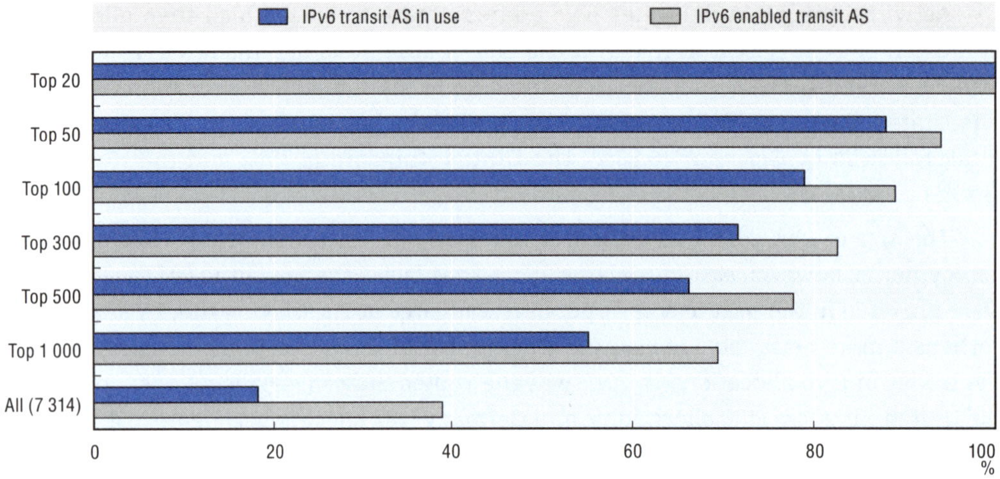

Source: Cisco.

StatLink http://dx.doi.org/10.1787/888932799190

of IPv6-capable AS. However, one must be cautious when comparing these ratios. Ten percent of IPv6 capable AS (representing 1 554 AS) in the United States (Figure 5.7, Table 5.9) amounts to much more in absolute terms than a small country, such as Luxembourg, where 44.7% of Autonomous Systems (21 AS) support IPv6.

Figure 5.7. **Share of ASN that are IPv6 ready, 2012**

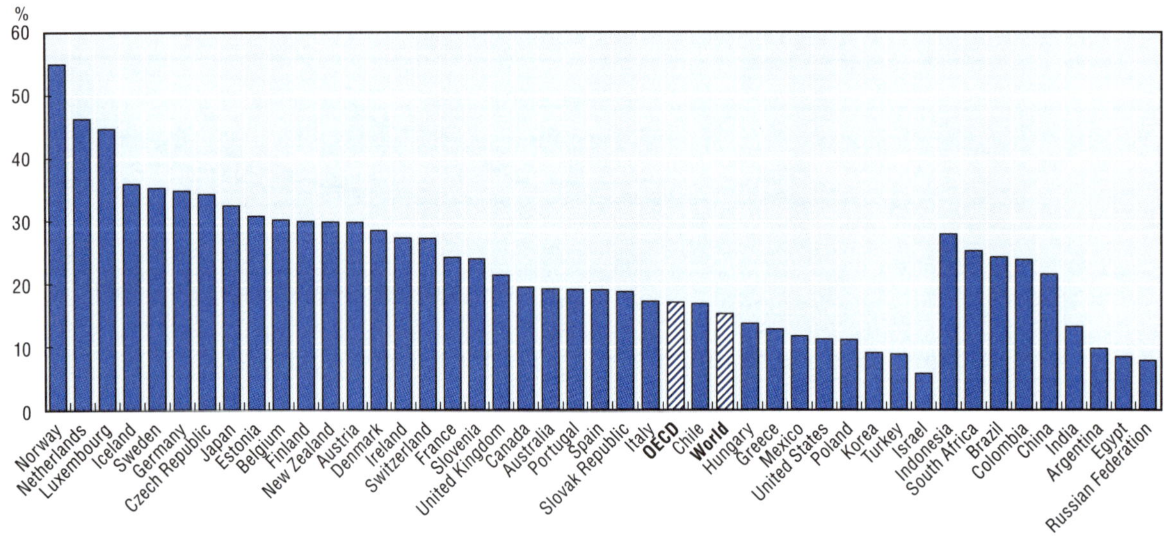

Source: Potaroo.net.

StatLink http://dx.doi.org/10.1787/888932799209

The percentage of announced AS numbers that route through IPv4, IPv6 or both for the year 2012 can be observed (Figure 5.8). Of the totality of announced Autonomous Systems in the world, almost 15% were both IPv4 and IPv6 capable at the end of 2012, up from 9% at the start of 2011 and 5.5% in early 2010. However, it is possible that not all IPv4 Autonomous Systems will convert into IPv6 because of historical routing reasons.

Figure 5.8. **Announced ASNs routing IPv4 and/or IPV6, 2012**

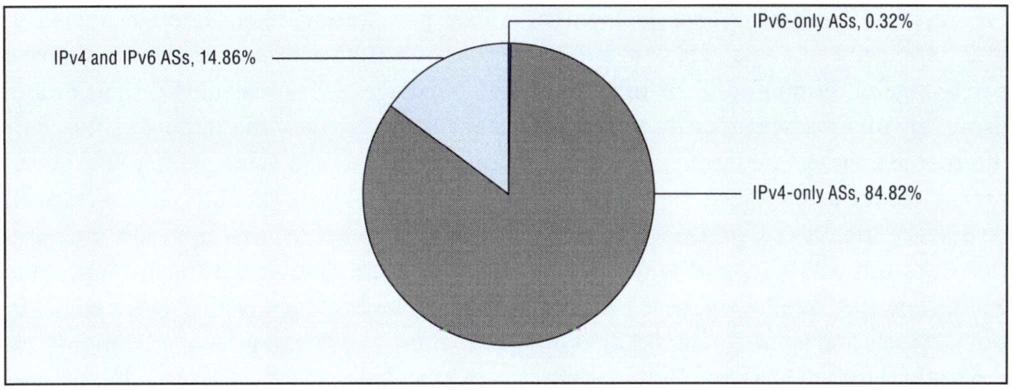

Source: Potaroo.net.

StatLink http://dx.doi.org/10.1787/888932799228

IPv6 web content

Earlier sections discussed the first two phases of IPv6 deployment: IPv6 prefix allocation and Transit and announced Autonomous Systems. Of equal importance are data regarding IPv6 adoption by content providers. As noted, some of the largest content providers globally participated in World IPv6 Launch in June 2012. Their content is now accessible to users around the globe who might already enjoy IPv6 connectivity.

Cisco has developed an indicator to measure the probable availability of web content to IPv6 users on a per country basis. This ratio takes into account the leading 500 webpages in each country ranked according to viewership by Alexa. In most countries, Facebook and Google rank highest; if they switch on content over IPv6, the ratio increases as they account for much of the content distributed within the country. Cisco estimates that on a global scale almost 30% of webpages were available using IPv6 in mid-2012. For some countries this ratio is even greater. For example, about half the webpages in Brazil and France could be accessed through IPv6 and almost half the webpages in the United States (Figure 5.9).

Figure 5.9. **Content available over IPv6, 2012**

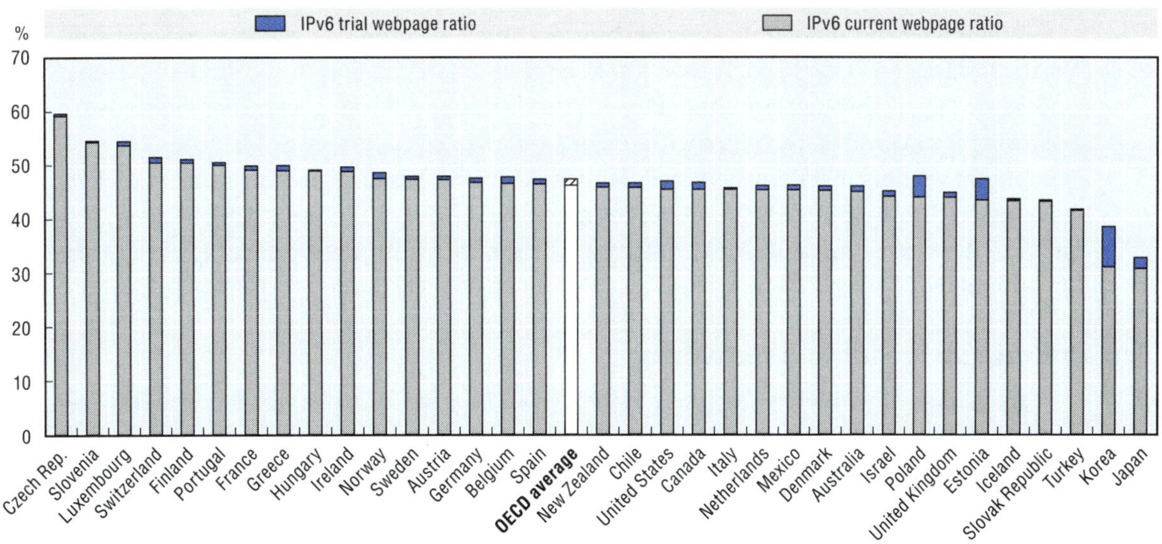

Source: Cisco.

StatLink http://dx.doi.org/10.1787/888932799247

IPv6 user penetration

Another measure of IPv6 deployment is user penetration rate. Cisco 6labs used the ratio provided by Google, which is similar to that of APNIC and essentially measures the percentage of terminal devices that "talk" IPv6 language. APNIC calculated their ratio by using YouTube's advertising distribution to reach a very significant sample of the entire Internet user base. The test measures the percentage of users in each country who show a preference for using IPv6 to download a dual-stack web object. This metric is termed the IPv6 user ratio. Data from Google reveal that the IPv6 global user penetration ratio grew from around 0.25% in mid-2009 to 0.71% in mid-2012. Google's data, in September 2009, showed the following user penetration ratios: France (1%) and the Netherlands, the United States and Japan (under 0.1%). In mid-2012 these ratios had grown to: France (4.7%), the Netherlands (0.32%), the United States (1.44%) and Japan (1.53%) (Figure 5.10).

Figure 5.10. **IPv6 user ratio, 2012**

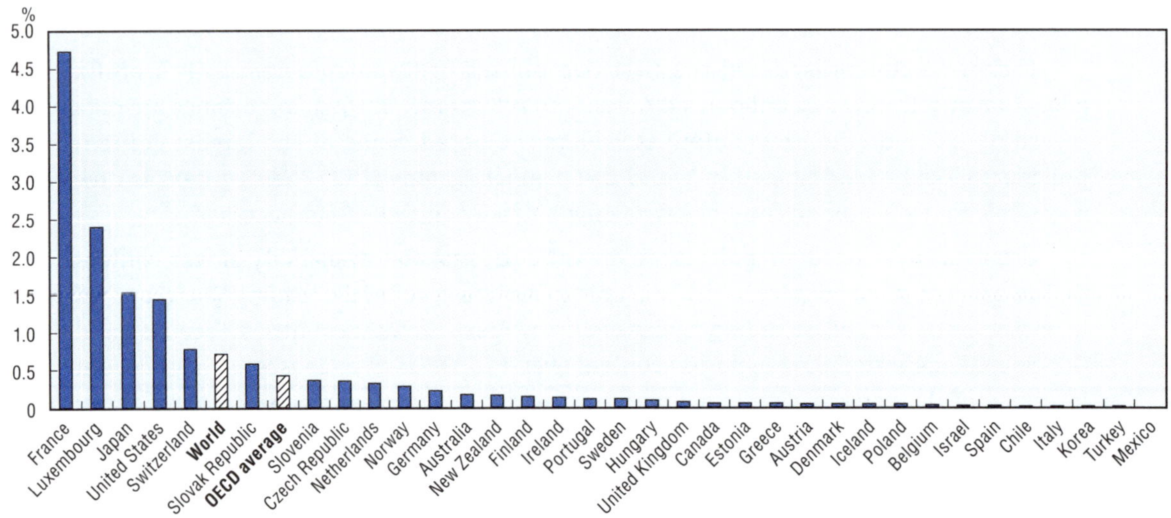

Source: Google.

StatLink http://dx.doi.org/10.1787/888932799266

While the penetration rate still appears very small, data on IPv6 take up appear more positive than a few years ago. It seems that co-ordination challenges are gradually being addressed largely through cross-industry efforts with support from government initiatives. Following IPv6 World Launch Day, with major sources of content now available in IPv6 and equipment vendors installing IPv6 by default in switching boxes, a growing number of ISPs have started to upgrade their networks. Increased IPv6 content availability may also provide incentives for additional network carriers to co-ordinate efforts towards this transition.

IPv6 and mobile broadband networks

As billions of wireless subscribers transition to Internet-capable mobile devices there is a growing need for IPv6 addresses in mobile networks, especially given the impending scarcity of IPv4 addresses. The huge rise in demand for mobile data traffic, including video services, has already provided an incentive for mobile network providers to upgrade their networks to "4G" technologies. CISCO VNI forecasts that mobile data traffic will grow at an average annual rate of 78% between 2011 and 2016. Thus, Long Term Evolution (LTE) could become a catalyst of IPv6 deployment because it is already IPv6 enabled, unlike other mobile technologies were enabling this feature could mean a significant increase in costs. It is noteworthy that some mobile network carriers are adopting transition strategies towards the implementation of IPv6. For example, Verizon Wireless, a key participant in the IPv6 World Launch Day, has required IPv6 support for all devices attached to its LTE network since 2009.

IPv6 policy initiatives

In addition to industry initiatives, OECD governments have continued to implement efforts to promote the deployment of IPv6 protocol. One major policy has been the coordination of IPv6 deployment within government networks, taking into account that governments are major consumers of network equipment. For example, the Czech Republic sought to ensure access to government websites and e-government services over both IPv4 and IPv6 protocols by the end of 2010.

The Australian Government Information Management Office (AGIMO) is overseeing the Australian government's transition to IPv6. At the end of 2012, the majority of agencies were reported to be IPv6 enabled with the remaining agencies expecting to complete the transition to IPv6 during 2013.

IPv6 has been mandatory in the Netherlands since 2010 for all new procurement by government bodies. By 2013, all governmental websites will be IPv6 ready. In Hungary, deployment of IPv6 in government infrastructure has begun and is proceeding apace. In Ireland, all public sector networks are to be IPv6 enabled and public sector ICT spending prioritises equipment that can run both IPv4 and IPv6. In Turkey, all public sector organisations must ensure IPv6 compatibility for IT infrastructure by the end of August 2013. The United States set a timeline for adopting IPv6 for use on public servers by the end of 2012.

The co-ordination challenges implicit in the IPv6 transition have led to large awareness-building efforts. These take the form of country task forces with governments seeking to promote common platforms of discussion between the distinct stakeholders of the Internet community. For example, in March 2011, in preparation for IPv6 Day the United Kingdom urged Internet companies to work together with 6UK, the organisation set up to aid the adoption of the new Internet protocol, to ensure a successful transition to IPv6. In New Zealand, an IPv6 task force drew members from the telecommunication industry, ISPs and government departments. The Danish Government is presently hosting meetings with ISPs and business associations regarding information for Danish enterprises about IPv6, with targeted information to SMEs planned for 1Q 2013.

There has been ongoing interest in IPv6 network research. In 2010 the Dutch IPv6 task force awarded research funds to monitor the implementation of IPv6. The Danish government has also initiated projects on IPv6 network research and established a test-bed (now discontinued) that allowed public agencies, universities and some companies to trial IPv6.

Considerable efforts towards the IPv6 transition have been made in countries outside the OECD area. China has implemented the Next Generation Internet project (CNGI) with the objective of ensuring a large proportion of the Internet is IPv6 enabled within five years. The country exhibited IPv6 network infrastructure at the 2008 Olympics using IPv6 to connect everything from security cameras to Olympic events. India has established an IPv6 Road Map: all major service providers (with at least 10 000 Internet customers) should be able to handle IPv6 traffic and offer IPv6 services by December 2011, and all central and state government ministries shall start using IPv6 services by March 2012. The Singapore government has set a goal of making government e-services and websites "IPv6 reachable" by September 2012.

Domain name system

Domain Name System (DNS) translates user-friendly host names (e.g. *www.oecd.org*) into IP addresses (e.g. 203.160.185.48). It works by asking the network the question: where is OECD.org located? The DNS answers using resolvers that query the data stored in a hierarchical and widely distributed sets of machines known as DNS servers. The DNS distributes this task among servers allocated to each domain and handles billions of requests daily. These servers are essential for the smooth functioning of the Internet.

The hierarchical syntax of a domain name is supported by the "dot" in the name and is read by the DNS server from right to left (.org is the Top Level Domain and .oecd is the sub-domain of this TLD and so on.) The top of the hierarchy is the "root", which acts like a "dot" at the end of an address, and is invisible to users. This root provides information for the resolvers to initiate a query for the Top Level Domain (such as .org, .com, .fr or .eu). Top Level Domains are divided into two classes: generic Top Level Domains (gTLDs), which include ".com" or ".org"; and country code Top Level Domains (ccTLDs), which consist of two-letter codes generally reserved for a country or a dependent territory (e.g. ".au" for Australia or ".fr" for France).

DNS management is in charge of the Internet Corporation of Assigned Names and Numbers (ICANN), among other Internet functions. As part of ICANN, IANA manages the DNS root, which involves delegating administrative responsibility of TLDs to sponsoring associations. For example, IANA allocates the management of ccTLDs to national registries (e.g. AfNIC in France or NIC.br in Brazil), which then establish different registration requirements and fees. These registries provide services to Registrars, for example, maintaining a centralised registry database for each TLD. Registrars (e.g. GoDaddy, Enom, Network Solutions), in turn, provide services to end-users wishing to purchase a certain domain (also called registrants) with whom they have contractual agreements.

Domain name registrations are one indicator of interest in adopting a presence on the World Wide Web.[1] There were around 233 million domain name registrations in mid-2012, up from 196 million in mid-2010. The growth in absolute numbers is aligned with a slight increase in the growth rate of domain name registrations from 14.4% in 2008-10 to 15.8% in 2010-12. This represents a drop in the growth rate from 37.5% in 2006-08. This tendency for more steady growth could reflect the beginning of saturation in the domain name market. Registrations under OECD-related ccTLDs grew on average by 22.1% annually during the 2000-12 period. Registrations under all ccTLDs worldwide grew by 23.9% annually over the same period, and registrations under major gTLDs grew by 18.4% annually (Table 5.6).

The introduction of country code Internationalised Domain Names (ccIDNs) in 2009 enabled the creation of ccTLDs in non-Latin languages or scripts. In October 2012 the root zone contained 328 top-level domains, of which 33 were assigned ccIDNs. Another 6 were defined but not assigned and 11 test ccIDNs were present in the rootzone file. ICANN has initiated a process to expand the number of gTLDs. In 2012, it received 1 960 applications for new gTLDs: some were for unique gTLDs and a further 751 were multiple applications for the same 230 gTLDs. Starting in 2013-14 as many as 1 439 new gTLDs could become available in the rootzone.

Registrations by domains

From 2010 to 2012 the number of registered domains increased by 18.7%. ccTLD registrations displayed growth almost equivalent to that of the world's major gTLD registrations over the same period (24.4% and 24.6% respectively). In mid-2012 the gTLDs .com and .net accounted for around 85% of all gTLD registrations and almost half of the world's total domain name registrations. Registrations under OECD area ccTLDs grew by an average of 9.5% per annum from mid-2010 to mid-2012, in contrast to growth of 11.4% in mid-2008 to mid-2010. Additionally, OECD area ccTLDs accounted for 26% of total domain name registrations in mid-2012 (Figure 5.11).

Figure 5.11. **Domain name registrations per type of top-level domain, 2000-12**

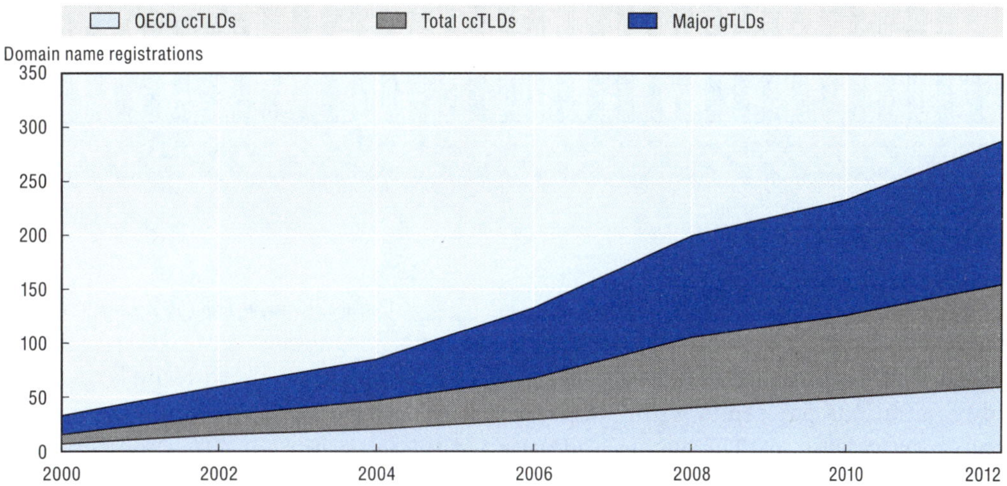

Source: OECD, from country NICs.

StatLink http://dx.doi.org/10.1787/888932799285

Out of more than 240 total ccTLDs, the 10 largest OECD countries (sorted by number of domain registrations) accounted for almost half of the global ccTLD market in mid-2012. Germany's ccTLD (.de) continues to be the largest with over 15 million names, followed by the United Kingdom's ccTLD (.uk) with a little over 10 million registrations. ccTLD registries are responsible for the policies and operation of the domain name (including implementation of requirements and fees), thus resulting in wide variations in the number of ccTLD registrations across countries. The high adoption of names under .de can be explained by several factors, including the large size of the country; and non-restrictive policies implemented by the registry, which allow registration of single-letter, two-letter and number-only domains, and accept all diacritics of the German language. The registry imposes only one restriction: an administrative contact must reside in Germany.

Registrations under China's ccTLD (.cn) dropped by almost 8.4 million between mid-2008 and mid-2012. This followed changes in registration requirements aimed at minimising anonymous registrations, which include paper application forms, a business license and an identity card. Additionally, individuals are not entitled to register .cn domains, and the registry operator has suspended .cn registrations through non-domestic registrars. China ranks fourth when compared to OECD area ccTLDs (preceded by Germany, United Kingdom and the Netherlands) with a total of 4 million registrations in mid-2012. The most dynamic OECD-related ccTLDs in terms of growth over the 2000-12 period were the United States (.us), Spain (.es), Canada (.ca) and Poland (.pl) (Figure 5.12).

Figure 5.12. **Average annual growth in domain name registrations by domain 2000-12 (%)**

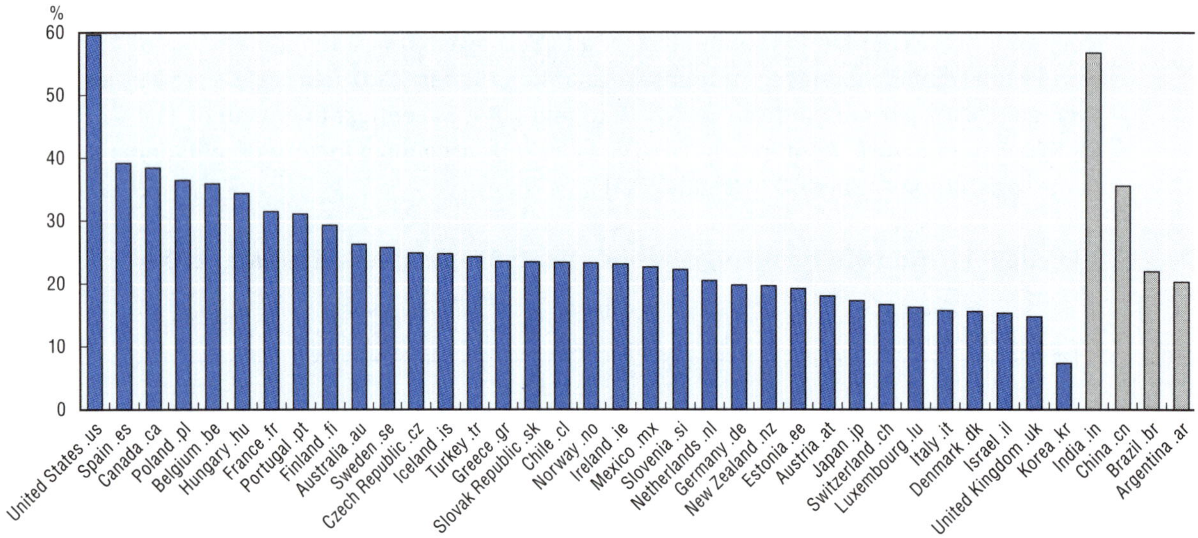

Source: OECD, based on country NICs.

StatLink http://dx.doi.org/10.1787/888932799304

In mid-2012, the Netherlands' ccTLD (.nl) was the country code with the largest ratio of registrations per capita with 301 per 1 000 inhabitants (or almost one third of the country's population). The Netherlands also led this ratio in mid-2010. Other countries with a high number of ccTLD registrations on a per capita basis were Switzerland (.ch), Denmark (dk.), Germany (.de) and the United Kingdom (.uk), all of which had over 150 domain names registered per 1 000 inhabitants (Figure 5.13).

Traffic on the Internet

Traffic on IP-networks is challenging to measure. With over 40 000 Autonomous Systems it is not possible to obtain data on all networks. Moreover, not all data traverses network borders and when traffic traverses networks it risks being double counted. The more reliable methods of measuring Internet traffic focus on the end-points in the network, where every communication starts and ends. Some countries request data on usage from network providers and these can provide an insight into traffic volumes per user. The difficulty here lies in obtaining a global overview, as the numbers are not easily extrapolated. Another approach is to look at core networks on the Internet or to extrapolate data from measurements in some networks and combine these with market trends.

The available data indicate that while the rate of IP traffic continues to show impressive growth, its pace has decreased in recent years. Such a decrease is to be expected when measured against a much larger base of Internet access penetration and broadband adoption in OECD countries. Most of the OECD area is now widely connected,

Figure 5.13. **OECD country-related ccTLD registrations per 1 000 inhabitants, mid-2012**

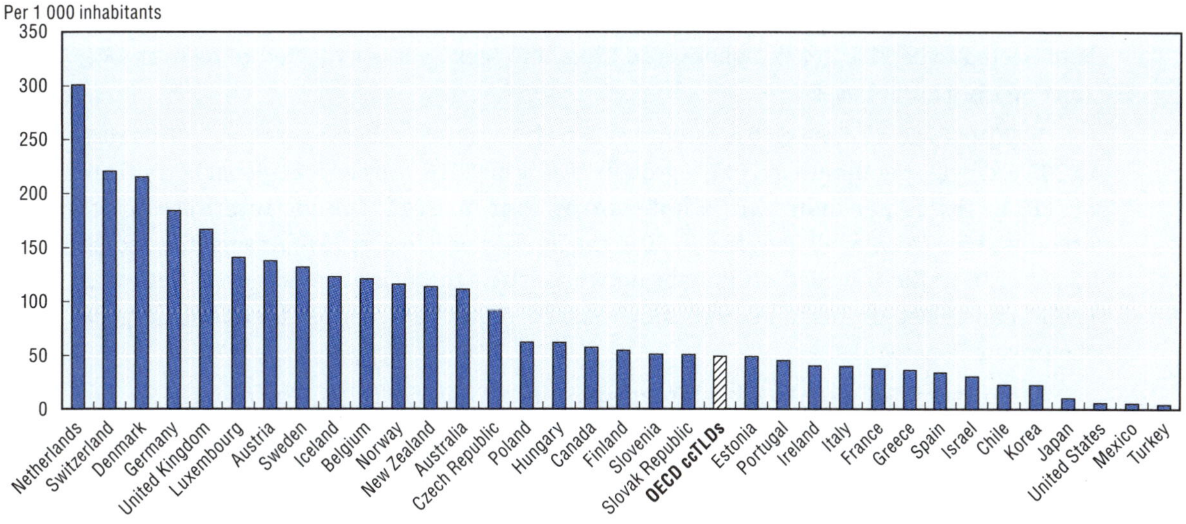

Source: OECD, based on country NICs.

StatLink http://dx.doi.org/10.1787/888932799323

which means that growth comes primarily from changes in Internet usage. This may be the result of applications that use more data, new forms of use or from greater use of the Internet for longer periods. All these factors are occurring, for example, through the increased use of video-on-demand and greater mobile Internet. In addition, business premises and households have an increasing number of devices through which employees and residents are connected over ever-increasing broadband speeds. At the same time there has been tremendous expansion in CDN infrastructure around the world over the last five to seven years. These companies distribute traffic in a more localised way with respect to where it is most in demand.

CDN infrastructure located around the world has played a major role in this trend, even at a time of increasing demand for content such as video. Other factors include greater use of peering and caching as well as overall economic conditions. In addition, many attributes of Internet traffic may not be captured, depending on methodologies used and different reporting and non-reporting practices, including the treatment of "on-net traffic" and "offloaded traffic". In the latter case, for example, traffic demand off-loaded from a cellular network to a fixed network, such as via Wi-Fi, may be reported differently by different operators. Caveats aside, the rate of Internet traffic growth during the global financial crisis has been substantial. While not all indicators may capture all factors they tell a similar story in relation to growth and the direction it is heading.

Cisco provides a widely cited indicator for trends in Internet traffic evolution. The company's Visual Network Index (VNI) consists of two elements. The first, "Forecast", gives an insight into global IP traffic based on projections of analysts on traffic use, traffic growth, applications, user growth and other elements. The VNI Forecast is an estimate and is not based on measurements. The second, "Usage", provides verification of their forecasts through usage data from 20 large broadband, mobile and cable networks across the globe with millions of end-customers. The company says that the prognosis of the Forecast is generally very close to the results of the VNI Usage study.

According to Cisco's VNI Forecast, global IP traffic volume has grown eightfold over the period 2006-11, but will increase only threefold over the period 2011-16 at a growth rate of 29% per year. In some countries, such as in France and the United States, growth until 2016 is projected to be 21% to 22%, whereas in Chile and Mexico it is expected to be over 45%, starting from lower bases.

In 2010, Cisco said the global volume of IP traffic was 20 Exabytes per month. This rose to 30.7 Exabytes per month in 2011, a growth of 40%. In 2016, they expect global IP traffic to be 110 Exabytes per month. It is noteworthy that, in 2011, "public Internet traffic" accounted for 77.8% of all IP traffic. In 2010 the figure was lower at 75%. At this time, some 22.2% of IP traffic travelled on private networks such as Business Wide Area Networks (WANs) and video-on-demand. According to Cisco, the public Internet, therefore, outgrew the private IP networks.

Cisco estimated global peak Internet traffic in 2011 at 153 Tbit/s. The average level of traffic, however, was half that rate at 73 Tbit/s. By 2015, the company estimates that peak Internet traffic will rise to 720 Tbit/s. As global peak Internet traffic is forecast to increase fivefold between 2011-16, compared to a threefold increase in volume, peaks at busy hours will become significantly higher relative to average traffic. Given that networks are built for peak capacity rather than average use, networks that wish to provide that level of service would need to be appropriately upgraded.

The various regions in the world showed different growth levels in 2011 from 2010. They also reflect the trend towards lower growth rates compared to the previous edition of the *Communications Outlook*.

- North America grew 41% to 10.3 Exabytes per month, up from 7.3 Exabytes.
- Latin America grew 53% to 1 Exabyte, up from 0.683 Exabytes.
- The Middle East and Africa grew 47% to 0.384 Exabytes, up from 0.261 Exabytes.
- Western Europe grew 40% to 7.3 Exabytes per month, up from 5.2 Exabytes.
- Central and Eastern Europe grew 45%, up from 0.800 Exabytes to 1.2 Exabytes.
- Asia Pacific grew 40% to 10.5 Exabytes, up from 7.5 Exabytes.

With the exception of Latin America, no region of the world had a growth rate above 50% during this period. Asia Pacific, North America and Western Europe close to doubled in the two-year period, a 40% increase.

While individual networks all have authoritative data on their own networks, there are little official data that describe national or regional trends. This makes it challenging to interpret available data and draw conclusions for what this may mean for a variety of potential outcomes, such as the level of investment required. In one sense this is a matter for each individual network and they are certainly best placed to make such judgments. On the other hand, policy makers and regulators are frequently called upon to make decisions that would likely have better outcomes if better information were available. Cisco's VNI is a valuable source for global trends. For some countries it can be complemented by data collected by regulatory authorities, such as in Germany and Sweden, or national statistical agencies as in Australia.

According to Cisco, mobile traffic is showing much higher growth rates than the overall growth of IP traffic, though naturally from a lower base than fixed networks. It grew 2.3 times in 2011 more than doubling from 2010. Cisco expects the growth rate to decline to below 100% per year in 2013. Given the higher growth rates for mobile networks, their share of overall IP traffic is expected to grow to 10% of all global traffic within about five years. In 2011 global mobile IP traffic was 0.6 Exabytes per month. Cisco expects this to rise to 10.8 Exabytes per month by 2016, which would amount to 9% of global IP traffic, up from 2% in 2011.

Currently, for those countries where national data are available, there are a wide variety of usage patterns for mobile IP traffic. In Germany, between 2011 and 2012, total mobile traffic grew by 42%. The predominant reason was a 34% growth in the number of users. Average usage per user grew by 5% to 270 Megabytes per month. In Japan, total traffic grew by 121%. The predominant reason was an increase in average use per subscriber, which grew 102% from 298 MB to 602 MB per month. The number of subscriptions grew 9% between 2011 and 2012. Also noteworthy in Japan is the availability of statistics on uploaded data. Over the same period there was 108% growth from 32 MB to 66 MB for uploaded data. Italy and the Netherlands had more even distribution between growth in number of users and average use per user. In Italy, total mobile traffic grew 36%, with traffic per user growing 20% to 180 MB per month, and the number of users growing 12%. Total mobile traffic in the Netherlands grew by 77%, traffic per device grew by 25% to 180 MB per month, and the number of users grew by 41%.

Patterns of use

While there has undoubtedly been an increase in the amount of traffic carried over the Internet, less public data are available which provide information on the composition of this traffic and the usage patterns of users. These data are, of course, available to every individual network but rarely available at an aggregated level to inform policy or regulatory decisions. In competitive markets, network operators and over-the-top service providers are understandably reluctant to share some types of data. Nevertheless, some of these actors simultaneously exert pressure on government to take decisions affecting issues such as spectrum policy, regulation, public investment and the Internet model for traffic exchange.

The increasing trend for regulatory authorities or other official statistical agencies to gather data and report it in an aggregated and anonymous format is welcome. In addition, new Internet tools allow users to generate "crowd-sourced" data on areas such as quality of service, while initiatives by network operators or other service providers make similar information available. Both provide opportunities for more informed decisions. However, some data just raise more questions instead of providing much-needed information for critical decisions.

For example, average usage as measured by data downloaded and uploaded may not indicate whether these data are on-net or off-net traffic. For example, it is not clear if the use of techniques such as Wi-Fi or femtocells for offloading traffic is counted as fixed or mobile traffic. Even data reported on average usage patterns of users does not necessarily provide information on associated tariffs or whether the traffic was generated at the busiest times of the day or in geographical locations with insufficient capacity to meet demand. Interpreting these data is challenging because some increases or decreases in the amount of traffic (against an overall increase in usage) could represent superior or inferior network and service provision in relation to, for example, transit and peering decisions, network offloading techniques from tools such as on-net Wi-Fi and so on, or the structure of tariffs or service choices available to end-users.

A limited number of studies have examined patterns of Internet use in ways that may inform issues relevant to policy makers and regulators, and draw on authoritative data direct from infrastructure and service providers. One example concerns the ongoing debate regarding the role of peer-to-peer networks in generating Internet traffic. In Japan, Kenjiro Cho, a senior researcher at IIJ,[2] has made annual studies of Japanese broadband usage, which reveal that use of peer-to-peer communications is declining in favour of direct-download sites. Another study undertaken in Europe by Maier, Feldmann et al. among 20 000 DSL-users drew similar conclusions in respect to peer-to-peer usage, and found that the Hypertext Transfer Protocol (HTTP) accounted for 60% of traffic.

These studies can help to inform debates that sometimes seemingly rely on data that do not allow for informed consideration of the issues involved. For example, Cho and others found that distribution of "heavy users" follows a power law distribution: a small proportion of users are responsible for generating a relatively large proportion of traffic. This is sometimes interpreted as "overuse" by a particular group to the detriment of other users. These data are, however, rarely made available alongside the tariffs paid (i.e. whether they include business customers, whether different tariffs are paid by these users, whether the traffic is on-net or off-net and so on). Moreover, the power law is visible even when the largest users are removed or when a random percentile is analysed. This analysis also shows that although there is a core group of heavy users, many users move in and out of the top 20 percentile. For example, if usage per day were recorded instead of per month, more users would feature in the top 20 percentile at least one day per month. Heavy use therefore seems to be less about "abuse" of a network and more about continued use during a certain period (e.g. over a month).

According to research undertaken by T-Mobile on some of its European mobile networks, similar power laws are visible not only with regard to the amount of data used, but also with regard to the amount of sites visited by users individually and collectively. A user on their network typically generates 50% of their traffic in a single cell, 30% of the traffic in another three cells, and the remaining 20% of the traffic in 28 cells. This, naturally, reflects the way most people live their day – at home, at work or leisure, and regularly travelling between two or three main locations.

In T-Mobile's experience only 2% of users exceed caps and only do so at the end of the month. The company has found that there is a 25% chance that a user who exceeds a fair use policy will be on one of the busiest cells during the busiest hour. Conversely, on 75% of the busiest sites no users exceeding the fair use policy will be in that cell site. During the busiest hour there will be, on average, 185 users in the busiest 20% of cell sites. Shutting off a heavy user only contributes 0.54% = (100%/185 users) extra bandwidth to all users. In other words, if a heavy user is discouraged from using the network or throttled, each individual other user on that cell site will receive an 0.5%/184 = 0.0028 percentage point increase.

Internet hosts

An Internet host is a machine or application connected to the Internet and uniquely identified with an IP address. As either servers (e.g. web, e-mail, and FTP servers) or clients, Internet hosts are viewed as basic elements of a TCP/IP network; therefore, surveys on Internet hosts, including that of the Internet Systems Consortium (ISC), can provide information on the size of the Internet and its growth. ISC's survey attempts to discover every visible host on the Internet by querying the domain system for the name assigned to every possible IP address. It should be noted that, for the purposes of the survey, domain names may be assigned at any level (i.e. not necessarily limited to registered domain names). However, this approach to measuring the size of the Internet by counting the number of hosts is gradually losing ground, as the one-to-one relationship between a host and an IP address is blurring due to the depletion of IP addresses.

Given that IPv4 addresses are no longer distributed in some regions and are near exhaustion in others, this factor will have an effect on the measurement of hosts and how such metrics need to be interpreted. The shortage of IPv4 addresses and slow diffusion of IPv6 addresses has resulted in large use of private addresses through network address translation (NAT) technology, which enables multiple hosts on a private network to access the Internet using a single public IP address. Almost all devices in a user's residence, for example, are hooked up behind NATs. At the same time, business users, ISPs, wireless operators and so on, often consider deploying Carrier Grade or large-scale NAT routers in their networks. NAT and other factors such as firewalls make many hosts in private and commercial networks undetectable, therefore host data given in the survey needs to be regarded as an estimate of the minimum size of the Internet.

As of January 2012, the number of Internet hosts worldwide reached 888 million, up 10% annually from 732 million in 2010. The increasing rate of host count in the last two years is quite low compared with the 26% compound annual growth rate (CAGR) throughout 2000 and 2010. Over half of all hosts (479 million) had a generic top-level-domain (gTLD) with over 99% of them under the .net (319 million) or .com (157 million) domains (Table 5.11).

Both .net and .com domains lack registration restrictions, as their use was originally intended for network and commercial operations, respectively. It is noteworthy that the share of .net hosts in gTLD total hosts has increased from 39% in 2000 to 67% in 2012, possibly reflecting the growing role of network operators, while .com hosts dropped from 58% to 33% over the same period.

In January 2012, 32% of all hosts (288 million) were connected under OECD-related country code Top Level Domains (ccTLD). The largest OECD country code domain in terms of hosts was Japan (.jp) followed by Italy (.it.), Germany (.de) and the United States (.us-related, i.e. .edu, .mil, .gov) (Figure 5.14). For historical reasons, most hosts in the United States are under gTLDs. The densest ccTLD in the OECD area was Iceland (.is) with 113 hosts per 100 inhabitants.

It can be noted that the number of hosts in the Slovenian ccTLD (.si) more than doubled. Conversely, the number of hosts in Germany (.de) has declined for three consecutive years. Hosts under the United Kingdom (.uk) domain also decreased from 2010 to 2011 by 12%. Between 2000 and 2012, the ccTLDs with the highest CAGR in the number of hosts were found in Poland (42.6%), the Slovak Republic (39.3%), Chile (37.6%), Portugal (36.1%) and Greece (36.0%).

Web servers

A web server is a computer or a program that transmits web page content to users using the World Wide Web's Hypertext Transfer Protocol (HTTP). As accessing web pages and exchanging e-mail, as well as downloading bulky files through File Transfer Protocol (FTP), are basic functions of the Internet, the number of web servers may serve as an

Figure 5.14. **Hosts by type of domain, 2012**

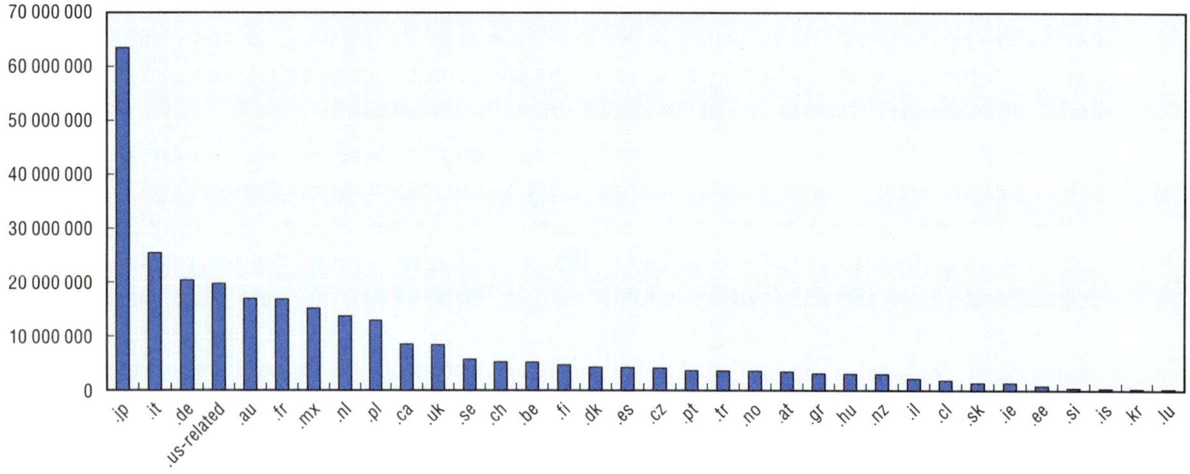

Source: ISC.

StatLink http://dx.doi.org/10.1787/888932799342

indicator of the degree of activity or popularity of the Internet in a specific region or domain.

E-soft (*www.securityspace.com*) conducts a monthly survey of web servers by running a general crawler that visits web pages on a subset of the Internet. It should be noted that this technique excludes about 90% of all websites, in particular those not linked to well-known sites (e.g. domain squatters, personal blogs and so on). In addition, site owners can easily deny web crawlers access to individual pages or to an entire site for privacy or performance reasons. As a result, the survey may underestimate the number of web servers.

In July 2012 there were 62 million websites in the world, representing a 32% increase since 2010. Hungary (46.9%) and Poland (45.1%) had the biggest growth of web servers in the OECD during this period. The lowest growth in number of web servers was in Estonia (19%) and Luxembourg (21.1%). The .edu and .gov domains showed even lower growth of 13.8%, however, these gTLDs are limited to educational institutions and government agencies in the United States. While they are not representative of the country as a whole, it does show that the use of web servers is still growing in the US public sector (Figure 5.15, Table 5.12).

Secure servers

Secure Socket Layer (SSL) is a security protocol used by Internet browsers and web servers to communicate sensitive information such as passwords, credit card numbers, purchases and so on, with users or clients. The SSL protocol consists of two sub-protocols: first, establishment of a secure connection between two sockets (parties) by setting parameters for use and authenticating both parties with the exchange of digital certificates; and second, the use of public key encryption to send private data over the secure connection. A certificate authority, such as those provided by companies like Symantec and GoDaddy, issues a digital certificate containing a public key and information about its owner, and confirms that a given public key belongs to a specific site. When the user enters a site with an SSL certificate, the key is received by the browser and used to

Figure 5.15. **Web servers by country, 2012**

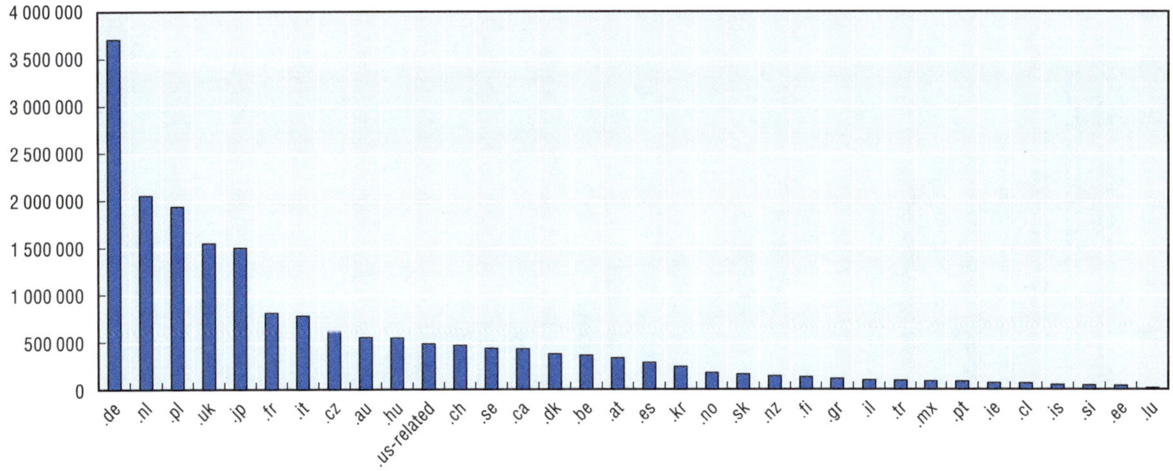

Source: E-soft.

StatLink http://dx.doi.org/10.1787/888932799361

encrypt the submitted information. The data can only be decrypted with the key issued to the certificated site, ensuring it will be readable only by the intended recipient.

Certificates are, of course, only useful if they are issued by an authoritative, reliable and trusted source. In July 2011, a hacker or hackers gained access to the network of DigiNotar, a company based in the Netherlands. The intruder(s) issued 531 fraudulent certificates for several popular private sites including Facebook, Google, Microsoft, Skype and the breach also affected some public sector sites. The false certificates coupled with so-called "Man-In-The-Middle" attacks were reportedly used for the surveillance of e-mail transmission. The security breach on this certificate authority was believed to affect at least 300 000 Gmail users and meant that many Internet users had to change passwords and update web browsers. These types of incident remain a relevant indicator for informing security trends on the Internet.

Netcraft carries out monthly secure server surveys on public secure websites (excluding secure mail servers, intranet and non-public extranet sites). The company has undertaken this survey for more than 15 years. According to the July 2012 survey, 2.2 million secure servers were implemented in the world, having grown by an average of 20% annually from 1.5 million since December 2010. This growth rate was slightly lower than the 24% rate in the 2008 to 2010 period, but was notably higher than the growth rate of total secure servers in the OECD area, at 11% between 2010 and 2012.

The number of secure servers in the OECD area was 1.2 million in July 2012, accounting for 53% of the total number of secure servers in the world. The United States had the largest number of secure servers (513 905), a 44% share of the OECD total. It was followed by Japan (105 868), the United Kingdom (105 541), and Germany (92 390) (Table 5.13). In terms of the number of secure servers per 100 000 inhabitants, Iceland (330), the Netherlands (308), Switzerland (238) and Denmark (240) were the leading countries (Figure 5.16). Since 2005, the countries with the most rapid adoption have been Poland (CAGR 47%), Czech Republic (41%), the Slovak Republic (41%), Korea (40%) and Turkey (38%). These countries started from a lower position and still have a relatively low number of secure web servers.

Figure 5.16. **Number of secure servers per 100 000 inhabitants, July 2012**

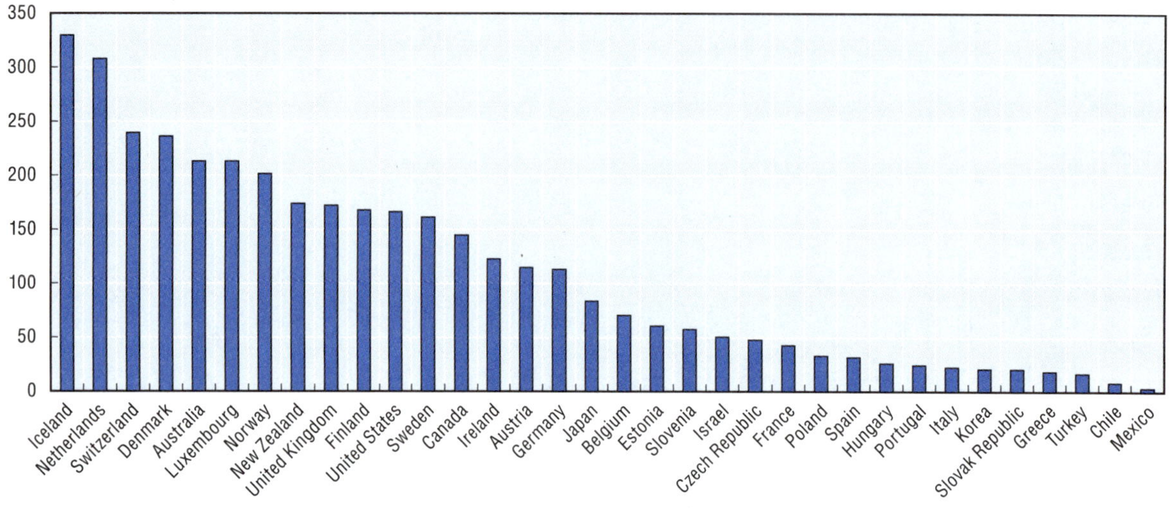

Source: Netcraft.

StatLink http://dx.doi.org/10.1787/888932799380

Notes

1. Since domain names are used in various networking contexts and services, they are not always or necessarily linked to the World Wide Web.
2. IJJ is a commercial ISP and board member of WIDE, a research consortium in Japan.

Table 5.1. **Routed autonomous systems by country, 1997-2012**

	1997	1998	1999	2000	2001	2002	2003	2004	2005	2006	2007	2008	2009	2010	2011	2012	CAGR 1997-2012	AS per 100 000 inhabitants
Australia	51	102	145	200	257	301	328	359	417	461	502	578	658	744	828	899	21.1	4.09
Austria	25	33	40	67	92	105	128	158	195	220	247	280	293	314	343	365	19.6	4.36
Belgium	14	16	20	27	38	45	59	73	87	102	116	124	138	144	152	159	17.6	1.47
Canada	93	127	180	289	344	403	444	508	560	601	641	673	729	778	835	898	16.3	2.66
Chile	4	9	9	25	43	52	57	57	60	60	70	79	88	98	107	113	24.9	0.67
Czech Republic	14	16	18	23	33	44	65	79	93	112	135	167	233	313	382	437	25.8	4.17
Denmark	8	10	18	26	34	41	48	58	75	80	107	126	136	161	172	191	23.6	3.46
Estonia	5	4	5	7	7	9	10	12	16	22	28	32	36	40	48	52	16.9	3.88
Finland	7	12	17	24	37	47	58	64	73	90	97	103	120	136	145	166	23.5	3.11
France	24	40	67	99	151	176	211	246	298	360	398	464	530	595	658	715	25.4	1.14
Germany	54	96	161	284	395	473	565	665	765	840	903	990	1 047	1 139	1 195	1 274	23.5	1.56
Greece	15	26	35	49	57	61	69	74	85	97	112	123	124	122	125	126	15.2	1.12
Hungary	24	24	31	37	51	57	66	75	83	97	114	134	147	162	168	169	13.9	1.69
Iceland	1	1	1	3	4	6	8	12	13	16	18	26	27	32	36	39	27.7	12.22
Ireland	4	5	7	13	12	15	24	35	40	50	69	87	96	98	108	119	25.4	2.67
Israel	6	11	19	36	55	63	73	81	100	110	128	147	167	178	188	195	26.1	2.61
Italy	50	67	89	132	199	230	258	292	313	345	382	427	469	509	546	587	17.8	0.98
Japan	14	15	13	16	24	30	37	40	47	54	54	58	58	62	65	73	11.6	0.06
Korea	30	33	32	43	40	48	62	67	66	73	74	75	72	68	69	67	5.5	0.14
Luxembourg	3	5	8	11	12	16	18	17	19	19	22	27	32	36	46	47	20.1	9.52
Mexico	35	41	45	69	87	91	106	112	129	135	143	152	166	170	175	171	11.2	0.16
Netherlands	21	29	34	60	84	112	152	187	232	270	299	332	363	407	450	490	23.4	2.96
New Zealand	6	17	27	40	52	58	63	76	90	99	116	148	169	193	212	237	27.8	5.49
Norway	11	12	14	24	34	37	46	50	58	63	69	80	96	113	132	149	19.0	3.09
Poland	13	22	35	83	134	168	217	310	401	522	657	771	885	1 048	1 299	1 529	37.4	4.01
Portugal	7	10	16	25	25	26	30	37	43	45	49	56	59	61	59	63	15.8	0.59
Slovak Republic	6	11	11	13	25	28	31	37	46	52	51	56	63	75	85	107	21.2	1.97
Slovenia	3	3	6	9	17	31	43	55	65	78	98	125	150	176	210	236	33.8	11.68
Spain	12	16	31	68	111	129	155	168	187	209	247	278	306	336	368	403	26.4	0.88
Sweden	35	39	47	56	76	99	124	152	183	206	244	284	315	338	367	399	17.6	4.29
Switzerland	22	42	59	97	135	153	168	204	228	266	302	336	378	409	434	450	22.3	5.81
Turkey	12	23	23	46	69	80	97	120	138	158	197	216	243	256	269	298	23.9	0.41
United Kingdom	62	82	133	205	293	365	463	566	651	757	842	963	1 066	1 158	1 225	1 320	22.6	2.17
United States	1 597	2 183	3 115	4 677	6 119	6 969	7 789	8 607	9 319	10 098	10 676	11 463	12 054	12 667	13 424	13 962	15.6	4.55
OECD	2 288	3 182	4 511	6 883	9 146	10 568	12 072	13 653	15 175	16 767	18 207	19 980	21 513	23 136	24 925	26 505	17.7	2.17
Rest of World	924	1 272	1 720	2 560	3 366	4 004	4 567	5 299	6 302	7 500	8 757	10 150	11 634	13 081	14 753	16 546	21.2	..
Selected Countries																		
Argentina	21	40	50	65	81	78	75	91	102	112	121	141	165	183	215	259	18.2	0.61
Brazil	19	41	68	102	129	150	162	182	192	231	273	287	293	290	290	289	19.9	0.14
China	7	10	8	13	27	31	50	51	54	63	64	69	69	71	81	85	18.1	0.01
Colombia	10	15	15	24	25	32	35	34	38	44	44	50	50	59	57	62	12.9	0.14
Egypt	4	4	5	6	9	14	16	18	24	28	31	36	39	44	46	48	18.0	0.06
India	4	8	14	26	50	63	73	75	103	126	155	211	297	368	420	495	37.9	0.04
Indonesia	9	11	11	21	31	63	68	71	74	80	88	91	97	106	115	122	19.0	0.05
Russian Federation	106	155	198	266	368	444	551	760	971	1 224	1 540	2 023	2 502	2 966	3485	3 987	27.4	2.89
South Africa	17	22	25	32	31	36	41	41	48	53	62	69	73	98	117	133	14.7	0.27
Total Selected countries:	197	306	394	555	751	911	1 071	1 323	1 606	1 961	2 378	2 977	3 585	4 185	4826	5 480	24.8	..
World	3 212	4 454	6 231	9 443	12 512	14 572	16 639	18 952	21 477	24 267	26 964	30 130	33 147	36 217	39678	43 051	18.9	..

Notes:

Data collected at the end of the year. UK data points INCLUDE data reported under GB.

Source: www.potaroo.net/reports/oecd, based on report files published by the RIRs.

StatLink http://dx.doi.org/10.1787/888932801717

Table 5.2. **Four views on Top 10 largest networks in the world, August 2012**

Customer Cone			IPV4 Adjacencies		
ASN	**Company name**	**Customer Cone**	**ASN**	**Name**	**Count**
AS3356	Level 3 Communications	0.83	AS174	Cogent Communications	3647
AS3549	Level 3 Communications	0.79	AS3356	Level 3 Communications, Inc.	3376
AS174	Cogent/PSI	0.77	AS7018	AT&T Services, Inc.	2423
AS2828	XO Communications	0.71	AS6939	Hurricane Electric, Inc.	2269
AS3320	Deutsche Telekom AG	0.68	AS3549	Level 3 Communications, Inc. (GBLX)	2235
AS3257	Tinet SpA	0.67	AS701	Verizon Business/UUnet	1700
AS26769	Bandcon	0.66	AS4323	tw telecom holdings, inc.	1622
AS12989	Eweka Internet Services B.V.	0.66	AS9002	ReTN.net	1592
AS1299	TeliaNet Global Network	0.62	AS209	Qwest Communications Company, LLC	1419
AS2914	NTT America, Inc.	0.59	AS39792	Anders Telecom Ltd.	1161
IPv4 Prefixes Announced			**IPv4 Addresses Originated**		
ASN	**Name**	**Count**	**ASN**	**Name**	**Count**
AS3356	Level 3 Communications, Inc.	120645	AS4134	China Telecom Backbone	109.8M
AS3257	Tinet SpA	53093	AS721	DoD Network Information Center	88.2M
AS3549	Level 3 Communications, Inc. (GBLX)	52386	AS7922	Comcast Cable Communications, Inc.	71.0M
AS6453	Tata Communications	41915	AS7018	AT&T Services, Inc.	58.0M
AS1299	TeliaNet Global Network	39325	AS4837	China Unicom Backbone	53.6M
AS174	Cogent Communications	32388	AS4766	Korea Telecom	45.5M
AS2914	NTT America, Inc.	31934	AS17676	Softbank BB Corp.	44.4M
AS701	Verizon Business/UUnet	30406	AS3356	Level 3 Communications, Inc.	44.3M
AS6939	Hurricane Electric, Inc.	30219	AS3320	Deutsche Telekom AG	34.6M
AS7018	AT&T Services, Inc.	28294	AS71	Hewlett-Packard Company	33.8M

Source: Route Views, CAIDA

StatLink http://dx.doi.org/10.1787/888932801736

Table 5.3. **IPv6 Cumulative Total Allocated Addresses by RIR**

	1999	2000	2001	2002	2003	2004	2005	2006	2007	2008	2009	2010	2011	2012
AfriNIC	0	0	0	0	0	1	4	21	35	45	54	157	312	4 515
APNIC	0	1	4	69	117	6 330	15 703	18 929	24 166	24 307	24 480	27 716	37 221	41 028
ARIN	1	8	16	28	70	109	166	218	354	14 840	15 087	15 664	17 942	19 618
LACNIC	0	0	0	0	5	9	61	77	116	65 865	65 895	65 941	66 592	70 917
Ripe NCC	1	2	5	107	290	7 377	24 879	31 360	32 617	33 251	33 864	35 752	38 199	41 917

Note: data collected on 8 February 2013.

Source: www.potaroo.net/reports/oecd, based on report files published by the

StatLink h'tp://dx.doi.org/10.1787/888932801755

Table 5.4. **IPv6 allocations by RIR, yearly basis**

	1999	2000	2001	2002	2003	2004	2005	2006	2007	2008	2009	2010	2011	2012
AfriNIC						1	3	17	14	10	9	103	155	4 203
APNIC	0	1	3	65	48	6 213	9 373	3 226	5 237	141	173	3 236	9 505	3 807
ARIN	1	7	8	12	42	39	57	52	136	14 486	247	577	2 278	1 676
LACNIC					5	4	52	16	39	65 749	30	46	651	4 325
Ripe NCC	1	1	3	102	183	7 087	17 502	6 481	1 257	634	613	1 888	2 447	3 718

Note: data collected on 8 February 2013.
the RIRs.

StatLink http://dx.doi.org/10.1787/888932801774

Table 5.5. **IPv4 and Ipv6 enabled ASNs per country, 2012 (year-end)**

	IPv4-only ASs	IPv4 and IPv6 ASs	IPv6-only ASs	Total
Australia	730	169	4	903
Austria	257	108	1	366
Belgium	111	48	0	159
Canada	725	173	2	900
Chile	94	19	0	113
Czech Republic	288	149	1	438
Denmark	138	53	2	193
Estonia	36	16	0	52
Finland	117	49	1	167
France	545	170	4	719
Germany	836	438	8	1 282
Greece	110	16	0	126
Hungary	146	23	0	169
Iceland	25	14	0	39
Ireland	88	31	2	121
Israel	184	11	0	195
Italy	488	99	2	589
Japan	50	23	1	74
Korea	61	6	0	67
Luxembourg	26	21	0	47
Mexico	152	19	1	172
Netherlands	266	224	5	495
New Zealand	167	70	1	238
Norway	68	81	2	151
Poland	1 363	166	3	1 532
Portugal	51	12	0	63
Slovak Republic	87	20	0	107
Slovenia	181	55	2	238
Spain	328	75	2	405
Sweden	259	140	1	400
Switzerland	329	121	2	452
Turkey	272	26	0	298
United Kingdom	1 042	278	5	1 325
United States	12 429	1 533	21	13 983
OECD	22 049	4 456	73	26 578
Selected Countries:				
Argentina	234	25	0	259
Brazil	219	70	0	289
BRICS	4 594	517	15	5 126
China	73	12	8	93
Colombia	48	14	1	63
Egypt	44	4	0	48
India	431	64	1	496
Indonesia	88	34	0	122
Russian Federation	3 682	305	4	3 991
South Africa	101	32	2	135
World	36 635	6 416	138	43 189

Source: Potaroo.net

StatLink http://dx.doi.org/10.1787/888932801793

Table 5.6. **Domain name registrations under top level domains, 2000-2012**

Domain		2000	2002	2004	2006	2008	2010	2012	Annual growth (%) 2000-12	Share of world domains (%)
Australia	.au	148 539	300 000	447 384	721 952	1 199 365	1 759 295	2 441 240	26.3	1.05
Austria	.at	157 387	252 441	341 841	548 060	759 033	939 951	1 151 136	18.0	0.49
Belgium	.be	32 709	206 989	348 401	1 056 976	802 287	1 044 492	1 302 338	35.9	0.56
Canada	.ca	39 127	299 391	438 559	714 607	1 059 536	1 451 167	1 940 141	38.4	0.83
Chile	.cl	31 191	77 099	105 296	153 985	215 211	285 593	387 510	23.4	0.17
Czech Republic	.cz	66 555	119 145	174 914	259 590	453 932	693 760	956 943	24.9	0.41
Denmark	.dk	208 300	397 552	528 886	708 693	930 904	1 070 525	1 188 868	15.6	0.51
Estonia	.ee	8 000	13 500	22 327	40 135	59 500	81 500	65 903	19.2	0.03
Finland	.fi	13 414	22 528	52 576	81 899	127 205	196 451	291 766	29.3	0.13
France	.fr	89 097	155 554	268 361	560 063	1 167 557	1 761 675	2 366 375	31.4	1.02
Germany	.de	1 732 994	5 666 269	7 799 823	10 013 686	12 148 809	13 723 381	15 090 458	19.8	6.48
Greece	.gr	32 518	65 747	87 324	168 923	252 358	335 130	410 898	23.5	0.18
Hungary	.hu	17 847	60 047	113 528	231 641	371 658	514 118	620 179	34.4	0.27
Iceland	.is	2 754	4 983	7 554	11 448	17 259	27 140	39 142	24.8	0.02
Ireland	.ie	14 863	29 384	39 438	62 710	105 054	145 653	180 111	23.1	0.08
Israel	.il	..	..	..	..	128 861	169 168	228 005	15.3	0.10
Italy	.it	417 609	735 156	909 241	1 236 918	1 566 390	1 932 090	2 412 735	15.7	1.04
Japan	.jp	190 709	482 644	587 412	845 603	1 033 412	1 170 965	1 295 731	17.3	0.56
Korea	.kr	470 544	474 948	612 740	687 714	926 865	1 086 439	1 102 014	7.3	0.47
Luxembourg	.lu	11 404	15 454	17 845	24 376	40 305	53 076	69 468	16.3	0.03
Mexico	.mx	49 947	71 590	91 559	174 490	266 896	430 259	576 718	22.6	0.25
Netherlands	.nl	532 596	748 510	1 005 292	1 991 799	3 027 731	3 976 244	4,973,038	20.5	2.13
New Zealand	.nz	56 765	107 046	149 269	221 433	341 490	402 331	489 832	19.7	0.21
Norway	.no	45 541	150 000	208 546	285 947	395 211	481 117	559 920	23.3	0.24
Poland	.pl	56 708	139 373	262 986	485 891	1 134 298	1 859 365	2 373 469	36.5	1.02
Portugal	.pt	18 739	38 048	57 546	118 452	222 293	322 843	481 634	31.1	0.21
Slovak Republic	.sk	22 081	57 091	64 100	97 811	158 128	220 364	276 050	23.4	0.12
Slovenia	.si	9 293	15 492	22 532	42 498	63 349	90 631	103 090	22.2	0.04
Spain	.es	29 590	43 476	85 309	507 874	1 082 757	1 247 978	1 563 449	39.2	0.67
Sweden	.se	78 449	101 553	255 633	461 721	741 208	968 405	1 223 853	25.7	0.53
Switzerland	.ch	267 425	445 230	609 426	785 406	1 169 074	1 454 660	1 708 958	16.7	0.73
Turkey	.tr	22 428	40 059	62 163	94 076	161 017	223 803	304 684	24.3	0.13
United Kingdom	.uk	1 938 740	3 635 585	3 802 885	5 141 040	6 941 940	8 587 726	10 153 840	14.8	4.36
United States	.us	6 468	269 233	875 016	1 003 212	1 397 964	1 715 000	1 772 222	59.6	0.76
OECD ccTLDs		6 830 514	*15 247 829*	*20 441 730*	*29 396 015*	*40 440 042*	50 220 772	60 101 718	19.9	25.79
China	.cn	103 203	126530	393974	1173330	12 364 615	7 246 686	3,984,188	35.6	1.71
Argentina	.ar	255 536	..	..	1 150 000	1 527 461	1 850 000	2 352 911	20.3	1.01
Brazil	.br	274 674	406 662	639 686	944 051	1 374 644	2 138 509	2 969 925	21.9	1.27
India	.in	2 319	..	7 000	170 000	389 858	510 000	510 000	56.7	0.22
Rest of world ccTLDs		1 806 964	*2 314 679*	*5 951 209*	*8 834 134*	*24 559 958*	26 079 228	34 798 282	27.9	14.93
Total ccTLDs		8 637 478	17 562 508	26 392 939	38 230 149	65 000 000	76 300 000	94 900 000	22.1	40.73
Major gTLDs		17 476 025	*27 113 371*	*38 278 040*	*65 242 646*	*94 202 651*	*106 660 193*	132 924 282	18.4	57.05
	.com	13 721 175	21 198 557	30 267 141	52 752 949	75 779 078	85 583 963	98 608 035	17.9	42.32
	.net	2 305 075	3 586 124	4 910 121	7 728 195	11 521 124	12 839 575	14092248	16.3	6.05
	.org	1 449 775	2 328 690	3 100 778	4 761 502	6 902 449	8 236 655	9642279	17.1	4.14
	.biz	..	700 962	1 028 314	1 423 179	1 973 994	2 028 703	2175761	12.0	0.93
	.info	..	864 457	1 235 485	3 132 195	4 851 813	6 163 763	7175073	23.6	3.08
	.name	..	78 041	99,509	205 326	284 692	243 337	205 025	10.1	0.09
	.mobi	..	..	..	..	924 690	975 568	1 025 861	2.6	0.44
Europe	.eu	..	..	..	2 000 218	2 832 437	3 211 948	3 745 976	11.0	1.61
World total		..	..	*64 500 000*	*105 000 000*	*168 000 000*	*196 300 000*	*233 000 000*	*23.9*	100.00

Note: Registrations at mid year (June), or nearest available count. Values in italics are estimates.

Source: OECD, compiled from country and generic NICs and from ZookNIC, July 2012.

StatLink http://dx.doi.org/10.1787/888932801812

Table 5.7. **Cumulative total of IPv4 address allocations by country, 1997-2012**

	1997	1998	1999	2000	2001	2002	2003	2004	2005	2006	2007	2008	2009	2010	2011	2012
Australia	18 237 184	18 508 800	18 848 000	19 787 520	20 634 368	21 165 312	21 920 256	23 401 728	25 081 088	28 847 104	31 539 456	34 605 312	37 164 800	46 198 784	47 592 704	47 778 048
Austria	3 089 408	3 317 248	3 737 472	4 421 248	5 204 640	5 488 544	5 966 752	7 145 120	7 901 600	8 707 808	9 365 984	9 672 672	10 157 280	11 187 936	11 465 440	11 679 840
Belgium	2 896 640	2 966 272	3 213 312	3 897 856	4 301 568	5 145 152	5 482 816	6 360 128	6 904 640	7 658 432	8 062 400	8 901 312	10 456 768	11 081 920	11 293 632	11 568 320
Canada	36 669 696	37 223 424	38 298 880	40 027 648	41 389 568	42 696 192	43 600 128	45 248 000	47 369 728	49 237 248	53 684 480	56 943 360	59 496 960	62 215 936	63 380 736	79 846 400
Chile	1 662 976	1 662 976	1 728 512	1 767 680	2 042 112	2 181 376	2 387 456	2 614 784	2 997 760	3 500 032	4 068 096	4 525 824	4 876 032	5 677 568	7 188 480	7 984 896
Czech Republic	1 421 056	1 470 208	1 527 552	1 650 432	1 770 240	1 967 360	2 280 704	3 329 280	3 931 136	5 100 672	5 734 016	6 672 768	7 229 056	7 684 736	8 248 960	8 984 192
Denmark	2 007 552	2 437 888	2 650 880	3 206 272	4 037 376	4 559 424	5 628 480	6 072 256	7 665 344	8 257 920	8 748 704	10 057 632	10 846 888	11 793 448	12 240 424	12 519 216
Estonia	131 072	147 456	180 224	184 320	241 664	299 264	372 992	476 672	707 584	950 016	986 880	1 015 832	1 046 040	1 143 608	1 201 976	1 263 416
Finland	6 032 896	6 402 048	6 615 040	6 968 064	7 322 368	7 909 504	8 267 648	10 023 680	11 189 760	12 163 072	12 405 248	12 615 552	12 777 728	13 042 240	13 436 800	13 682 304
France	27 311 360	27 693 056	28 331 264	29 118 432	30 536 288	32 107 520	37 133 184	40 088 192	53 113 024	66 104 544	79 487 072	80 118 272	83 964 800	87 948 864	93 401 424	95 692 048
Germany	43 949 064	45 333 384	47 405 704	51 723 400	57 620 776	58 502 696	59 611 880	67 525 864	74 065 832	84 741 192	93 819 720	101 086 792	104 617 704	111 438 952	116 281 960	119 668 584
Greece	979 200	1 162 240	1 210 624	1 493 760	1 737 216	1 760 768	1 832 192	2 080 512	2 329 344	2 917 632	3 701 760	4 309 248	4 657 664	5 363 456	5 747 712	6 005 504
Hungary	956 672	1 023 744	1 072 896	1 192 704	1 400 320	1 474 304	1 683 712	2 633 472	2 921 984	3 571 712	3 973 632	4 251 392	4 589 568	4 880 896	5 361 152	5 897 472
Iceland	205 056	213 248	213 248	250 112	274 688	336 128	422 144	508 160	637 184	712 960	725 248	742 144	742 656	762 624	818 176	822 784
Ireland	987 792	1 014 432	1 031 840	1 281 696	1 294 496	1 352 352	1 557 664	2 052 256	3 864 480	4 407 968	4 994 976	5 274 272	5 714 144	5 950 976	6 137 616	6 209 040
Israel	1 812 736	1 861 888	2 190 592	2 415 616	2 531 328	2 820 096	3 111 936	3 959 424	4 303 296	4 714 432	5 265 600	5 571 520	6 123 968	6 811 840	7 338 560	7 529 536
Italy	11 619 584	12 103 168	12 904 960	14 223 360	16 105 920	18 065 088	20 552 640	23 047 616	27 609 536	28 388 000	33 465 504	39 311 264	43 466 400	47 245 984	49 807 776	53 111 712
Japan	63 641 088	64 771 840	67 180 288	69 482 240	79 816 960	90 461 696	102 848 256	115 100 672	138 406 656	146 791 424	153 797 888	163 789 824	174 746 624	184 718 080	201 624 320	201 982 464
Korea	7 403 264	8 320 768	11 598 080	20 117 760	24 189 184	28 387 584	32 188 672	35 358 976	44 244 480	52 175 104	59 908 352	67 870 720	78 818 560	104 549 120	112 233 728	112 258 560
Luxembourg	742 912	751 104	768 768	781 312	807 168	830 208	833 280	963 328	1 020 672	1 023 744	1 085 696	1 161 984	1 268 480	1 302 016	1 315 072	1 327 104
Mexico	3 419 136	3 751 680	4 158 976	4 588 800	4 901 376	5 144 832	5 520 384	5 906 432	9 001 216	13 559 040	18 277 632	19 032 320	19 704 320	20 699 648	24 801 280	26 500 096
Netherlands	22 015 744	22 492 416	23 538 688	25 582 112	27 037 984	27 959 840	30 677 056	33 551 456	35 691 872	37 400 672	39 278 816	40 193 248	42 284 768	43 384 544	44 480 768	45 330 432
New Zealand	2 752 256	2 806 016	2 975 232	3 127 296	3 216 384	3 364 352	3 527 424	3 808 512	4 113 664	4 619 776	5 097 216	5 780 736	5 956 864	6 836 480	6 946 560	6 918 912
Norway	7 525 888	7 563 584	7 657 536	7 946 048	8 426 880	8 583 200	8 895 008	10 124 064	11 696 800	12 275 104	12 718 240	13 276 336	13 837 488	14 693 040	15 127 216	15 334 448
Poland	1 338 368	1 646 336	1 930 240	2 605 312	3 451 648	3 867 968	6 404 672	7 032 928	7 908 192	10 653 384	12 688 616	13 937 256	14 869 384	16 941 160	18 917 352	20 361 832
Portugal	806 144	823 552	1 102 848	1 231 104	1 321 216	1 485 824	1 712 640	2 662 144	2 783 232	3 443 712	3 725 824	4 377 888	4 995 616	5 426 464	5 757 728	5 933 088
Slovak Republic	787 456	848 896	866 560	905 472	959 488	1 018 368	1 052 160	1 164 288	1 417 728	1 589 760	1 766 656	1 854 208	2 130 944	2 400 768	2 502 144	2 562 816
Slovenia	481 536	497 920	555 264	576 512	602 368	648 960	701 952	780 672	901 504	1 164 672	1 254 272	1 437 824	1 652 096	2 021 248	2 528 384	2 559 872
Spain	3 048 448	3 286 016	4 254 720	5 683 456	7 821 056	9 426 944	11 137 568	15 251 488	18 181 920	20 630 048	22 445 984	23 742 368	24 824 992	26 297 504	27 911 584	28 660 352
Sweden	9 907 968	10 350 336	11 139 584	12 483 072	13 474 048	14 970 048	15 550 208	19 016 576	19 664 256	22 212 640	24 277 088	24 740 000	26 295 464	28 075 176	28 927 656	29 677 608
Switzerland	12 916 224	13 326 080	13 629 952	14 209 024	14 706 016	15 213 664	15 666 272	17 047 136	18 141 024	18 646 240	19 050 784	19 196 232	20 012 104	21 031 240	21 373 080	21 642 552
Turkey	705 792	992 512	1 345 024	1 900 800	2 019 840	2 345 472	2 464 768	3 772 928	5 953 280	6 087 424	8 691 200	10 365 184	10 980 096	12 221 440	15 081 728	16 224 768
United Kingdom	59 094 392	60 576 632	63 043 448	66 937 400	70 366 776	74 871 232	80 463 424	83 877 056	93 706 752	98 144 928	104 541 088	107 572 032	111 739 200	119 755 008	121 775 664	123 984 656
United States	1049 623 552	1103 166 720	1122 694 144	1147 712 256	1172 998 456	1193 115 576	1214 029 184	1243 407 232	1288 483 712	1333 043 328	1381 446 784	1435 229 568	1473 649 024	1515 942 016	1537 172 608	1562 862 976
OECD	1 406 180 112	1 470 513 888	1 509 600 352	1 569 480 096	1 634 561 784	1 689 526 848	1 755 485 512	1 845 393 032	1 983 910 280	2 103 441 744	2 230 080 912	2 339 232 896	2 435 694 480	2 566 724 720	2 649 420 400	2 714 365 848
Selected Countres																
Argentina	1 111 040	1 266 688	1 422 336	1 623 040	1 988 096	2 012 672	2 201 856	2 611 456	3 129 600	3 937 024	5 174 528	6 897 920	8 864 256	11 305 984	14 459 136	16 203 520
Brazil	10 290 688	10 290 688	10 290 688	10 290 688	10 814 976	10 814 976	11 863 552	12 912 128	17 106 432	19 203 584	23 397 888	29 689 344	33 883 648	40 175 104	46 466 560	54 855 168
China	3 751 936	5 406 720	7 553 024	13 258 752	21 785 600	29 473 792	41 632 768	60 325 888	74 334 208	97 976 320	134 941 440	181 370 624	232 032 256	277 104 384	330 179 328	330 011 136
Colombia	298 240	298 240	298 240	314 880	404 992	466 432	597 504	749 568	1 047 040	1 706 496	2 999 040	4 051 712	5 180 672	5 969 152	8 739 072	11 289 856
Egypt	82 432	82 432	98 816	496 128	692 736	758 272	840 192	1 104 384	1 247 232	1 390 848	2 146 560	2 431 232	4 247 296	5 652 736	8 482 816	9 393 408
India	384 256	490 752	681 216	1 250 560	1 621 248	2 210 560	2 380 544	4 731 904	5 750 016	7 902 976	13 507 072	17 718 784	19 272 448	28 618 752	34 633 984	34 817 792
Indonesia	323 840	349 184	420 864	625 664	728 576	891 392	975 872	1 140 992	1 596 928	1 857 792	2 676 480	6 830 848	8 408 064	10 231 040	17 325 056	17 425 152
Russian Federation	2 441 728	3 195 392	3 776 000	4 587 264	5 199 360	6 095 104	7 208 576	9 309 056	10 947 552	13 795 808	16 902 112	22 871 520	27 871 840	34 546 528	40 257 568	45 500 448
South Africa	8 254 208	8 295 168	8 311 552	8 360 704	8 391 424	8 421 632	8 443 648	8 486 400	8 896 000	10 094 848	14 133 248	14 570 240	16 469 760	17 950 976	20 356 352	24 435 456
Total Selected Countries:	26 938 368	29 675 264	32 852 736	40 807 680	51 627 008	61 144 832	76 144 512	101 371 776	124 055 008	157 865 696	215 878 368	286 432 224	356 230 240	431 554 656	520 899 872	543 931 936
World	1 451 037 712	1 519 729 120	1 564 952 416	1 636 633 824	1 718 636 888	1 787 511 840	1 873 942 856	1 996 828 136	2 170 314 128	2 336 537 496	2 539 321 656	2 741 724 456	2 929 546 168	3 172 135 704	3 372 812 376	3 483 331 008

Note: Data collected at the end of the year.

Source: www.potaroo.net/reports/oecd , based on report files published by the RIRs.

StatLink http://dx.doi.org/10.1787/888932801831

Table 5.8. **Routed IPv4 addresses by country, 1997-2012**

	1997	1998	1999	2000	2001	2002	2003	2004	2005	2006	2007	2008	2009	2010	2011	2012	CAGR 1997-2012
stralia	10 907 165	11 190 559	11 667 231	12 383 587	16 112 836	16 304 820	26 558 160	18 163 503	19 942 288	22 832 697	25 338 588	26 930 642	29 414 932	36 829 984	40 233 888	40 544 168	9.1
stria	3 108 865	3 339 009	3 461 377	3 770 372	4 691 215	5 184 524	6 105 101	6 306 575	7 240 722	7 932 984	8 539 936	8 816 640	9 253 888	10 160 128	10 572 544	10 831 616	8.7
lgium	2 178 816	2 311 683	2 518 274	2 819 077	3 595 784	4 086 792	7 385 103	5 250 318	5 772 563	6 431 510	6 842 880	7 438 592	9 129 728	9 662 720	9 962 752	9 992 960	10.7
ınada	22 818 594	23 863 340	25 278 577	29 512 606	30 685 873	32 025 269	37 583 256	34 748 579	36 684 493	36 674 786	40 321 792	42 751 744	45 304 064	47 545 344	48 116 736	47 409 152	5.0
ile	1 396 993	1 324 800	1 405 441	1 414 660	1 501 703	1 662 478	1 909 529	2 127 643	2 440 740	2 962 232	3 394 560	3 684 096	4 185 344	4 768 256	5 595 776	6 582 272	10.9
ech Republic	1 688 576	1 710 336	1 781 505	1 904 641	2 057 475	2 277 379	4 680 711	2 953 990	3 738 341	4 897 553	5 488 640	6 304 000	6 794 752	7 481 344	8 054 784	8 385 280	11.3
nmark	1 839 810	2 195 907	2 425 155	2 532 801	3 861 698	4 302 981	8 027 588	5 555 397	7 203 524	7 804 421	8 366 080	9 655 552	10 385 280	10 999 936	11 755 904	11 991 296	13.3
tonia	133 633	149 505	165 889	182 273	244 736	293 888	706 561	437 505	698 369	932 097	984 576	1 014 272	1 041 664	1 129 216	1 186 048	1 246 464	16.1
ıland	4 875 846	5 195 091	5 686 597	6 259 805	6 791 497	7 580 762	9 472 919	8 980 246	10 408 206	11 221 642	11 360 768	11 451 264	11 588 736	11 963 392	12 327 296	12 655 232	6.6
ance	25 806 087	26 148 869	26 738 221	27 177 904	29 028 761	30 482 330	55 391 386	33 987 481	40 672 291	50 397 865	61 399 684	64 837 120	66 941 952	67 179 008	73 472 000	76 496 128	7.5
rmany	39 806 347	41 670 571	42 572 851	45 166 521	53 388 576	53 431 082	73 304 961	60 663 626	66 615 386	76 700 379	86 964 480	94 270 976	97 737 728	103 764 736	108 031 488	111 841 024	7.1
eece	925 441	1 039 107	1 179 906	1 177 089	1 494 531	1 683 457	2 708 228	2 176 518	2 363 363	2 544 914	3 256 090	4 029 696	4 155 392	5 016 064	5 308 928	5 618 944	12.8
ngary	1 085 440	1 132 546	1 182 468	1 289 986	1 531 138	1 585 922	3 042 818	2 322 949	2 586 115	3 281 667	3 766 528	4 101 120	4 423 936	4 719 616	5 092 608	5 653 504	11.6
eland	197 120	214 528	280 064	308 992	340 992	402 432	475 904	426 752	559 360	647 424	650 752	681 472	676 096	687 616	728 320	751 360	9.3
land	937 473	896 784	977 664	1 119 489	1 317 888	1 439 744	4 188 672	1 910 275	3 663 107	4 173 317	4 662 784	4 915 968	5 296 384	5 590 528	5 819 136	5 854 720	13.0
ael	1 829 377	1 951 234	2 280 452	2 436 420	2 623 236	2 774 792	4 262 156	3 728 148	3 977 499	4 432 439	5 112 320	5 368 576	5 899 264	6 473 728	7 077 632	7 284 224	9.6
ly	6 566 499	12 114 279	12 723 301	13 204 587	15 756 404	16 979 994	37 052 441	20 482 075	23 488 295	26 472 177	30 308 544	36 428 032	40 133 376	42 744 064	46 816 000	50 872 832	14.6
pan	32 036 622	33 252 379	34 346 532	36 021 059	46 448 183	57 454 897	102 702 905	78 190 280	88 203 364	111 590 510	118 492 992	124 889 152	138 520 128	147 594 048	163 608 128	164 586 176	11.5
rea	7 238 147	7 972 876	9 807 334	16 808 006	23 154 918	26 767 746	57 526 637	33 701 240	41 509 324	50 422 822	57 787 584	64 122 344	74 605 920	97 060 896	103 637 280	105 416 960	19.6
xembourg	645 632	592 896	538 114	555 266	456 960	483 329	591 616	660 224	795 305	741 635	804 608	844 544	984 064	1 012 992	1 021 952	1 026 816	3.1
xico	3 873 544	4 186 374	4 348 938	4 603 412	4 898 845	5 160 998	10 148 904	5 660 465	7 064 332	12 919 889	15 237 888	17 639 168	18 512 896	17 900 800	20 015 616	21 040 128	11.9
therlands	19 035 141	19 662 082	20 242 183	21 356 073	24 237 077	19 050 784	24 730 906	23 389 230	28 244 318	30 282 307	32 197 380	33 074 692	34 850 308	35 969 408	37 032 448	37 665 024	4.7
w Zealand	2 377 995	2 260 483	2 232 068	2 270 474	2 284 058	2 450 453	3 757 070	2 852 876	3 010 583	3 348 007	4 018 944	4 289 792	4 894 976	5 385 984	5 569 280	5 590 016	5.9
rway	5 900 800	6 076 752	6 284 096	6 371 330	6 882 307	7 227 270	9 849 354	8 966 917	10 535 940	11 013 639	11 668 224	12 186 368	12 764 672	13 359 872	14 119 680	14 337 280	6.1
land	1 644 096	1 923 905	2 029 888	2 728 516	3 688 009	3 975 176	10 270 218	6 842 514	8 007 830	10 125 014	12 214 976	13 569 856	14 465 792	16 368 832	18 382 784	19 971 648	18.1
rtugal	617 761	661 793	843 042	1 064 739	1 216 544	1 396 994	2 758 915	1 768 196	2 687 753	3 226 892	3 538 944	3 911 680	4 852 480	5 231 360	5 657 856	5 836 288	16.2
ovak Republic	754 177	817 921	838 147	840 707	908 546	966 658	1 314 304	1 081 088	1 383 939	1 553 412	1 734 144	1 782 528	2 083 328	2 334 976	2 420 992	2 495 232	8.3
ovenia	420 096	428 288	518 400	603 394	629 504	647 424	1 060 354	717 569	880 545	1 129 732	1 208 832	1 369 088	1 552 384	1 818 624	2 349 824	2 463 232	12.5
ain	2 682 688	3 137 856	3 978 823	4 603 221	7 060 690	7 764 633	16 944 923	12 070 185	15 645 479	19 124 011	20 521 024	21 853 760	22 981 696	24 251 968	25 927 232	27 808 512	16.9
eden	5 734 728	6 184 846	6 912 333	7 813 774	10 008 661	11 409 182	15 733 788	15 027 740	15 976 729	18 522 986	20 332 800	21 106 176	21 573 888	23 755 776	25 187 136	26 126 976	10.6
itzerland	6 737 926	6 923 433	7 633 256	8 211 045	9 029 382	9 372 936	10 498 568	10 686 252	11 512 425	11 728 907	12 146 688	12 111 136	12 857 600	14 041 600	14 295 040	14 533 920	5.3
rkey	983 559	1 091 588	1 402 371	1 775 876	2 010 631	2 239 753	7 331 852	2 751 764	3 800 093	5 977 630	8 205 312	8 888 832	10 456 576	11 381 760	13 796 864	15 405 312	20.1
ited Kingdom	18 810 344	36 919 339	38 864 374	40 714 883	29 546 508	33 929 005	58 409 535	39 657 743	49 898 722	54 167 349	59 943 456	62 696 448	64 878 464	68 630 400	75 268 480	78 017 540	9.9
ited States	597 080 541	635 277 994	650 223 160	656 630 516	685 386 105	681 082 776	841 032 286	718 316 840	739 150 271	780 061 530	822 157 008	877 986 240	915 007 712	939 515 904	967 806 600	990 893 704	3.4
ECD	832 675 879	903 818 953	933 368 032	965 633 101	1 032 871 271	1 053 878 660	1 457 517 629	1 172 562 703	1 266 364 214	1 396 276 376	1 508 969 806	1 615 001 566	1 708 205 400	1 802 330 880	1 896 249 032	1 947 225 940	5.8
lected Countries																	
gentina	945 925	1 064 205	1 186 070	1 414 703	1 699 395	1 665 345	3 263 301	2 211 667	2 672 236	3 369 112	4 683 520	5 942 528	7 800 320	9 430 272	11 708 672	14 917 896	20.2
azil	5 892 619	6 503 433	7 300 872	7 450 373	8 748 820	9 463 342	31 312 979	11 464 279	13 343 869	16 583 585	20 709 888	25 019 648	30 234 624	34 792 192	41338368	50 059 008	15.3
ina	9 735 170	11 202 314	12 415 751	13 582 714	28 516 962	34 025 599	123 623 461	52 602 709	64 289 952	85 823 143	108 792 064	140 303 104	195 230 720	227 268 608	254379008	271 572 736	24.8
lombia	267 779	265 986	271 362	280 322	354 309	396 807	601 360	665 609	880 916	1 447 709	2 563 072	3 379 200	4 236 288	5 100 032	6971904	8 979 456	26.4
ypt	92 673	106 753	117 250	412 417	711 680	754 948	1 708 295	1 038 350	1 186 088	1 329 709	1 931 264	2 143 488	3 744 000	4 710 656	7803648	8 708 096	35.4
lia	649 473	734 469	975 878	1 519 129	1 652 554	2 083 939	8 366 164	2 858 648	4 968 609	6 260 227	8 655 104	10 196 736	12 530 688	17 138 432	23063040	28 441 088	28.7
lonesia	349 186	377 889	444 162	565 005	4 859 417	4 994 605	8 415 815	955 726	1 089 119	1 525 861	1 980 672	3 324 928	3 920 128	6 694 656	9844736	13 964 544	27.9
ssian Federation	2 742 033	3 619 848	4 138 248	4 936 717	5 941 005	6 531 346	17 623 910	8 362 760	10 264 443	12 517 573	15 303 552	20 567 680	25 499 648	30 758 656	36652416	43 047 552	20.2
uth Africa	6 308 356	6 495 498	6 607 624	6 292 486	6 534 152	6 409 227	8 808 716	6 337 805	6 644 494	7 684 375	9 307 648	11 670 272	12 991 488	14 899 712	17199104	19 093 504	7.7
tal Other countries:	26 983 214	30 370 395	33 457 217	36 453 866	59 018 294	66 325 158	203 724 001	86 497 553	105 339 726	136 541 294	173 926 784	222 547 584	296 187 904	350 793 216	408960896	458 783 880	20.8
orld	907 155 977	1 003 996 178	1 038 519 279	1 113 786 358	1 168 055 582	1 171 743 081	2 727 020 299	1 346 130 277	1 460 155 362	1 652 317 035	1 798 832 046	1 974 377 762	2 162 142 044	2 323 090 688	2 486 098 884	2 604 921 048	7.3

te: Data collected at the end of the year. UK data points INCLUDE data reported under GB.

urce: *www.potaroo.net/reports/oecd,* based on report files published by the RIRs.

StatLink http://dx.doi.org/10.1787/888932801850

Table 5.9. **Routed autonomous systems by country and type, 2012**

	Transit AS numbers with IPv4 support	Announced AS numbers with IPv4 support	Transit AS numbers with IPv6 support	Announced AS numbers with IPv6 support	Total announced AS numbers	IPv6 only AS	Percentage of IPv6 capable AS (%)
Australia	141	899	37	173	903	4	19.2
Austria	49	365	18	109	366	1	29.8
Belgium	28	159	12	48	159	0	30.2
Canada	147	898	31	175	900	2	19.4
Chile	16	113	6	19	113	0	16.8
Czech Republic	76	437	28	150	438	1	34.2
Denmark	20	191	9	55	193	2	28.5
Estonia	6	52	4	16	52	0	30.8
Finland	21	166	9	50	167	1	29.9
France	86	715	24	174	719	4	24.2
Germany	200	1 274	78	446	1 282	8	34.8
Greece	15	126	3	16	126	0	12.7
Hungary	28	169	6	23	169	0	13.6
Iceland	10	39	2	14	39	0	35.9
Ireland	22	119	5	33	121	2	27.3
Israel	20	195	2	11	195	0	5.6
Italy	77	587	22	101	589	2	17.1
Japan	18	73	10	24	74	1	32.4
Korea	9	67	3	6	67	0	9.0
Luxembourg	12	47	3	21	47	0	44.7
Mexico	32	171	4	20	172	1	11.6
Netherlands	86	490	43	229	495	5	46.3
New Zealand	50	237	20	71	238	1	29.8
Norway	33	149	22	83	151	2	55.0
Poland	185	1 529	29	169	1 532	3	11.0
Portugal	12	63	3	12	63	0	19.0
Slovak Republic	23	107	5	20	107	0	18.7
Slovenia	17	236	9	57	238	2	23.9
Spain	55	403	13	77	405	2	19.0
Sweden	62	399	25	141	400	1	35.3
Switzerland	45	450	19	123	452	2	27.2
Turkey	23	298	5	26	298	0	8.7
United Kingdom	204	1 320	60	283	1 325	5	21.4
United States	1 447	13 962	237	1 554	13 983	21	11.1
OECD	3 275	26 505	806	4 529	26 578	73	17.0
Selected Countries:							
Argentina	31	259	4	25	259	0	9.7
Brazil	84	289	20	70	289	0	24.2
BRICS	930	5 111	119	532	5 126	15	10.4
China	27	85	7	20	93	8	21.5
Colombia	7	62	5	15	63	1	23.8
Egypt	8	48	1	4	48	0	8.3
India	48	495	13	65	496	1	13.1
Indonesia	43	122	12	34	122	0	27.9
Russian Federati	705	3 987	62	309	3 991	4	7.7
South Africa	23	133	5	34	135	2	25.2
World	6 138	43 051	1 253	6 554	43 189	138	15.2

Note: Data collected at the end of the year.

Source: www.potaroo.net/reports/oecd , based on report files published by the RIRs.

StatLink http://dx.doi.org/10.1787/888932801869

Table 5.10. **Average routed IPv4 addresses per AS by country, 1997-2012**

	1997	1998	1999	2000	2001	2002	2003	2004	2005	2006	2007	2008	2009	2010	2011	2012
Australia	213 866	109 711	80 464	61 918	62 696	54 169	80 970	50 595	47 823	49 529	50 475	46 593	44 704	49 503	48 592	45 099
Austria	124 355	101 182	86 534	56 274	50 991	49 376	47 696	39 915	37 132	36 059	34 575	31 488	31 583	32 357	30 824	29 676
Belgium	155 630	144 480	125 914	104 410	94 626	90 818	125 171	71 922	66 351	63 054	58 990	59 989	66 157	67 102	65 544	62 849
Canada	245 361	187 900	140 437	102 120	89 203	79 467	84 647	68 403	65 508	61 023	62 905	63 524	62 145	61 112	57 625	52 794
Chile	349 248	147 200	156 160	56 586	34 923	31 971	33 501	37 327	40 679	49 371	48 494	46 634	47 561	48 656	52 297	58 250
Czech Republic	120 613	106 896	98 973	82 810	62 348	51 759	72 011	37 392	40 200	43 728	40 657	37 749	29 162	23 902	21 086	19 188
Denmark	229 976	219 591	134 731	97 415	113 579	104 951	167 241	95 783	96 047	97 555	78 188	76 631	76 362	68 323	68 348	62 782
Estonia	26 727	37 376	33 178	26 039	34 962	32 654	70 656	36 459	43 648	42 368	35 163	31 696	28 935	28 230	24 709	23 970
Finland	696 549	432 924	334 506	260 825	183 554	161 293	163 326	140 316	142 578	124 685	117 121	111 177	96 573	87 966	85 016	76 236
France	1 075 254	653 722	399 078	274 524	192 243	173 195	262 518	138 160	136 484	139 994	154 271	139 735	126 306	112 906	111 660	106 988
Germany	737 155	434 068	264 428	159 037	135 161	112 962	129 743	91 223	87 080	91 310	96 306	95 223	93 350	91 102	90 403	87 787
Greece	61 696	39 966	33 712	24 022	26 220	27 598	39 250	29 412	27 808	26 236	29 072	32 762	33 511	41 115	42 471	44 595
Hungary	45 227	47 189	38 144	34 864	30 022	27 823	46 103	30 973	31 158	33 832	33 040	30 605	30 095	29 133	30 313	33 453
Iceland	197 120	214 528	280 064	102 997	85 248	67 072	59 488	35 563	43 028	40 464	36 153	26 210	25 041	21 488	20 231	19 266
Ireland	234 368	179 357	139 666	86 115	109 824	95 983	174 528	54 579	91 578	83 466	67 577	56 505	55 171	57 046	53 881	49 199
Israel	304 896	177 385	120 024	67 678	47 695	44 044	58 386	46 027	39 775	40 295	39 940	36 521	35 325	36 369	37 647	37 355
Italy	131 330	180 810	142 958	100 035	79 178	73 826	143 614	70 144	75 042	76 731	79 342	85 312	85 572	83 977	85 744	86 666
Japan	2 288 330	2 216 825	2 642 041	2 251 316	1 935 341	1 915 163	2 775 754	1 954 757	1 876 678	2 066 491	2 194 315	2 153 261	2 388 278	2 380 549	2 517 048	2 254 605
Korea	241 272	241 602	306 479	390 884	578 873	557 661	927 849	503 004	628 934	690 724	780 913	854 965	1 036 193	1 427 366	1 501 990	1 573 387
Luxembourg	215 211	118 579	67 264	50 479	38 080	30 208	32 868	38 837	41 890	39 033	36 573	31 279	30 752	28 139	22 216	21 847
Mexico	110 673	102 107	96 643	66 716	56 309	56 714	95 744	50 540	54 765	95 703	106 559	116 047	111 523	105 299	114 375	123 042
Netherlands	906 435	678 003	595 358	355 935	288 537	170 096	162 703	125 076	121 741	112 157	107 684	99 623	96 006	88 377	82 294	76 867
New Zealand	396 333	132 970	82 669	56 762	43 924	42 249	59 636	37 538	33 451	33 818	34 646	28 985	28 964	27 907	26 270	23 587
Norway	536 436	506 396	448 864	265 472	202 421	195 332	214 116	179 338	181 654	174 820	169 105	152 330	132 965	118 229	106 967	96 223
Poland	126 469	87 450	57 997	32 874	27 522	23 662	47 328	22 073	19 970	19 397	18 592	17 600	16 346	15 619	14 151	13 062
Portugal	88 252	66 179	52 690	42 590	48 662	53 731	91 964	47 789	62 506	71 709	72 223	69 851	82 245	85 760	95 896	92 639
Slovak Republic	125 696	74 356	76 195	64 670	36 342	34 524	42 397	29 219	30 086	29 873	34 003	31 831	33 069	31 133	28 482	23 320
Slovenia	140 032	142 763	86 400	67 044	37 030	20 885	24 659	13 047	13 548	14 484	12 335	10 953	10 349	10 333	11 190	10 437
Spain	223 557	196 116	128 349	67 694	63 610	60 191	109 322	71 846	83 666	91 502	83 081	78 611	75 104	72 178	70 454	69 004
Sweden	163 849	158 586	147 071	139 532	131 693	115 244	126 885	98 867	87 305	89 917	83 331	74 318	68 489	70 283	68 630	65 481
Switzerland	306 269	164 844	129 377	84 650	66 884	61 261	62 491	52 384	50 493	44 094	40 221	36 045	34 015	34 332	32 938	32 298
Turkey	81 963	47 460	60 973	38 606	29 140	27 997	75 586	22 931	27 537	37 833	41 651	41 152	43 031	44 460	51 289	51 696
United Kingdom	303 393	450 236	292 213	198 609	100 841	92 956	126 155	70 067	76 649	71 555	71 192	65 105	60 862	59 266	61 444	59 104
United States	373 876	291 011	208 739	140 396	112 009	97 730	107 977	83 457	79 316	77 249	77 010	76 593	75 909	74 170	72 095	70 971
OECD	363 932	284 041	206 909	140 292	112 931	99 724	120 735	85 883	83 451	83 275	82 879	80 831	79 403	77 902	76 078	73 466
Selected Countries																
Argentina	45 044	26 605	23 721	21 765	20 980	21 351	43 511	24 304	26 198	30 081	38 707	42 146	47 275	51 532	54 459	57 598
Brazil	310 138	158 620	107 366	73 043	67 820	63 089	193 290	62 991	69 499	71 790	75 860	87 176	103 190	119 973	142 546	173 215
China	1 390 739	1 120 231	1 551 969	1 044 824	1 056 184	1 097 600	2 472 469	1 031 426	1 190 555	1 362 272	1 699 876	2 033 378	2 829 431	3 200 966	3 140 482	3 194 973
Colombia	26 778	17 732	18 091	11 680	14 172	12 400	17 182	19 577	23 182	32 902	58 252	67 584	84 726	86 441	122 314	144 830
Egypt	23 168	26 688	23 450	68 736	79 076	53 925	106 768	57 686	49 420	47 490	62 299	59 541	96 000	107 060	169 645	181 419
India	162 368	91 809	69 706	58 428	33 051	33 078	114 605	38 115	48 239	49 684	55 839	48 326	42 191	46 572	54 912	57 457
Indonesia	38 798	34 354	40 378	26 905	156 755	79 279	123 762	13 461	14 718	19 073	22 508	36 538	40 414	63 157	85 606	114 463
Russian Federation	25 868	23 354	20 900	18 559	16 144	14 710	31 985	11 004	10 571	10 227	9 937	10 167	10 192	10 370	10 517	10 797
South Africa	371 080	295 250	264 305	196 640	210 779	178 034	214 847	154 581	138 427	144 988	150 123	169 134	177 966	152 038	147 001	143 560
Selected Countries	136 971	99 250	84 917	65 683	78 586	72 805	190 218	65 380	65 591	69 628	73 140	74 756	82 619	83 822	84 741	83 720
World	282 427	225 414	166 670	117 948	93 355	80 411	163 893	71 028	67 987	68 089	66 712	65 529	65 229	64 144	62 657	60 508

Note: Data collected at the end of the year.

Source: www.potaroo.net/reports/oecd , based on report files published by the RIRs.

StatLink http://dx.doi.org/10.1787/888932801888

Table 5.11. **Internet hosts by domain, 1998-2012**

	Domain	1998	1999	2000	2001	2002	2003	2004	2005	2006	2007	2008	2009	2010	2011	2012	CAGR[2] 2000-2012 (%)
		Number of hosts, January of each year															
Australia	.au	665 403	792 351	1 090 468	1 615 939	2 288 584	2 564 339	2 847 763	4 820 646	6 039 486	8 529 020	10 707 139	11 337 838	12 666 849	15 796 154	16 952 039	25.7
Austria	.at	109 154	143 153	274 173	504 144	657 173	838 026	982 246	1 594 059	1 957 154	2 330 325	2 589 316	2 903 737	3 112 683	3 376 617	3 444 552	23.5
Belgium	.be	87 938	165 873	320 840	417 130	668 508	1 052 706	1 454 350	2 012 283	2 546 148	3 150 856	3 618 495	4 367 700	4 586 719	4 774 473	5 179 959	26.1
Canada	.ca	839 141	1 119 172	1 669 664	2 364 014	2 890 273	2 993 982	3 210 081	3 839 173	2 817 010	4 257 825	4 717 308	6 511 157	7 684 092	7 705 869	8 514 503	14.5
Chile	.cl	17 821	29 006	40 190	74 708	122 727	135 155	202 429	294 575	462 420	621 565	816 460	865 205	945 109	1 463 284	1 853 927	37.6
Czech Republic	.cz	52 498	73 770	112 748	153 902	213 803	239 885	315 974	724 631	993 778	1 502 537	2 093 497	3 126 690	3 302 696	3 831 296	4 140 126	35.0
Denmark	.dk	159 358	279 790	336 928	435 556	707 141	1 154 053	1 467 415	1 908 737	2 316 370	2 807 348	3 256 134	3 795 480	4 044 843	4 168 431	4 284 533	23.6
Estonia	.ee	14 299	21 991	29 682	40 094	68 729	109 643	113 154	237 461	355 015	449 036	564 608	656 053	713 772	811 017	848 009	32.2
Finland	.fi	450 044	546 244	631 248	771 725	944 670	1 140 838	1 224 155	1 915 506	2 505 805	3 187 643	3 728 551	4 041 241	4 358 465	4 480 613	4 699 507	18.2
France	.fr	333 306	488 043	779 879	1 229 763	1 670 694	2 157 628	2 770 836	4 999 770	6 863 156	10 335 974	14 356 747	13 650 159	14 828 048	15 549 341	16 839 991	29.2
Germany	.de	994 926	1 316 893	1 702 486	2 163 326	2 681 325	2 891 407	3 421 455	6 127 262	9 852 798	13 093 255	20 659 105	23 304 039	22 398 022	20 839 544	20 415 943	23.0
Greece	.gr	26 917	51 541	77 954	148 552	182 812	202 525	245 650	377 221	503 685	797 884	1 326 917	2 263 303	2 367 179	2 944 187	3 114 925	36.0
Hungary	.hu	46 082	83 530	113 695	158 732	210 804	254 462	313 576	611 887	894 800	1 176 592	1 689 456	2 168 182	2 523 024	2 790 778	3 063 519	31.6
Iceland	.is	17 450	21 894	29 598	44 040	61 682	68 282	106 296	144 636	191 528	209 071	229 916	268 100	271 042	338 872	360 910	23.2
Ireland	.ie	38 406	54 872	59 681	88 406	95 381	97 544	111 467	138 833	240 958	1 208 345	1 247 734	1 253 720	1 305 913	1 342 929	1 379 364	29.9
Israel	.il	64 233	102 090	139 946	180 263	223 012	230 167	634 001	1 004 141	1 212 264	1 311 769	1 397 740	1 522 217	1 612 648	2 185 740	2 179 398	25.7
Italy	.it	243 250	338 822	658 307	1 630 526	2 282 457	3 864 315	5 469 578	9 343 663	11 222 960	13 853 673	16 730 591	19 487 125	22 493 165	24 483 783	25 455 561	35.6
Japan	.jp	1 168 956	1 687 534	2 636 541	4 640 863	7 118 333	9 260 117	12 962 065	19 543 040	24 903 795	30 841 523	36 803 719	43 461 277	52 081 808	59 968 541	63 465 680	30.4
Korea[1]	.kr	121 932	186 414	283 459	397 809	439 859	407 318	253 242	213 045	245 566	304 113	342 178	331 231	297 268	293 644	293 862	0.3
Luxembourg	.lu	4 273	21 894	9 670	11 744	16 735	17 260	28 214	61 785	84 257	89 938	158 681	209 367	240 728	259 141	253 959	31.3
Mexico	.mx	41 659	112 620	404 873	663 553	918 288	1 107 795	1 333 406	1 868 583	2 555 047	6 697 570	10 071 370	12 515 249	12 471 162	13 817 502	15 165 150	35.2
Netherlands	.nl	381 172	564 129	820 944	1 309 911	1 983 102	2 415 286	3 419 182	6 443 558	7 258 159	9 014 103	10 540 083	11 682 001	12 512 224	13 596 075	13 715 032	26.4
New Zealand	.nz	169 264	137 247	271 003	345 107	408 290	432 957	474 395	651 065	971 900	1 355 534	1 687 494	1 772 571	2 458 678	2 750 179	3 027 638	22.3
Norway	.no	286 338	318 631	401 889	525 030	629 669	589 621	1 013 273	1 237 270	2 109 283	2 370 078	2 725 031	3 113 496	3 181 501	3 542 038	3 583 945	20.0
Poland	.pl	77 594	108 588	183 057	371 943	654 198	843 475	1 296 766	2 482 546	3 941 769	5 001 786	7 134 976	8 350 365	9 737 427	11 299 130	12 928 122	42.6
Portugal	.pt	39 533	49 731	90 757	177 828	263 821	291 355	299 923	605 648	1 378 817	1 510 958	1 643 768	1 919 035	2 208 550	3 384 374	3 664 339	36.1
Slovak Republic	.sk	11 836	17 953	25 906	36 680	68 972	80 660	98 788	188 352	322 753	486 020	695 520	759 106	1 003 248	1 323 163	1 387 022	39.3
Slovenia	.si	15 432	17 984	20 535	23 594	26 475	30 002	34 734	48 133	61 408	64 284	69 356	80 502	102 739	179 285	417 984	28.5
Spain	.es	168 913	264 245	415 641	663 553	1 497 450	1 694 601	1 127 366	1 304 558	2 459 614	2 929 627	3 085 513	3 325 990	3 706 492	3 996 519	4 231 738	21.3
Sweden	.se	319 065	431 809	594 627	764 011	1 141 093	1 209 266	1 539 917	2 668 816	2 817 010	3 039 770	3 513 170	3 868 362	3 999 207	4 741 830	5 727 196	20.8
Switzerland	.ch	114 816	224 350	306 073	461 456	613 918	723 243	1 018 445	1 785 427	2 125 269	2 570 891	3 308 684	3 464 568	3 974 072	5 062 397	5 249 367	26.7
Turkey	.tr	24 786	32 496	90 929	113 603	139 805	199 823	344 859	611 557	794 795	1 581 866	2 425 789	2 596 496	3 104 180	4 003 186	3 624 633	36.0
United Kingdom	.uk	987 733	1 423 804	1 901 812	2 291 369	2 462 915	2 583 753	3 715 752	4 449 190	5 778 422	6 650 334	7 727 550	8 980 515	9 430 637	8 340 756	8 408 891	13.2
United States		*6 618 382*	*8 746 846*	*10 490 416*	*12 052 491*	*12 579 595*	*11 683 370*	*11 422 195*	*13 872 605*	*14 831 525*	*14 896 066*	*15 758 584*	17 554 317	19 536 508	19 916 789	19 726 407	5.4
	.us	1 076 583	1 562 391	1 875 663	2 267 089	2 125 624	1 735 734	1 757 664	2 429 244	2 441 426	2 026 166	1 971 396	2 091 917	2 196 419	2 133 856	2 138 237	1.1
	.edu	3 944 967	5 022 815	6 085 137	7 106 062	7 754 038	7 459 219	7 576 992	8 992 398	9 806 021	10 177 586	10 659 326	11 812 901	12 316 029	12 388 204	12 587 684	6.2
	.mil	1 099 186	1 510 440	1 751 866	1 844 369	1 906 902	1 880 903	1 410 944	1 667 794	1 861 535	1 991 136	2 193 578	2 510 868	2 628 382	2 908 003	2 613 099	3.4
	.gov	497 646	651 200	777 750	834 971	793 031	607 514	676 595	783 169	722 543	701 178	934 284	1 138 631	2 395 678	2 486 726	2 387 387	9.8
OECD total		14 711 910	19 975 308	27 015 619	36 871 365	46 932 293	53 564 859	65 272 948	98 129 662	123 614 724	158 227 179	197 421 180	225 506 394	249 260 698	273 357 477	287 597 731	21.8
Europe	.eu	..	..	..	..	..	..	..	..	..	8696	25259	82072	131471	156978	173527	82.0*
gTLDs		*14 005 613*	*21 742 617*	*42 685 540*	*68 514 456*	*93 617 371*	*103 654 125*	*150 831 956*	*197 045 451*	*242 569 353*	*241 428 097*	*287 188 078*	*329 980 596*	*398 748 549*	*441 130 632*	*478 890 567*	22.3
	.com	8 201 511	12 140 747	24 863 331	36 352 243	44 520 209	40 555 072	48 688 919	56 428 268	69 578 775	76 984 153	95 448 209	123 324 475	142 526 322	151 953 306	156 860 679	16.6
	.net	5 283 568	8 856 687	16 853 655	30 885 116	47 761 383	61 945 611	100 751 276	139 057 448	171 346 396	162 929 985	190 267 719	204 683 342	253 853 098	286 276 469	319 311 234	27.8
	.org	519 862	744 285	959 827	1 267 662	1 321 104	1 116 311	1 332 978	1 459 335	1 516 898	1 396 498	1 333 870	1 711 252	1 994 366	2 117 249	2 078 160	6.6
	.int	672	898	8 727	9 435	11 048	11 594	13 625	13 120	15 756	16 808	16 484	18 944	18 487	19 739	19 435	6.9
	.biz	..	..	..	..	1 477	16 680	28 586	53 672	45 934	39 592	36 612	64 073	66 683	451 383	114 801	..
	.info	..	..	..	..	2 128	8 349	15 502	30 828	60 533	54 351	75 764	146 340	253 710	273 063	473 096	..
	.name	..	..	..	..	7	217	318	913	1 267	1 210	1 471	2 949	3 331	3 624	4 104	..
	.pro	..	..	..	..	2	2	5	15	36	46	61	164	418	1 103	1 719	..
	.aero	..	..	..	..	..	132	315	627	768	690	1 431	1 729	1 991	2 277	2 735	..
	.coop	..	..	..	..	9	148	417	1 191	2 953	4 705	6 354	23 262	25 272	27 501	19 671	..
	.museum	..	..	..	..	4	9	15	19	22	20	23	27	37	37	38	..
	.travel	..	..	..	..	..	..	..	15	**15**	39	80	4 039	4 834	4 881	4 895	..
World total	World	29 669 611	43 229 694	72 398 092	109 574 429	147 344 723	171 638 297	233 101 481	317 646 084	394 991 609	433 193 199	541 677 360	625 226 456	732 740 444	818 374 269	888 239 420	23.2

Notes: 1. Korea's actual number of hosts may be underestimated as the ISC survey methodology relies on ARPA zone information which is not reported by Korean network operators. 2. The Compound Average Growth Rate for .eu comprises of 5 years (2007-2012).

Source: Internet Software Consortium (*www.isc.org*). Data compiled July 2012.

StatLink http://dx.doi.org/10.1787/888932801907

Table 5.12. **Web servers by domain, 2000-12**

	Domain	Number of web servers, July of each year							Annual growth
		2000	2002	2004	2006	2008	2010	2012	
Australia	.au	26 119	66 605	121 004	163 737	268 387	380 857	549 740	28.9
Austria	.at	22 078	43 816	75 113	119 022	184 311	250 420	330 809	25.3
Belgium	.be	7 386	19 147	51 684	180 654	205 713	279 672	359 932	38.2
Canada	.ca	22 105	53 335	106 883	152 681	238 565	317 182	428 498	28.0
Chile	.cl	2 022	5 243	10 956	17 231	31 759	43 731	64 481	33.4
Czech Republic	.cz	12 626	35 600	69 120	116 240	261 879	414 375	610 243	38.2
Denmark	.dk	25 280	135 984	147 681	204 654	247 777	300 857	374 857	25.2
Estonia	.ee	4 803	11 777	15 645	20 531	25 785	32 885	38 758	19.0
Finland	.fi	9 836	16 708	25 284	37 762	59 465	88 202	130 315	24.0
France	.fr	20 471	47 200	55 981	155 163	411 471	613 391	809 058	35.9
Germany	.de	179 542	493 016	1 063 877	1 593 296	2 311 389	2 829 820	3 706 688	28.7
Greece	.gr	3 337	9 779	18 488	28 993	56 822	83 284	113 786	34.2
Hungary	.hu	5 392	15 919	41 556	118 214	263 090	379 531	545 343	46.9
Iceland	.is	1 199	2 914	7 243	9 731	21 385	36 525	48 548	36.1
Ireland	.ie	2 905	7 291	11 545	17 592	30 110	45 448	66 976	29.9
Israel	.il	8 387	10 277	14 605	20 681	44 648	68 021	96 160	22.5
Italy	.it	33 168	89 517	191 690	297 304	484 154	629 917	777 960	30.1
Japan	.jp	45 581	145 929	297 446	399 275	808 599	1 184 736	1 499 209	33.8
Korea	.kr	11 576	39 791	433 837	140 699	158 754	224 297	240 641	28.8
Luxembourg	.lu	1 409	2 467	3 747	5 321	8 559	10 767	14 073	21.1
Mexico	.mx	4 552	9 605	14 860	21 065	33 330	50 293	84 561	27.6
Netherlands	.nl	48 014	167 993	305 358	601 492	1 126 853	1 585 323	2 050 526	36.7
New Zealand	.nz	8 757	23 834	40 055	58 330	83 377	108 188	141 864	26.1
Norway	.no	10 531	26 646	48 471	69 061	104 585	137 574	175 729	26.4
Poland	.pl	22 265	133 501	373 468	524 888	741 599	1 340 977	1 934 579	45.1
Portugal	.pt	5 113	8 645	14 637	25 588	43 724	47 611	83 059	26.2
Slovak Republic	.sk	4 479	15 930	22 711	62 126	61 167	101 091	160 335	34.7
Slovenia	.si	2 632	9 411	10 665	10 140	18 124	26 457	41 858	25.9
Spain	.es	9 146	13 526	19 342	36 269	96 600	171 443	283 391	33.1
Sweden	.se	23 265	33 870	50 773	82 574	158 249	287 732	434 704	27.6
Switzerland	.ch	36 082	77 166	190 134	182 553	273 771	366 676	465 960	23.8
Turkey	.tr	4 897	9 546	14 227	19 918	37 650	54 537	91 990	27.7
United Kingdom	.uk	131 415	277 031	437 404	634 677	955 977	1 193 585	1 548 473	22.8
United States	.us-related	70 219	118 551	219 519	263 812	335 819	410 441	483 818	17.4
	.us	17 299	29 876	98 633	115 445	155 239	200 409	233 482	24.2
	.edu	46 272	78 213	106 244	129 458	156 845	182 515	218 918	13.8
	.gov	6 648	10 462	14 642	18 909	23 735	27 517	31 418	13.8
Total ccTLDs world						13 392 745	18 833 904		
Total gTLDs world						19 849 192	27 027 292		
	.com	992 618	4 689 003	7 239 594	8 884 634	14 782 393	20 587 353	27 597 775	31.9
	.net	106 613	534 214	1 078 762	1 293 624	2 138 109	2 888 408	3 646 001	34.2
	.org	124 150	451 254	791 389	1 081 603	1 628 373	2 124 981	2 831 192	29.8
World total	World	2 213 960	8 420 350	14 978 181	19 863 342	33 241 937	45 861 196	62 034 155	32.0

Source: Security Space *(www.securityspace.com)*

StatLink http://dx.doi.org/10.1787/888932801926

Table 5.13. **Secure servers in OECD countries, 1998-2012**

	1998	1999	2000	2001	2002	2003	2004	2005	2006	2007	2008	2009	2010	2011	2012
Australia	632	1 305	2 828	3 704	4 795	6 533	8 079	10 513	12 343	16 971	21 229	27 096	39 317	45 301	47 577
Austria	98	241	447	881	1 019	1 277	1 590	1 976	2 416	3 321	4 010	4 707	7 132	8 362	9 632
Belgium	52	159	268	431	458	658	912	1 284	1 549	2 041	2 678	3 523	5 329	6 600	7 655
Canada	929	1 789	3 896	6 050	8 497	12 623	15 166	18 946	21 383	26 789	30 200	33 826	42 258	47 194	49 414
Chile	..	..	131	..	197	253	..	353	378	482	589	694	904	1 165	1 408
Czech Republic	19	88	194	383	194	262	315	443	681	1 081	1 569	2 067	3 349	4 083	4 997
Denmark	44	112	289	523	794	1 284	1 681	2 679	3 441	4 397	5 698	6 598	10 336	12 155	13 099
Estonia	..	..	75	..	81	103	..	139	225	288	375	418	532	714	809
Finland	68	180	343	660	804	1 055	1 255	1 671	2 054	2 871	3 635	4 429	6 630	8 009	8 999
France	222	632	1 297	1 969	2 566	3 245	3 799	4 973	6 049	8 676	10 653	13 578	19 275	23 171	26 769
Germany	492	1 630	3 761	6 442	8 096	9 438	13 163	23 566	29 376	37 803	45 143	53 658	71 312	83 639	92 390
Greece	8	48	87	176	160	247	270	357	461	589	688	950	1 406	1 745	2 094
Hungary	18	26	90	165	101	162	199	312	370	593	838	1 190	1 662	2 193	2 601
Iceland	13	29	67	91	151	196	249	313	397	442	495	560	803	965	1 048
Ireland	56	97	245	467	639	959	1 201	1 542	1 790	2 408	3 001	3 372	4 472	5 180	5 492
Israel	..	..	306	..	430	660	..	1 142	1 308	1 637	1 993	2 222	3 024	3 651	3 837
Italy	167	432	795	1 264	1 196	1 669	1 977	2 696	3 236	4 330	5 552	6 733	9 340	11 599	13 480
Japan	429	1 170	2 900	7 952	9 196	15 044	19 610	34 379	43 960	51 013	60 235	66 917	82 823	95 087	105 868
Korea	38	106	243	397	626	793	878	991	1 180	4 704	5 005	6 008	8 195	9 076	10 562
Luxembourg	11	26	44	68	91	133	184	226	274	354	445	559	716	969	1 069
Mexico	26	58	176	310	358	508	605	899	1 054	1 404	1 667	1 890	2 337	3 079	3 636
Netherlands	127	306	541	1 064	1 466	2 553	3 595	5 519	6 945	12 786	18 173	24 398	37 828	45 911	51 171
New Zealand	90	227	482	778	1 054	1 427	1 668	2 111	2 524	3 503	4 184	4 635	6 506	7 037	7 590
Norway	55	130	273	491	599	865	1 122	1 474	1 864	2 929	4 030	5 062	8 073	8 966	9 848
Poland	23	61	188	467	358	447	557	865	1 518	2 104	3 229	4 999	8 049	10 306	12 590
Portugal	27	59	116	192	234	355	443	623	686	918	1 224	1 463	1 847	2 380	2 607
Slovak Republic	15		45	110	40	52	61	100	160	228	313	469	693	891	1 120
Slovenia	..	..	96	..	122	104	..	170	197	281	347	435	622	889	1 177
Spain	239	432	759	1 194	1 491	2 280	2 745	3 697	4 570	6 405	7 753	9 082	10 756	13 164	14 541
Sweden	145	406	811	1 261	1 302	1 860	2 826	3 134	3 831	5 495	7 119	8 103	11 895	13 720	15 119
Switzerland	152	401	854	1 370	1 668	2 179	2 826	3 622	4 486	6 137	7 456	8 882	14 614	16 901	18 741
Turkey	7	50	116	285	420	606	855	1 287	1 860	2 818	4 230	5 154	7 202	10 494	12 300
United Kingdom	714	1 735	4 404	7 916	10 853	16 061	20 339	29 100	34 548	46 011	55 564	66 051	86 878	99 758	105 541
United States	14 674	32 053	65 565	86 025	112 359	165 479	197 769	239 137	262 610	319 836	357 246	386 499	446 992	487 069	513 905
OECD	19 590	43 988	92 732	133 086	172 415	251 370	305 939	400 239	459 724	581 645	676 566	766 227	963 307	1 093 434	1 180 698
Unknown	..	..	1	..	..	98	..	37 335	78 876	175 484	291 302	441 434	561 776	775 866	954 033
World	20 300		119 020	140 841	182 678	261 094	324 816	452 630	556 022	779 920	996 172	1 242 270	1 522 128	1 931 460	2 209 148

Note: Data collected at the end of the year except for 2012 for which data were collected as of July.

Source: Netcraft *(www.netcraft.com)*

StatLink http://dx.doi.org/10.1787/888932801945

OECD Communications Outlook 2013

Chapter 6

Broadcasting and audiovisual content

Audio and audio-visual content is now delivered over an ever-increasing range of networks and services. This chapter traces recent developments in audio and audio-visual platforms and devices. It finds that the Internet, together with analogue audio broadcasting, has become the primary distribution method for audio content. The conversion to digital television is now all but completed in the OECD. In many countries broadcasters offer their content either live or via catch-up television over the Internet. Subscription video-on-demand services are seeing rapid adoption in the countries where they have been introduced.

Audio and video is now available on a wide number of digital platforms, available through many devices. Major trends include:

- Strong market position of analogue radio broadcasting, with very little adoption of digital radio broadcasting. Consumers are moving to subscription services for digital audio. Not all services are available across the OECD.
- The conversion to digital television is now all but concluded in the OECD. Analogue television remains on some cable and fibre networks.
- Internet has grown as a broadcast platform. Subscription video services have seen rapid adoption in the countries where they are available. Almost all countries have a catch-up television service via the Internet. IPTV has become a common part of broadband triple play services.
- Mobile television has not seen large adoption outside Japan and Korea.

The conversion to digital platforms characterized audiovisual services in the first decade of the 21st century. During the "digital switchover" analogue services were gradually replaced and then withdrawn in most OECD countries. In their place, many consumers now experience a wider range of over-the-air channels and enhanced quality in reception. The changeover has also freed spectrum for use by other service providers, such as for mobile communication.

Analogue radio still exists, however, and terrestrial digital audio broadcasting (T-DAB) has not become mainstream for many consumers. One reason may be the increasing variety of reception options available to consumers, both within and outside their traditional broadcasting footprints, including digitally via the Internet. At the same time, online developments from podcasts to streaming music services have given users greater choice as to how they consume their audio content, as well as making these markets much more competitive. Today, consumers can decide where, when and how they access content from a vast array of choices, in part as a result of the increasing range of devices such as smartphones and tablets that now have the capability to access and play audiovisual content.

These changes affected audio services first due to their lower requirement for bandwidth and computing power. The expansion of broadband access and concomitant increase in its capabilities is now also transforming visual services. Users today can consume content from an array of digital delivery platforms that include "free-to-air" (advertising or publicly funded) broadcasting, cable, managed IPTV and satellite services, as well as an increasing proliferation of over-the-top services (OTT) provided via broadband Internet access. As with radio services, new devices that make content available at any time and in any place, known as time and place-shifting, as well as cloud-based services that perform similar functions, allow users much greater choice in the content they consume.

A lively debate is occurring within the industry over which services will prevail in the market. Observers are asking whether audiovisual services will experience changes similar to those wrought in telecommunication markets as they converged with the Internet. For their part, service providers are adapting and changing their offers to meet the new demands of consumers, but face increasing competition for the leisure time of users. This includes time spent engaged in social networks, games and other media that may once have been spent watching scheduled broadcasting.

The wider availability of digital content, and devices to access that content, has created debate as to whether all consumption of audiovisual content will soon occur over IP networks. In 2012, the United Kingdom's House of Lords suggested: "It is likely that IPTV services will become ever more widespread, and eventually the case for transferring the carriage of broadcast content, including public service broadcasting, from spectrum to the Internet altogether will become overwhelming." Meanwhile, in the same year, the CEO of Swisscom said that he expected television to become a "free part" of a high-speed broadband package instead of a distinct paid-for part of a broadband subscription, with consumers moving seamlessly between ISP-supplied content services and OTT content services.

The company followed this up, in November 2012, by doubling the speed of their DSL offers and including 60 digital television channels, of which 22 are available in HD. The introductory offer is automatically included in DSL monthly subscriptions, providing customers with telephony, Internet and television for the same price. The company's existing television customers were upgraded to the extended tier with more than 110 channels, 27 of those in HD. These types of bundled offers are increasingly common across the OECD area. What made Swisscom's changes notable were the new offers in the area of mobile television in relation to the introduction of its 4G network, and the pricing of service by speed tiers. The companies mid-tier to high-speed offers enable access for users to watch more than 80 channels and live sporting events. Programmes can be watched "on the go" or recorded for later viewing on a different device such as a television. The high-priced speed tier plans allow customers to use features such as "live pause" and replay functions that have been features of set-top boxes for some years.

For some consumers with adequate fixed broadband network connectivity – and where competitive pricing has enabled consumption of the amounts of data required to replace over-the-air broadcasting with sufficient quality of service – the point at which this technology can act as a substitute for over-the-air broadcasting has largely been reached. In France, for example, Internet Protocol television (IPTV) provides all over-the-air digital stations as well as traditional cable or pay television channels and other OTT services. However, this is only the case where fibre-to-the-home (FTTH), DSL or cable networks provide coverage. For users without such coverage, as in other countries, over-the-air broadcasting continues to provide critical services and will likely do so for some time. In time, 4G offers may assist in this process, but such networks and services are just now being introduced and spectrum may be limited depending on the number of users looking to access video services.

This chapter examines developments surrounding audiovisual services in the OECD area. Major changes in how people consume audiovisual content clearly have major implications for business models in areas such as advertising. While many users continue to listen to the radio in their car, they may also be listening to a podcast, audiobook or an e-book read to them using text-to-speech software. In their residence, they are likely to be watching or listening to content over their home network and using an increasing array of devices from tablets to smartphones. While "on the go", the range of options to time-shift and place-shift content, and to access live content, is rapidly expanding.

The underlying business models, the availability of content and the technologies used, are all changing. While at present the traditional business models and technologies enabling audiovisual content production and distribution are still dominant, there will likely be an increasingly prevalent hybrid model combining traditional broadcasting channels with new IP-based channels for disseminating audiovisual content.

Audio

Users have traditionally consumed commercial audio content in physical formats (e.g. records, CDs) and through broadcasting (e.g. radio). The physical medium allowed users control over what was played with limitations on variety and the amount of content they could afford or carry with them. Broadcast offered greater variety and the ability to discover new content with additional services interspersed, such as news and weather, but with limited control and coverage tied to geographic location. The production and distribution of audio was for the most part the territory of professionals. Digitalisation and convergence have rendered these traditional differences less visible. Users can now create, disseminate and access a wide variety of audio sources largely irrespective of geography. The sale of physical products such as CDs for audio content has declined significantly over the last decade, overtaken by sales and subscription to digital content in OECD countries like the United Kingdom and the United States.

The availability of radio is near universal in OECD countries, though some geographical areas are likely to have more choices than others. When discussing radio a distinction needs to be made between the underlying transmission technology and the programmed broadcast. Traditionally, radio used short wave, AM or FM modulation techniques, and in recent years has employed other choices such as satellite delivery to expand coverage. Radio is still the most widely used means to access mass media in the world. Most households will have one or more AM/FM radio receivers, though this may no longer be the primary way to access programmes. At the same time, while vehicles come with a radio receiver as standard, they may also have the functionality to integrate this device with a smartphone.

A large number of mobile telephones have an integrated FM radio receiver. In 2012, out of 99 telephones listed by Darty (an electronic retailer in France) only two did not have FM radio, one notable exception being the iPhone. Radio broadcasting and programming have, of course, transcended traditional devices and are accessible through other means of reception. Radio stations have expanded the number of platforms their content is available on, away from traditional AM/FM and long wave transmission. Set-top boxes for satellite and terrestrial digital television have radio integrated with the signal transmitted over a digital television channel. Set-top boxes provided by ISPs generally integrate radio as part of the service, in some cases as part of their IPTV service and provided as a channel or via a menu accessed through the television. In other cases, the set-top box will connect directly to the stream made available over the Internet by the radio station.

Radio stations are also available through satellite television platforms. In Australia, for example, radio is provided via systems primarily put in place for pay television and offered as part of the bundle – a service especially popular in areas that would otherwise receive limited choice due to geographic coverage. In North America, SiriusXM provides a dedicated satellite radio across Canada and the continental United States. The distinguishing feature of dedicated satellite radio in North America is its accessibility on smaller receivers for vehicles and portable handheld devices, rather than via a satellite dish. SiriusXM, like most radio stations today, also makes its programmes available over the Internet or via apps for smartphones and tablets. Internet radio was originally provided through web browsers on desktop and laptop computers. Today, it is available via a range of devices such as stereo systems and alarm clocks.

Digital radio is becoming more important, with three main distribution platforms: digital radio over the Internet, cable television networks and Terrestrial Digital Audio Broadcasting. Internet radio is widely available across the OECD and the rest of the world. A recent study by the OECD, UNESCO and ISOC, on local content measured the availability of Internet radio stations per million inhabitants. Canada leads the world, followed by the United States, Australia, Norway and the Netherlands (Figure 6.1).

Figure 6.1. **Online radio stations per economy**

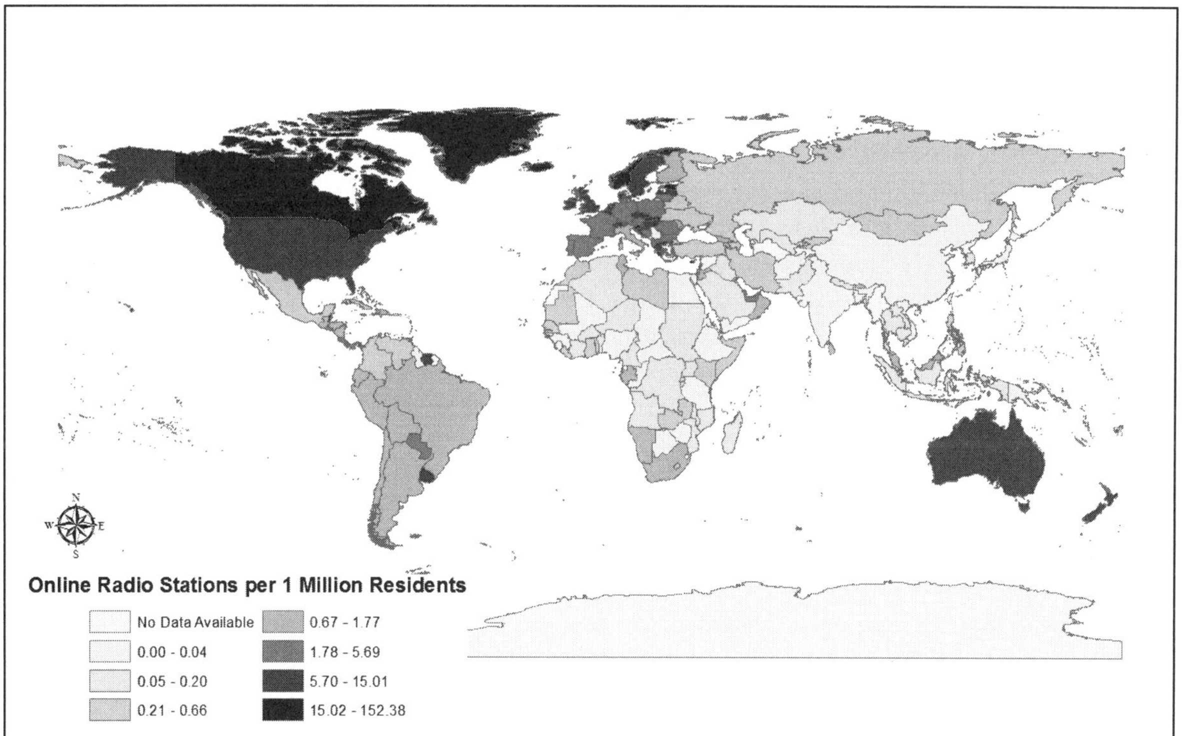

Note: This document and any map included herein are without prejudice to the status of or sovereignty over any territory, to the delimitation of international frontiers and boundaries and to the name of any territory, city or area.
Source: OECD, ISOC, UNESCO (2012).

The radio industry has experimented with Terrestrial-Digital Audio Broadcasting (T-DAB) for two decades. At first the standard known as DAB used MP2, the precursor to MP3 for audio encoding. Since 2007, DAB+ has become part of the standard. This approach uses High Efficiency Audio Coding (HE-AAC), which can give better audio quality or use less spectrum. Receivers that support DAB are not forward-compatible with DAB+. Some countries like Norway, Switzerland and the United Kingdom have seen significant consumer use of digital radio. Many other OECD countries have not gone further than trials or have no intention of using DAB to replace AM/FM radio (Table 6.1). So far, Norway is the only country to announce a switch-off date for FM broadcasting, though this is dependent upon meeting criteria such as market penetration; otherwise the switch off will be postponed. The United Kingdom has seen strong adoption of DAB with penetration of over 40% in 2012 and 19.4% of listening hours. It will therefore not move to DAB+. Mexico and the United States have adopted HD-Radio (NSRC-5B) as a digital broadcasting standard. The benefit of the standard is that the digital channel fits inside the traditional analogue channel where the broadcaster was broadcasting. The digital signal is broadcast together with and parallel to the analogue signal. To date, the technology has not seen widespread adoption; however, costs for broadcasting and reception equipment are decreasing, which may boost adoption.

Mexico has adopted voluntary use of HD Radio to be used on the AM and FM bands. The use of HD Radio technologies provides an alternative to digitalising radio, using the current bands where analogue services are provided. The result improves quality of service and provides multicasting of analogue and digital services in the FM band where the service has more reliability. As part of the transition to digital radio and in order to increase competition, Mexico established a policy in 2008 to facilitate the migration of AM stations to the FM band. Mexico had 473 FM commercial stations and 759 AM commercial stations. More than 500 AM commercial stations are in the process of migrating to FM.

Traditional linear broadcast radio is experiencing competition from new platforms that have become possible through the Internet. Like them, the new rivals share access to massive libraries of content, but have the advantage of being able to make these available to users on demand and with the ability to schedule (e.g. playlists) as they see fit. Professional and amateur content producers make the content available, which includes podcasts. They can be streamed or downloaded, for later use, from websites and are available through platforms such as iTunes or apps. Services like Pandora or Last.fm provide a subscription service equivalent to radio, playing songs based on the personal preferences of listeners. Spotify, Rdio and Deezer are subscription services providing access to a library of millions of songs. All three services use the same model, whereby users pay USD 4.99 per month for online access via a computer and USD 9.99 for offline and mobile use. Users can listen to individual songs, albums or make and share playlists. Some of these services have been integrated into user stereo sets to enable use independent of smartphones or laptops. Some users also use YouTube as a free radio service where they create their own playlists of available songs and listen (while not watching the videos).

Companies such as Amazon, Google and Apple allow users to upload purchased content to cloud services to access anywhere and at anytime. Content is also purchased and downloaded to devices from a variety of services. Many customers will use various services in combination, for example, podcasts, subscription services and cloud and device-based purchases.

The International Federation of the Phonographic Industry reports that all countries now have marketplaces for digital music. Not all audio content services, however, are equally available in OECD countries. In 2012, iTunes could not be used to purchase songs in Iceland, Israel, Turkey and Korea. Cloud services like iTunes Match and iCloud were not available in Denmark, Finland, Norway, Sweden and Iceland. At the same time, Spotify was only available in 16 countries, all in the OECD area. Meanwhile, Deezer was available in 100 countries, but not in Japan, South Korea or the United States. Rdio was not available in Italy, Mexico, the Netherlands or other OECD countries.

Although there are sometimes local alternatives to these services, the current situation highlights the lack of a global marketplace for content. Music is also not universally available on some new services. In 2012, for example, songs by Oasis and the Beatles are not available on any subscription-based service. Meanwhile, music by Pink Floyd is available on Deezer but not on Spotify. The main reason for this lack of global access to online audio services is that the intellectual property owners have not licensed the rights to the content for these services. In contrast, traditional radio services, even when accessible online as Internet radio or podcasts, are generally available without geographical restrictions, even though there may be a different copyright regime for online radio compared to offline radio. One exception is where rights to events such as sport have not been licensed (in these cases there may be separate licensing for Internet rights).

Developments in television

Television is almost universally available in OECD countries via one or more services including over-the-air, fixed broadband and satellite. The reported official and non-official statistics on the number of television households in this area may be inconsistent, however (Table 6.3). This is due to differences in methodologies in collecting data on households and number of television users across different countries. The wide availability of television may also mean that the indicator has become less relevant for policy makers with the result that less official data are collected and less often.

The collection of data in this area is also complicated by convergence. The gathering and treatment of data is not keeping pace with the expanding choice of networks to access television services. This can lead to double counting. In some countries, such as France and Switzerland, IPTV is bundled with a DSL/fibre subscription. At the same time, some DSL/fibre-customers may make use of satellite or cable networks for their television. Digital Terrestrial Television is also used in secondary locations where no other form of television is available, such as in a vehicle, on a boat or in a campsite. Furthermore, small and medium enterprises may subscribe to triple-play broadband offers at their business location, which may lead to an over count. The lack of statistics, broken down by technology, makes it challenging to follow and understand the rapid changes in television as a result of convergence.

Households in OECD countries watch between two and five hours of television per day. An exception is the United States where households are reported to watch over eight hours of television per day. From the statistics provided it is unclear whether this concerns only linear broadcasting, or also non-linear video-on-demand (VOD) and recorded television. Here again the data may record differences based on methodologies that have not been harmonised, but probably do provide a general indication of differences in consumption. Not every country has accurate and recent data on viewing hours per household per day. No data are available for Israel and Iceland. In other countries the data are not recent (Figure 6.2). Since 2000, average viewing hours have remained stable across the OECD at around three-and-a-half hours per household per day, though it is likely that more people are multi-tasking (e.g. surfing the Internet while televisions are on; see Chapter 8).

Figure 6.2. **Average viewing hours per household per day, 2011**

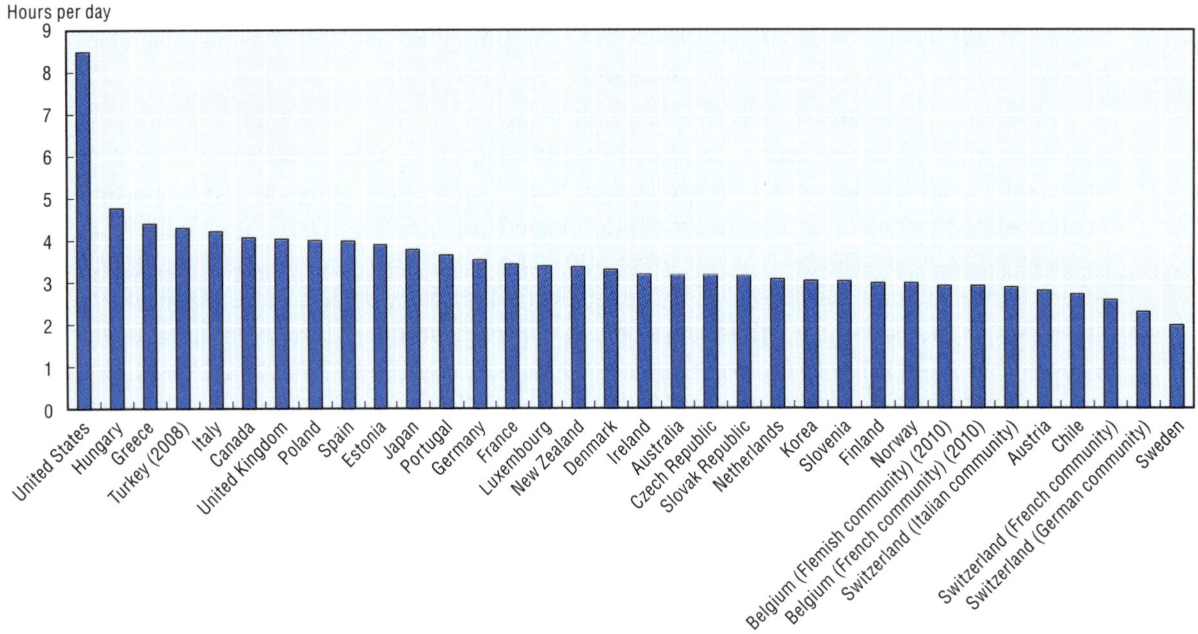

Source: OECD, IHS Screen Digest and European Audiovisual Observatory.

StatLink http://dx.doi.org/10.1787/888932799418

Since the turn of the century, the technologies used to make televisions have changed dramatically. During the previous century, the television screen was based on cathode ray tube (CRT) technology. There were competing standards for broadcasting, but television receivers themselves were square boxes – as wide and high, as they were deep. This changed, after 2000, when rapid advances in Liquid Crystal Display (LCD) and plasma technology radically transformed the market. LCD and plasma enabled much bigger and flatter screens with more pixels, resulting in a much higher image quality at a lower comparable price.

To illustrate the influence of LCD and plasma devices, DSG, a high street electronics chain with brands like Dixons in the United Kingdom, reported that during the fourth quarter of 2004, some 80% to 90% of its sales were traditional CRT screens. A year later, for the same period, CRT represented only 15-20% of the volume and 10% of the value, with LCD and plasma making up the rest. By 2006, CRT sales represented just 5%, and in 2008 Dixons ceased sales of CRT screens.

The market for the production of televisions has undergone similar rapid changes. Samsung and LG are now widely reported to have the leading market shares for the production of LCD televisions followed by Sony, Toshiba, Sharp and Panasonic. The market is extremely competitive. One traditional manufacturer, Philips, has exited the market, selling 70% of its television division to TPK of Hong Kong, China. Sony sold the manufacturing of LCD panels to its partner Samsung in 2012, focusing on design and assembly instead. Manufacturers from China are said to be quickly increasing their market share in the sale of televisions.

The shift to LCD screens has opened up research and development possibilities for larger screens with higher resolution than was previously possible with CRT. In addition, the advent of LCD and plasma technology made high definition (HD) television cost effective. Recent innovations have not only focused on making screens clearer, larger and faster, but have also concentrated on refining and making better use of higher resolutions. The technology for three-dimensional (3D) television became a practical reality with HD screens. The fourfold increase in resolution was used to show two slightly different pictures to both eyes, thereby creating a three-dimensional illusion. Availability of 3D content is gradually increasing.

In 2012, the first so-called "4K" televisions came onto the market. These televisions have twice the number of pixels, horizontally and vertically, than normal high definition television, resulting in a resolution four times higher. Digital movie production has shifted to 4K resolutions, as this is said to approach traditional celluloid film quality. The International Telecommunication Union (ITU) is working on new video-encoding standards for 4K and 8K resolutions.

One outcome of the rapid changes in televisions (and those forecast for the near future) is the changing expectations viewers have for content delivered over networks. Infrastructures need more bandwidth to broadcast television that makes use of these new resolutions, with 3D said to require 10 Mbit/s, 4K to require 25 Mbit/s and 8K to require 90 Mbit/s. It is notable that the mobile service offered by Swisscom, discussed earlier, markets speeds based on customers' requirement for video content. Swisscom recommends customers to subscribe to the 7.4 Mbit/s (shared downstream) mobile offer and 1 Mbit/s (upstream) for services such as YouTube and live television. This increases to 21 Mbit/s up and 2 Mbit/s down for viewing and uploading HD video over the mobile network, and the equivalent of 100 Mbit/s and 10 Mbit/s for saving data via the Internet.

The marketing of services by a mobile operator based on customer video requirements is an indication of the extent to which the market is shifting, based on convergence. This convergence is also evident in the underlying technology of televisions. A traditional television was a radio receiver with a screen. A modern television is a stand-alone computer with a screen, which is potentially far larger. These "connected" televisions now come with integrated networking and have the same type of processors as smartphones. Accordingly, some televisions today can be used to surf the Internet, though there is little evidence to date that this has proven popular with consumers relative to alternatives. What is far more evident is the use of dedicated applications on smart-televisions. Many of these apps allow the purchase and consumption of content on-demand and over the Internet. In the United States, for example, many connected televisions are sold with apps for Netflix and Hulu. In the United Kingdom similar apps for Netflix and Amazon's LOVEFiLM are sold with televisions, and in a growing number of countries there are local equivalents. Across most countries there are apps for services such as YouTube and Dailymotion. In other countries the catch-up and on-demand services of public and commercial broadcasters are available as apps, such as the BBC's iPlayer or Uitzending Gemist of the Netherlands' public broadcasters.

The peripheral devices connected to a television have also changed substantially. Before 2000, most households had one video recorder or DVD-player/recorder and maybe a game console connected to a television. The traditional video recorder/DVD recorder has been replaced. For digital television many networks now provide a set-top box, which may be equipped with digital video-recording functions. New DVD players all come equipped with Wi-Fi and smart-television options. Game consoles are now Internet connected and may offer video-on-demand services. Many televisions, set-top boxes, game consoles and DVD players may be able to connect to a home network using DLNA to access content.

Media players such as Roku, SlingPlayer, Apple TV, "Simple.TV" and Boxee are gaining wider acceptance and allow new functions for Internet-connected television. At the same time, attachments, apps or software, such as Elgato's EyeTV range of products, allow customers to extend the capabilities of smartphones, tablets and laptops as devices that receive television or interact with traditional television to record and display content stored or accessed by the user.

Television is now itself an application that can be accessed on any device that has Internet connectivity. Tablets, laptops and mobile telephones can all be television screens. The provider of the app can be a broadcaster, a provider of a broadcasting platform to which a user is subscribed, a third-party aggregator like Netflix or YouTube, or anyone else who has content available. The goal of the providers of apps is to make their content available at any time and in any place. Sometimes the same content can be accessed through the apps of various providers. As a result, the differences between market players for live and on-demand are diminishing.

Broadcasting platforms

Broadcasting platforms are the types of networks used to carry the television signal. In the 1950s television made use of analogue transmission technologies through antennas. Over time, satellite television and cable networks became important platforms. The introduction of digital television made it possible to fit at least four channels in the same spectrum as one analogue channel. Satellite broadcasting has changed fully to digital broadcasting because of the spectral savings and the better encryption options available. Both cable networks and terrestrial broadcasting are also moving to digital broadcasting. The Internet is emerging as a broadcasting platform and in some countries analogue television broadcasters no longer exist. Countries such as France, Finland and Italy, for example, have reached 100% penetration for digital television. Italy and Finland no longer have a broadcasting platform that supports analogue broadcasting. Notably, Finland passed a law that required all cable networks to shift to digital television. (Figure 6.3)

Terrestrial broadcast networks can, of course, use either analogue or digital broadcasting. All OECD countries have taken the decision to switch off analogue television broadcasts. The spectrum can then be repurposed for mobile communication, digital terrestrial broadcasts or for other uses. Digital terrestrial broadcasting offers a larger number of channels and, in principle, higher quality than analogue broadcasts. In 2006, Luxembourg and the Netherlands were the first countries in the OECD area to switch off analogue television. At the end of 2012, there were seven countries still in the process of switching over to all-digital television services. Only Chile has set no formal date as yet, though it is likely that the digital switchover will happen in 2017 or later, which would make it the final country in the OECD to switch off. Australia, Greece, New Zealand and the Slovak Republic all plan to end analogue service in 2013, whereas Mexico and Turkey will wait until at least 2015. When the spectrum has been cleared from analogue transmissions, it can take a while before countries issue the spectrum for new uses. For example, although the Netherlands was one of the first countries to clear the spectrum, the auction took place only six years later in 2012 (Table 6.2). In contrast, the United States auctioned the spectrum of the digital dividend before the digital switchover.

The use of Digital Terrestrial Broadcasting or Television (DTT) differs strongly by country, often because of the availability of competing infrastructures such as cable, IPTV

Figure 6.3. **Penetration of digital television by country as a percentage of television households, 2011**

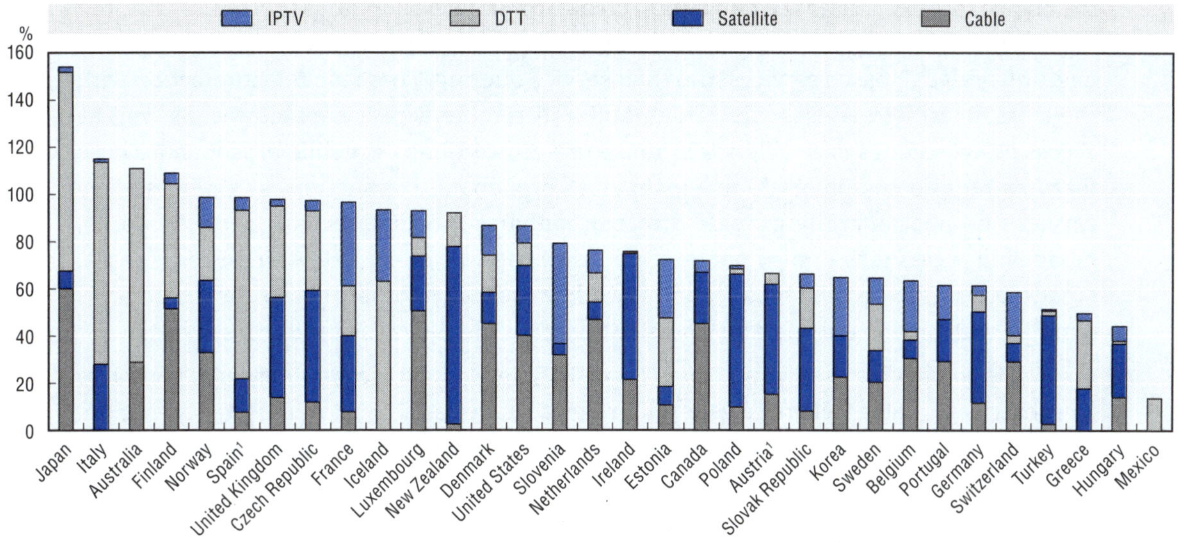

1. Cable includes IPTV.

Source: OECD, IHS Screen Digest and European Audiovisual Observatory.

StatLink http://dx.doi.org/10.1787/888932799437

and satellite. In 2011, the latest year for which data are available, some 24 countries had completed the digital switchover. For those countries where data is available, it is often unclear whether the data refer to the number of users or the number of DTT capable devices. Italy (86%), Japan (84%) and Australia (82%) have the highest penetration rates of DTT (Figure 6.4).

Figure 6.4. **Penetration of DTT by country as a percentage of television households, 2011**

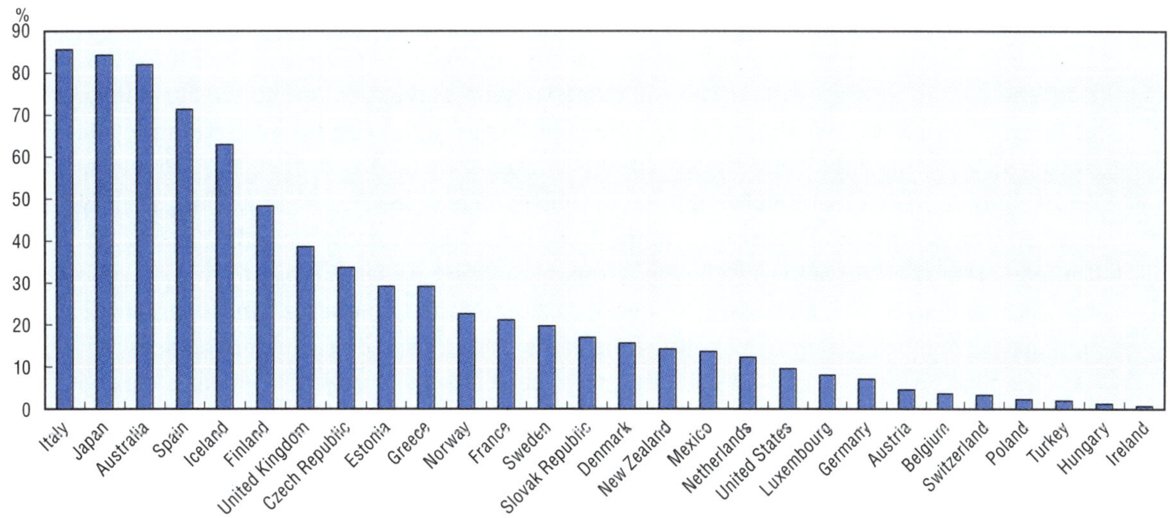

Source: OECD, IHS Screen Digest and European Audiovisual Observatory.

StatLink http://dx.doi.org/10.1787/888932799456

Direct Broadcast Satellite (DBS) services provide audio and video programming in a linear fashion with programming that may be free to viewers or available via subscriptions.

DBS coverage of population in a country is often close to 100%, though actual use varies greatly per country. The availability of alternative infrastructures and content influences the choice for DBS. In New Zealand and Poland 50% of television households use DBS, but less than 10% in Belgium and Finland. DBS broadcasters such as Dish, DirecTV and Sky have integrated IP-based services into their service offering. These include interaction with their broadcasting services via on-demand and remote access capabilities, as well as access on telephones and tablets that make use of fixed-line or mobile broadband. These features require customers to use a separate broadband connection, as their satellite services do not generally provide direct broadband Internet access. In other words, broadband access may or may not be supplied by the satellite broadcaster. In some cases, broadcasters have become broadband network operators; in others either their product is bundled with broadband operator offers or the consumer buys the products separately.

Cable companies use Hybrid Fibre-Coax (HFC) networks to deliver television and radio to subscribers. Cable networks generally offer a mix of analogue and digital television, except in Australia and Finland, where analogue broadcasts on cable networks were discontinued in 2007 and 2008. In countries that have implemented the digital switchover, analogue television remains available on cable networks for viewing by significant numbers of subscribers. In the Netherlands, for example, digital television on cable networks exceeded 50% of cable subscribers only in 2011. The use of Docsis 1.1 and Docsis 3.0 now enables cable networks to act as multiservice broadband networks that provide television, telephony and broadband Internet. Increasingly, IP is used as a carrier for new services, with some networks experimenting with using IPTV for less-watched channels, freeing up capacity when these channels are not in use. Most networks use IPTV for video-on-demand services.

Take up of cable services varies greatly across the OECD area, by country, and is dependent upon network coverage and maturity of the infrastructure. Belgium, Canada, Korea, the Netherlands, Switzerland and United States have the highest cable coverage: over 80% of the population can access cable television. In these countries cable is a viable second broadband infrastructure. In France, Italy, Mexico New Zealand, Spain and Turkey, cable coverage and therefore subscriptions do not exist or represent below 20% of households. The coverage ratio of cable networks provides an upper limit to the percentage of television viewers that subscribe to cable, but local market conditions determine the actual take-up of cable television. Of the countries that have over 80% coverage Switzerland leads with 88% subscriptions to cable, whereas the United States has slightly over 50% subscriptions (Figure 6.5). Some fibre networks, notably Verizon FiOS in the United States and Reggefiber in The Netherlands, use similar technology to that of cable networks to deliver television in the homes of consumers, and may serve a mix of analogue and digital. The technology is essentially the same as that used by cable networks, except that the fibre terminates in the home instead of at a cable head-end. The signal is carried over a separate colour on the fibre, next to the two colours that carry the upstream and downstream IP-packets. In official statistics these networks may however be registered differently.

Internet Protocol Television (IPTV) uses IP to deliver television to end-users. In recent years the market share of IPTV has increased. Broadband Internet providers using DSL and fibre networks were among the first to deploy IPTV to their customers for use in linear television. Some satellite and most cable networks have expanded their offers with IPTV to allow customers to access television on devices such as tablets, telephones and laptops, for

Figure 6.5. **Cable subscribers as a percentage of television households, 2011**

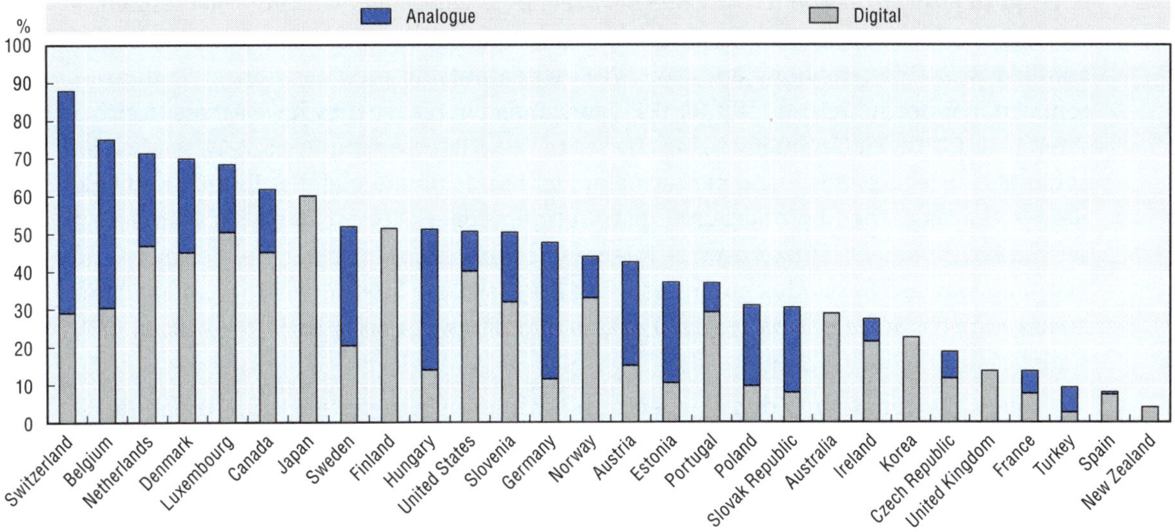

Source: OECD, IHS Screen Digest and European Audiovisual Observatory.

StatLink http://dx.doi.org/10.1787/888932799475

time and place-shifting type services. For on-demand services, IP has become the default platform used by all operators. As a result, official statistics on IPTV use may not accurately describe the use of IP to deliver the television viewing experience. France has among the highest percentage of IPTV subscribers out of television viewers in the OECD (Figure 6.6).

Figure 6.6. **IPTV subscribers as a percentage of television households, 2011**

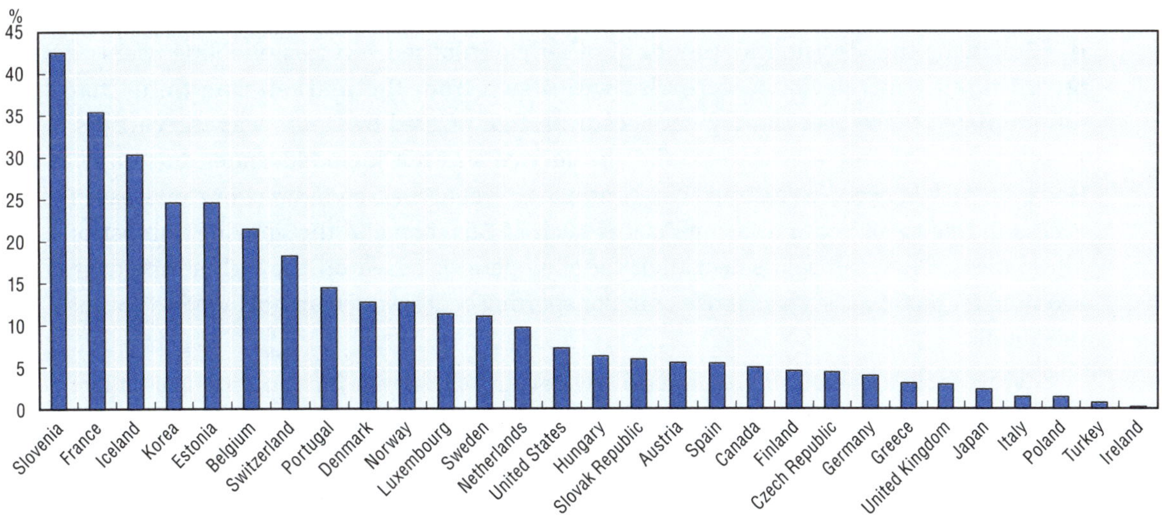

Source: OECD, IHS Screen Digest and European Audiovisual Observatory.

StatLink http://dx.doi.org/10.1787/888932799494

In France, most Internet users have access to IPTV via their broadband provider. Almost all broadband providers that use DSL offer IPTV as part of their package. IPTV typically refers to IP television provided over a managed network (as opposed to the public Internet). Conceptually, it is quite similar to multichannel broadcast distribution services provided over coaxial cable and DBS. Video offerings delivered over the public Internet are referred to as over-the-top (OTT) video services. IPTV is bundled with Internet and voice services and is seen by most consumers as a "free" part of the total offer (the offer comes bundled with free telephony and free Wi-Fi when outside their residence). If the DSL connection is not sufficiently fast for IPTV, consumers in France may have access to either satellite or DTV, which is integrated in the set-top box. In other countries, IPTV is often an optional choice in a broadband package. Canada has a commercial distribution network, NexTV, that is carried by traditional distribution networks (e.g. cable), but also uses a set-top box to broadcast packages of foreign channels from, for example, Germany and Viet Nam, but also from other countries using IPTV over the Internet and not an operator-managed network. Television received in this manner is not counted in the official IPTV statistics in Canada.

Given the constraints due to spectrum limitations associated with 3G networks, mobile operators have been cautious in promoting mobile television. There have been a few notable exceptions, but these have generally involved over-the-top options for customers. In the United Kingdom, for example, "3 UK" as far back as 2006 offered customers the ability to use Slingboxes on some tariff plans. Smartphone users can now also connect to a number of devices, such as those sold by EyeTV, to receive over-the-air television signals, thus not using mobile spectrum or incurring operator charges. Recently, some operators such as Optus in Australia have offered cloud-based services for the rebroadcast of free-to-air television, but these met with legal objections from a number of major sporting codes and copyright organisations. The sporting bodies argued that this restricted their ability to sell rights based on different technologies (e.g. free-to-air television, mobile or Internet rights), supported by the parties that had purchased those rights and copyright organisations that administered retransmission rights. The Australian courts ruled against Optus.

Following this decision a number of other cloud-based services enabling users to record free-to-air television were also withdrawn from the market. One thing that differentiated these over-the-top services from that offered by Optus was the carriers' ability to offer "unmetered access" (i.e. the service was not included under data caps). However, mobile users in Australia can continue to create their own direct or cloud-based access to free-to-air content, using devices such as Slingbox and the growing number of alternatives. They can also select their mobile operator based on the rights they have acquired for products, such as those of major sporting codes, which are included as part of the bundle.

While mobile operators are increasingly seeking the rights for content such as live sporting events, their experience with mobile television has been mixed. Broadcast television can be received on mobile telephones and other devices in a number of OECD countries (Japan, Korea and the United States). In Japan, all mobile operators offer their customers access to mobile broadcasting, using 1Seg-technology. A wide variety of smartphones are available and for those without a native receiver, such as the iPhone, an adaptor with an app is available. Korea uses DMB technology, which has become entrenched as a universal service thanks to the rapid diffusion of mobile devices. DMB can also be used in vehicles such as trains and automobiles, but watching DMB televisions while driving is prohibited.

In August 2012, "Dyle mobile TV" launched in selected markets in the United States. By November 2012, Dyle said its service was available from more than 90 stations in 35 markets, reaching approximately 55% of the national population. The Dyle application utilises the ATSC-Mobile DTV standard to enable live broadcast programming. The company works with EyeTV to provide the receiver and, as such, usage does not form part of the user's data allowance or incur data charges. Samsung has also developed a mobile telephone to access the Dyle services. Dyle is a joint venture operated by 12 major broadcast groups including Cox Media and Hearst Television, as well as Fox and NBC.

Dyle is one of a number of endeavours to introduce mobile television in the United States, earlier attempts having met with mixed results. Major mobile operators Sprint and T-Mobile use the MobiTV service to aggregate traditional broadcast and cable television content, which they rebrand. It is notable, however, that Dyle's initiative comes from a group of major media corporations rather than the "traditional" or four largest mobile operators. Dyle currently does not charge for the service (though they say this may change in the future) with the only cost to a user being the purchase of an appropriately equipped handset or attachment for a smartphone.

In European countries several mobile operators introduced DVB-H, a version of the digital television standard DVB-T for handheld devices. However, only Nokia supported the standard in its devices and consumer uptake of mobile television was limited. All European countries have now stopped DVB-H broadcasts and the spectrum has in many cases been returned to governments.

Consumption of audiovisual content on mobile devices is growing across the OECD. Some networks are reporting considerable amounts of traffic volume on LTE networks as a result of video usage from platforms such as YouTube and Vimeo, as well as their own branded services – more so than on 3G networks. There are, however, few official data available to establish the breakdown of on-net and off-net video traffic. They do show that LTE is a better fit for online video and, as a result, users use it more to access video. A growing number of video-streaming services, including Netflix, allow a user to shift the viewing of audiovisual content from the television to a mobile device. In other words, they can start watching on one device in one location and continue on another device elsewhere.

The Internet as a broadcasting platform

IPTV is often limited to an ISP's own network. However, the availability of high-speed broadband and the interconnections made possible by the Internet enable its use as a broadcasting platform. Users may experience variation in quality compared to ISP networks, but broadband access, together with the increasing use of content distribution networks (CDNs), means the experience can be indistinguishable from that of a managed ISP network. Data on the use of Internet as a broadcasting platform are difficult to interpret as most countries do not licence Internet broadcasting or video-on-demand, or apply regulation in the same manner as with traditional broadcasting.

Across the OECD area, broadcasters are increasingly using the Internet for live broadcasts and for on-demand television. In a growing number of countries, public broadcasters provide simultaneous Internet broadcasts of regular transmissions available on other platforms (Table 6.5). In more than 10 countries the largest commercial operator offers such services. In 2012, in at least eight countries, the main public broadcaster had not introduced simultaneous Internet broadcasts. In some countries there is regulation that precludes or makes it challenging for networks to offer simultaneous services on the Internet.

In Australia, Canada and the Netherlands, public broadcasters use the Internet for live broadcasting of special events (e.g. sport, elections) and for dedicated channels such as 24-hour news. In countries where there are geographical restrictions or content licensing, some cloud-based content services have had to restrict service to certain locations rather than providing national services.

In the Czech Republic, the Authors Act sets a restriction on the public broadcaster, that does not allow it to broadcast via the Internet. In Japan, the public broadcaster NHK provides simultaneous Internet broadcasts of "NHK World TV" and "NHK World Radio JAPAN" for abroad and domestic radio programmes on a trial basis. In the United Kingdom, live broadcasts are available, but proof of payment of the TV license fee is required for access. The United Kingdom and other countries may have geographical restrictions regarding who can access live broadcasts. Austria has six broadcasters that are registered as exclusive to the Internet.

While regulation precludes or restricts public or private broadcasters from offering some types of services over the Internet, in some countries they do increasingly offer "catch-up" television. Catch-up television is available in almost all OECD countries from public as well as commercial broadcasters. In some countries, where these services are not offered, the reason may be commercial rather than regulatory. In the Czech Republic, for example, only the public broadcaster offers catch-up services. In the United States, both commercial and non-commercial broadcast channels, and some cable channels, make programming available on a catch-up basis on their individual websites. Broadcast channels ABC, FOX and NBC also make episodes of their programming available through Hulu, which they jointly own. In addition, many networks have agreements in place with cable, satellite and IPTV services to make programming available via video-on-demand or online as part of their video subscriptions.

In most countries catch-up services offer one week to one month of television, often limited to their own productions, and as streaming products rather than downloads. Some countries, however, offer much longer periods; for example, the public broadcasters of the Netherlands offer over 10 years of their own productions. In the United Kingdom, the public operator allows users to download programmes to store on their own devices for up to 30 days. This enables users to transfer programmes to their chosen devices (e.g. smartphones, tablets), use content when they do not have a direct Internet connection, and take this content with them to different locations (e.g. foreign countries).

In most countries the content provided for the catch-up services of public broadcasters has no additional charge (i.e. it is not covered by licence fees or general revenue to which they contribute). In Japan, viewers have to pay USD 11.60 per month for the use of catch-up television. For the broadcasts of commercial broadcasters the situation can vary between countries and can also depend upon whether the broadcasters themselves or third parties produce the content. In Slovenia, the largest commercial broadcaster charges USD 6.40 for 30 days of access, while smaller broadcasters do not charge a fee.

Most OECD countries have video-on-demand services (VOD). The provision of VOD is regulated differently between countries. Australia, Ireland and Japan do not regulate VOD services. In Canada, VOD delivered by Broadcast Distribution Units (cable) is regulated and subject to a 5% charge for the production of Canadian content, but VOD over the Internet is not subject to this charge. In Austria, the Czech Republic and Mexico, a registration is required to offer services. In the United Kingdom, there is a co-regulatory system with the Authority for Television on Demand (ATVOD) in which both the government and the industry participate. Ofcom, the communications and media regulator, has delegated certain of its functions to this organisation. In other countries various forms of regulation exist. In general, it can be said that VOD is more lightly regulated than traditional broadcast television.

The number of VOD providers is not known in many countries because there is no requirement to register. In 2012, for the countries for which data are available, the United Kingdom has 198 registered VOD providers, the Czech Republic has 118 and Austria has 64. In the Czech Republic a large number of registered VOD providers aimed services at countries other than the Czech Republic, as in the case of HBO Bulgaria. Estonia, Germany and the Netherlands have lower numbers of registered VOD services with 1, 6 and 14, respectively.

The type of VOD services and the content available over them vary greatly by OECD country. There are four main business models: advertising supported (YouTube, Dailymotion), download to own (iTunes, Google Video), download to rent (iTunes, cable companies) and subscription for a limited number of videos or for an unlimited number of videos (Hulu Plus, Netflix, HBO). In many OECD countries, the availability of VOD is still limited to local productions and less popular titles. Subscription services, such as HBO, Hulu Plus and Netflix, have proven very popular in countries where they are available. However, these services are not available universally in OECD countries and often have no local equivalent.

Obtaining accurate data on the use of video-on-demand services is a challenge. However, there are some unofficial statistics and indications available. One is from IHS Screen Digest, which estimates that in 2010 some 60.8% of online movie revenues in the United States went to iTunes for movies that consumers downloaded to rent or own, while the subscription service of Netflix accounted for 0.5%. In 2011, Netflix claimed a 44% share of online movie expenditure, surpassing iTunes at 32.3%.

Netflix is said to be the largest generator of Internet traffic in countries where it operates. In the first eight months of operation in the United Kingdom and Ireland, Netflix reached 1 million subscribers. It launched in Scandinavia in October 2012 and within a month upgraded its peering connections at the Netnod Internet Exchange Point in Sweden from 10 Gbit/s to 40 Gbit/s. One information point Netflix provides for its customers concerns the performance of hundreds of millions of long-duration, high-definition video streams, which are delivered over the Internet by different ISPs. These data, first provided to customers in North America, will make for a thought-provoking set of metrics as the company expands to more markets. Data reported in 2011 for North America revealed two cable networks as having the highest performance: Charter led for United States streams with a 2.7 Mbit/s average over the period; and Rogers led, at the time, in Canada with a 3 Mbit/s average.

Statistics from Netflix indicate that there may be a positive correlation between the speeds broadband networks offer and how much customers use Netflix. In some countries with caps for data downloads, Netflix offers their customers the ability to opt for a lower quality service. In Canada, for example, a Netflix customer can choose to watch 30 hours of video with a typical usage of 30 GB (70 GB for HD). Alternatively they can opt for lower quality, which may use 9 GB.

In the United Kingdom, Ofcom found in its 2012 Update to the Infrastructure Report that broadband speeds seem to be a significant constraint or enabler on how much data consumers can use on the Internet. Ofcom has found that the amount of data downloaded and uploaded by consumers increases steadily as broadband speeds rise, up to around 8 Mbit/s. Above that level, they report, it remains essentially constant until speeds are reached where much faster services are being used. They hypothesize that consumers with lower broadband speeds are being deterred from using data-hungry services such as HD Internet television or larger files such as those associated with VOD.

Table 6.1. **Digital audio broadcasting in OECD countries**

	Remarks
Australia	Public and commercial stations available in major regions, Adelaide, Brisbane, Melbourne, Perth, Sydney, Canberra and Darwin.
Austria	No services on air, only test transmissions.
Belgium	Only public radio stations.
Chile	No digital radio broadcasting.
Czech Republic	Tendered licenses for 2011-12. Some broadcasts available.
Denmark	Started trials in 1995. Availability of public and commercial stations. Considering analogue switch off.
France	Market has not yet launched. Some trials.
Germany	Full national coverage by 2014.
Hungary	Only tests.
Ireland	Public broadcaster broadcasts in Cork, Dublin and Limerick. Trials in other locations ongoing.
Italy	Regular service started in 2012 in the Trentino region.
Netherlands	Only public broadcaster, not available everywhere. Aims for full coverage in 2017.
New Zealand	Ongoing trials
Norway	Switch off of analogue radio in January 2017 if conditions are met.
Poland	One trial.
Korea	Researching digital audio broadcasting standard.
Spain	Started in 1998, testing DAB+
Sweden	Started in 1995, trialling DAB+
Switzerland	Over 1 million radios sold, testing coverage in tunnels.
United Kingdom	DAB accounts for 19.4% of listening hours. Digital radio on all platforms accounts for 31.5% of radio listening hours, no plans for DAB+
United States	Adopted HD-Audio standard, no wide-scale use.

Source: OECD and WorldDMB.

Table 6.2 **Analogue television broadcast switch off**

	Target date	Remark
Australia	10-12-2013	started in 30-06-2010
Austria	07-06-2011	
Belgium	01-03-2011	03-11-2008 for Flanders
Canada	31-08-2011	exceptions for remote communities
Chile	no date	
Czech Republic	11-11-2011	
Denmark	31-10-2009	
Estonia	01-07-2010	
Finland	31-08-2008	analogue broadcasts on cable stopped 29-02-2008
France	30-11-2011	
Germany	01-06-2009	
Greece	31-12-2013	
Hungary	31-12-2012	
Iceland	01-01-2013	
Ireland	24-10-2012	
Israel	31-03-2011	
Italy	30-06-2012	
Japan	24-07-2011	3 prefectures hit by earthquake postponed untill 31/03/2012
Korea	31-12-2012	Started on June 29, 2011
Luxembourg	01-09-2006	
Mexico	31-12-2015	
Netherlands	11-12-2006	
New Zealand	01-12-2013	Started on 30-09-2012 on the South Island
Norway	01-12-2009	Started in 2007
Poland	31-07-2012	
Portugal	26-04-2012	
Slovak Republic	31-03-2013	
Slovenia	01-12-2010	
Spain	03-04-2010	
Sweden	15-10-2007	started in 2005
Switzerland	25-02-2008	
Turkey	31-12-2015	
United Kingdom	24-10-2012	started in 2007
United States	12-06-2009	01-09-2015 for low power stations

StatLink http://dx.doi.org/10.1787/888932801983

Table 6.3. **Television households**

Thousands

	1995	2000	2001	2002	2003	2004	2005	2006	2007	2008	2009	2010	2011
Australia	6 500	7 177	7 293	7 431	7 569	7 706	7 842	7 702	7 850	8 001	8 154	8 311	8 470
Austria	2 648	3 185	3 220	3 184	3 196	3 328	3 475	3 508	3 537	3 566	3 598	3 624	3 650
Belgium	3 794	4 176	4 179	4 181	4 275	4 300	4 330	4 363	4 414	4 506	4 506	4 557	4568
Canada	10 485	11 575	11 796	11 924	12 067	12 276	12 474	12 660	12 855	13 032	13 198	..	13,321
Chile	..	..	..	4 048	4 048	..	4 250	4 250	..	4 592	4 592	..	..
Czech Republic	3 213	3 804	4 164	4 151	3 095	3 086	3 263	3 329	3 389	3 939	3 996	4 079	4 276
Denmark	2 061	2 349	2 379	2 364	2 402	2 402	2 429	2 429	2 457	2 443	2 439	2 847	2 847
Estonia	..	..	..	547	546	545	547	553	563	588	588	588	565
Finland	1 915	2 160	2 183	2 163	2 166	2 197	2 198	2 220	2 265	2 236	2 256	2 257	2 264
France	21 557	22 580	22 840	23 060	23 300	23 650	24 120	24 541	26 263	25 516	26 612	26 733	26790
Germany	32 634	36 790	37 110	37 365	38 165	36 190	36 500	36 800	36 900	37 412	37 412	37 714	39311
Greece	3 332	3 500	3 510	3 520	3 530	3 612	3 622	3 646	3 667	4 191	4 275	4 313	3920
Hungary	3 773	3 740	3 729	3 717	3 701	3 810	3 900	3 962	3 962	3 686	3 945	3 901	3 910
Iceland	91	98	99	101	101	101	115	110	110	117	117	119	122
Ireland	991	1 204	1 194	1 262	1 329	1 359	1 379	1 452	1 491	1 540	1 592	1 589	1 586
Israel	..	36 236	37 953	..	..	..	..	..	..	..	..	..	..
Italy	..	20 660	20 900	20 693	22 053	22 053	22 262	22 609	23 356	23 595	24 002	24 255	24 525
Japan	..	..	..	43 230	43 670	44 120	43 710	44 660	44 910	45 040	45 590	45 970	47 800
Korea	14 517	15 113	15 500	15 854	16 380	16 708	16 944	17 113	17 462	17 666	17 978	18 259	18 569
Luxembourg	155	168	170	170	177	180	179	181	185	188	190	192	195
Mexico	16 000	18 471	20 705	22 938	23 410	23 883	23 654	24 860	25 038	25 885	26 514	26 834	..
Netherlands	5 850	6 685	6 757	6 823	6 905	7 000	7 000	7 075	7 000	7 113	7 175	7 241	7270
New Zealand	1 145	1 395	1 413	1 431	1 454	1 480	1 501	1 520	1 537	1 555	1 572	1 607	1 619
Norway	1 582	1 980	1 990	1 946	1 958	1 970	1 996	2 024	2 062	2 100	2 127	2 158	2 146
Poland	11 996	9 026	8 917	8 902	8 780	8 805	8 605	7 745	7 488	12 699	12 959	13 007	12959
Portugal	3 191	3 503	3 561	3 532	3 561	3 547	3 547	3 820	3 829	3 865	3 899	3 915	3926
Slovak Republic	1 742	1 858	1 881	1 883	1 869	1 879	1 881	1 885	1 938	1 702	1 745	1 742	2238
Slovenia	..	..	..	..	..	..	..	317	387	454	461	484	502
Spain	11 683	12 961	13 805	13 962	14 120	1 473	14 774	15 792	16 033	16 700	17 076	17 172	16377
Sweden	3 368	4 219	4 232	4 074	4 000	4 000	4 130	4 140	4 550	4 590	4 620	4 670	4 900
Switzerland	2 435	2 661	2 702	2 760	2 778	2 658	2 682	2 693	2 717	3 127	3 161	3 161	3195
Turkey	11 500	13 770	14 257	..	14 690	15 700	16 700	17 640	17 640	17 955	17 955	19 111	19411
United Kingdom	20 736	24 100	24 300	24 670	24 620	24 840	25 135	25 253	25 348	25 450	25 319	26 190	26 109
United States	95 300	102 200	104 400	106 700	108 400	109 600	110 200	111 400	112 800	114 500	114 900	115 900	114 700

Note: Data for Australia are estimates.

Source: OECD, IHS Screen Digest and European Audiovisual Observatory

StatLink http://dx.doi.org/10.1787/888932802002

Table 6.4. **Digital television households by platform**

Thousand of TV households

	Cable				Satellite (DTH)				DTT				DSL (IPTV)			Total Digital HH			
	2008	2009	2010	2011	2008	2009	2010	2011	2008	2009	2010	2011	2009	2010	2011	2008	2009	2010	2011
Australia	2 334	2 409	2 424	2 440	..	..	..	..	..	5 024	6 464	6 945	..	..	..	5 024	..	..	..
Austria	220	278	391	550	172	146	168	1 708	1 465	1 670	1 695	1 708	69	130	200	1 857	2 094	2 254	2 425
Belgium	898	1 271	1 461	1390	73	100	329	361	58	66	121	165	623	822	980	1 476	2 096	2 803	2 896
Canada	3 933	4 562	5 498	6 018	..	2 800	2 862	2 877	..	10 540	..	..	300	417	657	..	..	8 777	9 552
Chile	892	1 003	1 076	1 138	570	661	853	930	..	..	..	..	..	..	..	..	..	..	..
Czech Republic	310	418	470	500	494	515	1 747	2032	355	880	1 330	1440	171	170	186	1 541	2 004	3 717	4 158
Denmark	167	590	..	1 079	418	384	347	317	290	353	380	373	182	233	304	1 316	1 668	1 699	2 073
Estonia	17	16	29	59	35	40	47	45	30	50	170	164	101	123	139	160	207	368	408
Finland	922	1 091	1 131	1 164	78	146	111	106	1 111	1 129	1 139	1 093	9	42	102	2 197	2 256	2 257	2 264
France	1 558	1 989	2 213	2146	6 645	6 823	6 877	6 549	9 137	12 863	16 098	16344	4 944	6 378	8240	16 819	20 860	24 480	26 601
Germany	3 326	3 820	4 623	4526	11 673	12 300	13 417		..	4 370	2 452	2775	864	1 173	1548	21 608	22 879	21 665	24 100
Greece	0	0	0	0	448	480	530	690	99	392	768	1140	130	121	118	925	879	1 450	1 948
Hungary	..	327	432	545		717	837	893	..	..	..	49	78	142	245	..	1 122	1 411	1 732
Iceland	12	..	..	..	11	..	..	10	..	..	..	77	0	37	37	70	..	77	..
Ireland	316	351		340	..	..	..	845	..	..	..	11	..	..	3	803	936	1 103	1 189
Israel	..	..	..	..	..	..	..	..	..	..	..	..	..	..	..	..	..	..	..
Italy	0	0	0	0	6 507	6 659	6 854	6 858	5 079	9 497	16 729	20 990	429	366	337	11 976	16 585	23 949	28 185
Japan	31 302	32 642	33 965	28 668	3 567	3 554	3 549	3 602	37 804	38 202	39 751	40 274	695	900	1 092	..	..	..	..
Korea	1 913	2 675	3 423	4 186	2 338	2 457	2 826	3 262	..	..	..	..	1 741	3 086	4 570	..	..	..	..
Luxembourg	131	134	..	98	45	45	45	45	..	..	11	16	9	14	22	184	190	..	..
Mexico	..	..	..	..	..	..	..	..	..	5 252	3 647	..	..	..	..	..	..	..	..
Netherlands	1 979	2 491	2 940	3407	800	895	689	539	..	879	820	892	154	445	705	3 273	4 419	4 894	5 543
New Zealand	70	60	50	40		890	1 004	1 220	50	160	202	230	..	..	..	..	..	..	..
Norway	532	608	661	706	714	691	666	655	322	449	478	484	178	225	272	1 893	1 885	..	1 877
Poland	492	772	1 088	1406	5 405	6 920	7 685	8 161	87	151	181	330	151	181	196	5 353	6 951	9 015	10 093
Portugal	562	788	1 074	1 147	586	645	670	699	..	176	..	605	377	498	567	1 350	1 810	2 242	..
Slovak Republic	50	98	138	178	521	614	793	790	0	5	50	380	89	117	132	518	550	1 090	1 480
Slovenia	18	43	102	160	16	21	24	24	..	85	..	..	180	200	213	173	245	327	397
Spain	1 112	1 169	1 221	1220	2 035	1 846	1 773	1 756	1 460	1 442	1 508	1464	798	856	913	11 062	15 966	15 672	16 149
Sweden	468	576	620	982	781	786	798	648	689	665	636	949	244	321	531	3 936	3 936	..	2 769
Switzerland	..	323	798	1 093	472	484	262	253	..	106	..	105	236	407	582	..	1 477	..	..
Turkey	0	3	129	518	3 400	2 656	8 938	8 938	..	5	50	380	89.0	117.0	132	..	2 696	9 142	9 819
United Kingdom	3 279	3 175	3 339	3 558	9 472	9 804	10 979	11 132	9 387	9 588	9 308	10 092	550	650	747	22 588	23 117	24 276	25 529
United States	40 432	42 646	44 746	46 045	31 321	32 678	33 366	33 910	..	11 830	11 080	10 970	5 137	6 908	8 299	74 856	92 291	96 100	99 224

Note: Austria: Cable includes IPTV.

Source: OECD, IHS Screen Digest and European Audiovisual Observatory

StatLink http://dx.doi.org/10.1787/888932802021

Table 6.5. **Availability of live, catch-up and on-demand television over the Internet**

	Live	Public catch-up	Public how long back	Costs	Commercial live	Commercial catch-up	Commercial catch-up how long back	Costs2	Commercial Video on Demand	Regulation.	Further comments
Australia	No, except for news broadcasts	Yes, ABC iView and SBS on Demand	+/- 2 weeks	No	No, except for A-PAC news	Yes, The Seven Network, The Nine Network and Network Ten	+/- 2 weeks	No	Yes, but unknown	Not regulated	
Austria	Yes	Yes	Up to 7 days, Sports 24 hours	No	Yes	Yes	No restrictions	No restrictions	64 registered providers of VOD	Registration	Six registered Internet broadcasters of television
Canada	No, except for special events like Olympics	Yes	Varying	No	Yes, a wide variety including Internet only broadcasters like NexTV and JumpTV	Yes	Varying	Varying	Yes, 33 Broadcast Distribution Unit (cable) VOD and many Internet based	BDU VOD is regulated, Internet based is not.	Video on Demand as part of cable subscription subject to 5% of gross revenues contribution to Canadian programme production. VOD over Internet exempt. Many forms of place shifting type services
Chile	Yes	Yes	Unknown	No	Yes	Yes	Unknown	Bazuca and Netflix and others			
Czech Republic	No, live broadcasts prohibited by Authors Act	Yes	Unknown	No	No	No	No	No	Yes, 118 registered, also aimed at other countries. Like HBO Go Netherlands/Bulgaria	Registration	
Denmark	Yes	Yes	Unknown	Unknown	Yes	Yes	Unknown	Unknown	Yes	Lighter regulation	
Estonia	Yes, ETV and ETV2	Yes	Unknown	No	No	Yes, limited to own production	Kanal 2 08/01/2010 and TV3 one month back	No	One operator,Elion, second operator in Q4	Unknown	
Germany	Yes	Yes	Different depending on content	No	Yes	Yes	Unknown	No	7 registered	Yes	
Ireland	Yes	Yes	Varying	No	Yes	Yes	Varying	No	Yes	No regulation	
Japan	Yes, except for domestic TV programs	Yes	One to three weeks and some under Past Programme Service	945 yen	Unknown	Yes	2005	Unknown	Yes various	Not regulated	
Luxembourg	Yes	Yes	Unknown	No	Yes	Yes	Unknown	No	itunes, numericable	Yes	
Mexico	Unknown	Unknown	Unknown	Unknown	Unknown	Unknown	Unknown	Unknown	Yes, mostly by cable providers	Is considered add-on, only registration at Cofetel necessary	
Netherlands	Not of main programming, but of politics, sports and news events	Yes, Uitzending Gemist	10 years	No	No	Yes	7 days for RTL, SBS longer for own content	Yes and no	Yes, 14 registered providers	Yes, commercial VoD needs to be registered	
New Zealand	No, only via place shifting services	Yes	One month	No	Only via place shifting services	Yes	One month	3 providers Sky, Quickflix and Telstra clear			
Portugal	Yes	Yes	Unknown	No	No	Yes	Unknown	No	5 registered in 2010	Yes	
Slovenia	Yes	Yes	Unknown	No	No	Yes	Unknown	The biggest commercial broadcaster asks 5 euro per 30 days access	30 registered services	Yes, light regime	
Switzerland	Yes, both for regional and national public broadcasters	Yes	5 days	After the 5 days limit it may demand a fee to cover costs	Yes	Yes	Unknown	Yes			
Turkey	Yes	Yes	Unknown	No	Yes	Yes	Unknown	No	Yes		
United Kingdom	Yes, only to United Kingdom TV license payers	Yes	7 days	No	Yes	Yes, ITV, Channel 4and Channel 5	7 days and sometimes longer	No	Yes, 198 registered	Regulation viat ATVOD.co.uk.	
United States	No, though there are place shifting services	No	No	No	No, though there are place shifting Services	No	No	No	Yes		

StatLink http://dx.doi.org/10.1787/888932801299

Chapter 7

Main trends in pricing

Prices for fixed telephony and, more markedly, for mobile voice services have decreased from 2010 to 2012, showing significant declines across all consumption patterns, with the exception of fixed business services. The chapter conducts, for the first time, a comprehensive benchmarking exercise of mobile broadband prices for laptops, tablets and smartphones, based on the recently approved OECD wireless broadband baskets.

The statistical data for Israel are supplied by and under the responsibility of the relevant Israeli authorities. The use of such data by the OECD is without prejudice to the status of the Golan Heights, East Jerusalem and Israeli settlements in the West Bank under the terms of international law.

1. Note by Turkey: The information in this document with reference to "Cyprus" relates to the southern part of the Island. There is no single authority representing both Turkish and Greek Cypriot people on the Island. Turkey recognises the Turkish Republic of Northern Cyprus (TRNC). Until a lasting and equitable solution is found within the context of the United Nations, Turkey shall preserve its position concerning the "Cyprus issue".
2. Note by all the European Union Member States of the OECD and the European Union: The Republic of Cyprus is recognised by all members of the United Nations with the exception of Turkey. The information in this document relates to the area under the effective control of the Government of the Republic of Cyprus.

Consumers need clear, transparent and readily accessible information on communication services to make informed decisions. Any comprehensive list of metrics developed by policy makers and regulators should therefore include indicators on prices. Access to this information empowers consumers, provides useful insights about the level of competition in a given market, and complements other metrics used to assess the sector's overall efficiency and performance. Benchmarking these prices across cities, different geographical regions and the world allows stakeholders to evaluate progress towards their objectives.

The OECD uses a set of telecommunication price benchmarking indicators based on a basket approach. It selects the least costly options among surveyed offers, thereby providing a tool to compare prices available to consumers and businesses. This chapter reports pricing data for fixed voice, mobile voice and leased line services. In addition to fixed broadband service prices this edition of *Communication Outlook* highlights mobile broadband prices. Prices are reported in exchange rate and purchase power parity (PPP) terms to provide a comprehensive view on prices for telecommunication services. Graphs are also based on purchase power parity terms, which provide a better view of actual prices faced by consumers relative to domestic prices for goods and services.

Higher speed and broadband mobile services are rapidly gaining a larger share of the wireless and overall market for communication services. Today, mobile operators provide numerous different tariff options for connection to the Internet. As users tend to have different consumption patterns based on the communication device they use, this chapter reports on wireless broadband prices across different allowances for smartphones, tablets and laptops separately.

Consumers broadly purchase fixed network communication services as bundles, usually double and triple-play packages that include Internet access, telephony and television. Quadruple-play services that integrate mobile services as part of a seamless bundle are less common across OECD countries, although there have been relevant developments. While discounts across services from the same provider are available, different services within the quadruple bundle are generally billed separately (e.g. as two separate lines on a bank statement). Nonetheless, an increasing number of offers bundle mobile services across multiple users (e.g. shared data plans for couples or families).

The prices for fixed residential voice services decreased between 5% and 20% from 2010 to 2012, depending on consumption patterns, while business prices remained stable, accounting for inflation. Mobile voice services have experienced notable price declines, up to 40% for higher usage baskets, and the trend towards paying a higher share of costs as a fixed charge is consolidating. This has been made possible by significant reductions in mobile termination rates in most OECD countries, which make these markets more competitive. Given that mobile operators have an effective monopoly over the termination of calls to their customers, any reductions imposed by regulators provide a greater role for market forces in determining the prices operators charge. This trend is already proving effective in increasing usage, lowering end-user charges and enabling alternative providers to challenge established operators, thus increasing the level of competition.

Between 2010 and 2012, fixed broadband prices increased by 6.76% per annum for DSL and 2.48% for cable, based on monitoring of specific connections from leading DSL and cable providers. These price increases were accompanied by notable speed upgrades with advertised speeds increasing by CAGR 43.81% for incumbent DSL/fibre and CAGR 27.49% for cable broadband providers. Given that mobile broadband services constitute a relatively new market compared to fixed broadband, there tends to be greater experimentation in wireless markets. Moreover, the evolution of the smartphone ecosystem has resulted in a complex array of stakeholders who determine these prices. The tariff options for fixed services are largely independent of the devices used to connect to these networks. For some mobile services, consumers choose a device then select a plan based upon expected usage and device capabilities.

A number of operators have introduced 4G (LTE) plans to meet customer demand for increased data usage. The associated tariffs in some cases resemble those offered by fixed networks. For example, some operators of LTE networks are offering tariffs based on access speeds rather than data caps. Others are offering shared data plans across multiple users. This is similar in some respects to the way in which users treat a fixed broadband connection for a single household. The major trend is the bundling of voice and SMS with data into a single service, billed by data usage with caps or at flat rates irrespective of the application generating the data.

As in all markets the interplay between supply and demand is most evident in the pricing of services. In the case of mobile broadband supply factors include the level of investment required to meet increasing demand for data, and the effectiveness of competition in applying discipline to prices in a market where spectrum limitations restrict the number of operators. Some operators argue that certain types of pricing constitute appropriate network management practices, designed to make best use of available spectrum and levels of investment, while enabling an attractive return. In a competitive market, these types of pricing schemes are disciplined by customer acceptance and expectations. Customers in competitive markets can pay for services that suit their requirements and will take into account factors such as network coverage and performance.

During the monopoly era, operators priced services using factors such as time of day or day of the week to determine which customers would pay higher prices and which were more price sensitive, employing efficiency-enhancing effects to manage peak loads on their networks. They also employed factors such as distance for much the same reason, regardless of what could have been dictated by directly attributable costs. Following liberalisation, competition increased and these types of pricing attributes largely disappeared in many markets. In competitive markets, users make calls irrespective of factors such as time of day or distance and traditional charges such as initial connection fees have long disappeared on wireless networks. Some question whether pricing should be used to reveal the value users place on specific services rather than treating access and use as a commodity. Others say that pricing can be a tool allowing users to self govern usage, thereby ensuring efficient use of networks. This is not necessarily a simple matter as congestion may occur only at certain locations or a certain times of day, with a particular group of users in a cell generating higher levels of traffic rather than the average level of use becoming too high by intense use of some services. For this reason, it is important that policy and regulatory authorities use effective competition to provide discipline to pricing rather than backing specific tariff schemes.

Some industry players argue that intensive users of over-the-top (OTT) services (i.e. from third-party providers over the Internet) may place a heavy load on their networks, requiring increased levels of investment. Regardless, competitive and effective markets will establish the required balance to address these issues, and it is important to facilitate commercial negotiations while ensuring that no party exerts monopoly power unchecked by competition. Moreover, it is crucial that market forces ensure efficient outcomes to enable all stakeholders to benefit from increased wireless broadband uptake.

A further influence on the development of broadband wireless prices is the interplay between different actors in the value chains around smartphones. With regard to handset devices, markets are witnessing competitive dynamics between smartphone makers such as Apple, Nokia, HTC or Samsung and mobile operators. On the one hand, popular smartphones remain a widely used tool for consumer attraction and retention and are frequently sold jointly with mobile plans, at lower upfront prices than available elsewhere. On the other hand, the bargaining power of mobile operators may be diminished as a result of these dynamics.

In at least one case the runaway popularity of a particular brand of smartphones gave the manufacturer a considerable say in the make-up of the ongoing tariffs and the capabilities associated with its use. Over time, as might be expected, market competition for handset supply has increased with far greater choice for consumers. As a result, manufacturers may have less influence in determining upfront discounts and ongoing pricing. In some markets, for example, certain operators have reduced the amount of upfront discounts against the same length of contract, or even eliminated the practice while still offering monthly repayments (i.e. a more transparent form of credit). Others are now charging fees to upgrade handset models before expiration of the contract period, whereas previously a customer could upgrade by extending the contract, without paying an upfront charge.

Service bundling

Broadband services are frequently sold as mixed bundles that allow users to choose among stand-alone offers or bundled services. Broadband bundles are typically sold with a significant reduction in relation to stand-alone prices. Indeed, in many markets the majority of consumers subscribe to broadband services as part of a broader bundle. While the benefits and possible drawbacks for end-users resulting from the popularity of bundled services have been discussed extensively, evidence shows that bundles are increasingly popular and therefore any systematic analysis of the telecommunications sector should consider them and the issues they raise.

The OECD report *Fixed and Mobile Networks: Substitution, Complementarity and Convergence* (2012) noted that quadruple-play offers are relatively rare in OECD countries, but are now flourishing in certain markets, for example, Austria (A1), Belgium (Mobistar), France (Bouygues, Free, Orange, SFR) and Korea (KT). Integrated quadruple-play offers have been used as a competitive tool in countries such as France or Spain to retain customers that may want to switch to a less expensive mobile phone provider. In Spain, the recent launch of "Fusión" by Telefonica is viewed as an attempt to tackle the company's current reduction in market share in fixed and mobile services.

These trends may become more widespread as traditional fixed operators with no current spectrum resources are awarded spectrum from the digital dividend, or as smaller mobile players use more spectrum resources. This may increase incentives among some mobile operators to establish alliances with fixed operators (e.g. small cable providers) to provide convergent services.

Double and triple-play offers remain widespread across OECD countries. As a recent electronic communication household survey in the European Union (Eurobarometer, June 2012) showed, more than 4 in 10 households in the European Union area buy bundle packages, rising to 64% of households in the Netherlands (Figure 7.1). Moreover, the use of bundles has increased slightly since the last survey by approximately 1% year on year, with significant increases in France (6%), Portugal (7%) and, from a lower base, Finland (5%).

Figure 7.1. **Percentage of households with contracted bundles of communication services – December 2011**

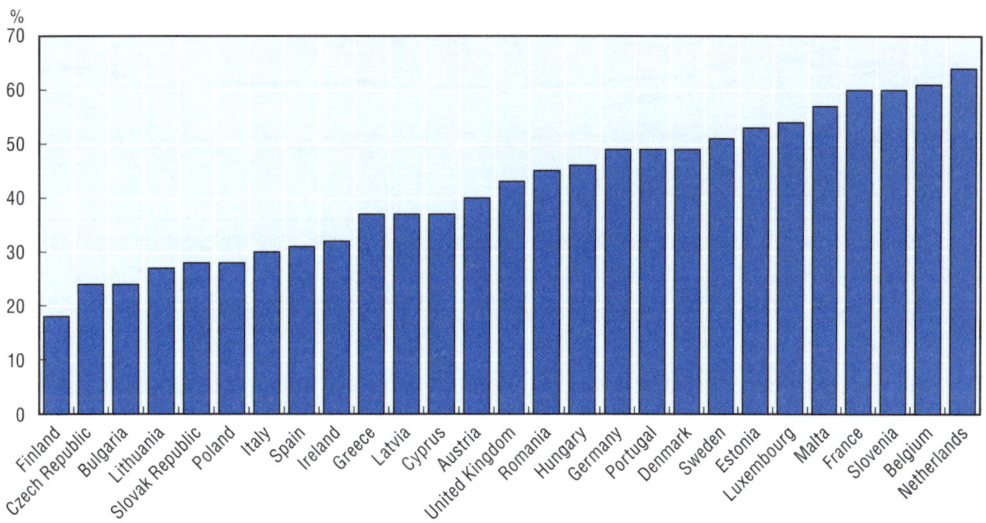

Source: EUROSTAT, European Union household survey, June 2012.

StatLink http://dx.doi.org/10.1787/888932799513

Fixed pricing trends

Tracking prices for stand-alone fixed voice services has become increasingly challenging, or less relevant in some cases, as these services are now frequently a component of a larger bundle, typically including Internet access and television services. Moreover, Voice over Internet Protocol (VoIP) services or pricing schemes that encourage mobile substitution have become readily available. As a result, unmetered or flat rate pricing for local calls in countries such as Australia, Canada, New Zealand and the United States is now common across all services, including for some of the most competitive markets, such as international telephony. VoIP services provided by over-the-top (OTT) operators, such as Skype, or in the form of "managed services" through a broadband connection provided directly by DSL or cable operators, typically offer unlimited or large buckets of minutes. This has contributed over the years, and where it can still be measured separately, to lower unit prices for fixed telephony.

The OECD fixed-line baskets provide a tool for comparing prices, as they refer to established consumption patterns rather than to prices of calling packages or per-minute tariffs. The results reported here clearly show that fixed telephony prices have continued to decrease over the 2010-12 period by 5% to 20% for residential services. Prices for business fixed-line services have remained overall stable, with a slight nominal increase of 2% to 3% in two years, which may well account for inflation in the overall consumer price index. In contrast, the overall trend already accounts for the effects of inflation (Figure 7.2).

Figure 7.2. **Time series for residential telephone charges, 1992-2012, OECD average**

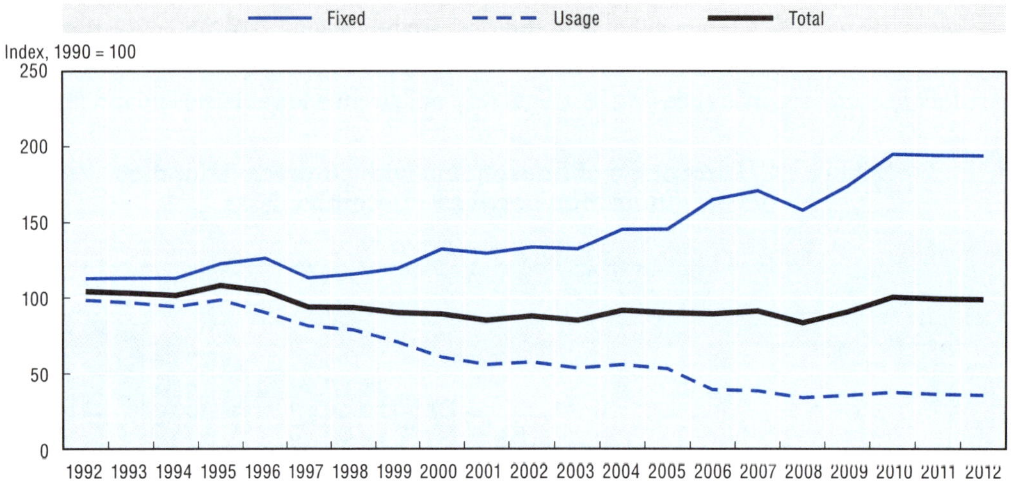

Source: OECD and Teligen.

StatLink http://dx.doi.org/10.1787/888932799532

The past two decades have seen a shift in consumers' telecommunication expenses from usage to fixed components. When telephony was the primary service this practice was known as rebalancing. Historically, operators used usage charges, particularly long-distance and international, to "cross-subsidise" the cost of a consumer's line. Irrespective of the merits of this practice it can be noted that liberalisation initially made usage much more competitive than fixed components. In fact, the trend increased considerably around 1998 when a considerable number of countries first opened their markets to competition (e.g. France, Germany).

Following liberalisation, consumers increasingly expressed a desire for fixed prices to ensure predictable outcomes at the end of billing cycles. The reduction of termination rates for fixed telephony services and, more recently, the significant decrease in mobile termination rates have contributed to this trend. Some argue that in countries with high mobile termination charges, fixed networks cross-subsidise the development of mobile networks. To the extent that this is so, regulatory authorities around the world have reduced the ability of all networks to exert monopoly power in the area of termination, and this is reflected in more competitive usage charges. Moreover, this development has triggered the inclusion of minutes with mobile phones in fixed subscription packages, a relatively rare practice in most countries until recently. Once again, Canada and the United States were outliers in this respect, with long-standing practices of no termination charges or low termination charges respectively. As a result, these countries record no difference in price for calling fixed and mobile numbers from fixed lines, as has been the case in most other OECD countries.

Since 1990, residential telephone charges have remained stable. Taking into account the effect of inflation, residential telephone charges have not changed since 1990 (Figure 7.2, Table 7.1). Business telephone charges, however, have stabilised at approximately 70% of their 1990 level, reflecting greater market competition (Figure 7.3). A common pattern can be drawn from the overall time-series for all residential and business baskets: the percentage of fixed charges has increased constantly and now lies at 90% (residential) and 70% (business) above the original 1990 level. Usage-based charges have dropped accordingly, to 35.5% and 44.4% of their respective 1990 levels.

Figure 7.3. **Time series for business phone charges, 1990-2012, OECD average**

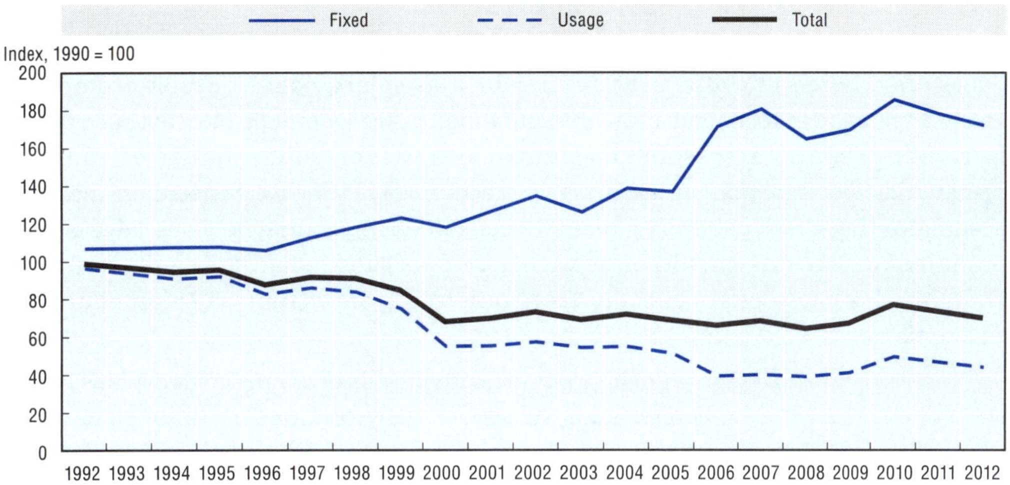

Source: OECD and Teligen.

StatLink http://dx.doi.org/10.1787/888932799551

Mobile pricing trends

The overall reduction in mobile termination rates in the OECD area has undoubtedly played a major role in shaping mobile pricing trends. Flat rates or, alternatively, large buckets of minutes are now more widespread and inexpensive. This trend is likely to continue in the coming years, triggered by declining termination rates in many, mostly European, OECD countries, but with still some way to go. The rapid drop in prices for higher consumption mobile voice baskets compared to other baskets, in a context of overall decreasing mobile voice prices, supports this view. Golan Telecom is the fifth operator to enter the mobile market in Israel and the first new brand to do so since 1999. Taking advantage of regulated termination rates below USD 0.02, Golan Telecom launched a set of inexpensive offers, resembling those of Free Mobile in France (unlimited service for USD 25 per month – ILS 99 in local currency terms– and an entry-level offer for USD 2.5 – ILS 9.9).

The results from the mobile baskets presented here also support the conclusion that flat-rate mobile tariffs are gaining prominence in OECD markets. In August 2010, 9 out of 34 countries were represented in the 300 calls mobile basket by a single fixed charge. In other words, the charge covered all 300 calls and the other services such as SMS included in that basket. In 2012, this number rose to 21 countries. In consequence, all these countries had a fixed charge tariff that included at least 300 calls.

Looking instead at the 900 calls basket, 11 countries in August 2010 had a fixed-charge tariff best suited to these baskets. In other words they included at least 900 calls. This number increased to 25 countries by August 2012. Given the high number of calls included in the 900 calls basket it can be reasonably assumed that these 25 countries had a flat rate, or a flat rate-like tariff, at best available prices selected for this basket, hence the least expensive options among the operators surveyed.

Mobile broadband services have, to date, largely presented different pricing structures from those of fixed broadband services. While most fixed broadband services are marketed by speed, mobile broadband services mainly advertise the amount of data included, assigning speed a relatively minor role. The OECD basket methodology for wireless broadband services acknowledged this point, but it was noted that in countries launching LTE services some operators were making reference to speeds as a marketing tool, even if they did not form the principle differentiator of tariffs.

Today, the amount of data in a tariff plan remains the principal way by which mobile operators differentiate offers for devices such as smartphones or tablet computers. Frequent references to a trend away from unlimited usage ignore the fact that such options were not common across OECD countries. However, in a number of countries communication regulators or consumer protection agencies have insisted on the use of accurate terms when publicising plans to consumers. Accordingly, plans advertised as unlimited or flat rate should offer such services. Thus, what was perceived as the withdrawal of these offers in some cases reflected more the introduction of accurate descriptions.

There have been a number of high-profile withdrawals of unlimited data plans for new subscribers, though even here existing subscribers may continue on these plans. In the United States, AT&T and Verizon discontinued unlimited access to data, in 2010 and 2011 respectively, for new subscribers. This practice remained rare in other countries, even where advertising indicated similar options without examination of the terms and conditions. Nonetheless, as might be expected in a competitive market, the third-largest operator, Sprint, promotes some of its offers as including unlimited access to data.

Competition is continuing to drive further change in the United States mobile market as operators respond to the repositioning of telephony and SMS services as data applications for smartphone users. This is most evident in the plans launched by AT&T and Verizon, which offer shared usage across different users and devices on a single account. Verizon's "Share Everything" plans include unlimited voice minutes, texts and picture messaging. A data allowance is shared among up to 10 Verizon wireless devices. The plan also includes a mobile hotspot service without extra charge. The shared data allowance starts at 1 GB (USD 50) and goes up to 10 GB (USD 100) with an additional charge per device added (USD 40 for smartphones, USD 30 for basic phones, USD for notebooks/netbooks and USD 10 for tablets).

The new tariffs acknowledge the fact that OTT services make separate charging for telephony and SMS less feasible for smartphone or tablet users. In practice they should also decrease tensions between network operators and OTT service providers, as the attractions of OTT services drive demand for smartphones, their capabilities and willingness among consumers to subscribe to plans.

Market popularity will be the ultimate test of shared data plans with specific allowances and the ability to purchase additional data. In a competitive market consumers will choose among different plans taking into consideration factors such as network coverage. The key role for regulators is to ensure the presence of sufficient competition. In this regard, the decision in 2012 to reject the purchase of the fourth-largest operator (T-Mobile) by the second-largest (AT&T) is notable.

In Switzerland, the leading mobile operator has taken a different approach to most its OECD peers as one of the first to offer LTE services according to speed. In mid-2012, the Swiss incumbent operator, Swisscom, launched a new set of mobile plans differentiated by speed, all with unlimited data consumption (Figure 7.4). As in the United States, Swisscom's plans include unlimited telephony and SMS while removing the tension between customer demand for OTT services and incentives among operators to meet this demand. Swisscom reports that the new plans have proven popular with both customers and capital markets (in the latter case inferred from rising stock prices). Swisscom has said that customers are more familiar with tariff plans based on speed than those based on data, where outcomes are less predictable in terms of overage charges or not utilising full allowances. LTE networks may introduce prices similar to those of fixed networks, but as before, the final test remains market performance in a context of sufficient competition. Moreover, the degree to which mobile broadband providers deliver speeds closer to those advertised remains a key issue when undertaking international speed comparisons.

Figure 7.4. **Swisscom plans for mobile services**

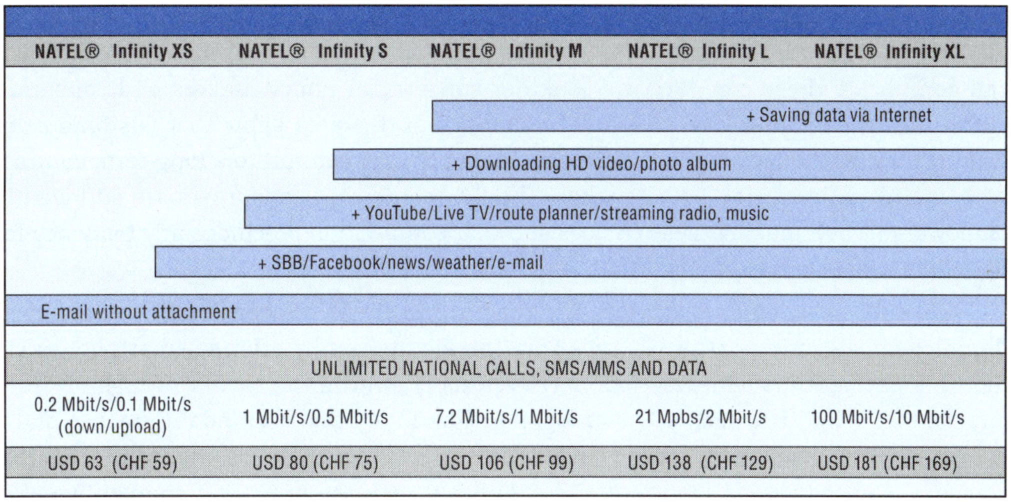

Source: Swisscom.

Consumer reaction to new tariff plans that do not meet their requirements can be swift and effective, where alternatives are available. In 2012, the Swedish operator TeliaSonera floated the idea of charging mobile customers separately for use of VoIP. Its customers did not welcome the proposal and one competitor (Three), which had previously welcomed the use of Skype, refused to follow the move. In September 2012, Telia announced that mobile IP telephony would form part of all its subscriptions, with the exception of one (Telia Flex Bas) designed for customers primarily using traditional voice services with limited need for data communication. The company adjusted its data tariffs accordingly, raising the maximum fee charged per day from USD 1.36 to USD 2.88, and increasing the amount of data included from 0.5 MB to 2 MB. The company confirmed that customers could continue to use Skype and that the adjustments and price changes reflected growing demand for data communication.

The tariff options introduced by companies such as TeliaSonera and Swisscom imply that OTT providers can be potential allies for operators: they develop services attractive that are users and encourage them to pay for transport. However, the necessary adjustments are not easy for all mobile operators, as billing practices have been based around telephony for more than a century. For regulators it is noteworthy that the integration of apps and the degree of separation provided by "app stores" has produced outcomes not dissimilar from those expected from "structural separation" or "unbundling" driven by regulators. Regulators did, of course, contribute to these developments by ensuring effective competition, which has opened markets to the innovation witnessed around the apps market. In addition, spectrum has been auctioned with conditions designed to increase competition and consumer choice. The FCC, for example, imposed "open platform conditions" on buyers of one block of 700-megahertz spectrum in the United States.

Issues surrounding the tied sale of handset devices have also gained significance. In a context of increasing smartphone and wireless broadband penetration, tied sales largely drive mobile market dynamics, as many consumers choose to commit to long-term contracts (e.g. 12, 24 or 36 months) if offered popular smartphones at a lower upfront charge. OECD research shows that, in some instances, these marketing and sales strategies can involve a higher final price for consumers if both the handset device and the mobile plan are considered. However, varying levels of network quality and handset availability can complicate these comparisons. Another concern for policy makers and consumer authorities is customer lock-in. Market dynamics in this area show that customers are usually reluctant to face high up-front payments but may commit to a long-term contract with their mobile carrier. Policy makers should ensure that consumers are sufficiently empowered when making these choices and that competition precludes any tendency for operators to abuse market power.

More recently, some operators are reacting to these concerns and to perceived diminished bargaining power vis-à-vis smartphone makers by offering smartphones via monthly payment instalments. This may eventually remove the barrier of high up-front payments. In countries such as Australia, Korea and Italy this is standard practice and a component of the monthly mobile plan bill corresponds to handset payments. In 2012, this choice was introduced in France after a significant increase in market competitiveness, while T-Mobile in the United States has introduced an instalment payment plan for smartphone users.

Price basket methodologies

Fixed and mobile voice baskets

The OECD's methodology for measuring prices for communication services is based on consumption patterns or "baskets". These include fixed, mobile and leased line communication services across a wide range of baskets and prices, and are collected from multiple operators with the largest market shares in each country. There are six fixed-line baskets (two targeting residential users and two business users) and six mobile baskets.

The OECD has recently developed a new set of baskets for broadband services, both for fixed broadband (adopted in 2009) and wireless broadband (2012). This chapter focuses on the newly adopted baskets, providing only a brief summary of fixed, mobile and leased line baskets. The annexed tables provide all the underlying data for these baskets including some breakdown of the data.

The trend towards a higher proportion of fixed subscription costs and lower usage costs has continued, for both fixed and mobile baskets, since 2010. All fixed voice baskets have followed this trend. By way of example, the proportion of fixed charges has increased from 45.4% to 60.35% for the 140 calls baskets, and from 37.3% to 48.35% for the 420 calls basket. This confirms the perceived trend towards a larger share of fixed charges and a more significant presence of flat-rate tariffs. Analysis of the evolution of average, maximum and minimum prices of the fixed and mobile baskets is provided here (Tables 7.1, 7.5 to 7.15). These data take stock of average prices for a given basket across all OECD countries, as well as the price trend from 2010 to 2012. They do not reflect leased line prices, which given the scarcity of publicly available data (not all OECD countries have data) are only represented in the annexed tables.

The overall trend for the fixed baskets shows a reduction of average prices, with the exception of a slight increase in the business baskets (between 2% and 3% over two years). Price falls are quite significant for the 20 calls basket (5.4% over two years), 60 calls baskets (7.7%), the 140 calls basket (14.8%) and the 420 calls basket (20.6%). The prices for residential baskets range from USD 23.82 PPP (20 calls) to USD 75.67 PPP (420 calls) per month. Business prices have increased slightly: the 100 calls and 260 calls business baskets cost on average USD 42.59 and USD 86.91 per month in PPP terms, although this trend is likely to reflect inflation rather than real increases in rates.

The prices of mobile baskets have fallen more markedly. For example, the price of the 900 calls mobile basket has dropped 39.3% in two years, down to USD 74.44 PPP per month, and the 400 SMS basket has dropped by 33.01% to USD 15.30 PPP per month, which may reflect the competitive constraints posed by data-based application such as WhatsApp.

Table 7.1. **Summary of the OECD communication price benchmarking baskets (USD PPP)**

Basket	OECD average	% 2010-2012	OECD cheapest	% 2010-2012	OECD most expensive	% 2010-2012
Fixed line – Residential call baskets						
20 calls	23.82	–5.4	6.80	–46.1	34.60	–9.4
60 calls	33.89	–7.7	10.83	–38.2	48.41	–2.9
140 calls	50.28	–14.8	15.90	–49.5	79.81	–12.4
420 calls	75.67	–20.6	27.77	–26.2	193.43	+9.4
Fixed line – Business call baskets						
100 calls	42.59	+3.4	16.92	–17.9	74.63	+4.1
260 calls	86.91	+2.3	41.32	+32.1	147.57	–4.7
Mobile call baskets						
30 calls	13.75	–18.3	5.07	–7.8	28.38	–12.4
100 calls	25.00	–24.2	8.60	–30.5	51.56	–12.3
300 calls	43.02	–31.6	8.60	–45.7	107.72	–23.2
900 calls	74.44	–39.3	16.90	–46.7	272.47	–10.5
Prepaid 40 calls	19.01	–12.6	5.42	–6.2	68.20	4.6
400 SMS	15.30	–33.0	5.15	–32.0	42.31	–36.0

Source: OECD elaboration based on Teligen.

StatLink http://dx.doi.org/10.1787/888932802059

The monthly price of the mobile baskets starts at USD 13.75 PPP per month (30 call basket) up to USD 74.44 PPP per month for the 900 mobile calls basket. Consumers pay on average USD 25 PPP for the 100 calls basket and USD 43.02 PPP for the 300 calls basket.

At the same time, the price of the least expensive country has fallen across all baskets. This has also been the trend for the maximum prices, although there were some slight increases for specific baskets, such as the 420 calls fixed basket, where the price paid in the most expensive country increased by 9.4%, or the 100 calls fixed business basket, where it increased by 4.1%.

Leased line baskets

Even though comprehensive data on leased line prices are not available across OECD countries, data are presented here from selected OECD countries on prices for a national 34 Mbit/s leased line (Figure 7.5, Table 7.17). Furthermore, the overall price evolution for the 1992-2012 period for a 2 Mbit/s leased line is reported (Table 7.18). Leased line prices declined sharply from 1992 until approximately 2003. They seem to have stabilised since 2005 at approximately one third of their 1992 price level.

Figure 7.5. **National OECD leased lines basket, 34 Mbit/s, August 2012, per month, VAT excluded**

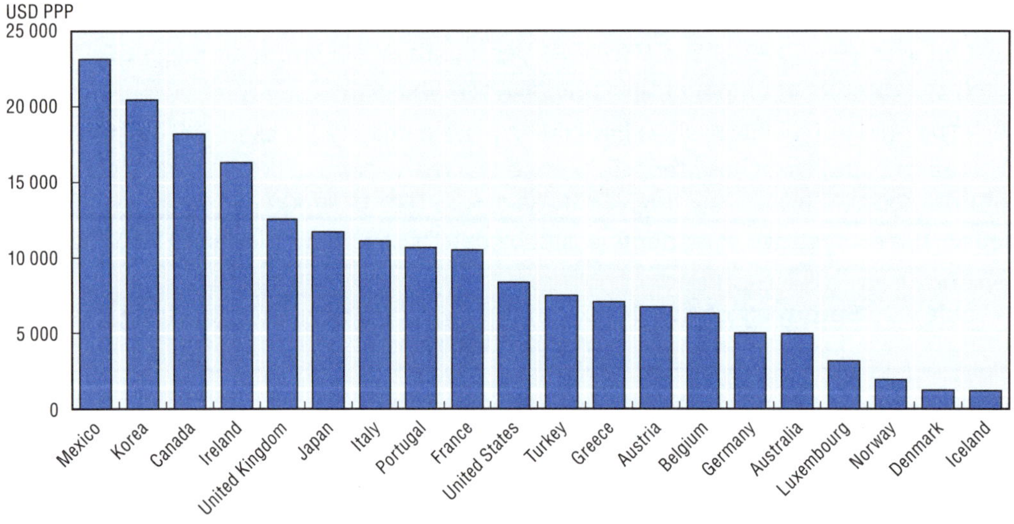

Source: OECD and Teligen.

StatLink http://dx.doi.org/10.1787/888932799589

Fixed broadband baskets

The OECD adopted the revised fixed broadband baskets in 2009 and has undertaken regular reporting since 2011. Basket specification is largely driven by speed ranges and data usage per month. As a result, certain speed thresholds have been established for the development of fixed broadband baskets, each of which has a low and a high usage profile (Table 7.2). For example, the baskets with the lowest speeds, 256 Kbit/s or above, are split into two different usage profiles: up to 2 GB/month (or lower) and 6 GB per month. The different profiles only have an impact on the final price of the basket for those countries or operators that billed their customers on the basis of usage. Those operators offering unlimited download capacity have an identical price for low and high usage profile

Table 7.2. **Fixed broadband basket profile, by advertised download capacity**[1]

	Usage per month			
	Lower use		Higher use	
Speed range	GB	Hours	GB	Hours
> 0.25 Mbit/s	2	10	6	30
> 2.5 Mbit/s	6	15	18	45
> 15 Mbit/s	11	20	33	60
> 30 Mbit/s	14	25	42	75
> 45 Mbit/s	18	30	54	90

1. In some countries, such as Chile or Mexico, the uptake of high-speed broadband offers is limited.
Source: OECD and Teligen.

StatLink http://dx.doi.org/10.1787/888932799608

baskets. Even though number of hours of use is a parameter in the fixed broadband baskets, the number of offers that bill customers by the hour was very low. Among the nearly 700 offers originally included in the data collection, only a handful of entry-level offers in Italy, Portugal and Spain followed a time-based charging model and, consequently, these were not retained for the final dataset. Offers listed for a country may not be widely available within that country, although it is a requirement that they are available in the country's largest city. Reported speeds are those advertised and may not be accurate representations of speeds actually received by customers of the service.

The current fixed wired broadband price data collection includes 691 standalone broadband offers across the 34 OECD countries (Tables 7.19 to 7.28). If stand-alone broadband is not available from a given operator, then the least expensive bundled package was selected and included in the comparison. For the sake of consistency, redundant offers in terms of broadband service provided have been removed from the list, for example, if the same broadband service is provided under different bundles. For cases where differences exist between rates offered to existing and new customers, the dataset generally includes those offers available to new customers. The dataset does not include offers addressed to special groups or users fulfilling additional requirements such as broadband offers for students. The comparison includes line charges, that is, charges consumers may face when subscribing to a broadband service, but that do not strictly correspond to the broadband service. For example, if subscribers must buy a telephone line or a pay television service to obtain a broadband subscription, those charges are accounted for as "line charges".

The lowest speed tier (> 256 Kbit/s) of the fixed broadband baskets has two levels of usage, 2 GB and 6 GB, but the resulting prices are fairly similar. The average price across OECD for this basket was USD 27.19 PPP for the 2 GB profile and USD 28.39 PPP for the 6 GB profile. The two most expensive countries for these two different baskets also coincide: Spain (USD 47.70 PPP) and Norway (USD 43.09 PPP), as do two of the three least expensive countries: Estonia (USD 14.39 PPP) and Korea (USD 16.35 PPP). By way of contrast, Hungary has the least expensive offer in the OECD area for the 2 GB data allowance (USD 12.85 PPP), but only the sixth least expensive for the 6 GB allowance within this speed tier, as the initial offer was limited to 5 GB data capacity (Figures 7.6 and 7.7).

Figure 7.6. **OECD Fixed Broadband basket Low 1: 2 GB, 0.250 Mbit/s and above, Sept. 2012**

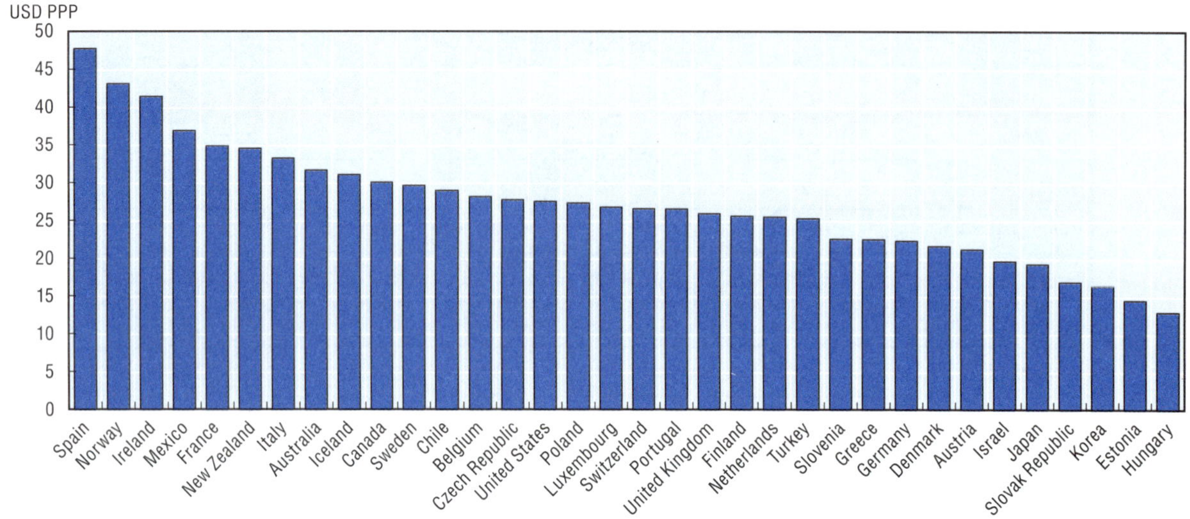

Source: OECD and Teligen.

StatLink http://dx.doi.org/10.1787/888932799627

Figure 7.7. **OECD Fixed Broadband basket High 1: 6 GB, 0.250 Mbit/s and above, Sept. 2012**

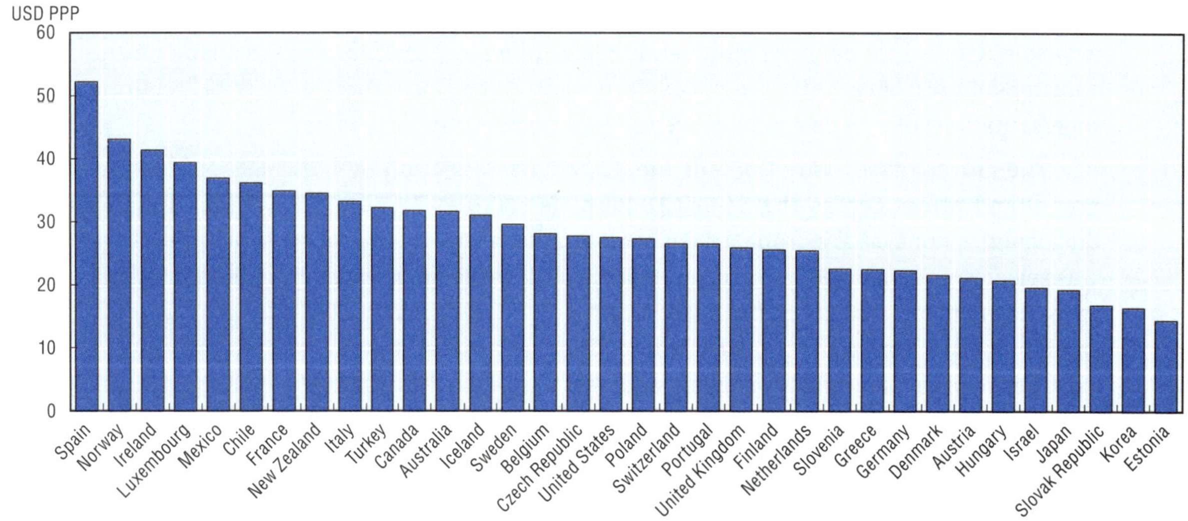

Source: OECD and Teligen.

StatLink http://dx.doi.org/10.1787/888932799646

A similar observation is valid for the broadband baskets with speeds higher than 2.5 Mbit/s of advertised download speed. Most countries retain their relative positions with a few experiencing slight variations explained by the usage-based billing of their services. For baskets including data allowances of 6 GB (low usage) and 18 GB (high usage) the resulting average prices for the OECD were USD 30.23 PPP and USD 31.36 respectively. Estonia (USD 19.23), Slovak Republic (USD 16.83) and Korea (16.35) are the least expensive countries for both levels of usage, while Spain, Chile and Norway have the highest prices for the 6 GB allowance (between USD 46 and USD 52). Turkey is, however, the most expensive country if the 18 GB usage level is considered, with prices as high as USD 58.66 in PPP terms (Figures 7.8 and 7.9).

Figure 7.8. **OECD Fixed Broadband basket Low 2: 6 GB, 2.5 Mbit/s and above, Sept. 2012**

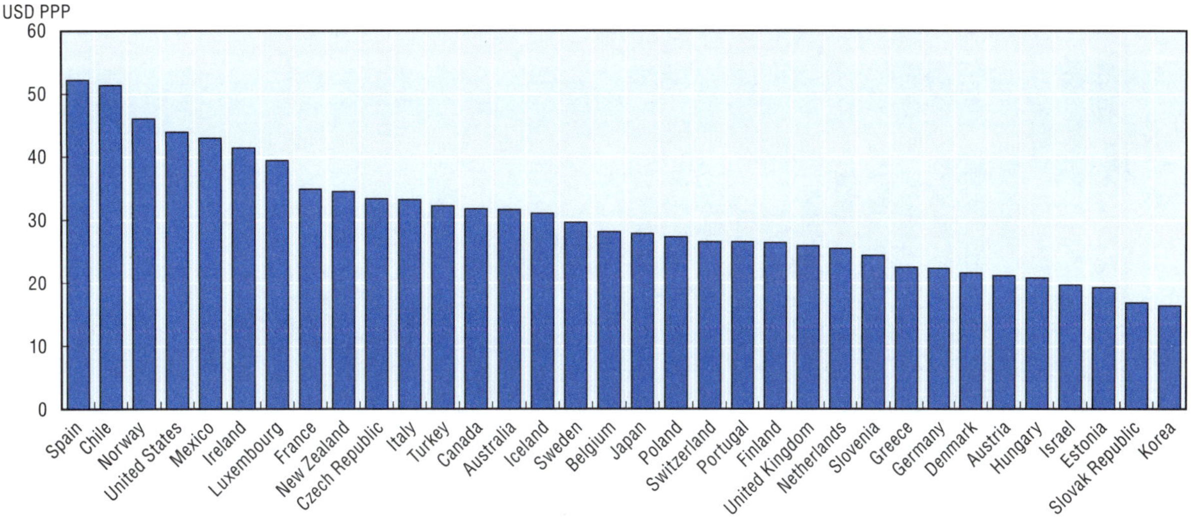

Source: OECD and Teligen.

StatLink http://dx.doi.org/10.1787/888932799665

Figure 7.9. **OECD Fixed Broadband basket High 2: 18 GB, 2.5 Mbit/s and above, Sept. 2012**

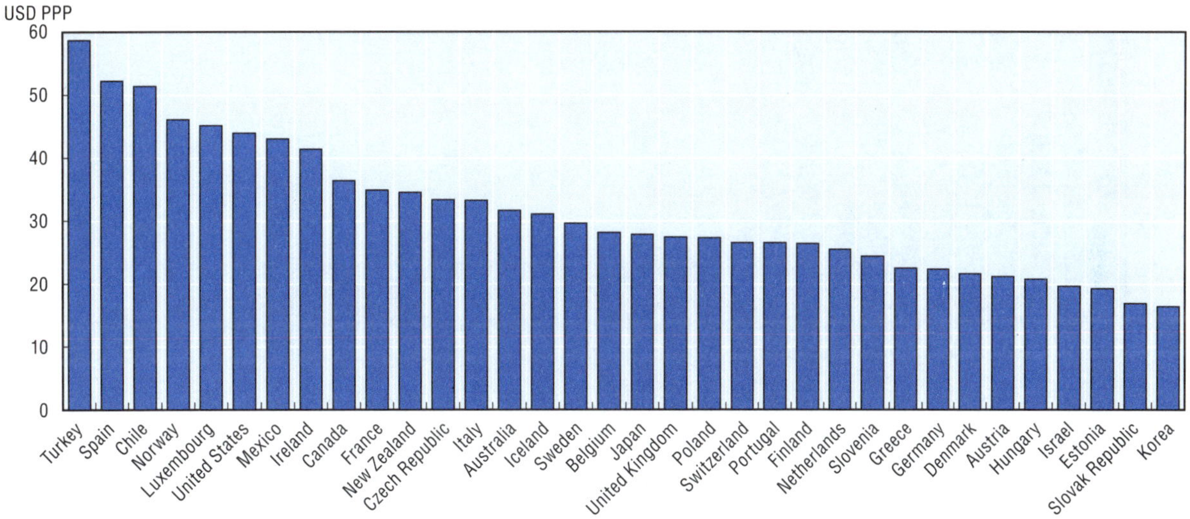

Source: OECD and Teligen.

StatLink http://dx.doi.org/10.1787/888932799684

For speeds higher than 15 Mbit/s, across two usage levels of 11 GB and 33 GB, the average price across the OECD is USD 35.00 PPP for lower usage and USD 38.06 for higher use. Chile (USD 66.62) and Mexico (66.02) have the highest prices for this speed tier for the 11 GB profile. The price in Mexico, however, is much higher for the 33 GB profile – up to 132.45 – which reflects high overage charges above a certain limit (10 GB). Chile (66.62) and Turkey (USD 63.57) have the second and third-highest charges for the 33 GB profile (Figures 7.10 and 7.11).

It can be noted, however, that distribution of broadband subscriptions across speed tiers varies significantly across countries. In some cases, newly launched high-speed offers may have higher prices than is the case for speed tiers with higher adoption rates, and may

Figure 7.10. **OECD Fixed Broadband basket Low 3: 11 GB, 15 Mbit/s and above, Sept. 2012**

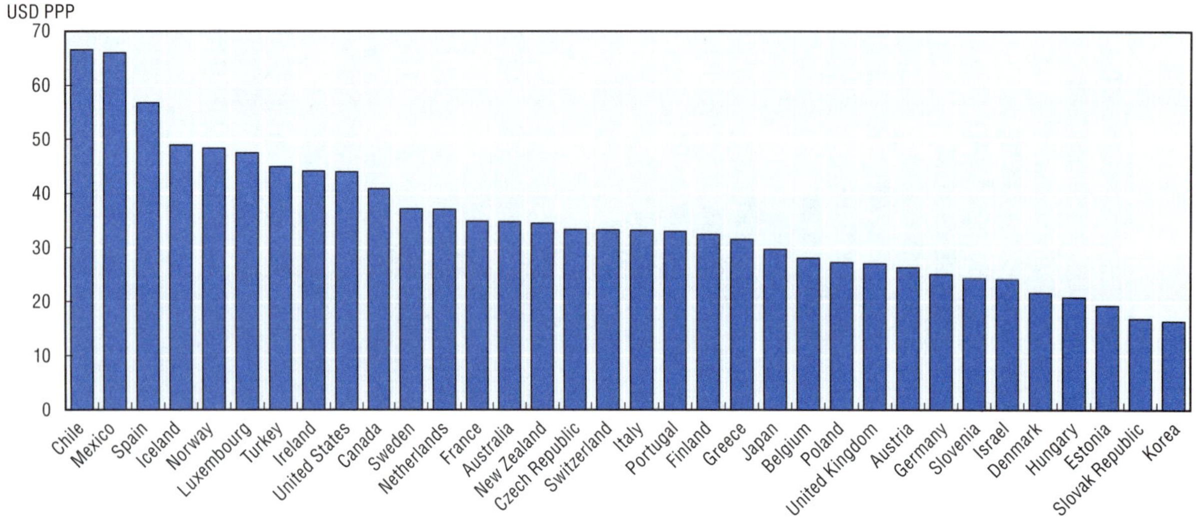

Source: OECD and Teligen.

StatLink http://dx.doi.org/10.1787/888932799703

Figure 7.11. **OECD Fixed Broadband basket High 3: 33 GB, 15 Mbit/s and above, Sept. 2012**

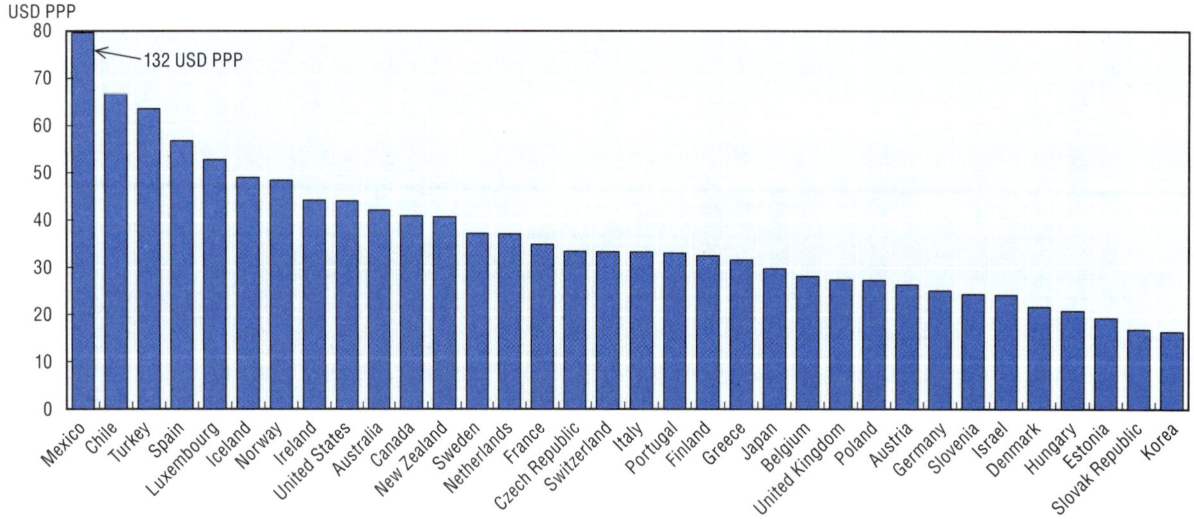

Source: OECD and Teligen.

StatLink http://dx.doi.org/10.1787/888932799722

be subject to increased competitive pressure. In the case of Chile, in 2012, only 18% of broadband customers enjoyed advertised speeds higher than 10 Mbps. In this country, geographic and demographic conditions also shape high-speed broadband supply limiting network deployment and economies of scale outside Santiago. In Chile's capital, broadband offers in the range of 100 Mbps of download speed were launched in 2011 when the largest cable operator started offering a 120 Mbps plan. This was the fastest speed in South America and was associated with a premium price for that market.

The next speed tier accounts for offers at 30 Mbit/s or higher speeds with two different data allowances: 14 GB and 42 GB. The average price across the OECD is USD 44.66 PPP (lower profile) or USD 50.15 for the higher profile (Figures 7.12 and 7.13). The convergence

Figure 7.12. **OECD Fixed Broadband basket Low 4: 14 GB, 30 Mbit/s and above, Sept. 2012**

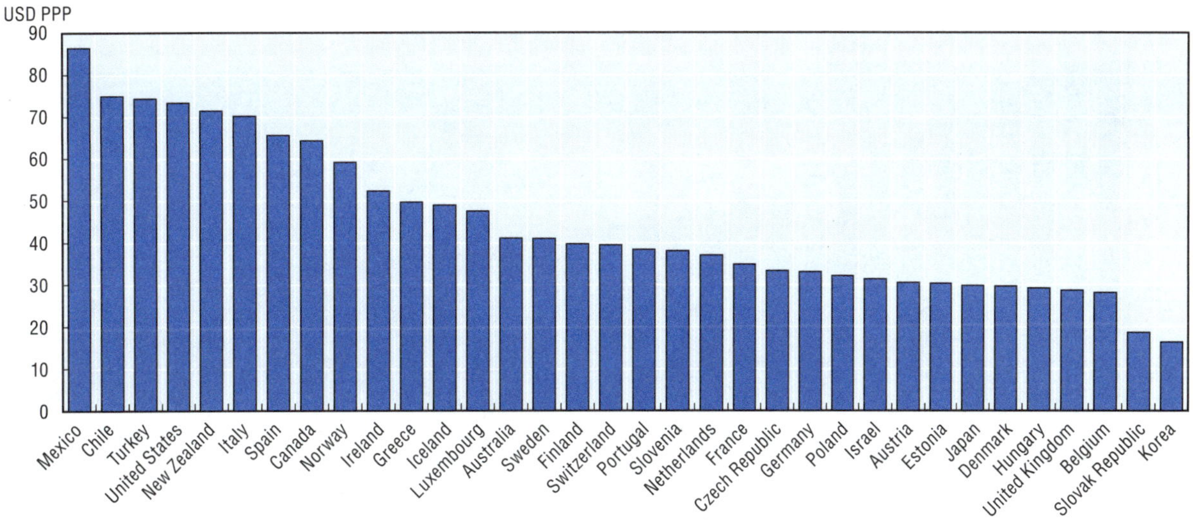

Source: OECD and Teligen.

StatLink http://dx.doi.org/10.1787/888932799741

Figure 7.13. **OECD Fixed Broadband basket High 4: 42 GB, 30 Mbit/s and above, Sept. 2012**

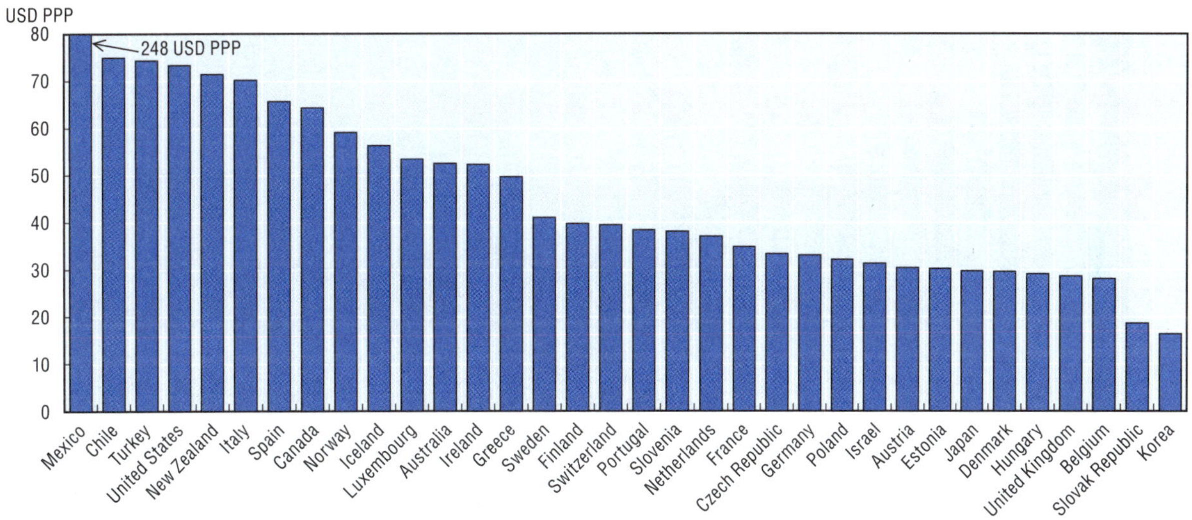

Source: OECD and Teligen.

StatLink http://dx.doi.org/10.1787/888932799760

of average prices for both usage patterns means that unlimited offers are more important for higher speeds. Six countries (Chile, Italy, Mexico, New Zealand, Turkey and the United States) have prices higher than USD 70, whereas the least expensive countries are Belgium (28.10), Slovak Republic (USD 18.62) and Korea (USD 16.35). In the case of Mexico, high-speed offers (at 50 Mbps) are limited by a 20 GB cap. Additional GB may be added at an extra charge, which causes a dramatic increase in the price paid for the 42 GB basket (close to USD 250 PPP).

Finally, the group of connections with the highest speeds has an advertised download speed higher than 45 Mbit/s for usage profiles of 18 GB or 54 GB (Figures 7.14 and 7.15). The average price across OECD countries was USD 52.01 PPP for the lower usage profile (USD 60 for the higher profile). As happened with the 30 Mbit/s speed tier, most prices do not

Figure 7.14. **OECD Fixed Broadband basket Low 5: 18 GB, 45 Mbit/s and above, Sept. 2012**

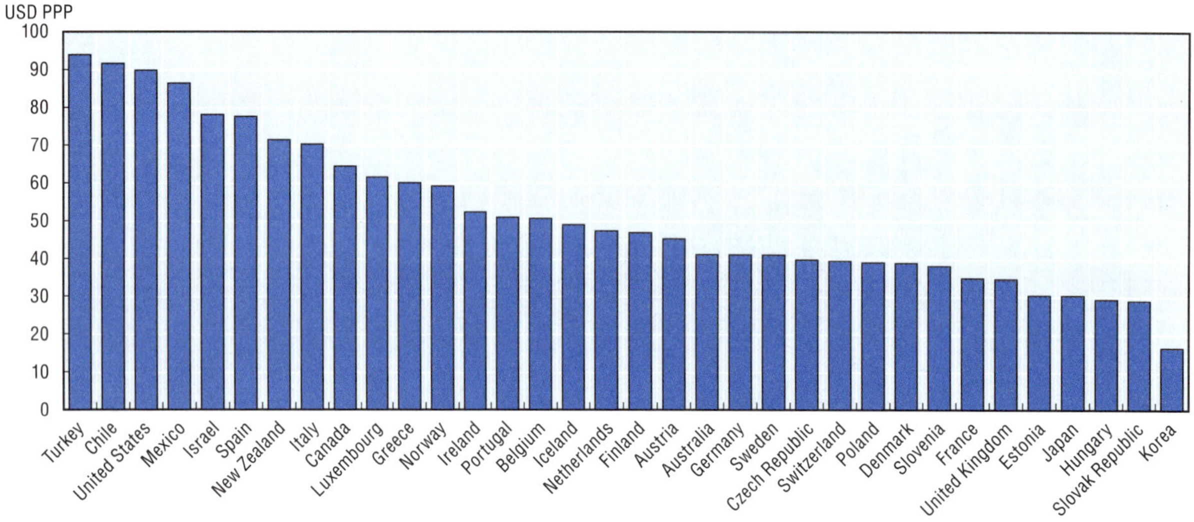

Source: OECD and Teligen.

StatLink http://dx.doi.org/10.1787/888932799779

Figure 7.15. **OECD Fixed Broadband basket High 5: 54 GB, 45 Mbit/s and above, Sept. 2012**

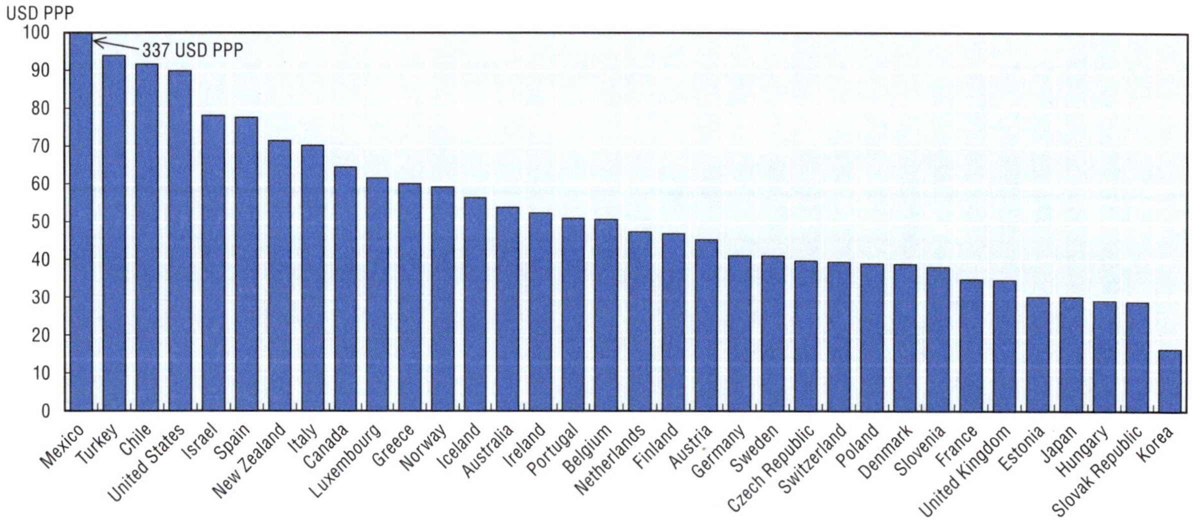

Source: OECD and Teligen.

StatLink http://dx.doi.org/10.1787/888932799798

change across usage profiles, the most expensive countries being Turkey (USD 93.96 PPP), Chile (USD 91.62) and the United States (USD 89); and the least expensive being Hungary (USD 29.13), Slovak Republic (28.80) and Korea (USD 16.35). The price paid for higher data requirements in Mexico jumps to over USD 300 PPP.

A useful indicator of the prices consumers are paying for broadband connections is the range of prices. Even though some of the selected offers for this comparison may cover bundles, such as in France or Spain where very little stand-alone broadband is sold, these provide a fair idea of entry-level prices for a broadband connection, especially if countries are ranked by minimum price. These prices take into account line charges, as discussed above (Figure 7.16).

Figure 7.16. **Broadband subscription price ranges, Sept. 2012, including line charge, USD PPP**

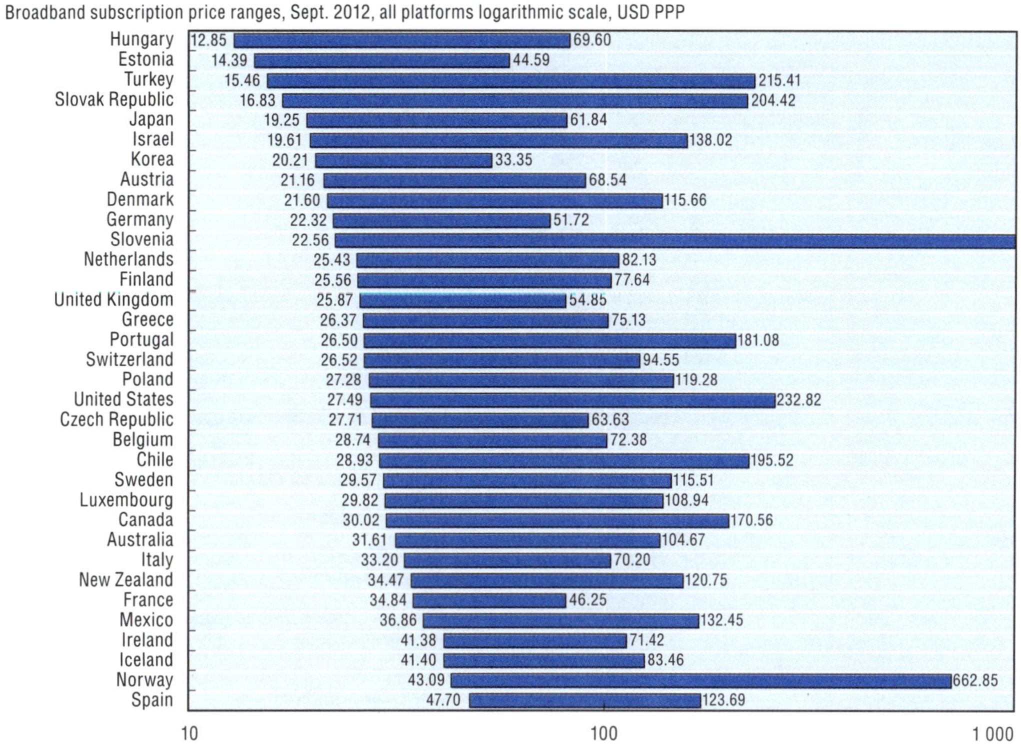

Source: OECD and Teligen.

StatLink http://dx.doi.org/10.1787/888932799817

Hungary (USD 12.85 PPP) and Estonia (USD 14.39) have the lowest entry price for a broadband subscription, while the average price for the OECD was USD 27.55 PPP. For most countries, entry-level broadband subscriptions are within the USD 20-30 range, with the most expensive countries being Iceland (USD 41.40), Norway (43.09) and Spain (47.70).

Another way to measure broadband prices is to consider the price per megabit per second of advertised speed, also including line charges (Figure 7.17). This comparison favours plans with higher speeds, however, as they usually provide lower prices per Mbit/s. Countries with very high speed offers, such as Japan and Sweden, lead the OECD ranking accordingly for this metric. Conversely, countries where operators offer lower speeds are also characterised by higher prices per Mbit/s, at least if the lowest price is considered. Mexico has the highest price per Mbit/s of advertised speeds, if countries are ranked by lowest available price. At least USD 1.69 per Mbit/s is paid in Mexico, down from USD 11.77 in September 2010, followed by Greece (USD 1.25) and Turkey (USD 1.12). At the other end of the ranking, Japan (USD 0.04), the Netherlands (USD 0.08) and Sweden (USD 0.11) are the least expensive countries according to this metric.

Price-speed evolution of broadband offers in OECD countries

The following data show the evolution of broadband speeds and prices in OECD countries between 2010 and 2012 (Figures 7.18 and 7.19, Tables 7.29 and 7.30). Speeds have experienced a steep increase in some cases at two-digit rates, while broadband prices have slightly increased. This is due to the fact that operators have upgraded their broadband offers, while trying to extract more revenues from enhanced connections. The graphs above (Figures 7.18 and 7.19) were developed by tracking a given offer from the incumbent

Figure 7.17. **Broadband subscription prices ranges per megabit per second of advertised speed, with line charges, Sept. 2012, USD PPP**

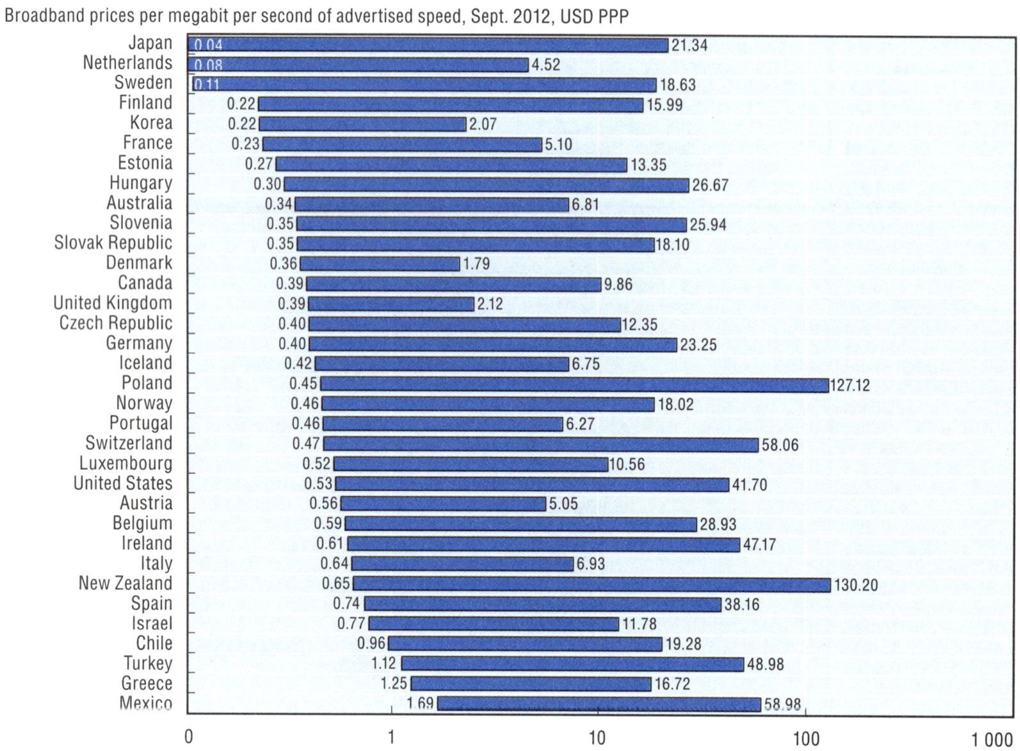

StatLink http://dx.doi.org/10.1787/888932799836

Figure 7.18. **Incumbent broadband price and speed changes, ADSL or fibre, Sept. 2010-Sept. 2012**

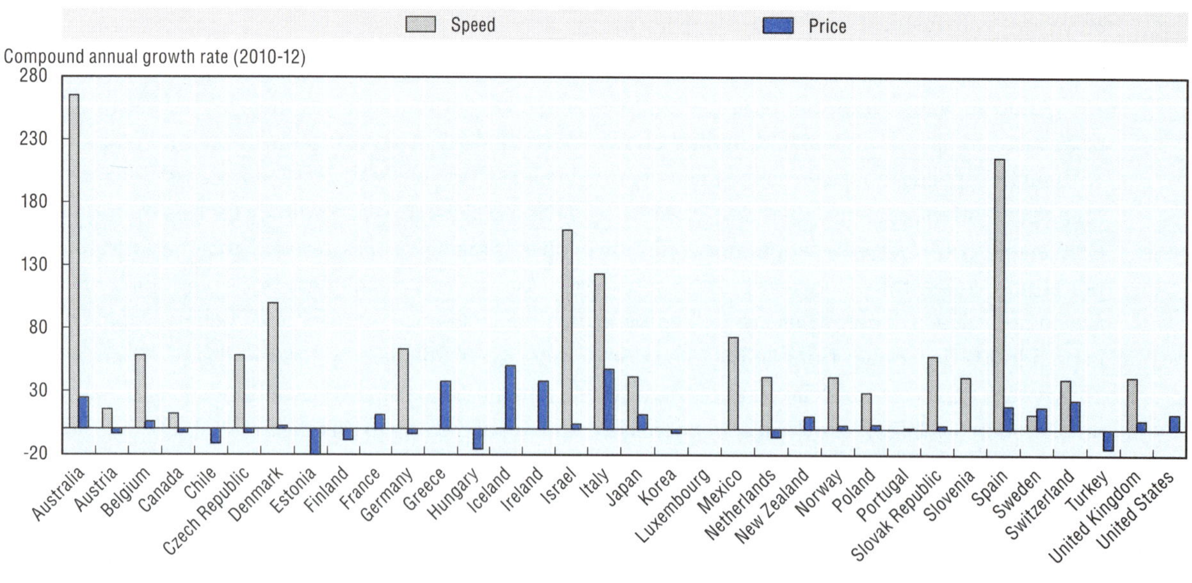

StatLink http://dx.doi.org/10.1787/888932799855

Figure 7.19. **Cable broadband price and speed changes, Sept. 2010-Sept. 2012**

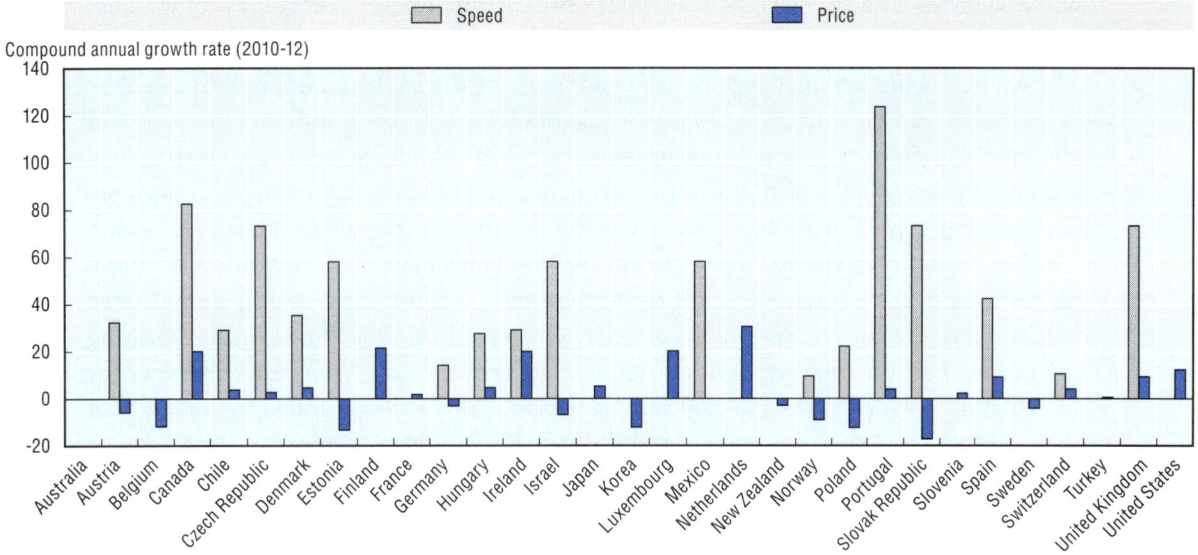

Source: OECD and Teligen.

StatLink http://dx.doi.org/10.1787/888932799874

operator or leading cable provider. While the most popular broadband package in a given country may or may not correspond to the selected offers, this methodology provides a consistent way of tracking broadband prices from a given operator over time. All prices have been expressed in local currency terms, without considering line charges. This means that emphasis needs to be placed on actual price/speed evolution, rather than on the absolute values that underpin this comparison.

Price comparison conducted using the above methodology showed an increase between 2010 and 2012 of CAGR 6.76% for incumbent providers and 2.48% for cable operators. Concurrently, the speed of broadband connection increased by 43.81% per annum for incumbent operators and by 27.49% for cable operators. Customer upgrades that augmented download speeds at a slight price increase were the principle cause of this trend. Some operators, such as Telstra in Australia, UPC in the Slovak Republic and Ono in Spain, have provided customers with a twofold or threefold upgrade in download speeds. In some countries, costs associated with international connectivity form a significant element of retail broadband prices. Countries that are more distant from larger international traffic routes or which have less developed regional infrastructure to exchange traffic, such as Chile, may face prices several times higher than in Europe or the United States. This is currently being addressed by a regional initiative to deploy a fibre backbone ring in South America, which will interconnect the countries in the region to two new submarine cables, from Fortaleza (Brazil) to peering centres in the Northern hemisphere.

Wireless broadband baskets

The OECD's fixed line, mobile and leased line baskets were revised in 2009 to include a fixed broadband basket for the first time. This was followed, in 2012, by the introduction of a wireless broadband basket reflecting the growing significance and use of this service and the development of an agreed methodology.

Given the very distinct usage and pricing patterns of fixed wireless services (e.g. fixed WiMAX, satellite), wireless broadband baskets focus only on mobile technologies, such as

UMTS, HSPA/HSPA+, CDMA2000, IEEE 802.16e or LTE, which represent close to the totality of wireless broadband services, as fixed wireless accounts for, on average, 1% of wireless broadband subscriptions in OECD countries.

The wireless broadband usage patterns adopted for the baskets were underpinned by data gathered and consultation with all stakeholders, including industry players and regulators. This process revealed that usage patterns for handsets (smartphones), tablets and laptops present substantial differences, especially between handsets and laptops/ tablets. At that stage, for example, data usage for handsets was always bundled with a voice plan, or more generally with a plan including voice calls, SMS and other services. That is still largely the case although new offers may emerge. As a result, the wireless broadband baskets for handsets are currently conceived as an addition to the existing OECD mobile baskets and are defined using their terminology. Therefore, the cost for 500 MB of data is added to the 100 calls mobile basket, 1 GB of data is added to the 300 calls basket, and so on. Reference to the mobile baskets should be understood as per the established OECD methodology (Table 7.3).

Table 7.3. **Mobile broadband basket proposal: tablet, laptop and handset use**

Laptop use (data volumes)	Tablet use (data volumes)	Handset use (data volumes + voice/SMS basket)
500 MB	250 MB	100 MB + 30 calls basket
1 GB	500 MB	500 MB + 100 calls basket
2 GB	1 GB	1 GB + 300 calls basket
5 GB	2 GB	2 GB + 900 calls basket
10 GB	5 GB	2 GB + 100 calls basket

Source: OECD and Teligen.

StatLink http://dx.doi.org/10.1787/888932799893

Laptop and tablet-based service provides connectivity using a laptop/tablet and a USB modem, dongle, and so on, without explicit voice capability. Handset-based service is provided through a smartphone or other type of mobile device with explicit voice capability. While many plans are available for laptop and tablet use, in many cases operators offer tablet-specific plans. Moreover, the evidence suggests that tablet usage is significantly lower than that of laptop, which is reflected in the baskets adopted.

Another important element of this methodology is the absence of speed-based baskets. At the time, evidence showed that speeds are not usually marketed and that actual speeds may be significantly lower than advertised speeds, with the gap considerably higher than for fixed broadband services. The situation may change for further network upgrades, such as LTE technology, and some operators may now include speed parameters in their marketing strategies and even, as in the case of Swisscom, a tariff differentiator. In order to capture this, high speeds are flagged in the results (Tables 7.31 to 7.45).

The presentation of the price benchmarking results for mobile broadband services first covers services provided over a handset or smartphone. As available evidence confirms that users purchase those services in conjunction with other mobile services, such as calls or SMS, the baskets are bundled and take existing mobile voice baskets as a reference. The handset-based mobile baskets are largely based on mobile baskets, presented earlier in this chapter, and both are presented again here for the sake of reference. Going forward, the mobile baskets would normally include mobile data services, and only one set of mobile baskets would need to be presented.

The basket with the lowest usage of data is the 30 call mobile basket, complemented by 100 MB of wireless broadband data, which costs on average USD 19.74 PPP per month (Figure 7.20). Japan had the most expensive offer (over USD 60) with Chile, Hungary and the United States following (around USD 30 PPP). It should be noted that mobile broadband plans in Japan focus particularly on higher usage with an overall adoption of flat-rate plans, visible also in the other baskets (e.g. Figures 7.25, 7.26, 7.30 and 7.31). Three countries, Austria, Estonia and the United Kingdom, had the least expensive offers (all under USD 10). At this point, some analysis could be undertaken on the share of costs corresponding to data charges but, given that many offers have some amount of data in their fixed charges, an extensive analysis of variable charges and its share allocated to data services would not be meaningful.

Figure 7.20. **OECD 30 calls + 100 MB mobile basket, August 2012, VAT included**

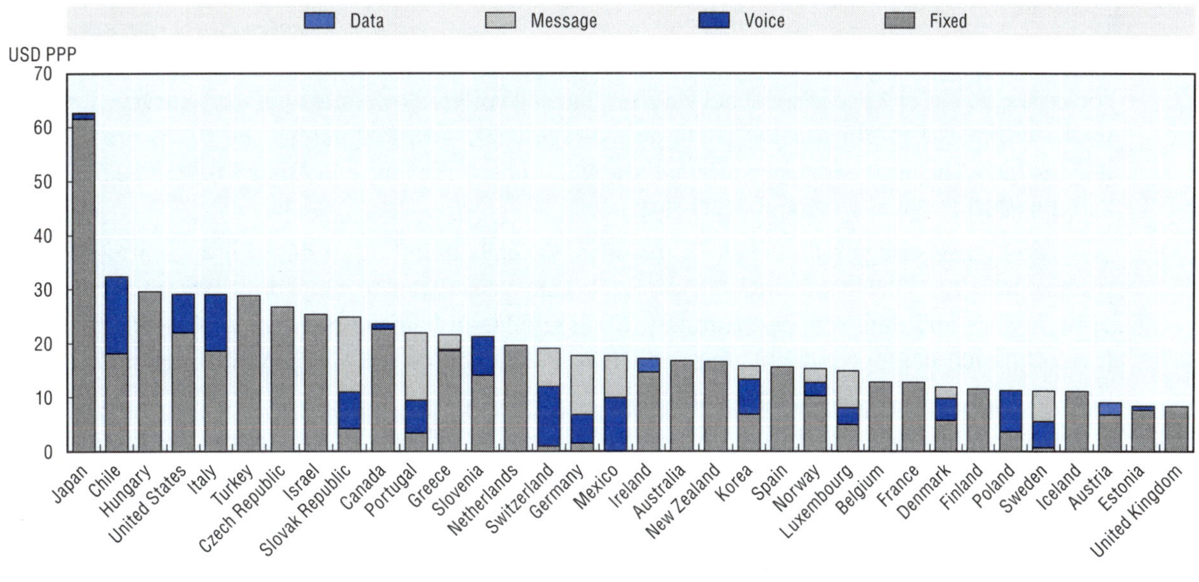

Source: OECD and Teligen.

StatLink http://dx.doi.org/10.1787/888932799912

The OECD 500 MB basket matches average smartphone usage in many OECD countries. It includes 100 calls, 140 SMS and 500 MB of data and costs an average of USD 33.63 PPP across the OECD area (Figure 7.21). Austria, Estonia and the United Kingdom were the least expensive countries for this consumption pattern (just above USD 10 per month), while the most expensive countries were Japan (USD 74.40), Chile (USD 62.90) and the United States (USD 55.00). It should be noted that, in this case, mobile broadband plans in Japan focus particularly on higher usage with an overall adoption of flat-rate plans. This is one reason why plans in Japan are particularly expensive for lower usage baskets compared to other OECD countries.

The next highest basket in terms of usage intensity is the 300 calls basket plus 1 GB of wireless data. As was the case with the 300 calls baskets without a data component, most

Figure 7.21. **OECD 100 calls + 500 MB mobile basket, August 2012, VAT included**

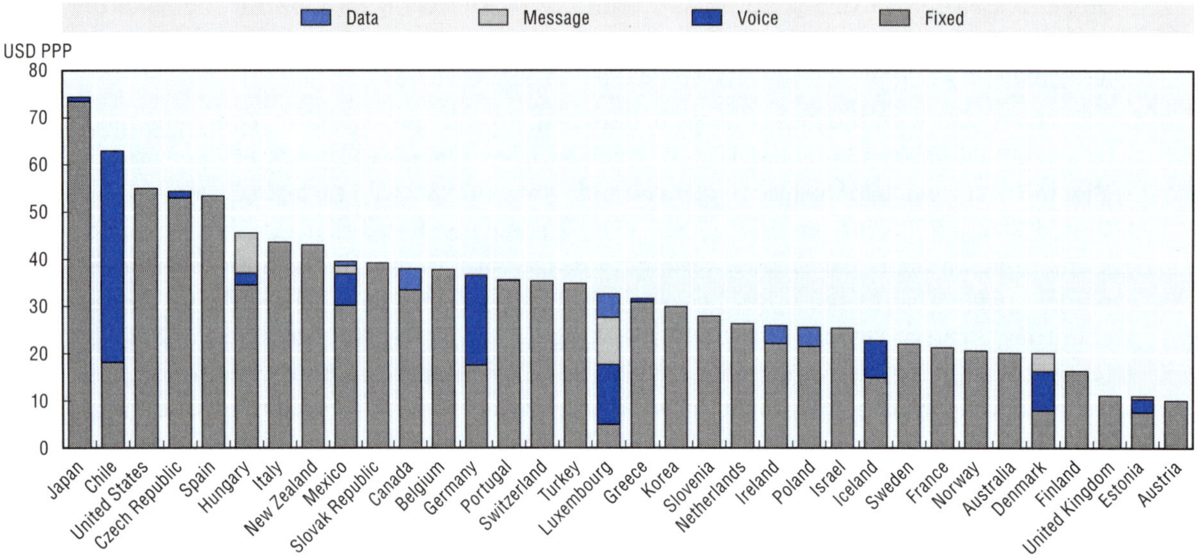

Source: OECD and Teligen.

StatLink http://dx.doi.org/10.1787/888932799931

countries (24 out of 33 countries) have a fixed package offered by some operators surveyed in the comparison. This basket costs on average USD 54.58 PPP across the OECD area (Figure 7.22). The least expensive offers in the OECD area were found in Austria (USD 10.08), Estonia (USD 15.30) and Finland (USD 19.25).

The highest usage mobile basket includes 900 calls and 2 GB of data for an average price of USD 90.70 PPP per month (Figure 7.23). This basket was developed for larger users of both voice and data services. Hungary, Chile and Japan had the most expensive offers (over USD 200 PPP per month). Austria (USD 17.83), France (USD 21.25) and Australia (USD 26.83) had the least expensive offers for this basket.

Finally, the OECD wireless broadband baskets also include a usage pattern of 2 GB combined with 100 calls, which aims to capture data intensive usage patterns with a moderate amount of voice services consumption (Figure 7.24). This basket costs on average USD 44.17 per month, with prices ranging from just above USD 80 (Germany) or just above USD 70 (Japan), down to just above USD 10 (Austria and Estonia).

Wireless broadband baskets for laptop and tablets

Unlike the handset-based wireless broadband baskets, the baskets for laptop and tablet use are stand-alone, meaning that voice and SMS elements are not counted in the calculations. This is in accordance with evidence of consumption patterns revealed during the development of the price benchmarking methodology. Hence, five baskets were developed for laptop use (with data allowances from 500 MB to 10 GB) and five for tablet use (from 250 MB to 5 GB). The data collection included over 2 000 offers for residential customers in the 34 OECD countries.

The first laptop-based wireless broadband basket includes offers within the 500 MB allowance range (Figure 7.25). The average cost across the OECD was USD 13.04 in PPP

Figure 7.22. **OECD 300 calls + 1 GB mobile basket, August 2012, VAT included**

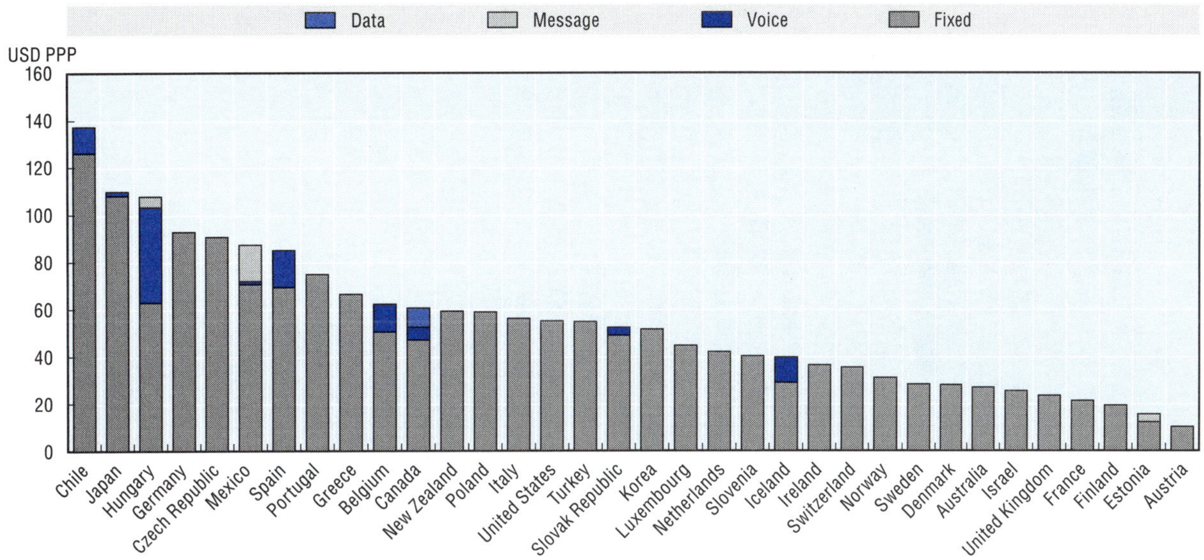

Source: OECD and Teligen.

StatLink http://dx.doi.org/10.1787/888932799950

Figure 7.23. **OECD 900 calls + 2 GB mobile basket, August 2012, VAT included**

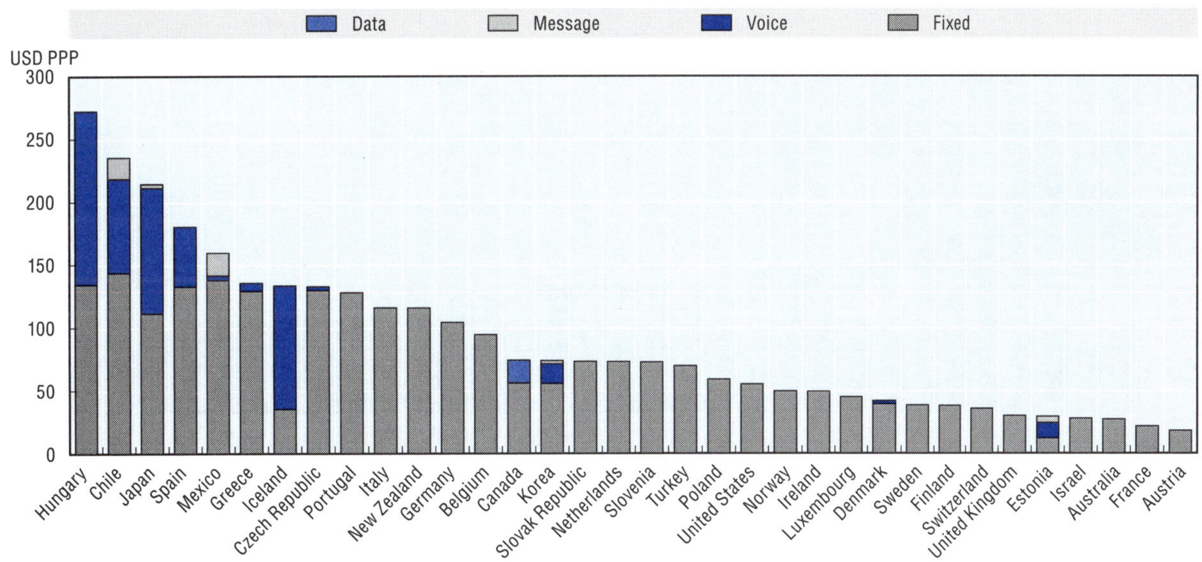

Source: OECD and Teligen.

StatLink http://dx.doi.org/10.1787/888932799969

terms, although countries with the highest costs, such as Japan whose plans are based on flat-rate tariffs, and the United States, were in the range of USD 30. This is in line with previous statements which confirm that average consumption in some countries, including these two, is higher and that operators focus on wireless broadband plans which provide higher allowances of data. By way of contrast, four countries, Austria, Denmark, Finland and the United Kingdom, have offers within the USD 5-6 interval for this consumption pattern.

Figure 7.24. **OECD 100 calls + 2 GB mobile basket, August 2012, VAT included**

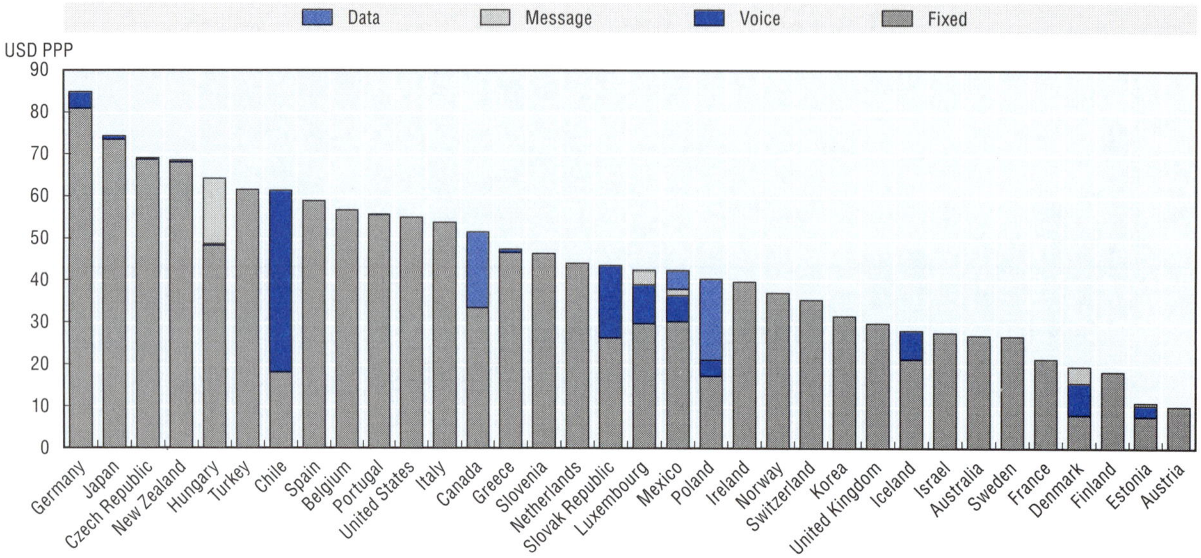

Source: OECD and Teligen.

StatLink http://dx.doi.org/10.1787/888932799988

Figure 7.25. **OECD Wireless Broadband basket, OECD Laptop 500 MB, September 2012**

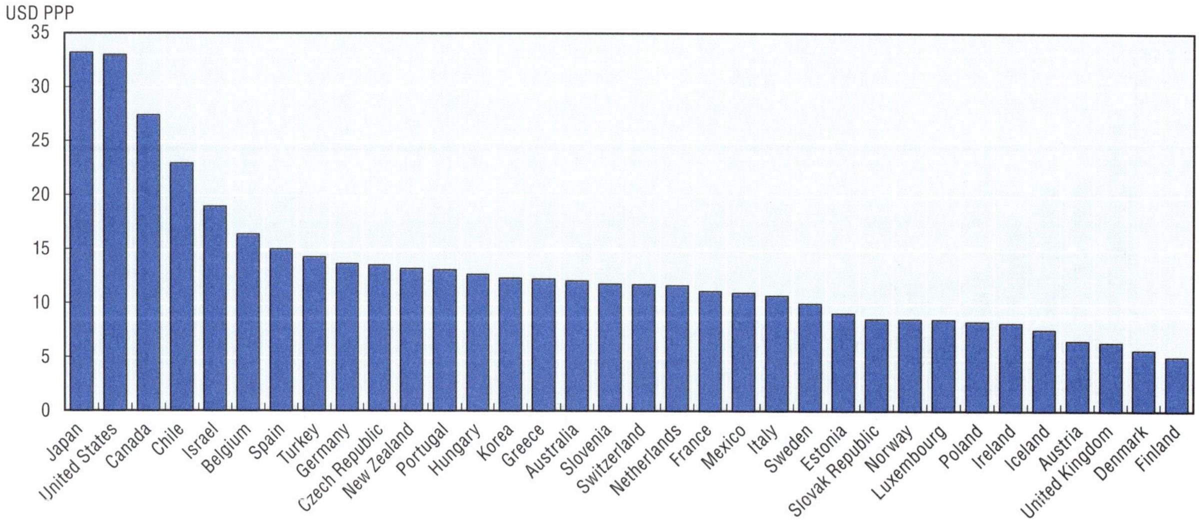

Source: OECD and Teligen.

StatLink http://dx.doi.org/10.1787/888932800007

The second lowest laptop broadband basket in terms of data allowance is the 1 GB basket (Figure 7.26). The average cost for this basket is slightly higher than that of the 500 MB basket: USD 15.26 per month. The highest and lowest tariff interval for this data allowance lies very close to the 500 MB basket. The least expensive countries such as Austria, Finland and Iceland are in the USD 5-7 range. The higher level of tariffs again remains between USD 25 and USD 33, giving a clear indication that many plans are valid both for the 500 MB and the 1 GB data allowance.

Figure 7.26. **OECD Wireless Broadband basket, OECD Laptop 1 GB, September 2012**

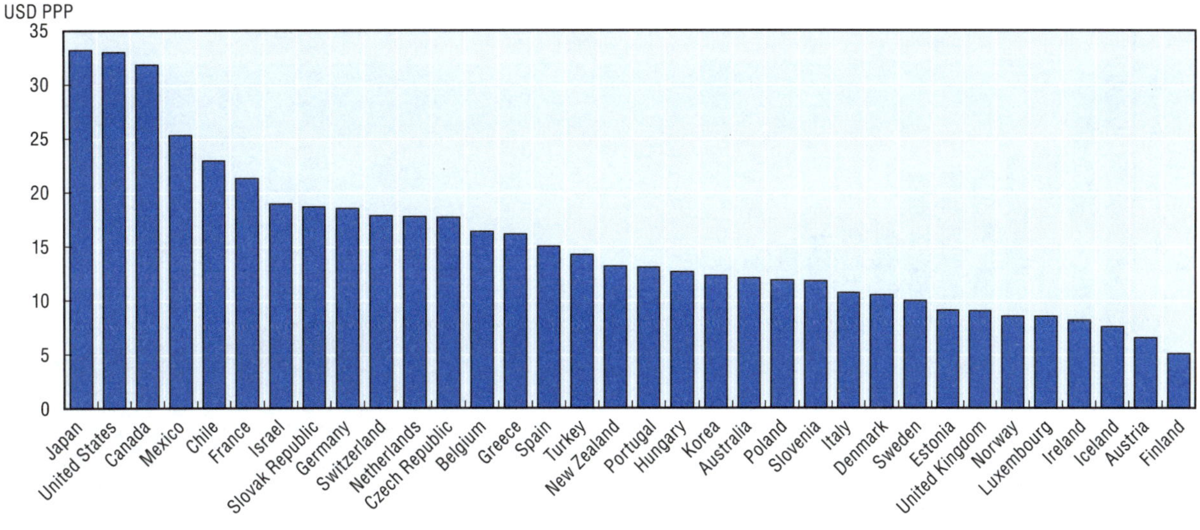

Source: OECD and Teligen.

StatLink http://dx.doi.org/10.1787/888932800026

The average price for the 2 GB wireless broadband basket for laptop is significantly higher than the two previous baskets, with an average cost across the OECD of nearly USD 20 PPP per month or USD 19.13 (Figure 7.27). The lowest prices were found in Finland (USD 7.11), Luxembourg (USD 8.48) and Estonia (USD 9.04), whereas the most expensive countries were Spain (USD 34.45), the United States (USD 33.61) and Japan (USD 33.20).

Figure 7.27. **OECD Wireless Broadband basket, OECD Laptop 2 GB, September 2012**

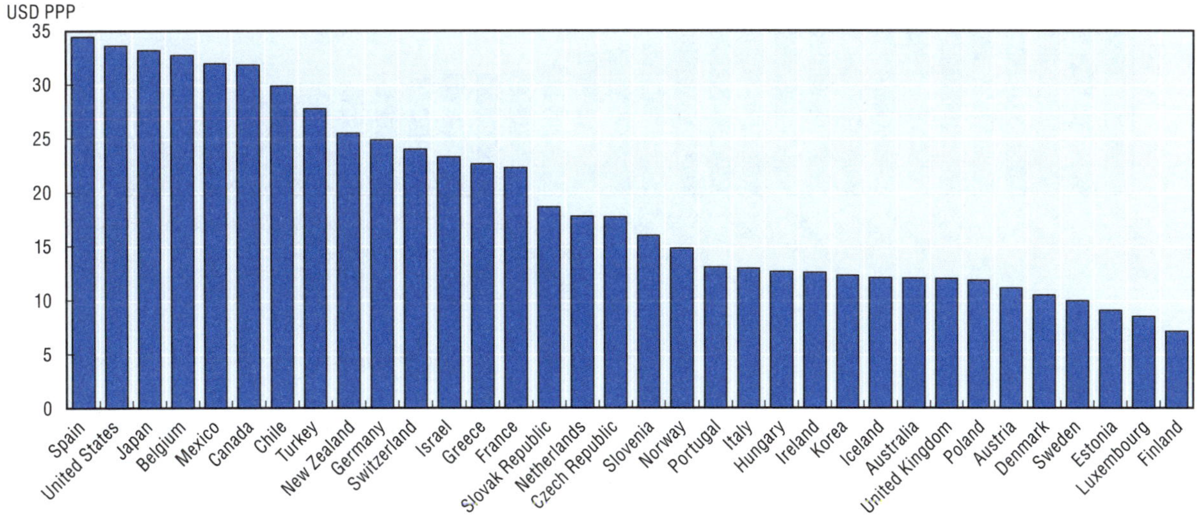

Source: OECD and Teligen.

StatLink http://dx.doi.org/10.1787/888932800045

Finally, the largest broadband baskets correspond to data allowances of 5 GB and 10 GB for laptop use over a month's period (Figures 7.28 and 7.29). Average expenditure across the OECD for these baskets was USD 26.70 for the 5 GB basket and USD 37.15 for the 10 GB basket. Some countries (Estonia, Finland) remain the least expensive for these two baskets,

Figure 7.28. **OECD Wireless Broadband basket, OECD Laptop 5 GB, September 2012**

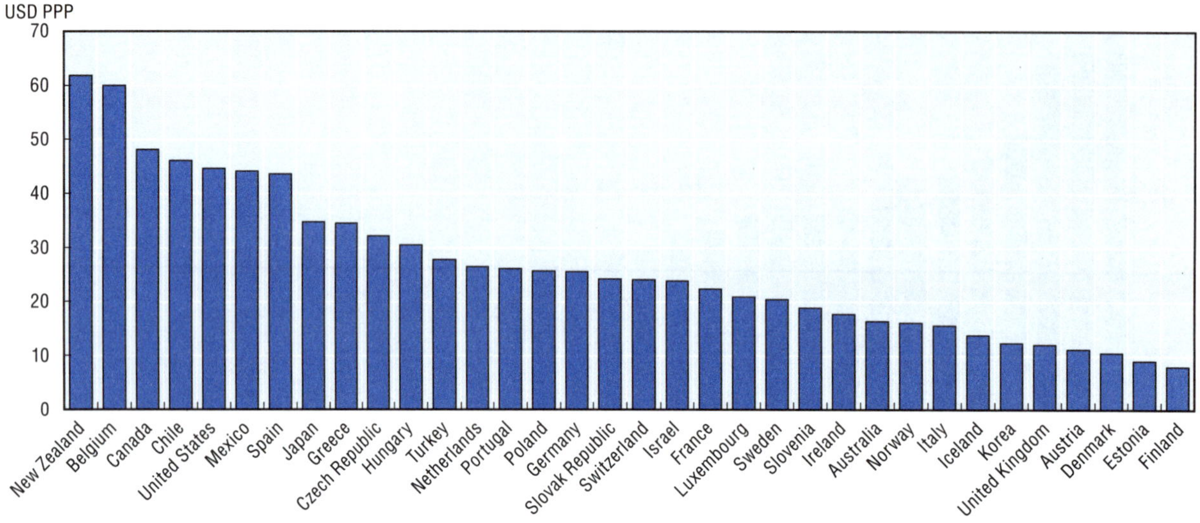

Source: OECD and Teligen.

StatLink http://dx.doi.org/10.1787/888932800064

Figure 7.29. **OECD Wireless Broadband basket, OECD Laptop 10 GB, September 2012**

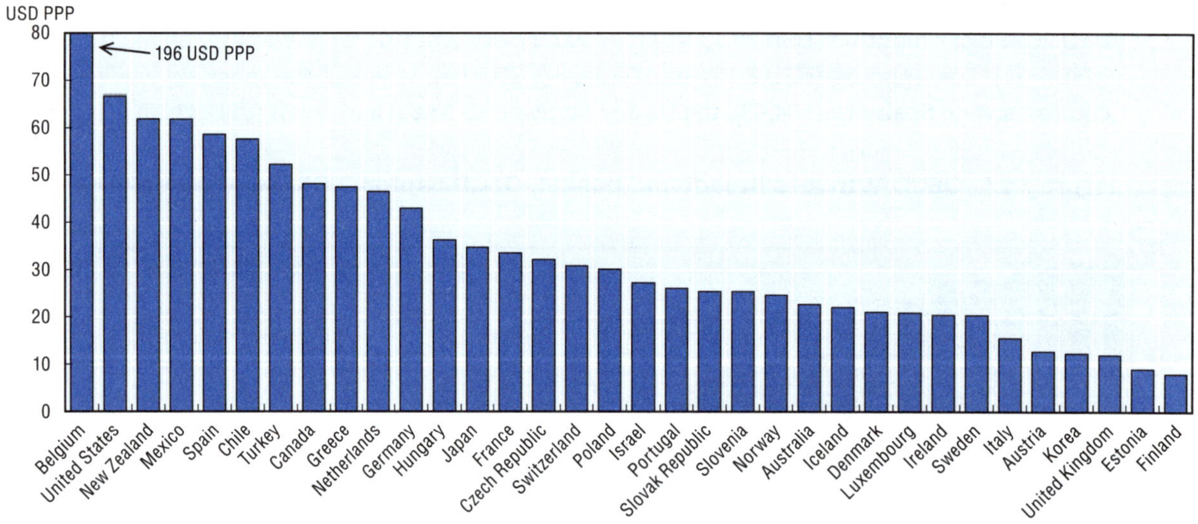

Source: OECD and Teligen.

StatLink http://dx.doi.org/10.1787/888932800083

with prices below USD 10. It should be noted, however, that these entry-level offers correspond to low download speeds. The most expensive 5 GB wireless broadband plans are those in New Zealand (USD 61.84), Belgium (60.01) and Canada (USD 48.11), whereas the highest prices for 10 GB were found in Belgium (USD 196.39), the United States (USD 66.61) and New Zealand (USD 61.84). The fact that Belgium has no plans with allowances higher than 4 GB, as data consumption beyond that level is subject to overage charges, makes wireless broadband services especially expensive for the 5 GB and 10 GB baskets. To a lesser extent, a similar finding is valid for New Zealand.

Even though the basket definitions for wireless broadband service do not take into account advertised download speeds, they may be an important factor for marketing purposes in some countries. For example, the least expensive plans for the 10 GB laptop basket correspond to low speeds (1 Mbit/s in Estonia and 400 Kbps in Finland), but some countries have tariffs with higher download speeds. Most of the offers selected for this basket (least expensive in their countries) advertise a download speed of between 2 Mbit/s and 20 Mbit/s, as is the case in 19 out of 34 countries. Two countries (Canada and Slovak Republic) offer 100 Mbit/s, while Norway's Netcom provides 80 Mbit/s and seven countries (France, Germany, Greece, Japan, Korea, Poland, Slovenia) advertised speeds in the range of 40 Mbit/s. This implies that advertised download speeds, while not a parameter in the basket definition, may be informative for consumers in relation to the service.

With regard to tablet-based wireless broadband baskets, it should be emphasized that in many cases plans for laptop and tablets are interchangeable. Notwithstanding this, usage patterns are believed to be distinct, with data consumption for tablets accounting for about half of that of laptops.

The lowest usage tablet broadband basket includes 250 MB of data for an average price of USD 11.02 per month across the OECD area (Figure 7.30). Countries in the highest range provide this data allowance for about between USD 15 and 20 (e.g. Chile, the United States) with the exception of Japan. Operators in this country, as highlighted above, promote flat-rate tariffs, which make entry-level offers relatively expensive. In this case, the lowest 250 MB plan for tablets costs USD 33.20 PPP. The least expensive offers are offered in Estonia, Denmark, Finland, Sweden and the United Kingdom, all between about USD 5 and USD 6.

Figure 7.30. **OECD Wireless Broadband basket, OECD Tablet 250 MB, September 2012**

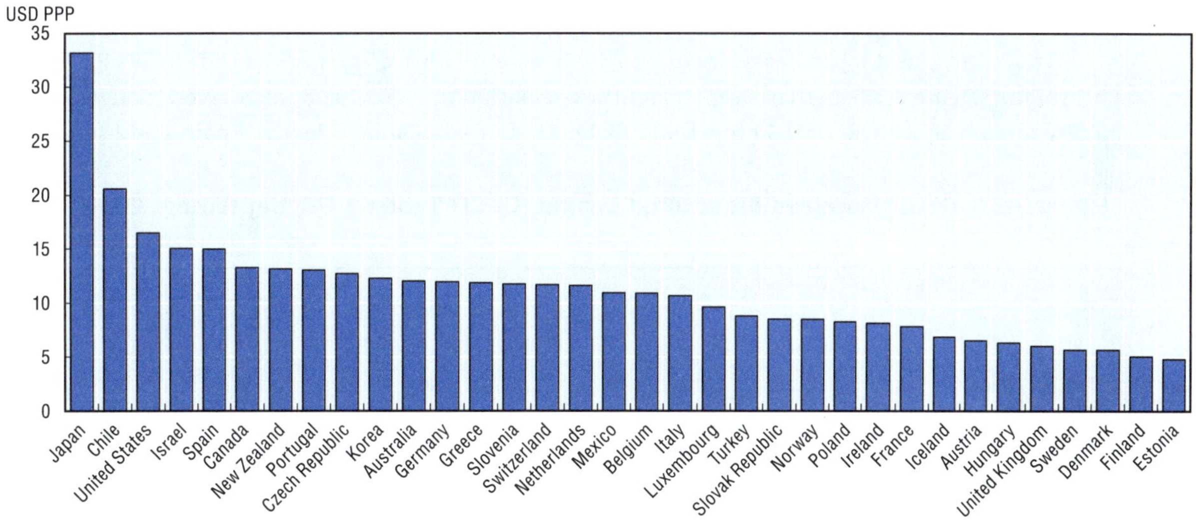

Source: OECD and Teligen.

StatLink http://dx.doi.org/10.1787/888932800102

The second lowest usage tablet broadband basket includes 500 MB of data. The average price in OECD countries is USD 11.96, just above the average price of the 250 MB basket (Figure 7.31). Denmark, Finland, Sweden and the United Kingdom remain the four least-expensive countries, whereas Japan (USD 33.20) with plans based on flat-rate tariffs, the United States (USD 32.99) and Chile (USD 20.60) have the highest costs for this consumption pattern.

Figure 7.31. **OECD Wireless Broadband basket, OECD Tablet 500 MB, September 2012**

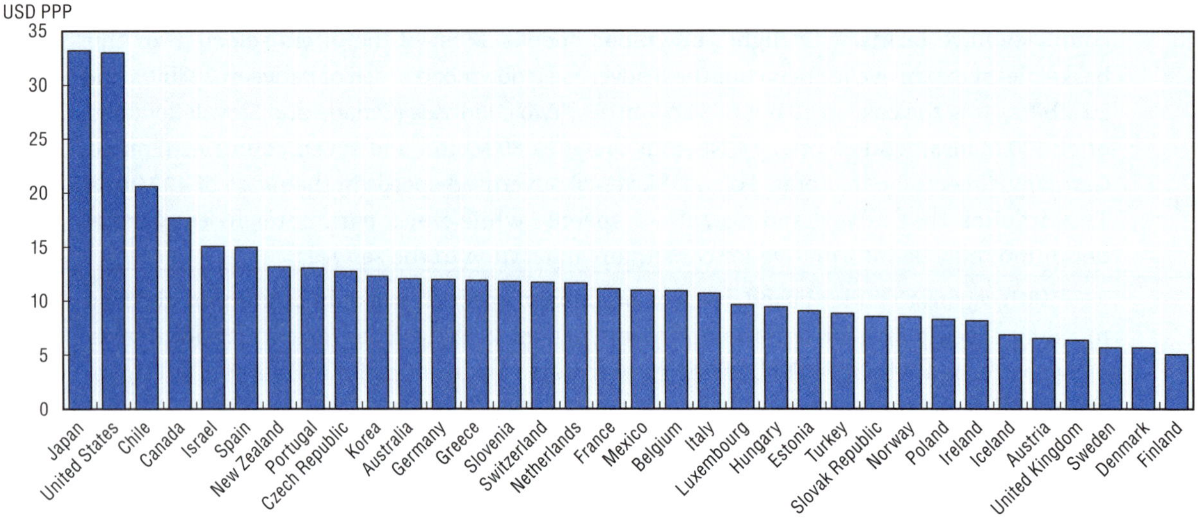

Source: OECD and Teligen.

StatLink http://dx.doi.org/10.1787/888932800121

The average cost for the 1 GB basket increases to USD 14.58 PPP per month, but most results only change slightly, as the majority of countries with the most expensive offers remain at the USD 25-35 level, while the least expensive countries provide the service at prices very similar to those retained for the previous baskets, between USD 5 and USD 7 (Figure 7.32).

The next highest tablet basket includes 2 GB of data for an average price of USD 18.58 PPP per month (Figure 7.33). In this case, 11 countries lie within the USD 15-25 range, even though a few remain above the USD 30 threshold (Belgium, Canada, Japan, Mexico, Spain and the

Figure 7.32. **OECD Wireless Broadband basket, OECD Tablet 1 GB, September 2012**

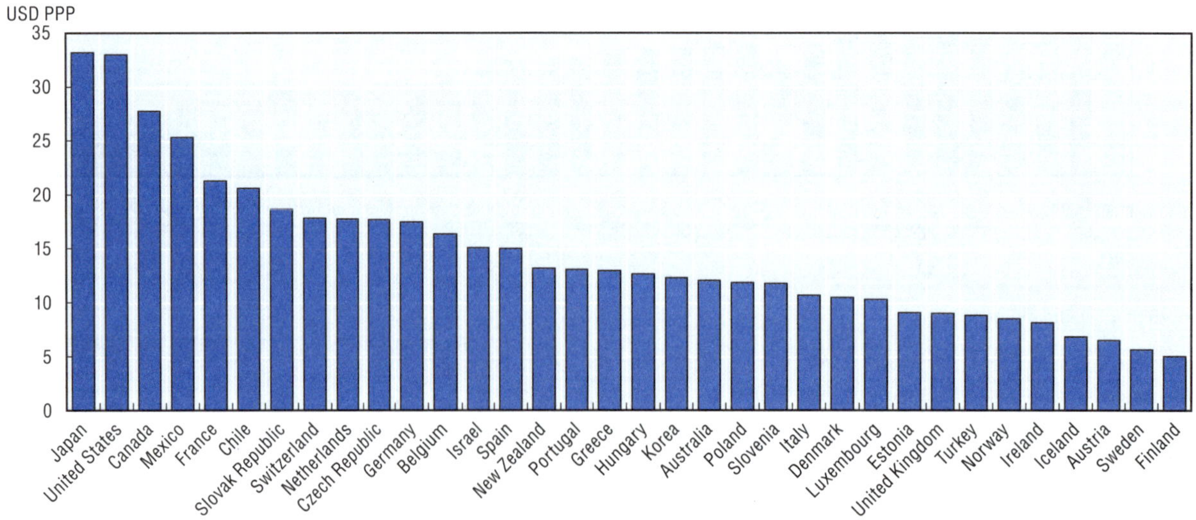

Source: OECD and Teligen.

StatLink http://dx.doi.org/10.1787/888932800140

Figure 7.33. **OECD Wireless Broadband basket, OECD Tablet 2 GB, September 2012**

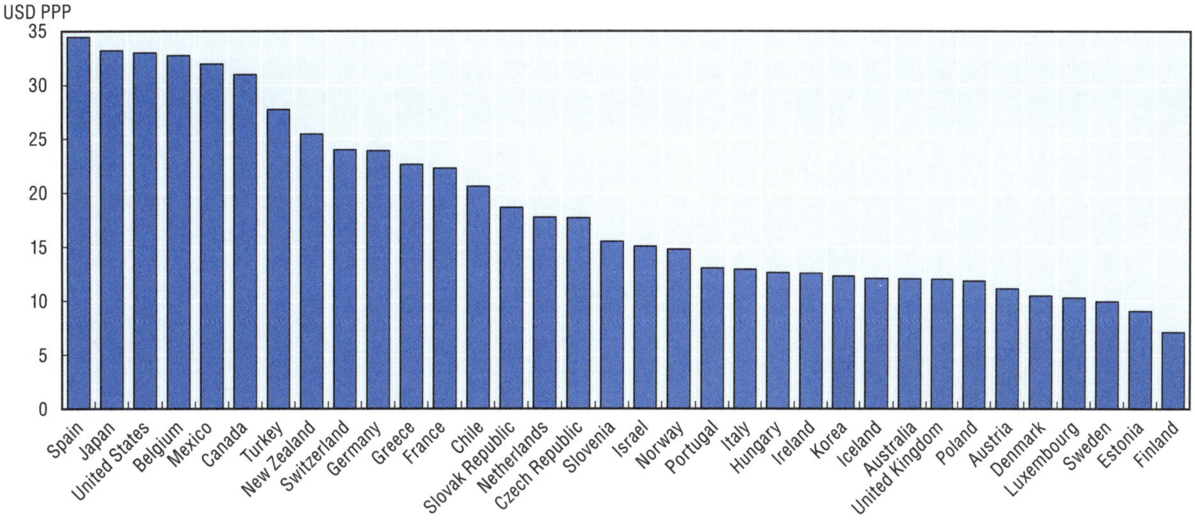

Source: OECD and Teligen.

StatLink http://dx.doi.org/10.1787/888932800159

United States) and some below USD 10 (Estonia, Finland, Sweden). Some of the least expensive offers, as happened with laptops, only offer low connectivity speeds (Estonia, Finland).

Finally, 5 GB is the highest consumption basket for tablets (Figure 7.34). The average cost across OECD countries is USD 24.88, distributed across a broad range of prices, starting from USD 7.98 (Finland) to USD 61.84 (New Zealand), with more than 10 countries offering prices between USD 20 and USD 30. Regarding the relationship between speeds and prices, the statement for laptops also applies to tablet baskets. Offers with lower prices usually correspond to unlimited or high data allowances at low speeds, as is the case of Estonia and Finland for the 5 GB tablet basket. There are, however, a few available high-speed plans at very reasonable rates, such as Denmark's Telenor (USD 10.47 at 16 Mbit/s) and Norway's Netcom (USD 16 at 80 Mbit/s advertised download speeds).

Figure 7.34. **OECD Wireless Broadband basket, OECD Tablet 5 GB, September 2012**

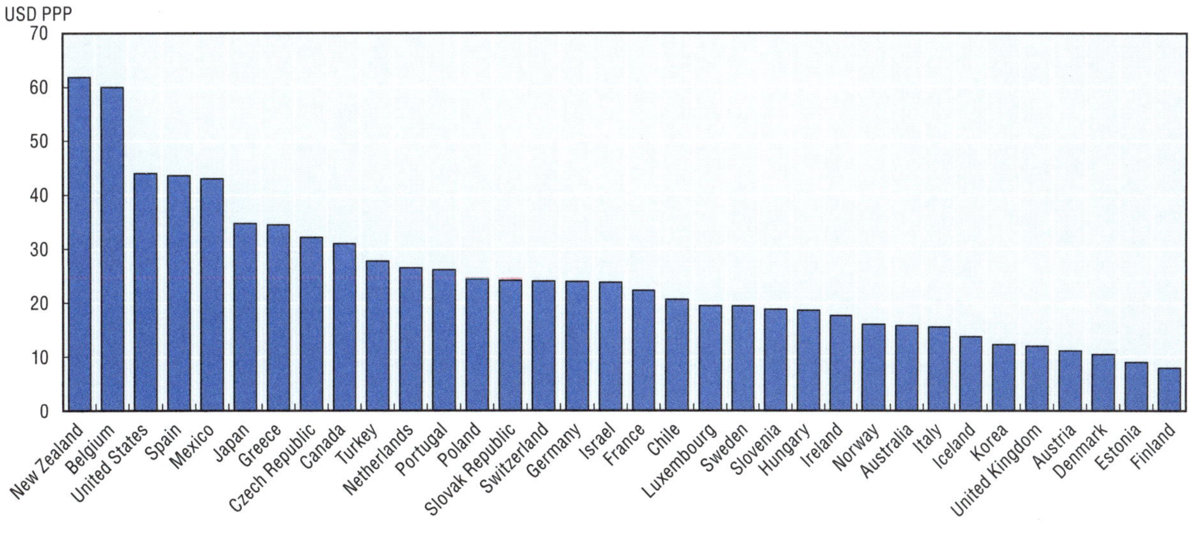

Source: OECD and Teligen.

StatLink http://dx.doi.org/10.1787/888932800178

Table 7.4. **OECD time series for telephone charges**

	1990	1991	1992	1993	1994	1995	1996	1997	1998	1999	2000	2001	2002	2003	2004	2005	2006	2007	2008	2009	2010	2011	2012
Residential																							
Fixed	100	109.2	112.7	112.8	112.8	122.4	125.9	113.0	115.5	119.3	132.0	129.1	133.3	132.2	145.2	145.6	165.0	170.8	158.1	173.8	195.0	194.2	194.2
Usage	100	104.2	98.4	96.8	94.1	98.6	90.1	81.3	78.7	70.5	60.6	55.8	57.5	53.5	55.7	53.2	39.0	38.3	33.9	35.4	37.3	36.2	35.5
Total	100	106.2	104.1	103.2	101.6	108.1	104.4	94.0	93.4	90.0	89.2	85.1	87.8	85.0	91.5	90.1	89.4	91.3	83.6	90.7	100.4	99.4	99.0
Business																							
Fixed	100	104.3	107.4	107.6	108.0	108.1	106.4	113.1	118.7	123.4	118.6	126.9	135.0	126.5	139.1	137.2	171.8	181.1	165.3	170.2	186.1	178.5	172.5
Usage	100	103.5	96.9	94.2	91.3	92.5	83.3	86.5	84.3	75.2	55.5	55.5	57.7	54.6	55.2	51.6	39.7	40.4	39.3	41.5	49.9	47.1	44.4
Total	100	103.7	99.0	96.9	94.6	95.6	87.9	91.8	91.2	84.8	68.1	69.8	73.2	69.0	72.0	68.8	66.1	68.5	64.5	67.3	77.1	73.4	70.0

Source: OECD and Teligen.

StatLink http://dx.doi.org/10.1787/888932802040

Table 7.5. **OECD basket of residential telephone charges, 20 calls, VAT included, August 2012**

	Fixed		Usage		Total	
	USD	USD PPP	USD	USD PPP	USD	USD PPP
Australia	29.37	18.71	16.56	10.55	45.93	29.26
Austria	23.93	21.56	4.03	3.63	27.96	25.19
Belgium	26.57	22.71	5.23	4.47	31.80	27.18
Canada	39.88	31.40	3.36	2.64	43.23	34.04
Chile	15.89	21.77	1.99	2.73	17.88	24.50
Czech Republic	14.50	18.12	11.59	14.49	26.09	32.62
Denmark	24.74	16.60	4.47	3.00	29.21	19.60
Estonia	8.87	10.56	2.42	2.88	11.29	13.44
Finland	15.89	12.52	10.21	8.04	26.10	20.55
France	20.78	18.07	4.52	3.93	25.30	22.00
Germany	23.48	21.94	3.36	3.14	26.84	25.08
Greece	19.47	19.47	5.53	5.53	25.00	25.00
Hungary	16.56	23.65	0.43	0.61	16.98	24.26
Iceland	6.75	5.72	1.28	1.08	8.02	6.80
Ireland	33.20	26.56	5.27	4.21	38.47	30.78
Israel	15.05	12.86	1.90	1.63	16.95	14.49
Italy	22.42	22.65	2.31	2.34	24.74	24.98
Japan	33.58	22.53	7.10	4.77	40.68	27.30
Korea	6.03	7.54	3.89	4.86	9.92	12.40
Luxembourg	23.78	19.18	3.02	2.43	26.80	21.61
Mexico	15.70	24.53	4.09	6.38	19.79	30.92
Netherlands	13.96	12.58	11.82	10.65	25.78	23.23
New Zealand	38.19	30.31	5.41	4.29	43.60	34.60
Norway	15.99	9.99	1.50	0.94	17.49	10.93
Poland	15.84	26.85	0.86	1.45	16.70	28.30
Portugal	20.93	21.14	4.63	4.67	25.56	25.82
Slovak Republic	11.17	14.70	3.66	4.81	14.83	19.51
Slovenia	20.45	22.72	1.56	1.73	22.01	24.46
Spain	22.83	22.17	9.22	8.96	32.06	31.12
Sweden	23.74	18.54	4.52	3.53	28.25	22.07
Switzerland	26.66	15.32	6.28	3.61	32.93	18.93
Turkey	12.82	22.89	0.81	1.45	13.63	24.35
United Kingdom	26.32	19.94	7.66	5.80	33.97	25.74
United States	24.42	24.42	4.44	4.44	28.86	28.86
OECD average	20.88	19.42	4.85	4.40	25.72	23.82

Notes: The OECD basket of residential telephone charges includes fixed access and 20 calls (broken down according to distance, destination [fixed, mobile and international], and time of day) over a one-month period. USD purchasing power parities (PPP) are used to aid in international comparisons.

Source: OECD and Teligen.

StatLink http://dx.doi.org/10.1787/888932802078

Table 7.6. **OECD basket of residential telephone charges, 60 calls, VAT included, August 2012**

	Fixed		Usage		Total	
	USD	USD PPP	USD	USD PPP	USD	USD PPP
Australia	47.25	30.09	27.42	17.46	74.67	47.56
Austria	23.93	21.56	15.27	13.76	39.20	35.32
Belgium	36.52	31.21	0.13	0.12	36.66	31.33
Canada	39.88	31.40	10.85	8.54	50.72	39.94
Chile	20.01	27.42	3.78	5.18	23.80	32.60
Czech Republic	34.06	42.58	3.75	4.68	37.81	47.26
Denmark	24.74	16.60	16.25	10.91	40.99	27.51
Estonia	8.87	10.56	10.82	12.88	19.69	23.44
Finland	15.89	12.52	34.10	26.85	49.99	39.37
France	25.57	22.23	6.23	5.42	31.80	27.65
Germany	25.94	24.24	7.57	7.07	33.51	31.31
Greece	25.29	25.29	5.51	5.51	30.80	30.80
Hungary	16.56	23.65	2.74	3.91	19.29	27.56
Iceland	6.75	5.72	6.03	5.11	12.78	10.83
Ireland	33.20	26.56	20.62	16.50	53.82	43.06
Israel	15.05	12.86	8.14	6.96	23.19	19.82
Italy	22.42	22.65	14.90	15.05	37.32	37.70
Japan	33.58	22.53	30.42	20.41	63.99	42.95
Korea	6.03	7.54	15.43	19.29	21.46	26.82
Luxembourg	23.78	19.18	11.86	9.57	35.64	28.74
Mexico	15.70	24.53	15.04	23.50	30.74	48.03
Netherlands	31.22	28.13	10.73	9.67	41.95	37.79
New Zealand	43.05	34.16	14.34	11.38	57.38	45.54
Norway	15.99	9.99	7.49	4.68	23.48	14.68
Poland	15.84	26.85	8.61	14.60	24.45	41.44
Portugal	20.93	21.14	17.27	17.45	38.20	38.59
Slovak Republic	17.52	23.06	10.12	13.31	27.64	36.37
Slovenia	20.45	22.72	7.34	8.16	27.79	30.88
Spain	27.29	26.50	22.57	21.91	49.86	48.41
Sweden	28.15	21.99	8.99	7.02	37.14	29.02
Switzerland	34.63	19.90	10.47	6.02	45.10	25.92
Turkey	15.60	27.86	3.98	7.11	19.58	34.97
United Kingdom	35.65	27.01	6.73	5.10	42.38	32.11
United States	24.42	24.42	12.53	12.53	36.95	36.95
OECD average	24.46	22.78	12.00	11.11	36.46	33.89

Notes: The OECD basket of residential telephone charges includes fixed access and 60 calls (broken down according to distance, destination [fixed, mobile and international], and time of day) over a one-month period. USD purchasing power parities (PPP) are used to aid in international comparisons.

Source: OECD and Teligen.

StatLink http://dx.doi.org/10.1787/888932802097

Table 7.7. **OECD basket of residential telephone charges, 140 calls, VAT included, August 2012**

	Fixed		Usage		Discount		Total	
	USD	USD PPP	USD	USD PPP	USD	USD PPP	USD	USD PPP
Australia	57.76	36.79	67.54	43.02			125.30	79.81
Austria	23.93	21.56	37.81	34.07			61.75	55.63
Belgium	54.95	46.96	0.00	0.00			54.95	46.96
Canada	51.30	40.40	0.17	0.13			51.47	40.53
Chile	33.49	45.88	9.08	12.44			42.57	58.31
Czech Republic	37.47	46.83	9.65	12.06			47.12	58.89
Denmark	35.47	23.80	21.50	14.43			56.97	38.23
Estonia	8.87	10.56	27.64	32.90			36.51	43.46
Finland	15.89	12.52	83.53	65.77			99.42	78.29
France	50.14	43.60	0.37	0.32			50.51	43.92
Germany	25.94	24.24	26.98	25.22			52.92	49.46
Greece	42.73	42.73	0.44	0.44			43.18	43.18
Hungary	16.56	23.65	6.32	9.03			22.88	32.68
Iceland	12.55	10.63	6.22	5.27			18.77	15.90
Ireland	60.84	48.67	6.01	4.80			66.85	53.48
Israel	18.26	15.61	14.27	12.19			32.53	27.80
Italy	22.42	22.65	40.27	40.68			62.69	63.32
Japan	34.92	23.44	71.60	48.05			106.52	71.49
Korea	6.03	7.54	37.83	47.29			43.86	54.83
Luxembourg	23.78	19.18	29.69	23.95			53.47	43.12
Mexico	31.24	48.81	11.68	18.24	- 0.02	- 0.03	42.90	67.03
Netherlands	43.50	39.19	5.31	4.78			48.81	43.97
New Zealand	47.91	38.02	32.54	25.83			80.45	63.85
Norway	32.39	20.25	7.86	4.91			40.25	25.16
Poland	24.88	42.17	10.21	17.30			35.09	59.47
Portugal	20.93	21.14	40.85	41.27			61.79	62.41
Slovak Republic	23.65	31.12	16.92	22.27			40.58	53.39
Slovenia	20.45	22.72	18.15	20.17			38.60	42.89
Spain	70.01	67.97					70.01	67.97
Sweden	33.30	26.02	15.27	11.93			48.58	37.95
Switzerland	34.63	19.90	25.45	14.62			60.08	34.53
Turkey	12.82	22.89	20.00	35.71			32.82	58.60
United Kingdom	42.17	31.94	3.66	2.77			45.82	34.71
United States	29.88	29.88	24.39	24.39			54.27	54.27
OECD average							53.83	50.16

Notes: The OECD basket of residential telephone charges includes fixed access and 140 calls (broken down according to distance, destination [fixed, mobile and international], and time of day) over a one-month period. USD purchasing power parities (PPP) are used to aid in international comparisons.

Source: OECD and Teligen.

StatLink http://dx.doi.org/10.1787/888932802116

Table 7.8. **OECD basket of residential telephone charges, 420 calls, VAT included, August 2012**

	Fixed		Usage		Discount		Total	
	USD	USD PPP	USD	USD PPP	USD	USD PPP	USD	USD PPP
Australia	120.80	76.94	9.72	6.19			130.52	83.14
Austria	23.93	21.56	104.44	94.09			128.37	115.65
Belgium	38.46	32.87	25.31	21.63			63.77	54.50
Canada	57.04	44.92	20.46	16.11			77.50	61.03
Chile	35.56	48.72	68.72	94.13			104.28	142.85
Czech Republic	37.47	46.83	73.99	92.49			111.46	139.32
Denmark	35.47	23.80	28.30	18.99			63.77	42.80
Estonia	8.87	10.56	53.26	63.40			62.13	73.96
Finland	15.89	12.52	229.76	180.91			245.66	193.43
France	50.14	43.60	0.62	0.54			50.76	44.14
Germany	38.22	35.72	18.36	17.16			56.58	52.88
Greece	42.73	42.73	1.17	1.17			43.90	43.90
Hungary	16.56	23.65	10.79	15.42			27.35	39.07
Iceland	12.55	10.63	40.98	34.73			53.52	45.36
Ireland	60.84	48.67	10.16	8.13			71.00	56.80
Israel	18.26	15.61	37.29	31.87			55.56	47.48
Italy	34.19	34.53	42.23	42.65			76.41	77.18
Japan	34.92	23.44	159.96	107.36			194.88	130.79
Korea	6.03	7.54	92.39	115.48			98.42	123.02
Luxembourg	36.07	29.09	22.92	18.49			58.99	47.57
Mexico	47.21	73.77	36.81	57.51	- 2.91	- 4.55	81.10	126.73
Netherlands	43.50	39.19	8.85	7.97			52.35	47.17
New Zealand	52.76	41.88	60.20	47.78			112.96	89.65
Norway	32.39	20.25	12.04	7.53			44.44	27.77
Poland	33.01	55.96	12.31	20.87			45.33	76.82
Portugal	33.45	33.79	38.30	38.69			71.75	72.48
Slovak Republic	24.43	32.14	30.12	39.63			54.54	71.77
Slovenia	20.45	22.72	21.92	24.35			42.37	47.08
Spain	70.01	67.97					70.01	67.97
Sweden	33.30	26.02	46.53	36.35			79.84	62.37
Switzerland	36.68	21.08	54.29	31.20			90.96	52.28
Turkey	34.59	61.77	12.41	22.16			47.00	83.93
United Kingdom	42.17	31.94	5.78	4.38			47.95	36.32
United States	73.69	73.69	5.47	5.47			79.15	79.15
OECD average							79.25	75.19

Notes: The OECD basket of residential telephone charges includes fixed access and 420 calls (broken down according to distance, destination [fixed, mobile and international], and time of day) over a one-month period. USD purchasing power parities (PPP) are used to aid in international comparisons.

Source: OECD and Teligen.

StatLink http://dx.doi.org/10.1787/888932802135

Table 7.9. **OECD basket of business telephone charges, 100 calls, VAT excluded, August 2012**

	Fixed		Usage		Discount		Total	
	USD	USD PPP	USD	USD PPP	USD	USD PPP	USD	USD PPP
Australia	47.30	30.13	61.68	39.29			108.99	69.42
Austria	35.85	32.29	1.83	1.65			37.68	33.94
Belgium	36.52	31.21	11.75	10.04			48.27	41.25
Canada	38.36	30.20	9.50	7.48			47.86	37.68
Chile								
Czech Republic	31.82	39.78	27.88	34.85			59.70	74.63
Denmark	30.93	20.76	37.01	24.84			67.94	45.60
Estonia	10.68	12.71	22.27	26.51			32.94	39.22
Finland	15.89	12.52	50.64	39.87			66.53	52.39
France	37.85	32.91	0.11	0.10			37.96	33.01
Germany	25.94	24.24	12.83	11.99			38.77	36.23
Greece	19.47	19.47	22.78	22.78			42.26	42.26
Hungary	23.47	33.53	14.11	20.15			37.58	53.68
Iceland	7.58	6.42	12.38	10.49			19.96	16.92
Ireland	52.55	42.04	3.19	2.55			55.74	44.59
Israel	15.05	12.86	8.11	6.93			23.16	19.79
Italy	35.92	36.28	31.02	31.33			66.94	67.61
Japan	45.66	30.65	37.45	25.14			83.12	55.78
Korea	6.03	7.54	17.39	21.73			23.42	29.27
Luxembourg	23.78	19.18	17.05	13.75			40.83	32.92
Mexico	22.55	35.24	9.30	14.54			31.86	49.78
Netherlands	23.85	21.49	26.98	24.31			50.83	45.79
New Zealand	49.24	39.08	22.49	17.85			71.73	56.92
Norway	31.76	19.85	27.28	17.05			59.04	36.90
Poland	16.54	28.03	8.14	13.79			24.68	41.82
Portugal	25.29	25.54	27.34	27.62			52.63	53.16
Slovak Republic	17.09	22.49	30.89	40.65			47.98	63.13
Slovenia	20.45	22.72	13.03	14.48			33.48	37.20
Spain	27.29	26.50	27.41	26.61			54.70	53.11
Sweden	27.14	21.20	27.71	21.65			54.85	42.85
Switzerland	26.66	15.32	32.19	18.50			58.85	33.82
Turkey	15.60	27.86	6.53	11.66			22.14	39.53
United Kingdom	34.50	26.13	5.15	3.91			39.65	30.04
United States	25.93	25.93	11.97	11.97			37.90	37.90
OECD average	26.60	24.47	19.86	18.12			46.47	42.59

Notes: The OECD basket of business telephone charges includes fixed access and 100 calls (broken down according to distance, destination [fixed, mobile and international], and time of day) over a one-month period. USD purchasing power parities (PPP) are used to aid in international comparisons.

Source: OECD and Teligen.

StatLink http://dx.doi.org/10.1787/888932802154

Table 7.10. **OECD basket of business telephone charges, 260 calls, VAT excluded, August 2012**

	Fixed		Usage		Discount		Total	
	USD	USD PPP	USD	USD PPP	USD	USD PPP	USD	USD PPP
Australia	141.94	90.41	55.17	35.14			197.11	125.55
Austria	42.24	38.06	34.90	31.44			77.14	69.50
Belgium	38.46	32.87	50.61	43.26			89.07	76.13
Canada	38.36	30.20	40.00	31.50			78.36	61.70
Chile								
Czech Republic	31.82	39.78	85.15	106.44			116.98	146.22
Denmark	30.93	20.76	112.67	75.62			143.60	96.37
Estonia	10.68	12.71	66.05	78.63			76.73	91.34
Finland	15.89	12.52	160.13	126.08			176.02	138.60
France	37.85	32.91	31.34	27.25			69.19	60.17
Germany	25.94	24.24	40.52	37.87			66.46	62.11
Greece	34.59	34.59	40.13	40.13			74.72	74.72
Hungary	27.94	39.91	47.45	67.78			75.39	107.70
Iceland	7.58	6.42	41.41	35.09			48.98	41.51
Ireland	52.55	42.04	25.88	20.70			78.42	62.74
Israel	15.05	12.86	40.14	34.31			55.18	47.17
Italy	35.92	36.28	105.69	106.76			141.61	143.04
Japan	45.66	30.65	174.21	116.92			219.88	147.57
Korea	6.03	7.54	80.21	100.27			86.24	107.80
Luxembourg	23.78	19.18	52.81	42.59			76.58	61.76
Mexico	60.27	94.17	22.65	35.39	-0.13	-0.21	82.79	129.36
Netherlands	23.85	21.49	79.15	71.31			103.00	92.80
New Zealand	49.24	39.08	68.70	54.52			117.94	93.60
Norway	31.76	19.85	79.68	49.80			111.44	69.65
Poland	23.89	40.49	21.75	36.87			45.64	77.36
Portugal	25.29	25.54	96.60	97.57			121.88	123.11
Slovak Republic	34.22	45.02	65.72	86.47			99.94	131.49
Slovenia	20.45	22.72	36.84	40.93			57.29	63.66
Spain	46.61	45.25	65.69	63.77			112.30	109.03
Sweden	27.14	21.20	79.61	62.19			106.75	83.40
Switzerland	26.66	15.32	90.20	51.84			116.86	67.16
Turkey	26.18	46.76	20.62	36.83			46.81	83.59
United Kingdom	34.50	26.13	20.05	15.19			54.54	41.32
United States	25.93	25.93	41.52	41.52			67.45	67.45
OECD average							96.74	89.54

Notes: The OECD basket of business telephone charges includes fixed access and 260 calls (broken down according to distance, destination [fixed, mobile and international], and time of day) over a one-month period. USD purchasing power parities (PPP) are used to aid in international comparisons.

Source: OECD and Teligen.

StatLink http://dx.doi.org/10.1787/888932802173

Table 7.11. **OECD basket of mobile telephone charges, 30 calls, VAT included, August 2012**

		Fixed		Usage		Messages		Grand total		Contract type*
		USD	USD PPP	USD	USD PPP	USD	USD PPP	USD	USD PPP	
Australia, Optus	Optus Pre-paid - Connect 4 Less, $10	0.29	0.19	13.59	8.66	10.39	6.62	24.27	15.46	PP
Austria, T-Mobile	Tele.ring - Masta Mini , SIM only	7.50	6.76	0.00	0.00	0.00	0.00	7.50	6.76	
Belgium, Mobistar	Dauphin 15 , 12 months	14.74	12.60	0.00	0.00	0.00	0.00	14.74	12.60	
Canada, Bell Mobility	Virgin Prepaid $10 , 5000 SMS	17.22	13.56	8.23	6.70	0.00	0.00	25.45	20.26	PP
Chile, Entel Movil	Raimundo Plan , 140 SMS, Calls to 1 sel.num.	4.98	6.82	10.37	14.21	0.00	0.00	15.35	21.02	PP / SD
Czech Rep., T-Mobile	Podnikatel Plus 250 HIT , 24 months	14.60	18.25	0.00	0.00	0.00	0.00	14.60	18.25	
Denmark, Telenor	Selvhenter.dk	0.23	0.15	5.32	3.57	3.22	2.16	8.77	5.89	PP
Estonia, Tele2	Mini , 100 SMS	5.80	6.90	0.59	0.70	0.00	0.00	6.39	7.61	
Finland, Sonera	TeleFinland - Perus , 100 SMS	3.40	2.68	5.62	4.43	0.00	0.00	9.03	7.11	
France, Orange	Forfait M6 1h , SIM only, 12 months	12.16	10.57	0.00	0.00	0.00	0.00	12.16	10.57	
Germany, T-Mobile	Xtra Click , SIM only	0.34	0.32	5.70	5.33	11.69	10.92	17.73	16.57	PP
Greece, Cosmote	What's Up , Call Them All Plus	8.60	8.60	0.00	0.00	0.00	0.00	8.60	8.60	PP
Hungary, T-Mobile	Domino Aktív , 200 SMS	10.13	14.47	7.63	10.90	0.00	0.00	17.76	25.37	PP
Iceland, Vodafone	Risafrelsi Eitt verð , on-net calls & SMS	8.54	7.24	0.00	0.00	0.00	0.00	8.54	7.24	PP
Ireland, O2	O2 Choices 300 , 24 months	18.33	14.66	0.00	0.00	0.00	0.00	18.33	14.66	
Israel, Cellcom	Prepaid - Stockmann Basic , 120 SMS	8.73	7.46	14.10	12.05	0.00	0.00	22.83	19.51	PP
Italy, TIM	TIM 9 , 1000 SMS, TIMx2 Calls to 1 sel.num.	19.10	19.29	8.99	9.08	0.00	0.00	28.09	28.38	SD
Japan, KDDI au	Plan S Simple , Everybody Discount with 24M, 24 months	21.86	14.67	1.46	0.98	0.24	0.16	23.56	15.81	
Korea, KT	EvergreenMobile Zero , SIM only	0.59	0.74	5.23	6.54	1.95	2.43	7.77	9.71	
Luxembourg, Tango	Knock-out , + Tango Family (disc calcs for on-net only)	0.00	0.00	3.85	3.11	8.50	6.85	12.35	9.96	SD
Mexico, Telcel	Amigo Plus , $200 Top-up, $60 bonus, Calls and SMS to 3 sel.num.	0.00	0.00	2.88	6.63	4.67	5.16	7.54	11.79	PP / SD
Netherlands, Vodafone	SIM only Scherp 110 , 100 SMS, SIM only, 24 months	12.69	11.43	0.00	0.00	0.00	0.00	12.69	11.43	
New Zealand, Telecom	Smart Prepaid , TalkText Value Pack	16.31	12.94	3.99	3.17	0.00	0.00	20.30	16.11	PP
Norway, Telenor	Prat , SIM only, Frifamilie Calls and SMS to 7 sel.num.	0.00	0.00	4.01	2.51	4.11	2.57	8.12	5.07	SD
Poland, Polkomtel	Na_Karte , Unlimited SMS 3000 SMS, SIM only	1.54	2.60	4.48	7.60	0.00	0.00	6.02	10.20	PP
Portugal, TMN	UZO , Minimum usage	0.00	0.00	6.22	9.35	12.21	9.26	18.42	18.61	PP
Slovak Rep., T-Mobile	Podľa seba 1 , Any-net + SMS, SIM only, 24 months	12.61	16.60	0.00	0.00	0.00	0.00	12.61	16.60	
Slovenia, Si.mobil	SIMPL , 100 SMS	6.55	7.28	6.42	7.13	0.00	0.00	12.97	14.41	PP
Spain, Vodafone	Tarjeta Internet y SMS Gatuitas , Top-up €10	16.10	15.63	0.00	0.00	0.00	0.00	16.10	15.63	PP / SD
Sweden, Tele 2 Comviq	Comviq Kontant Smart , SEK 69 Top-up	0.00	0.00	6.18	4.83	7.21	5.64	13.39	10.46	PP
Switzerland, Swisscom	M-Budget Mobile	0.56	0.32	14.02	8.06	10.05	5.77	24.63	14.16	PP
Turkey, Vodafone	Hesabini Bilen Mini , 100 SMS	11.64	20.78	0.00	0.00	0.00	0.00	11.64	20.78	
United Kingdom, T-Mobile	Pay Monthly 7 , 24 months	10.99	8.32	0.00	0.00	0.00	0.00	10.99	8.32	
United States, AT&T	Pay As You Go - 10c / min , 200 SMS	5.49	5.49	7.10	7.10	0.00	0.00	12.59	12.59	PP
OECD average		7.99	7.86	4.29	4.19	2.18	1.69	14.47	13.75	

Notes: The OECD basket of mobile telephone charges includes subscription and usage (30 voice calls and 100 SMS messages, distributed between peak and off-peak hours and based on an average call duration) over a one-month period. Calling patterns were all determined through extensive discussions with carriers across the OECD. USD purchasing power parities (PPP) are used to aid international comparisons. (*) Contact type : PP = Pre-paid plan ; SD = Including selective discounts.

Source: OECD and Teligen.

StatLink http://dx.doi.org/10.1787/888932802192

Table 7.12. **OECD basket of mobile telephone charges, 100 calls, VAT included, August 2012**

		Fixed		Usage		Messages		Grand total		Contract type*
		USD	USD PPP	USD	USD PPP	USD	USD PPP	USD	USD PPP	
Australia, Telstra	Pre-Paid Beyond Talk $30 top-up	31.60	20.13	0.00	0.00	0.00	0.00	31.60	20.13	PP
Austria, T-Mobile	Tele.ring - Masta Max , SIM only	11.19	10.08	0.00	0.00	0.00	0.00	11.19	10.08	
Belgium, Mobistar	Panthere Pro 30 , 12 months	29.47	25.19	0.00	0.00	0.00	0.00	29.47	25.19	
Canada, Bell Mobility	Virgin Choice $25 , 5000 SMS, Voicemail option	42.48	33.45	0.00	0.00	0.00	0.00	42.48	33.45	
Chile, Entel Movil	Pablo Plan , Top-up $10000, 140 SMS, Calls to 1 sel.num.	4.98	6.82	32.67	44.75	0.00	0.00	37.64	51.56	PP / SD
Czech Rep., T-Mobile	S námi 790 + HIT , 24 months	30.75	38.44	1.11	1.39	0.00	0.00	31.86	39.83	
Denmark, Telenor	Selvhenter.dk , Mine nærmeste Calls to 10 sel.num.	3.37	2.26	9.35	6.28	4.55	3.05	17.27	11.59	PP / SD
Estonia, Tele2	Nuti 4.99 , SIM only	6.40	7.62	2.39	2.84	0.49	0.58	9.28	11.05	
Finland, Elisa	Saunalahti Satakieli 2000 , 300 SMS, 12 months	10.35	8.15	7.55	5.95	0.00	0.00	17.90	14.09	
France, SFR	Forfait Carre Bloques 2h , SIM only, 24 months, Calls to 2 sel.num.	22.11	19.23	2.29	1.99	0.00	0.00	24.40	21.21	SD
Germany, Vodafone	CallYa Smartphone Fun , Top up EUR 15	12.61	11.79	20.43	19.09	0.00	0.00	33.04	30.88	PP
Greece, Cosmote	What's Up , Call Them All Plus	8.60	8.60	0.00	0.00	0.00	0.00	8.60	8.60	PP
Hungary, Telenor	Red Bull Mobile All-in , SIM only, 24 months	24.16	34.51	1.77	2.53	5.98	8.55	31.91	45.59	
Iceland, Vodafone	Risafrelsi Eitt verð , on-net calls & SMS	8.54	7.24	9.27	7.85	0.00	0.00	17.81	15.09	PP
Ireland, O2	O2 Choices 300 , SIM only, 1 months	18.33	14.66	0.00	0.00	0.00	0.00	18.33	14.66	
Israel, Cellcom	Unlimited 119	29.67	25.36	0.00	0.00	0.00	0.00	29.67	25.36	
Italy, Vodafone	Vodafone 12 , Smart 200, Calls to 1 sel.num.	30.71	31.02	1.32	1.34	10.11	10.21	42.14	42.57	PP / SD
Japan, KDDI au	Plan M Simple , Everybody Discount with 24M, 24 months	39.61	26.59	1.40	0.94	0.00	0.00	41.01	27.53	SD
Korea, KT	EvergreenMobile Donation Zero SMS 200 , SIM only	16.64	20.80	0.00	0.00	0.78	0.97	17.42	21.77	
Luxembourg, Tango	Knock-out , + Tango Family (disc calcs for on-net only)	0.00	0.00	15.78	12.72	12.37	9.98	28.15	22.70	SD
Mexico, Telcel	Amigo Óptimo , $200 Top-up, $60 bonus, 1000 SMS	7.47	11.67	10.69	16.71	0.00	0.00	18.16	28.37	PP / SD
Netherlands, Vodafone	SIM only Scherp 225 , 200 SMS, SIM only, 24 months	16.78	15.12	0.00	0.00	0.00	0.00	16.78	15.12	
New Zealand, Vodafone	Smart $65 , 12 months, 1 x NZBestmates Calls and SMS to 1 sel.num.	44.54	35.35	0.00	0.00	0.00	0.00	44.54	35.35	SD
Norway, Telenor	Prat 100 , SIM only, Frifamilie Calls and SMS to 7 sel.num.	11.43	7.15	0.00	0.00	5.45	3.40	16.88	10.55	SD
Poland, Polkomtel	Na_Karte , Top-Up ZL 150, Unlimited SMS 3000 SMS, SIM only	12.10	20.50	0.00	0.00	0.00	0.00	12.10	20.50	PP / SD
Portugal, TMN	Pack Negocios Rede 12 , 200 SMS	30.67	30.98	0.12	0.13	0.00	0.00	30.79	31.11	
Slovak Rep., T-Mobile	Podľa seba 1 , SMS + Redu Chrg, SIM only, 24 months	12.61	16.60	13.63	17.93	0.00	0.00	26.24	34.52	
Slovenia, Si.mobil	Paket PODJETNI Start 2	22.40	24.88	0.00	0.00	0.00	0.00	22.40	24.88	
Spain, Vodafone	Tarjeta Llamadas Gatuitas , Top-up €10	35.42	34.39	6.36	6.18	0.00	0.00	41.78	40.57	PP / SD
Sweden, Tele 2 Comviq	Comviq Kontant Smart , SEK 169 Top-up	24.88	19.44	0.00	0.00	0.00	0.00	24.88	19.44	PP
Switzerland, Swisscom	Natel Infinity XS	61.47	35.33	0.00	0.00	0.00	0.00	61.47	35.33	
Turkey, Vodafone	Hesabini Bilen Mini , 500 SMS	12.75	22.77	0.00	0.00	0.00	0.00	12.75	22.77	
United Kingdom, T-Mobile	Business Pay Monthly 500 Mins - Internet Booster , SIM only, 12 months	14.65	11.10	0.00	0.00	0.00	0.00	14.65	11.10	
United States, AT&T	Pay As You Go , $25 Monthly Plan	27.50	27.50	0.00	0.00	0.00	0.00	27.50	27.50	PP
OECD average		21.07	19.55	4.00	4.37	1.17	1.08	26.24	25.00	

Notes: The OECD basket of mobile telephone charges includes subscription and usage (100 voice calls and 140 SMS messages, distributed between peak and off-peak hours and based on an average call duration) over a one-month period. Calling patterns were all determined through extensive discussions with carriers across the OECD. USD purchasing power parities (PPP) are used to aid international comparisons. (*) Contact type : PP = Pre-paid plan ; SD = Including selective discounts

Source: OECD and Teligen.

StatLink http://dx.doi.org/10.1787/888932802211

Table 7.13. **OECD basket of mobile telephone charges, 300 calls, VAT included, August 2012**

		Fixed		Usage		Messages		Grand total		Contract type*
		USD	USD PPP	USD	USD PPP	USD	USD PPP	USD	USD PPP	
Australia, Optus	Boost Pre-paid - Super Cap $40	42.12	26.83	0.00	0.00	0.00	0.00	42.12	26.83	PP
Austria, T-Mobile	Tele.ring - Masta Max , SIM only	11.19	10.08	0.00	0.00	0.00	0.00	11.19	10.08	
Belgium, Mobistar	Kangourou 40 , 12 months	39.31	33.59	8.43	7.20	15.35	13.12	63.09	53.92	
Canada, Bell Mobility	Virgin Choice $40 A , 5000 SMS, Calls to 5 sel.num.	59.70	47.01	6.77	5.33	0.00	0.00	66.48	52.34	SD
Chile, Entel Movil	Planes Red 1500 , 250 SMS	57.19	78.35	8.26	11.31	0.00	0.00	65.45	89.66	
Czech Rep., T-Mobile	Tarify Grand v Siti , 24 months	72.50	90.63	0.00	0.00	0.00	0.00	72.50	90.63	
Denmark, TDC Mobil	Mobil 6hrs , Fri SMS, SIM only, 6 months	33.31	22.35	0.00	0.00	0.00	0.00	33.31	22.35	
Estonia, Tele2	Nuti 7.99 , SIM only	10.09	12.01	0.00	0.00	2.76	3.29	12.85	15.30	
Finland, Sonera	Sopiva Fiksu , SIM only	24.44	19.25	0.00	0.00	0.00	0.00	24.44	19.25	
France, Orange	Sosh - 24/7 2GB , SIM only	24.44	21.25	0.00	0.00	0.00	0.00	24.44	21.25	
Germany, Vodafone	CallYa Smartphone M , Top up EUR 15	19.09	17.84	31.69	29.61	0.00	0.00	50.78	47.46	PP
Greece, Cosmote	What's Up , Call Them All Plus	8.60	8.60	0.00	0.00	0.00	0.00	8.60	8.60	PP
Hungary, Telenor	Red Bull Mobile All-in + , SIM only, 24 months	43.96	62.79	28.26	40.37	3.19	4.56	75.40	107.72	
Iceland, Vodafone	Risafrelsi Eitt verð , 5 numbers + on-net calls & SMS	25.11	21.28	12.63	10.70	0.00	0.00	37.74	31.98	PP / SD
Ireland, O2	O2 Choices 800 (600 mins & 200 SMS) , SIM only, 12 months	45.45	36.36	0.00	0.00	0.00	0.00	45.45	36.36	
Israel, Cellcom	Unlimited 119	29.67	25.36	0.00	0.00	0.00	0.00	29.67	25.36	
Italy, TIM	Tutto Compreso 500 , 1000 SMS, TIMx2 Calls to 1 sel.num.	54.48	55.03	0.00	0.00	0.00	0.00	54.48	55.03	SD
Japan, KDDI au	Plan LL Simple , Everybody Discount with 24M, 24 months	91.03	61.10	2.70	1.82	0.00	0.00	93.74	62.91	
Korea, KT	Control 540 w/o SMS pack , KT 50% Discount, 24 months	36.09	45.11	0.00	0.00	0.00	0.00	36.09	45.11	
Luxembourg, LuxGSM	Relax , LuxGSM + SMS + Landlines + Tango	36.85	29.72	0.00	0.00	0.00	0.00	36.85	29.72	
Mexico, Telcel	Telcel Plus 300 , Calls to 2 sel.num.	26.33	41.13	0.81	1.26	9.96	15.56	37.09	57.95	SD
Netherlands, KPN	Mobiel Instap - Call-SMS 600	37.87	34.12	0.00	0.00	0.00	0.00	37.87	34.12	
New Zealand, Vodafone	Smart $100 , 12 months, 3 x NZBestmates Calls	64.79	51.42	0.00	0.00	0.00	0.00	64.79	51.42	SD
Norway, Telenor	Prat 400 , SIM only, Frifamilie Calls and SMS to 7 sel.num.	24.69	15.43	0.00	0.00	9.70	6.06	34.39	21.49	SD
Poland, Orange	Rozmowy bez limitu 88.88 , SIM only, 24 months	29.05	49.23	0.00	0.00	0.00	0.00	29.05	49.23	
Portugal, Vodafone	Plano Pro 91 Net 240	73.98	74.73	0.00	0.00	0.00	0.00	73.98	74.73	
Slovak Rep., T-Mobile	Happy M , 24 months	29.81	39.22	9.86	12.97	0.00	0.00	39.67	52.20	
Slovenia, Mobitel	Paketi 2012 , Calls to 3 sel.num.	25.12	27.91	0.00	0.00	0.00	0.00	25.12	27.91	SD
Spain, Vodafone	Tarifa @M , Calls to 2 sel.num.	59.45	57.72	16.19	15.72	0.00	0.00	75.64	73.44	SD
Sweden, Tele 2 Comviq	Comviq fastpris	36.06	28.18	0.00	0.00	0.00	0.00	36.06	28.18	
Switzerland, Swisscom	Natel Infinity XS	61.47	35.33	0.00	0.00	0.00	0.00	61.47	35.33	
Turkey, Vodafone	Hesabini Bilen Midi , 500 SMS	18.32	32.71	3.40	6.07	0.00	0.00	21.72	38.79	
United Kingdom, T-Mobile	Business Pay Monthly 500 Mins - Landline Booster	14.65	11.10	0.00	0.00	0.00	0.00	14.65	11.10	
United States, AT&T	Pay As You Go , Monthly Unltd Plan	55.00	55.00	0.00	0.00	0.00	0.00	55.00	55.00	PP
OECD average		38.86	37.58	3.79	4.19	1.20	1.25	43.86	43.02	

Notes: The OECD basket of mobile telephone charges includes subscription and usage (300 voice calls and 225 SMS messages, distributed between peak and off-peak hours and based on an average call duration) over a one-month period. Calling patterns were all determined through extensive discussions with carriers across the OECD. USD purchasing power parities (PPP) are used to aid international comparisons. (*) Contact type : PP = Pre-paid plan ; SD = Including selective discounts.

Source: OECD and Teligen.

StatLink http://dx.doi.org/10.1787/888932802230

Table 7.14. **OECD basket of mobile telephone charges, 900 calls, VAT included, August 2012**

		Fixed		Usage		Messages		Grand total		Contract type*
		USD	USD PPP	USD	USD PPP	USD	USD PPP	USD	USD PPP	
Australia, Optus	Boost Pre-paid - Super Cap $40	42.12	26.83	0.00	0.00	0.00	0.00	42.12	26.83	PP
Austria, A1 Telekom Austria	E-Free Smart+	18.76	16.90	0.00	0.00	0.00	0.00	18.76	16.90	PP
Belgium, Mobistar	Kangourou Unlimited , 3000 SMS, 12 months	104.41	89.24	0.00	0.00	0.00	0.00	104.41	89.24	
Canada, Bell Mobility	Virgin Choice $50 , 5000 SMS, Voicemail option	71.19	56.05	0.00	0.00	0.00	0.00	71.19	56.05	
Chile, Entel Movil	Planes Red 1500 , 250 SMS	57.19	78.35	54.53	74.70	12.44	17.04	124.16	170.08	
Czech Rep., O2	Smart Neon XL , 200 SMS, 24 months	72.15	90.18	5.02	6.27	0.00	0.00	77.16	96.46	
Denmark, TDC Mobil	Mobil 12hrs , Fri SMS, SIM only, 6 months	41.56	27.90	3.84	2.58	0.00	0.00	45.41	30.47	
Estonia, Tele2	Nuti 7.99 , SIM only	10.09	12.01	10.03	11.93	4.30	5.12	24.41	29.06	
Finland, Elisa	Saunalahti Nopsa Paketti + , 500 SMS, 12 months	47.79	37.63	0.00	0.00	0.00	0.00	47.79	37.63	
France, Orange	Sosh - 24/7 2GB , SIM only	24.44	21.25	0.00	0.00	0.00	0.00	24.44	21.25	
Germany, Vodafone	SuperFlat Internet Allnet Spezial , SIM only, 24 months	86.94	81.25	0.00	0.00	0.00	0.00	86.94	81.25	
Greece, Vodafone	Vodafone Unlimited 100 , SIM only	110.55	110.55	0.00	0.00	0.00	0.00	110.55	110.55	
Hungary, T-Mobile	Mozaik XXL , Extra SMS 500 SMS	94.07	134.38	96.66	138.09	0.00	0.00	190.73	272.47	SD
Iceland, Vodafone	R safrelsi Eitt verð , 5 numbers + on-net calls & SMS	25.11	21.28	115.75	98.09	0.00	0.00	140.85	119.37	PP / SD
Ireland, O2	O2 Choices 1200 (800 mins & 400 SMS) , SIM only, 12 months	55.27	44.22	0.00	0.00	0.00	0.00	55.27	44.22	
Israel, Cellcom	Unlimited 119	29.67	25.36	0.00	0.00	0.00	0.00	29.67	25.36	
Italy, TIM	Tutto Compreso 1500 , 1000 SMS, TIMx2 Calls to 1 sel.num.	103.61	104.66	0.00	0.00	0.00	0.00	103.61	104.66	SD
Japan, KDDI au	Plan LL Simple , Everybody Discount with 24M, 24 months	96.02	64.44	148.79	99.86	4.66	3.13	249.47	167.43	SD
Korea, KT	Style 550 , Style SMS 200 SMS, 24 months	39.49	49.37	12.42	15.52	1.95	2.43	53.86	67.32	
Luxembourg, LuxGSM	Relax , LuxGSM + SMS + Landlines + Tango	36.85	29.72	0.00	0.00	0.00	0.00	36.85	29.72	
Mexico, Telcel	Telcel Plus 1000 , Calls to 7 sel.num.	67.81	105.96	2.42	3.79	11.62	18.15	31.85	127.89	SD
Netherlands, KPN	Mobiel Instap - Call-SMS Unlimited	50.15	45.18	0.00	0.00	0.00	0.00	50.15	45.18	
New Zealand, Vodafone	Business Smart $200 , 12 months, FCN Calls to 1 sel.num.	129.58	102.84	0.00	0.00	0.00	0.00	129.58	102.84	SD
Norway, Telenor	Prat 1200 , 250 SMS, SIM only	64.29	40.18	0.00	0.00	0.00	0.00	64.29	40.18	SD
Poland, Orange	Rozmowy bez limitu 88.88 , SIM only, 24 months	29.05	49.23	0.00	0.00	0.00	0.00	29.05	49.23	
Portugal, Vodafone	Plano Pro Premium	126.64	127.92	0.00	0.00	0.00	0.00	126.64	127.92	
Slovak Rep., Orange	Wow Infinity 3	51.94	68.34	0.00	0.00	0.00	0.00	51.94	68.34	
Slovenia, Si.mobil	L Zate	43.28	48.09	0.00	0.00	0.00	0.00	43.28	48.09	
Spain, MoviStar	Con Tu Negocios 24 Horas y Navega 80	119.77	116.28	0.00	0.00	0.00	0.00	119.77	116.28	
Sweden, Tele 2 Comviq	Comviq fastpris	36.06	28.18	0.00	0.00	0.00	0.00	36.06	28.18	
Switzerland, Swisscom	Natel Infinity XS	61.47	35.33	0.00	0.00	0.00	0.00	61.47	35.33	
Turkey, Vodafone	Ceo Avantaj Her Yöne , 500 SMS	37.25	66.52	0.00	0.00	0.00	0.00	37.25	66.52	
United Kingdom, O2	On & On SIMplicity , SIM only, 12 months	31.39	23.78	0.00	0.00	0.00	0.00	31.39	23.78	
United States, AT&T	Pay As You Go , Monthly Unltd Plan	55.00	55.00	0.00	0.00	0.00	0.00	55.00	55.00	PP
OECD average		60.91	59.83	13.22	13.26	1.03	1.35	75.16	74.44	

Notes: The OECD basket of mobile telephone charges includes subscription and usage (900 voice calls and 350 SMS messages, distributed between peak and off-peak hours and based on an average call duration) over a one-month period. Calling patterns were al determined through extensive discussions with carriers across the OECD. USD purchasing power parities (PPP) are used to aid international comparisons. (*) Contact type : PP = Pre-paid plan; SD = Including selective discounts.

Source: OECD and Teligen.

StatLink http://dx.doi.org/10.1787/888932802249

Table 7.15. **OECD basket of mobile telephone charges, 40 calls pre-paid, VAT included, August 2012**

		Fixed		Usage		Messages		Grand total		Contract type*
		USD	USD PPP	USD	USD PPP	USD	USD PPP	USD	USD PPP	
Australia, Telstra	Pre-Paid Simplicity $30 top-up	0.06	0.04	13.46	8.57	7.57	4.82	21.09	13.43	PP
Austria, A1 Telekom Austria	bob wertkarte 6.8 cent	0.51	0.46	7.08	6.38	5.01	4.51	12.60	11.35	PP
Belgium, Mobistar	Pay & Go Partner - Plug €10 , 500 SMS	18.42	15.75	3.81	3.26	0.00	0.00	22.24	19.01	PP
Canada, Bell Mobility	Virgin Prepaid $10 , 5000 SMS	17.22	13.56	11.74	9.49	0.00	0.00	28.97	23.05	PP
Chile, Entel Movil	Raimundo Plan , 140 SMS, Calls to 1 sel.num.	4.98	6.82	13.79	18.89	0.00	0.00	18.76	25.70	PP / SD
Czech Rep., T-Mobile	Combi 500 , 24 months	0.00	0.00	20.70	25.88	3.63	4.54	24.33	30.41	PP
Denmark, Telenor	Selvhenter.dk , Mine nærmeste Calls to 10 sel.num.	3.37	2.26	2.82	1.89	1.89	1.27	8.07	5.42	PP / SD
Estonia, Tele2	Smart kõnekaart Hinnaliider	0.05	0.06	6.89	8.21	2.93	3.49	9.88	11.76	PP
Finland, Elisa	Saunalahti Prepaid	0.34	0.27	6.09	4.80	4.86	3.83	11.30	8.89	PP
France, SFR	La Carte , EUR 65 Top-up, 60 SMS	26.27	22.85	1.49	1.29	0.00	0.00	27.76	24.14	PP / SD
Germany, T-Mobile	Xtra Click , SIM only	0.34	0.32	7.52	7.03	6.98	6.53	14.85	13.88	PP
Greece, Cosmote	What's Up , Call Them All Plus	8.60	8.60	0.00	0.00	0.00	0.00	8.60	8.60	PP
Hungary, T-Mobile	Domino Aktív , 50 SMS	10.04	14.35	8.34	11.92	0.81	1.16	19.20	27.42	PP / SD
Iceland, Vodafone	Risafrelsi Eitt verð , on-net calls & SMS	8.54	7.24	0.00	0.00	0.00	0.00	8.54	7.24	PP
Ireland, O2	Speak Easy - o2 Experience Plus , EUR 20 Top-up	24.57	19.65	0.00	0.00	4.50	3.60	29.07	23.26	PP
Israel, Orange Israel	Big Talk , Bonus Card 70	17.45	14.92	0.00	0.00	10.17	8.69	27.62	23.61	PP
Italy, TIM	TIM 9 , TIMx2 Calls to 1 sel.num.	5.90	5.96	12.90	13.03	6.58	6.64	25.38	25.64	PP / SD
Japan, KDDI au	au Prepaid , (NO VOICEMAIL)	1.49	1.00	96.29	64.62	3.84	2.58	101.62	68.20	PP
Korea, SK Telecom	PPS General (No MMS)	0.00	0.00	19.13	23.91	1.17	1.46	20.30	25.37	PP
Luxembourg, Tango	Pronto , + Tango Family (disc calcs for on-net only)	0.00	0.00	7.93	6.40	5.92	4.77	13.85	11.17	PP / SD
Mexico, Telcel	Amigo Óptimo , $150 Top-up, Calls and SMS to 3 sel.num.	11.31	17.68	0.00	0.00	0.00	0.00	11.31	17.68	PP / SD
Netherlands, Vodafone	Vodafone Meerwaarderen , Bonus Plus with €20 Top-up	24.57	22.13	0.19	0.17	0.04	0.04	24.79	22.34	PP
New Zealand, Telecom	OneRate Prepaid , TalkText Value Pack	26.03	20.66	4.32	3.43	0.00	0.00	30.35	24.09	PP / SD
Norway, Telenor	djuice Easy	0.23	0.14	12.38	7.74	1.79	1.12	14.40	9.00	PP
Poland, Polkomtel	Na_Karte , Top-Up ZL 100, Unlimited SMS 3000 SMS	6.58	11.16	0.58	0.98	0.00	0.00	7.16	12.14	PP
Portugal, TMN	UZO , Minimum usage	0.00	0.00	13.21	13.34	5.22	5.27	18.42	18.61	PP
Slovak Rep., T-Mobile	Easy Free New , Top-up EUR 10	12.28	16.16	0.35	0.46	0.19	0.24	12.82	16.86	PP
Slovenia, Si.mobil	SIMPL , 100 SMS	6.55	7.28	9.78	10.87	0.00	0.00	16.34	18.15	PP
Spain, Vodafone	Tarjeta Internet y SMS Gatuitas	16.10	15.63	1.82	1.77	0.00	0.00	17.92	17.40	PP / SD
Sweden, Tele 2 Comviq	Comviq Kontant Smart , SEK 69 Top-up	0.00	0.00	8.48	6.63	4.33	3.38	12.81	10.01	PP
Switzerland, Swisscom	M-Budget Mobile	0.56	0.32	21.18	12.17	6.04	3.47	27.78	15.97	PP
Turkey, Vodafone	Cebine Göre Tarife , (TL 20 Card), 100 SMS	15.03	26.85	0.00	0.00	0.00	0.00	15.03	26.85	PP
United Kingdom, O2	Simplicity Pay & Go 100 Mins + 500 Texts	16.48	12.48	1.26	0.95	0.00	0.00	17.74	13.44	PP
United States, AT&T	Pay As You Go - 10c / min , 200 SMS	5.49	5.49	10.43	10.43	0.00	0.00	15.92	15.92	PP
OECD average		8.51	8.53	9.53	8.37	2.45	2.10	20.49	19.00	

Notes: The OECD basket of mobile telephone charges includes subscription and usage (40 pre-paid voice calls, distributed between peak and off-peak hours and based on an average call duration) over a one-month period. Calling patterns were all determined through extensive discussions with carriers across the OECD. USD purchasing power parities (PPP) are used to aid international comparisons. (*) Contact type : PP = Pre-paid plan ; SD = Including selective discounts.

Source: OECD and Teligen.

StatLink http://dx.doi.org/10.1787/888932802268

Table 7.16. **OECD basket of mobile telephone charges, 400 messages, VAT included, August 2012**

		Fixed		Usage		Messages		Grand total		Contract type*
		USD	USD PPP	USD	USD PPP	USD	USD PPP	USD	USD PPP	
Australia, Optus	Optus Pre-paid - Connect 4 Less, $10 , 500 SMS	10.81	6.88	3.55	2.26	0.00	0.00	14.36	9.15	PP
Austria, T-Mobile	Tele.ring - Masta Max , SIM only	11.19	10.08	0.00	0.00	0.00	0.00	11.19	10.08	
Belgium, Proximus	Generation MTV €10 , 24 months	12.28	10.50	0.00	0.00	0.00	0.00	12.28	10.50	
Canada, Bell Mobility	Virgin Prepaid $0 , 5000 SMS	5.74	4.52	7.09	5.60	0.00	0.00	12.83	10.12	PP
Chile, Entel Movil	Raimundo Plan , 250 SMS, Calls to 1 sel.num.	7.46	10.22	2.68	3.67	15.55	21.30	25.69	35.20	PP / SD
Czech Rep., T-Mobile	Podnikatel Plus 250 HIT , 100 SMS, 24 months	19.21	24.01	0.19	0.24	3.65	4.56	23.06	28.82	
Denmark, TDC Mobil	Om Click Minut , Fri SMS 3000 SMS	8.54	5.73	1.24	0.83	0.00	0.00	9.79	6.57	PP
Estonia, Tele2	Smart kõnekaart Tavahind Omadega Soodsalt 30 days	7.42	8.84	1.51	1.79	0.00	0.00	8.93	10.63	PP
Finland, Sonera	TeleFinland - Tekstari	9.96	7.84	2.35	1.85	0.00	0.00	12.31	9.70	
France, Orange	Forfait M6 1h , SIM only, 12 months	12.16	10.57	0.00	0.00	0.00	0.00	12.16	10.57	
Germany, T-Mobile	Xtra Card , Xtra SMS Flat 1000 allnet 1000 SMS, SIM only	12.56	11.74	1.10	1.02	0.00	0.00	13.66	12.77	PP
Greece, Cosmote	What's Up , Call Them All Plus	8.60	8.60	0.00	0.00	0.00	0.00	8.60	8.60	PP
Hungary, T-Mobile	Domino SMS , 200 SMS	10.31	14.72	3.89	5.56	15.42	22.02	29.61	42.31	PP / SD
Iceland, Vodafone	Risafrelsi Eitt verð , on-net calls & SMS	8.54	7.24	0.00	0.00	0.00	0.00	8.54	7.24	PP
Ireland, O2	Speak Easy - o2 Experience More , EUR 20 Top-up	24.57	19.65	0.00	0.00	0.00	0.00	24.57	19.65	PP
Israel, Cellcom	Unlimited 119	29.67	25.36	0.00	0.00	0.00	0.00	29.67	25.36	
Italy, Vodafone	Vodafone Al Secondo , Vodafone Smart Square, Calls to 1 sel.num.	18.42	18.61	2.93	2.96	0.00	0.00	21.35	21.57	PP / SD
Japan, KDDI au	Plan SS Simple , Everybody Discount with 24M, 24 months	18.57	12.46	0.69	0.46	0.00	0.00	19.26	12.93	SD
Korea, KT	EvergreenMobile Zero , SIM only	0.59	0.74	1.57	1.96	7.78	9.73	9.94	12.43	
Luxembourg, LuxGSM	Relax Youz , SMS	12.28	9.91	1.95	1.57	0.00	0.00	14.24	11.48	
Mexico, Telcel	Amigo Plus , $50 Top-up, 1000 SMS, Calls and SMS to 3 sel.num.	7.47	11.67	3.77	5.89	0.00	0.00	11.24	17.56	PP / SD
Netherlands, Vodafone	SIM only Scherp 110 , 400 SMS, SIM only, 24 months	16.37	14.75	0.00	0.00	0.00	0.00	16.37	14.75	
New Zealand, Telecom	Smart Prepaid , 2500 SMS	9.83	7.80	6.12	4.86	0.00	0.00	15.95	12.66	PP
Norway, Telenor	Prat , SIM only, Frifamilie Calls and SMS to 7 sel.num.	0.00	0.00	0.60	0.37	9.85	6.15	10.44	6.53	SD
Poland, Orange	POP , Grosz za SMS 600 SMS	1.83	3.11	1.20	2.04	0.00	0.00	3.04	5.15	PP
Portugal, Vodafone	YORN Power SMS , 100 SMS, Calls to 10 sel.num.	8.05	8.13	5.06	5.11	10.51	10.62	23.61	23.85	PP / SD
Slovak Rep., Orange	Wow Base PAYU , Infinite SMS & MMS 3000 SMS	8.96	11.79	2.22	2.92	0.00	0.00	11.18	14.71	
Slovenia, Si.mobil	U ORTO NULO	15.15	16.83	0.78	0.86	0.00	0.00	15.93	17.70	
Spain, Vodafone	Tarjeta Rate XS 8 , Mensamanía a Todos 32500 SMS	20.56	19.96	2.79	2.70	0.00	0.00	23.35	22.67	PP / SD
Sweden, Telia	Refill Till vänner , - Top-Up SEK 50, 5000 SMS	6.12	4.78	7.36	5.75	0.00	0.00	13.48	10.53	PP
Switzerland, Sunrise	Sunrise Sunflat 1 , SIM only	31.81	18.28	1.69	0.97	0.00	0.00	33.51	19.26	
Turkey, Turkcell	Tarife Saniye prepaid , 2000 SMS	6.24	11.15	2.09	3.72	0.00	0.00	8.33	14.87	PP
United Kingdom, T-Mobile	Business Pay Monthly 500 Mins - Texts Booster	14.65	11.10	0.00	0.00	0.00	0.00	14.65	11.10	
United States, AT&T	Pay As You Go - 10c / min , 1000 SMS	10.99	10.99	2.08	2.08	0.00	0.00	13.07	13.07	PP
OECD average		11.97	11.13	1.96	1.97	1.85	2.19	15.77	15.30	

Notes: The OECD basket of mobile telephone charges includes subscription and usage (400 SMS messages, distributed between peak and off-peak hours and based on an average call duration) over a one-month period. Calling patterns were all determined through extensive discussions with carriers across the OECD. USD purchasing power parities (PPP) are used to aid international comparisons. (*) Contact type: PP = Pre-paid plan; SD = Including selective discounts.

Source: OECD and Teligen.

StatLink http://dx.doi.org/10.1787/888932802287

Table 7.17. **OECD basket of national leased line charges, monthly price, August 2012, VAT excluded**

	2 Mbit/s		34 Mbit/s	
	USD	USD PPP	USD	USD PPP
Australia	4 365	2 780	7 773	4 951
Austria	1 103	994	7 480	6 738
Belgium	1 424	1 217	7 384	6 311
Canada	3 602	2 836	23 091	18 182
Czech Republic	3 755	4 694		
Denmark	590	396	1 851	1 242
Finland				
France	1 942	1 688	12 087	10 510
Germany	1 231	1 150	5 330	4 981
Greece	952	952	7 075	7 075
Hungary				
Iceland	448	380	1 423	1 206
Ireland	1 723	1 378	20 384	16 307
Italy	1 720	1 737	10 972	11 082
Japan	4 845	3 251	17 441	11 705
Korea	3 007	3 759	16 345	20 432
Luxembourg	929	749	3 920	3 161
Mexico	1 862	2 909	14 803	23 130
Netherlands	1 444	1 301		
New Zealand				
Norway	902	564	3 079	1 924
Poland	693	1 174		
Portugal	1 329	1 342	10 564	10 671
Slovak Republic	5 433	7 149		
Spain				
Sweden	586	458		
Switzerland				
Turkey	668	1 192	4 203	7 505
United Kingdom	2 753	2 086	16 554	12 541
United States	1 659	1 659	8 373	8 373

Source: OECD and Teligen.

StatLink http://dx.doi.org/10.1787/888932802306

Table 7.18. **Trends in leased line pricing over different distances, 1992-2012**

OECD average	1992	1993	1994	1995	1996	1997	1998	1999	2000	2001	2002	2003	2004	2005	2006	2007	2008	2009	2010	2011	2012
2 Mbit/s																					
2 km	100	99.7	106.3	108.1	105.6	101.5	94.8	59.8	58.3	56.6	53.4	50.3	47.4	44.5	47.7	41.0	40.9	41.4	45.7	44.8	43.9
50 km	100	98.3	89.3	84.5	78.2	72.3	59.6	40.1	43.2	39.8	37.8	34.5	31.3	27.6	34.5	33.4	34.3	32.2	33.6	33.2	32.8
200 km	100	98.8	94.6	87.9	77.3	73.1	60.7	41.7	45.4	38.9	35.9	33.5	30.7	26.3	32.3	32.9	33.9	31.4	30.9	30.5	30.2

Source: OECD/Teligen.

StatLink http://dx.doi.org/10.1787/888932802325

Table 7.19. **OECD Fixed Broadband basket Low 1 : 2 GB, 0.250 Mbit/s and above, September 2012**

	ISP	Package	Speed Up/Down	Total		ADSL, Fibre or Cable Modem
				USD	USD PPP	
Australia	Internode	NBNTP Bronze 30	1024 / 12288	51.52	31.61	F
Austria	A1 Telekom Austria	A1 Breitband 8 Mbit/s	768 / 8192	23.07	21.16	A
Belgium	Belgacom	Internet Start	1500 / 30720	32.32	28.10	A
Canada	Bell	Essential Plus	800 / 2048	38.73	30.02	A
Chile	VTR	Banda Ancha Mega 1	512 / 1024	21.69	28.93	A
Czech Republic	O2	O2 Internet Start	250 / 2048	21.89	27.71	A
Denmark	Stofa.dk	Bredband 15Mbit. / 2Mbit.	2000 / 15000	31.53	21.60	C
Estonia	STV	STV Internet 1	512 / 1024	11.94	14.39	C
Finland	Saunalahti	Cable Modem 2/2 M	2048 / 2048	31.70	25.56	C
France	Free	Zone Dégroupées ADSL	1024 / 28672	39.02	34.84	A
Germany	Kabel Deutschland	Internet & Telefon 8	600 / 8000	23.43	22.32	C
Greece	CYTA	2play 4Mbps	512 / 4096	21.82	22.49	A
Hungary	UPC	Fiber Power 10	1000 / 10000	8.99	12.85	F
Iceland	Hringiðan	ADSL 10Mb/20GB	1024 / 10240	36.92	31.02	A
Ireland	Vodafone	Value Broadband	512 / 8192	50.90	41.38	A
Israel	Hot/Smile	5MB Cable Modem	768 / 5120	22.36	19.61	C
Italy	Fastweb	Joy	1024 / 20480	32.20	33.20	A
Japan	BB Excite/NTT-East	BB Excite ADSL	512 / 1024	28.68	19.25	A
Korea	KT	Special/Light + PSTN	.. / 100000	13.24	16.35	F
Luxembourg	P&T	LuxDSL Junior with INTEGRAL	512 / 5120	32.62	26.74	A
Mexico	Axtel	Acceso Universal Banda Ancha 512Kbps	128 / 512	25.07	36.86	A
Netherlands	Ziggo	Internet Z1	1024 / 8192	27.47	25.43	C
New Zealand	TelstraClear	InHome LightSpeed 20G Naked Broadbanc	2048 / 15360	44.82	34.47	C
Norway	Telenor	Bredband 2	400 / 2048	68.94	43.09	A
Poland	UPC	Fiber Power 15Mb/s	1500 / 15000	16.09	27.28	F
Portugal	PT	SAPO ADSL 12Mb - Zonas sapo	1024 / 12000	25.71	26.50	A
Slovak Republic	Swan/ MAX Multimedia	Klasik 10/10	10240 / 10240	12.46	16.83	F
Slovenia	T-2	VDSL 1Mb (Total)	256 / 1024	20.08	22.56	A
Spain	Telefonica	Servicio Universal de Banda Ancha	256 / 1024	48.18	47.70	A
Sweden	ComHem	Medium 10 (12M)	1024 / 10240	38.45	29.57	C
Switzerland	Cablecom	Fiber Power Internet 10	1000 / 10000	45.08	26.52	C
Turkey	TTNet	1 Mbps / 2 GB	256 / 1024	14.13	25.23	A
United Kingdom	BT	Broadband and Evening & Weekend Calls	.. / 16384	34.15	25.87	A
United States	Verizon	High Speed Internet	512 / 1024	27.49	27.49	A
OECD average				30.08	27.19	

Notes: The OECD basket of Fixed Broadband services includes the total charges for a subscription with a minimum speed of 0.25 Mbit/s and 2GB usage per month. USD purchasing power parities (PPP) are used to aid international comparisons.

Source: OECD and Teligen.

StatLink http://dx.doi.org/10.1787/888932802344

Table 7.20. **OECD Fixed Broadband basket Low 2 : 6 GB per month. 2.5 Mbit/s and above, September 2012**

	ISP	Package	Speed Up/Down	Total		ADSL, Fibre or Cable Modem
				USD	USD PPP	
Australia	Internode	NBNTP Bronze 30	1024 / 12288	51.52	31.61	F
Austria	A1 Telekom Austria	A1 Breitband 8 Mbit/s	768 / 8192	23.07	21.16	A
Belgium	Belgacom	Internet Start	1500 / 30720	32.32	28.10	A
Canada	Bell	Fibe 5	1024 / 5120	40.92	31.72	F
Chile	Telmex - Claro Chile	Plan Internet Estandar (6Mbps)	512 / 6144	38.53	51.38	A
Czech Republic	UPC	UPC Fiber Power 30	1024 / 30720	26.33	33.32	F
Denmark	Stofa.dk	Bredband 15Mbit. / 2Mbit.	2000 / 15000	31.53	21.60	C
Estonia	STV	STV Internet 5	1024 / 5120	15.96	19.23	C
Finland	Saunalahti	ADSL 4M/1M	1024 / 4096	32.71	26.38	A
France	Free	Zone Dégroupées ADSL	1024 / 28672	39.02	34.84	A
Germany	Kabel Deutschland	Internet & Telefon 8	600 / 8000	23.43	22.32	C
Greece	CYTA	2play 4Mbps	512 / 4096	21.82	22.49	A
Hungary	UPC	Fiber Power 25	1500 / 25000	14.52	20.74	F
Iceland	Hringiðan	ADSL 10Mb/20GB	1024 / 10240	36.92	31.02	A
Ireland	Vodafone	Value Broadband	512 / 8192	50.90	41.38	A
Israel	Hot/Smile	5MB Cable Modem	768 / 5120	22.36	19.61	C
Italy	Fastweb	Joy	1024 / 20480	32.20	33.20	A
Japan	Yahoo! BB/NTT-East	8M	922 / 8192	41.43	27.81	A
Korea	KT	Special/Light + PSTN	.. / 100000	13.24	16.35	F
Luxembourg	P&T	LuxDSL Junior with INTEGRAL	512 / 5120	48.04	39.38	A
Mexico	Telmex	Paquete Conectes	128 / 3072	29.25	43.01	A
Netherlands	Ziggo	Internet Z1	1024 / 8192	27.47	25.43	C
New Zealand	TelstraClear	InHome LightSpeed 20G Naked Broadband	2048 / 15360	44.82	34.47	C
Norway	Telenor	Bredband 5	500 / 5120	73.73	46.08	A
Poland	UPC	Fiber Power 15Mb/s	1500 / 15000	16.09	27.28	F
Portugal	PT	SAPO ADSL 12Mb - Zonas sapo	1024 / 12000	25.71	26.50	A
Slovak Republic	Swan/ MAX Multimedia	Klasik 10/10	10240 / 10240	12.46	16.83	F
Slovenia	Amis	BS3 Optika - Enka 15/15	15000 / 15000	21.67	24.35	F
Spain	Telefonica	ADSL Libre 10Mb	800 / 10240	52.73	52.21	A
Sweden	ComHem	Medium 10 (12M)	1024 / 10240	38.45	29.57	C
Switzerland	Cablecom	Fiber Power Internet 10	1000 / 10000	45.08	26.52	C
Turkey	TTNet	Net 6 / 6GB	1024 / 8192	18.03	32.19	A
United Kingdom	BT	Broadband and Evening & Weekend Calls	.. / 16384	34.15	25.87	A
United States	Comcast	Economy Plus	768 / 3072	43.95	43.95	C
OECD average				32.95	30.23	

Notes: The OECD basket of Fixed Broadband services includes the total charges for a subscription with a minimum speed of 2.5 Mbit/s and 6GB of usage per month. USD purchasing power parities (PPP) are used to aid international comparisons.

Source: OECD and Teligen.

StatLink http://dx.doi.org/10.1787/888932802363

Table 7.21. **OECD Fixed Broadband basket Low 3 : 11 GB per month. 15 Mbit/s and above, September 2012**

	ISP	Package	Speed Up/Down	Total USD	Total USD PPP	ADSL, Fibre or Cable Modem
Australia	Internode	NBNTP Silver 30	5120 / 25600	56.67	34.77	F
Austria	A1 Telekom Austria	A1 Breitband Gigaspeed 16 Mbit/s	1024 / 16384	28.70	26.33	A
Belgium	Belgacom	Internet Start	1500 / 30720	32.32	28.10	A
Canada	Bell	Fibe 15	10240 / 15360	52.71	40.86	F
Chile	Movistar	Banda Ancha Hogar 15Mega /15Mega	15360 / 15360	49.97	66.62	A
Czech Republic	UPC	UPC Fiber Power 30	1024 / 30720	26.33	33.32	F
Denmark	Stofa.dk	Bredband 15Mbit. / 2Mbit.	2000 / 15000	31.53	21.60	C
Estonia	STV	STV Saturn Internet 20/10	10240 / 20240	15.96	19.23	F
Finland	Saunalahti	Full Rate 20M/1M	1024 / 20480	40.23	32.45	A
France	Free	Zone Dégroupées ADSL	1024 / 28672	39.02	34.84	A
Germany	Kabel Deutschland	Internetanschluss 16	1000 / 16000	26.36	25.10	C
Greece	CYTA	2play 24Mbps	1024 / 24576	30.60	31.55	A
Hungary	UPC	Fiber Power 25	1500 / 25000	14.52	20.74	F
Iceland	Vodafone	Ljósleiðari - Fibre Internet 40	50000 / 50000	58.23	48.93	F
Ireland	UPC	Fibre Power 25MB + Anywhere 100	2560 / 25600	54.27	44.12	F
Israel	Bezeq/Smile	15MB	800 / 15360	27.58	24.19	A
Italy	Fastweb	Joy	1024 / 20480	32.20	33.20	A
Japan	BB Excite/NTT-East	BB Excite ADSL (FLET More II (40 Mbps Type))	1024 / 40960	44.30	29.73	A
Korea	KT	Special/Light + PSTN	.. / 100000	13.24	16.35	F
Luxembourg	Numericable	TV + NET	1024 / 30720	57.93	47.48	F
Mexico	Telmex	Infinitum 20 Mbps	762 / 20480	44.90	66.02	A
Netherlands	Ziggo	Internet Z2	4096 / 40960	40.02	37.05	C
New Zealand	TelstraClear	InHome LightSpeed 20G Naked Broadband	2048 / 15360	44.82	34.47	C
Norway	Lyse	Internett 50 Opptill 25/25	25600 / 25600	77.35	48.34	F
Poland	UPC	Fiber Power 15Mb/s	1500 / 15000	16.09	27.28	F
Portugal	PT	SAPO ADSL 24Mb - Zonas sapo	1024 / 24000	31.98	32.97	A
Slovak Republic	Swan/ MAX Multimedia	Klasik 20/1	1024 / 20480	12.46	16.83	F
Slovenia	Amis	BS3 Optika - Enka 15/15	15000 / 15000	21.67	24.35	F
Spain	Orange	ADSL máxima velocidad	1024 / 20480	57.36	56.79	A
Sweden	ComHem	Large 25 (12M)	1024 / 25000	48.24	37.11	C
Switzerland	Cablecom	Fiber Power Internet 25	2500 / 25000	56.57	33.28	C
Turkey	TTNet	HiperNET 12 Mb LİMİTSİZ	1024 / 20480	25.17	44.94	F
United Kingdom	BT	Broadband and Evening & Weekend Calls	.. / 16384	35.74	27.07	A
United States	Verizon	High Speed Internet Enhanced	1126.4 / 15360	43.99	43.99	A
OECD average				37.91	35.00	

Notes: The OECD basket of Fixed Broadband services includes the total charges for a subscription with a minimum speed of 15 Mbit/s and 11GB of usage per month. USD purchasing power parities (PPP) are used to aid international comparisons.

Source: OECD and Teligen.

StatLink http://dx.doi.org/10.1787/888932802382

Table 7.22. **OECD Fixed Broadband basket Low 4: 14 GB per month. 30 Mbit/s and above, September 2012**

	ISP	Package	Speed Up/Down	Total		ADSL, Fibre or Cable Modem
				USD	USD PPP	
Australia	Internode	NBNTP Gold 30	20480 / 51200	66.99	41.10	F
Austria	UPC Telekabel	Fiber Power	4096 / 35840	33.15	30.41	C
Belgium	Belgacom	Internet Start	1500 / 30720	32.32	28.10	A
Canada	Shaw	Broadband 50	3072 / 51200	83.02	64.36	C
Chile	Movistar	Banda Ancha Hogar 30Mega /15Mega	15360 / 30720	56.22	74.95	A
Czech Republic	UPC	UPC Fiber Power 30	1024 / 30720	26.33	33.32	F
Denmark	Stofa.dk	Bredband 30Mbit. / 3Mbit.	3000 / 30000	43.22	29.60	C
Estonia	Elion	Homemade Solution up to 50Mbit/s	10240 / 51200	25.09	30.23	A
Finland	Saunalahti	Cable Modem 30/2 M	2048 / 30720	49.26	39.73	C
France	Free	Zone Dégroupées Fibre	51200 / 102400	39.02	34.84	F
Germany	Kabel Deutschland	Internet & Telefon 32	2000 / 32000	34.72	33.07	C
Greece	CYTA	2play 35Mbps	3072 / 35840	48.17	49.66	A
Hungary	UPC	Fiber Power 50	5000 / 50000	20.39	29.13	F
Iceland	Vodafone	Ljósleiðari - Fibre Internet 40	50000 / 50000	58.23	48.93	F
Ireland	UPC	Fibre Power 50MB + Anywhere 100	5120 / 51200	64.30	52.28	F
Israel	Hot/018	30 MB	1500 / 30000	35.70	31.31	A
Italy	Telecom Italia - Alice	Fiber	10240 / 102400	68.09	70.20	F
Japan	BB Excite/NTT-East	BB Excite ADSL (FLET More II (40 Mbps Type))	1024 / 40960	44.30	29.73	A
Korea	KT	Special/Light + PSTN	.. / 100000	13.24	16.35	F
Luxembourg	Numericable	TV + NET	1024 / 30720	57.93	47.48	F
Mexico	Telmex	Infinitum 50 Mbps	1024 / 51200	58.76	86.41	A
Netherlands	Ziggo	Internet Z2	4096 / 40960	40.02	37.05	C
New Zealand	TelstraClear	InHome WarpSpeed 100G Naked Broadband	10240 / 102400	92.88	71.44	C
Norway	Lyse	Internett 120 Opptill 60/60	61440 / 61440	94.58	59.11	F
Poland	UPC	Fiber Power 30Mb/s	3072 / 30720	18.97	32.15	F
Portugal	PT	SAPO Fibra 30Mb	3072 / 30720	37.21	38.36	F
Slovak Republic	UPC	Fiber Power 30	3072 / 30720	13.78	18.62	C
Slovenia	T-2	Fiber 100/10	10240 / 102400	33.88	38.06	F
Spain	Telefonica	Imagenio VDSL Familiar hasta 30Mb + DVR	1024 / 30720	66.31	65.65	A
Sweden	Telia	Bredband Fiber 100/10	10000 / 100000	53.31	41.01	F
Switzerland	Cablecom	Fiber Power Internet 50	5000 / 50000	67.02	39.42	C
Turkey	TTNet	HiperNET 35 Mb Unlimited	1024 / 35840	41.64	74.35	F
United Kingdom	Virgin Media	Up to 30 Mbps	3072 / 30720	37.80	28.64	F
United States	Comcast	Blast Plus (special offer)	6144 / 30720	73.32	73.32	C
OECD average				47.92	44.66	

Notes: The OECD basket of Fixed Broadband services includes the total charges for a subscription with a minimum speed of 30 Mbit/s and 14 GB of usage per month. USD purchasing power parities (PPP) are used to aid international comparisons.

Source: OECD and Teligen.

StatLink http://dx.doi.org/10.1787/888932802401

Table 7.23. **OECD Fixed Broadband basket Low 5: 18 GB per month. 45 Mbit/s and above, September 2012**

	ISP	Package	Speed Up/Down	Total		ADSL, Fibre or Cable Modem
				USD	USD PPP	
Australia	Internode	NBNTP Gold 30	20480 / 51200	66.99	41.10	F
Austria	UPC Telekabel	Super Fit	5120 / 51200	49.33	45.25	A
Belgium	Telenet	Fibernet	4096 / 61440	58.14	50.56	F
Canada	Shaw	Broadband 50	3072 / 51200	83.02	64.36	C
Chile	Movistar	Banda Ancha Hogar 80Mega /15Mega	15360 / 81920	68.71	91.62	A
Czech Republic	UPC	UPC Fiber Power 60	6144 / 61440	31.38	39.72	F
Denmark	Stofa.dk	Bredband 50Mbit. / 5Mbit.	5000 / 50000	56.71	38.84	C
Estonia	Elion	Homemade Solution up to 50Mbit/s	10240 / 51200	25.09	30.23	A
Finland	Saunalahti	Cable Modem 100/2 M	2048 / 102400	58.04	46.81	C
France	Free	Zone Dégroupées Fibre	51200 / 102400	39.02	34.84	F
Germany	Kabel Deutschland	Internet & Telefon 100	6000 / 100000	43.09	41.04	C
Greece	CYTA	2play 50Mbps	10240 / 51200	58.20	60.00	A
Hungary	UPC	Fiber Power 50	5000 / 50000	20.39	29.13	F
Iceland	Vodafone	Ljósleiðari - Fibre Internet 40	50000 / 50000	58.23	48.93	F
Ireland	UPC	Fibre Power 50MB + Anywhere 100	5120 / 51200	64.30	52.28	F
Israel	Hot/Smile	100MB Cable Modem	1000 / 102400	89.03	78.10	C
Italy	Telecom Italia - Alice	Fiber	10240 / 102400	68.09	70.20	F
Japan	BB Excite/NTT-East	BB Excite ADSL (FLET'S ADSL MORE III (47 Mbps Type))	5120 / 48128	44.97	30.18	A
Korea	KT	Special/Light + PSTN	.. / 100000	13.24	16.35	F
Luxembourg	Numericable	Fiberboost 70Mega	5120 / 71680	75.00	61.48	F
Mexico	Telmex	Infinitum 50 Mbps	1024 / 51200	58.76	86.41	A
Netherlands	KPN	Glasvezel internet en Bellen Instap	51200 / 51200	51.06	47.28	F
New Zealand	TelstraClear	InHome WarpSpeed 100G Naked Broadband	10240 / 102400	92.88	71.44	C
Norway	Lyse	Internett 120 Opptill 60/60	61440 / 61440	94.58	59.11	F
Poland	UPC	Fiber Power 60Mb/s	6144 / 61440	22.99	38.97	F
Portugal	PT	SAPO Fibra 100Mb	10240 / 102400	49.34	50.87	F
Slovak Republic	UPC	Fiber Power 60	6144 / 61440	21.31	28.80	C
Slovenia	T-2	Fiber 100/10	10240 / 102400	33.88	38.06	F
Spain	Ono	Banda Ancha 50Mb + Phone	3072 / 51200	78.37	77.60	F
Sweden	Telia	Bredband Fiber 100/10	10000 / 100000	53.31	41.01	F
Switzerland	Cablecom	Fiber Power Internet 50	5000 / 50000	67.02	39.42	C
Turkey	TTNet	HiperNET 50 Mb Unlimited	1024 / 51200	52.62	93.96	F
United Kingdom	Virgin Media	Up to 60 Mbps	6144 / 61440	45.71	34.63	F
United States	Verizon	FiOS Internet up to 50/25Mbps	25000 / 51200	89.82	89.82	F
OECD average				55.37	52.01	

Notes: The OECD basket of Fixed Broadband services includes the total charges for a subscription with a minimum speed of 45 Mbit/s and 18 GB of usage per month. USD purchasing power parities (PPP) are used to aid international comparisons.

Source: OECD and Teligen.

StatLink http://dx.doi.org/10.1787/888932802420

Table 7.24. **OECD Fixed Broadband basket High 1: 6 GB per month. 0.250 Mbit/s and above, September 2012**

	ISP	Package	Speed Up/Down	Total		ADSL, Fibre or Cable
				USD	USD PPP	
Australia	Internode	NBNTP Bronze 30	1024 / 12288	51.52	31.61	F
Austria	A1 Telekom Austria	A1 Breitband 8 Mbit/s	768 / 8192	23.07	21.16	A
Belgium	Belgacom	Internet Start	1500 / 30720	32.32	28.10	A
Canada	Bell	Fibe 5	1024 / 5120	40.92	31.72	F
Chile	Telmex - Claro Chile	Plan Internet Entrada (2Mbps)	512 / 2048	27.08	36.10	A
Czech Republic	O2	O2 Internet Start	250 / 2048	21.89	27.71	A
Denmark	Stofa.dk	Bredband 15Mbit. / 2Mbit.	2000 / 15000	31.53	21.60	C
Estonia	STV	STV Internet 1	512 / 1024	11.94	14.39	C
Finland	Saunalahti	Cable Modem 2/2 M	2048 / 2048	31.70	25.56	C
France	Free	Zone Dégroupées ADSL	1024 / 28672	39.02	34.84	A
Germany	Kabel Deutschland	Internet & Telefon 8	600 / 8000	23.43	22.32	C
Greece	CYTA	2play 4Mbps	512 / 4096	21.82	22.49	A
Hungary	UPC	Fiber Power 25	1500 / 25000	14.52	20.74	F
Iceland	Hringiðan	ADSL 10Mb/20GB	1024 / 10240	36.92	31.02	A
Ireland	Vodafone	Value Broadband	512 / 8192	50.90	41.38	A
Israel	Hot/Smile	5MB Cable Modem	768 / 5120	22.36	19.61	C
Italy	Fastweb	Joy	1024 / 20480	32.20	33.20	A
Japan	BB Excite/NTT-East	BB Excite ADSL (FLET'S ADSL Entries 1Mbps Type)	512 / 1024	28.68	19.25	A
Korea	KT	Special/Light + PSTN	.. / 100000	13.24	16.35	F
Luxembourg	P&T	LuxDSL Junior with INTEGRAL	512 / 5120	48.04	39.38	A
Mexico	Axtel	Acceso Universal Banda Ancha 512Kbps	128 / 512	25.07	36.86	A
Netherlands	Ziggo	Internet Z1	1024 / 8192	27.47	25.43	C
New Zealand	TelstraClear	InHome LightSpeed 20G Naked Broadband	2048 / 15360	44.82	34.47	C
Norway	Telenor	Bredband 2	400 / 2048	68.94	43.09	A
Poland	UPC	Fiber Power 15Mb/s	1500 / 15000	16.09	27.28	F
Portugal	PT	SAPO ADSL 12Mb - Zonas sapo	1024 / 12000	25.71	26.50	A
Slovak Republic	Swan/ MAX Multimedia	Klasik 10/10	10240 / 10240	12.46	16.83	F
Slovenia	T-2	VDSL 1Mb (Total)	256 / 1024	20.08	22.56	A
Spain	Telefonica	ADSL Libre 10Mb	800 / 10240	52.73	52.21	A
Sweden	ComHem	Medium 10 (12M)	1024 / 10240	38.45	29.57	C
Switzerland	Cablecom	Fiber Power Internet 10	1000 / 10000	45.08	26.52	C
Turkey	TTNet	Net 6 / 6GB	1024 / 8192	18.03	32.19	A
United Kingdom	BT	Broadband and Evening & Weekend Calls	.. / 16384	34.15	25.87	A
United States	Verizon	High Speed Internet	512 / 1024	27.49	27.49	A
OECD average				31.17	28.39	

Notes: The OECD basket of Fixed Broadband services includes the total charges for a subscription with a minimum speed of 0.25 Mbit/s and 6 GB of usage per month. USD purchasing power parities (PPP) are used to aid international comparisons.

Source: OECD and Teligen.

StatLink *http://dx.doi.org/10.1787/888932802439*

Table 7.25. **OECD Fixed Broadband basket High 2: 18 GB per month. 2.5 Mbit/s and above, September 2012**

	ISP	Package	Speed Up/Down	Total		ADSL, Fibre or Cable Modem
				USD	USD PPP	
Australia	Internode	NBNTP Bronze 30	1024 / 12288	51.52	31.61	F
Austria	A1 Telekom Austria	A1 Breitband 8 Mbit/s	768 / 8192	23.07	21.16	A
Belgium	Belgacom	Internet Start	1500 / 30720	32.32	28.10	A
Canada	Bell	Performance	1024 / 6144	46.90	36.36	A
Chile	Telmex - Claro Chile	Plan Internet Estandar (6Mbps)	512 / 6144	38.53	51.38	A
Czech Republic	UPC	UPC Fiber Power 30	1024 / 30720	26.33	33.32	F
Denmark	Stofa.dk	Bredband 15Mbit. / 2Mbit.	2000 / 15000	31.53	21.60	C
Estonia	STV	STV Internet 5	1024 / 5120	15.96	19.23	C
Finland	Saunalahti	ADSL 4M/1M	1024 / 4096	32.71	26.38	A
France	Free	Zone Dégroupées ADSL	1024 / 28672	39.02	34.84	A
Germany	Kabel Deutschland	Internet & Telefon 8	600 / 8000	23.43	22.32	C
Greece	CYTA	2play 4Mbps	512 / 4096	21.82	22.49	A
Hungary	UPC	Fiber Power 25	1500 / 25000	14.52	20.74	F
Iceland	Hringiðan	ADSL 10Mb/20GB	1024 / 10240	36.92	31.02	A
Ireland	Vodafone	Value Broadband	512 / 8192	50.90	41.38	A
Israel	Hot/Smile	5MB Cable Modem	768 / 5120	22.36	19.61	C
Italy	Fastweb	Joy	1024 / 20480	32.20	33.20	A
Japan	Yahoo! BB/NTT-East	8M	922 / 8192	41.43	27.81	A
Korea	KT	Special/Light + PSTN	.. / 100000	13.24	16.35	F
Luxembourg	Visual Online	Vodsl Starter Kombi	512 / 5120	55.05	45.12	A
Mexico	Telmex	Paquete Conectes	128 / 3072	29.25	43.01	A
Netherlands	Ziggo	Internet Z1	1024 / 8192	27.47	25.43	C
New Zealand	TelstraClear	InHome LightSpeed 20G Naked Broadband	2048 / 15360	44.82	34.47	C
Norway	Telenor	Bredband 5	500 / 5120	73.73	46.08	A
Poland	UPC	Fiber Power 15Mb/s	1500 / 15000	16.09	27.28	F
Portugal	PT	SAPO ADSL 12Mb - Zonas sapo	1024 / 12000	25.71	26.50	A
Slovak Republic	Swan/ MAX Multimedia	Klasik 10/10	10240 / 10240	12.46	16.83	F
Slovenia	Amis	BS3 Optika - Enka 15/15	15000 / 15000	21.67	24.35	F
Spain	Telefonica	ADSL Libre 10Mb	800 / 10240	52.73	52.21	A
Sweden	ComHem	Medium 10 (12M)	1024 / 10240	38.45	29.57	C
Switzerland	Cablecom	Fiber Power Internet 10	1000 / 10000	45.08	26.52	C
Turkey	TTNet	NETLİMİTSİZ / unlimited	1024 / 8192	32.85	58.66	A
United Kingdom	Sky	Broadband Unlimited with Sky Talk Freetime	1300 / 20480	36.13	27.37	A
United States	Comcast	Economy Plus	768 / 3072	43.95	43.95	C
OECD average				33.83	31.36	

Notes: The OECD basket of Fixed Broadband services includes the total charges for a subscription with a minimum speed of 2.5 Mbit/s and 18 GB of usage per month. USD purchasing power parities (PPP) are used to aid international comparisons.

Source: OECD and Teligen.

StatLink http://dx.doi.org/10.1787/888932802458

Table 7.26. **OECD Fixed Broadband basket High 3: 33 GB per month. 15 Mbit/s and above, September 2012**

	ISP	Package	Speed Up/Down	Total		ADSL, Fibre or Cable Modem
				USD	USD PPP	
Australia	Optus	120GB Broadband + Phone	640 / 20480	68.55	42.06	A
Austria	A1 Telekom Austria	A1 Breitband Gigaspeed 16 Mbit/s	1024 / 16384	28.70	26.33	A
Belgium	Belgacom	Internet Start	1500 / 30720	32.32	28.10	A
Canada	Bell	Fibe 15	10240 / 15360	52.71	40.86	F
Chile	Movistar	Banda Ancha Hogar 15Mega /15Mega	15360 / 15360	49.97	66.62	A
Czech Republic	UPC	UPC Fiber Power 30	1024 / 30720	26.33	33.32	F
Denmark	Stofa.dk	Bredband 15Mbit. / 2Mbit.	2000 / 15000	31.53	21.60	C
Estonia	STV	STV Saturn Internet 20/10	10240 / 20240	15.96	19.23	F
Finland	Saunalahti	Full Rate 20M/1M	1024 / 20480	40.23	32.45	A
France	Free	Zone Dégroupées ADSL	1024 / 28672	39.02	34.84	A
Germany	Kabel Deutschland	Internetanschluss 16	1000 / 16000	26.36	25.10	C
Greece	CYTA	2play 24Mbps	1024 / 24576	30.60	31.55	A
Hungary	UPC	Fiber Power 25	1500 / 25000	14.52	20.74	F
Iceland	Vodafone	Ljósleiðari - Fibre Internet 40	50000 / 50000	58.23	48.93	F
Ireland	UPC	Fibre Power 25MB + Anywhere 100	2560 / 25600	54.27	44.12	F
Israel	Bezeq/Smile	15MB	800 / 15360	27.58	24.19	A
Italy	Fastweb	Joy	1024 / 20480	32.20	33.20	A
Japan	BB Excite/NTT-East	BB Excite ADSL (FLET More II (40 Mbps Type))	1024 / 40960	44.30	29.73	A
Korea	KT	Special/Light + PSTN	.. / 100000	13.24	16.35	F
Luxembourg	Numericable	Fiberboost 30Mega	1024 / 30720	64.34	52.74	F
Mexico	Telmex	Paquete Todo Mexico Sin Limites	256 / 20480	90.07	132.45	A
Netherlands	Ziggo	Internet Z2	4096 / 40960	40.02	37.05	C
New Zealand	TelstraClear	InHome LightSpeed 40G Naked Broadband	2048 / 15360	52.83	40.64	C
Norway	Lyse	Internett 50 Opptill 25/25	25600 / 25600	77.35	48.34	F
Poland	UPC	Fiber Power 15Mb/s	1500 / 15000	16.09	27.28	F
Portugal	PT	SAPO ADSL 24Mb - Zonas sapo	1024 / 24000	31.98	32.97	A
Slovak Republic	Swan/ MAX Multimedia	Klasik 20/1	1024 / 20480	12.46	16.83	F
Slovenia	Amis	BS3 Optika - Enka 15/15	15000 / 15000	21.67	24.35	F
Spain	Orange	ADSL máxima velocidad	1024 / 20480	57.36	56.79	A
Sweden	ComHem	Large 25 (12M)	1024 / 25000	48.24	37.11	C
Switzerland	Cablecom	Fiber Power Internet 25	2500 / 25000	56.57	33.28	C
Turkey	TTNet	Ultranet unlimited bundle	4096 / 16384	35.60	63.57	A
United Kingdom	Sky	Broadband Unlimited with Sky Talk Freetime	1300 / 20480	36.13	27.37	A
United States	Verizon	High Speed Internet Enhanced	1126.4 / 15360	43.99	43.99	A
OECD average				40.33	38.06	

Notes: The OECD basket of Fixed Broadband services includes the total charges for a subscription with a minimum speed of 15 Mbit/s and 33 GB of usage per month. USD purchasing power parities (PPP) are used to aid international comparisons.

Source: OECD and Teligen.

StatLink http://dx.doi.org/10.1787/888932802477

Table 7.27. **OECD Fixed Broadband basket High 4: 42 GB per month. 30 Mbit/s and above, September 2012**

	ISP	Package	Speed Up/Down	Total		ADSL, Fibre or Cable
				USD	USD PPP	
Australia	BigPond/Telstra	BigPond Ultimate Cable 50GB	1024 / 30720	85.50	52.45	C
Austria	UPC Telekabel	Fiber Power	4096 / 35840	33.15	30.41	C
Belgium	Belgacom	Internet Start	1500 / 30720	32.32	28.10	A
Canada	Shaw	Broadband 50	3072 / 51200	83.02	64.36	C
Chile	Movistar	Banda Ancha Hogar 30Mega /15Mega	15360 / 30720	56.22	74.95	A
Czech Republic	UPC	UPC Fiber Power 30	1024 / 30720	26.33	33.32	F
Denmark	Stofa.dk	Bredband 30Mbit. / 3Mbit.	3000 / 30000	43.22	29.60	C
Estonia	Elion	Homemade Solution up to 50Mbit/s	10240 / 51200	25.09	30.23	A
Finland	Saunalahti	Cable Modem 30/2 M	2048 / 30720	49.26	39.73	C
France	Free	Zone Dégroupées Fibre	51200 / 102400	39.02	34.84	F
Germany	Kabel Deutschland	Internet & Telefon 32	2000 / 32000	34.72	33.07	C
Greece	CYTA	2play 35Mbps	3072 / 35840	48.17	49.66	A
Hungary	UPC	Fiber Power 50	5000 / 50000	20.39	29.13	F
Iceland	Vodafone	Ljósleiðari - Fibre Internet 80	50000 / 50000	67.03	56.33	F
Ireland	UPC	Fibre Power 50MB + Anywhere 100	5120 / 51200	64.30	52.28	F
Israel	Hot/018	30 MB	1500 / 30000	35.70	31.31	A
Italy	Telecom Italia - Alice	Fiber	10240 / 102400	68.09	70.20	F
Japan	BB Excite/NTT-East	BB Excite ADSL (FLET More II (40 Mbps Type))	1024 / 40960	44.30	29.73	A
Korea	KT	Special/Light + PSTN	.. / 100000	13.24	16.35	F
Luxembourg	Numericable	Triple Play	1024 / 30720	65.21	53.45	F
Mexico	Telmex	Infinitum 50 Mbps	1024 / 51200	168.85	248.31	A
Netherlands	Ziggo	Internet Z2	4096 / 40960	40.02	37.05	C
New Zealand	TelstraClear	InHome WarpSpeed 100G Naked Broadband	10240 / 102400	92.88	71.44	C
Norway	Lyse	Internett 120 Opptill 60/60	61440 / 61440	94.58	59.11	F
Poland	UPC	Fiber Power 30Mb/s	3072 / 30720	18.97	32.15	F
Portugal	PT	SAPO Fibra 30Mb	3072 / 30720	37.21	38.36	F
Slovak Republic	UPC	Fiber Power 30	3072 / 30720	13.78	18.62	C
Slovenia	T-2	Fiber 100/10	10240 / 102400	33.88	38.06	F
Spain	Telefonica	Imagenio VDSL Familiar hasta 30Mb + DVR	1024 / 30720	66.31	65.65	A
Sweden	Telia	Bredband Fiber 100/10	10000 / 100000	53.31	41.01	F
Switzerland	Cablecom	Fiber Power Internet 50	5000 / 50000	67.02	39.42	C
Turkey	TTNet	HiperNET 35 Mb Unlimited	1024 / 35840	41.64	74.35	F
United Kingdom	Virgin Media	Up to 30 Mbps	3072 / 30720	37.80	28.64	F
United States	Comcast	Blast Plus (special offer)	6144 / 30720	73.32	73.32	C
OECD average				52.17	50.15	

Notes: The OECD basket of Fixed Broadband services includes the total charges for a subscription with a minimum speed of 30 Mbit/s and 42 GB of usage per month. USD purchasing power parities (PPP) are used to aid international comparisons.

Source: OECD and Teligen.

StatLink http://dx.doi.org/10.1787/888932802496

Table 7.28. **OECD Fixed Broadband basket High 5: 54 GB per month. 45 Mbit/s and above, September 2012**

	ISP	Package	Speed Up/Down	Total		ADSL, Fibre or Cable Modem
				USD	USD PPP	
Australia	Internode	NBNTP Gold 300	20480 / 51200	87.61	53.75	F
Austria	UPC Telekabel	Super Fit	5120 / 51200	49.33	45.25	A
Belgium	Telenet	Fibernet	4096 / 61440	58.14	50.56	F
Canada	Shaw	Broadband 50	3072 / 51200	83.02	64.36	C
Chile	Movistar	Banda Ancha Hogar 80Mega /15Mega	15360 / 81920	68.71	91.62	A
Czech Republic	UPC	UPC Fiber Power 60	6144 / 61440	31.38	39.72	F
Denmark	Stofa.dk	Bredband 50Mbit. / 5Mbit.	5000 / 50000	56.71	38.84	C
Estonia	Elion	Homemade Solution up to 50Mbit/s	10240 / 51200	25.09	30.23	A
Finland	Saunalahti	Cable Modem 100/2 M	2048 / 102400	58.04	46.81	C
France	Free	Zone Dégroupées Fibre	51200 / 102400	39.02	34.84	F
Germany	Kabel Deutschland	Internet & Telefon 100	6000 / 100000	43.09	41.04	C
Greece	CYTA	2play 50Mbps	10240 / 51200	58.20	60.00	A
Hungary	UPC	Fiber Power 50	5000 / 50000	20.39	29.13	F
Iceland	Vodafone	Ljósleiðari - Fibre Internet 80	50000 / 50000	67.03	56.33	F
Ireland	UPC	Fibre Power 50MB + Anywhere 100	5120 / 51200	64.30	52.28	F
Israel	Hot/Smile	100MB Cable Modem	1000 / 102400	89.03	78.10	C
Italy	Telecom Italia - Alice	Fiber	10240 / 102400	68.09	70.20	F
Japan	BB Excite/NTT-East	BB Excite ADSL (FLET'S ADSL MORE III)	5120 / 48128	44.97	30.18	A
Korea	KT	Special/Light + PSTN	.. / 100000	13.24	16.35	F
Luxembourg	Numericable	Fiberboost 70Mega	5120 / 71680	75.00	61.48	F
Mexico	Telmex	Infinitum 50 Mbps	1024 / 51200	228.90	336.62	A
Netherlands	KPN	Glasvezel internet en Bellen Instap	51200 / 51200	51.06	47.28	F
New Zealand	TelstraClear	InHome WarpSpeed 100G Naked Broadband	10240 / 102400	92.88	71.44	C
Norway	Lyse	Internett 120 Opptill 60/60	61440 / 61440	94.58	59.11	F
Poland	UPC	Fiber Power 60Mb/s	6144 / 61440	22.99	38.97	F
Portugal	PT	SAPO Fibra 100Mb	10240 / 102400	49.34	50.87	F
Slovak Republic	UPC	Fiber Power 60	6144 / 61440	21.31	28.80	C
Slovenia	T-2	Fiber 100/10	10240 / 102400	33.88	38.06	F
Spain	Ono	Banda Ancha 50Mb + Phone	3072 / 51200	78.37	77.60	F
Sweden	Telia	Bredband Fiber 100/10	10000 / 100000	53.31	41.01	F
Switzerland	Cablecom	Fiber Power Internet 50	5000 / 50000	67.02	39.42	C
Turkey	TTNet	HiperNET 50 Mb Unlimited	1024 / 51200	52.62	93.96	F
United Kingdom	Virgin Media	Up to 60 Mbps	6144 / 61440	45.71	34.63	F
United States	Verizon	FiOS Internet up to 50/25Mbps	25000 / 51200	89.82	89.82	F
OECD average				61.24	59.96	

Notes: The OECD basket of Fixed Broadband services includes the total charges for a subscription with a minimum speed of 45 Mbit/s and 54 GB of usage per month. USD purchasing power parities (PPP) are used to aid international comparisons.

Source: OECD and Teligen.

StatLink http://dx.doi.org/10.1787/888932802515

Table 7.29. **Changes in DSL/fibre offerings, September 2005 to 2012**

DSL		Speed (kbit/s)								Price (local currency)								Bitcap (MB)								CAGR (2010-12)	
		2005	2006	2007	2008	2009	2010	2011	2012	2005	2006	2007	2008	2009	2010	2011	2012	2005	2006	2007	2008	2009	2010	2011	2012	Speed	Price
Australia	Bigpond	1 536	1 536	1 536	1 536	1 536	1 536	20 480	20 480	129.40	109.95	69.95	69.95	79.95	39.95	95.43	61.90	10 000	10 000	12 000	12 000	12 000	2 000	5 000	5 000	265.15	24.48
Austria	A1 Telekom Austria	2 048	2 048	2 048	2 048	2 048	6 144	8 192	8 192	54.90	54.90	59.90	39.90	39.90	19.90	31.98	18.38	12 000	15 000	20 000	..	..	..	..	..	15.47	-3.89
Belgium	Belgacom	4 096	4 096	4 096	4 096	4 096	12 288	15 360	30 720	54.95	54.95	57.05	57.05	31.55	32.50	34.31	34.86	30 000	30 000	30 000	60 000	4 000	50 000	100 000	150 000	58.11	3.57
Canada	Bell Canada	5 120	5 120	7 168	7 168	12 288	12 288	12 288	15 360	50.00	46.95	49.95	47.96	51.95	51.95	59.64	52.17	..	..	60 000	60 000	50 000	50 000	50 000	75 000	11.80	0.21
Chile	Movistar	..	..	..	..	..	2 048	2 048	2 048	..	..	..	..	..	18 490	12 990	14 990	..	..	..	..	..	..	..	..	0.00	-9.96
Czech Republic	O2	1 024	2 048	2 048	8 192	8 192	8 192	8 192	20 480	3 568	713	475	475	500	750	739	700	..	8 000	10 000	..	..	..	..	..	58.11	-3.39
Denmark	TDC	4 096	4 096	4 096	4 096	4 096	5 120	20 480	20 480	499	474	319	194	244	249	260.06	260.06	..	15 000	..	..	..	..	..	..	100.00	2.20
Estonia	Elion	..	..	..	..	..	102 400	102 400	102 400	..	..	..	..	..	49.82	24.00	24.00	..	..	..	..	..	..	..	..	0.00	-30.59
Finland	Sonera	24 000	24 000	24 576	24 576	24 576	24 576	10 240	24 576	68.9	59	49	24.5	49	39.9	39.31	33.2	..	..	..	..	..	..	..	..	0.00	-8.78
France	France Telecom	18 432	18 432	18 432	18 432	18 432	20 480	20 480	20 480	39.90	34.90	34.90	34.90	29.90	29.90	36.07	36.90	..	..	..	..	..	..	..	..	0.00	11.09
Germany	T-Com	6 144	6 144	6 144	6 144	6 144	6 144	16 384	16 384	34.94	34.98	28.58	39.95	39.95	39.95	33.55	36.89	..	..	..	..	..	..	..	..	63.30	-3.91
Greece	OTE	1 024	1 024	1 024	1 024	2 048	2 048	2 048	2 048	32.90	28.50	21.50	16.50	16.50	17.06	31.12	32.31	..	..	..	..	..	..	..	..	0.00	37.62
Hungary	T-Com	2 048	2 048	4 096	8 192	15 360	15 360	15 360	15 360	22 188	15 600	6 900	3 990	6 150	6 900	4 850	4 880	..	..	..	..	..	..	..	..	0.00	-15.90
Iceland	Simmin	6 144	8 192	8 192	8 192	8 192	12 188	12 188	12 288	5 790	5 990	5 990	6 190	4 400	3 050	5 189	6 879	..	..	..	80 000	..	10 000	10 000	10 000	0.41	50.18
Ireland	Eircom	2 048	2 048	2 048	3 072	3 072	3 072	3 072	3 072	54.45	29.99	39.99	29.99	30.11	29.99	56.16	57.11	16 000	20 000	20 000	30 000	..	30 000	30 000	30 000	0.00	38.00
Israel	Bezeq	..	..	..	..	..	1 536	1 536	10 240	..	..	..	..	..	86.9	80.8	93.9	..	..	..	..	..	..	..	..	158.20	3.95
Italy	Alice	4 000	20 480	20 480	20 480	20 480	20 480	20 480	102 400	41.9	36.95	36.95	24.95	24.95	24.95	51.66	54.27	..	..	..	..	..	..	..	..	123.61	47.48
Japan	NTT	102 400	102 400	102 400	102 400	102 400	102 400	204 800	204 800	4 064	3 612	2 930	3 255	3 518	3 115	3 863	3 873	..	..	..	..	..	..	..	..	41.42	11.51
Korea	KT	102 400	102 400	102 400	102 400	102 400	102 400	102 400	102 400	36 000	36 000	36 000	34 200	34 200	32 400	30 600	30 600	..	..	..	..	..	..	..	..	0.00	-2.82
Luxembourg	EPT	3 072	3 072	15 360	15 360	15 360	20 480	20 480	20 480	90.50	79.00	79.00	79.00	79.00	79.00	79.00	79.00	25 000	..	..	..	..	..	..	..	0.00	0.00
Mexico	Telmex	1 024	1 024	1 024	1 024	1 024	1 024	1 024	3 072	599.00	401.35	399.00	599.00	389.00	389.00	389.00	389.00	..	..	..	..	..	..	..	..	73.21	0.00
Netherlands	KPN	8 192	6 144	6 144	20 480	20 480	40 960	40 960	81 920	74.95	49.95	50.00	50.00	50.00	50.00	44.17	44.17	..	..	..	..	..	..	..	..	41.42	-6.01
New Zealand	TCNZ	2 048	2 048	24 576	24 576	24 576	24 576	24 576	24 576	69.95	39.95	69.95	79.95	69.95	69.95	83.00	85.00	1 000	1 000	10 000	15 000	20 000	20 000	10 000	15 000	0.00	10.23
Norway	Telenor	4 096	6 144	6 144	6 144	8 192	8 192	16 384	16 384	549	499	499	499	449	449	455	478	..	..	..	..	..	..	..	..	41.42	3.18
Poland	TP	6 144	6 144	6 144	6 144	6 144	6 144	10 240	10 240	291.58	156.00	156.00	109.00	124.00	99.90	91.57	107.46	50 000	..	..	..	..	..	..	..	29.10	3.71
Portugal	Portugal Telecom	8 192	8 192	8 192	16 384	24 576	24 576	24 576	24 576	59.99	49.50	35.58	35.28	24.99	24.99	24.68	25.49	8 000	30 000	50 000	..	..	..	..	..	0.00	1.00
Slovak Republic	Slovak Telecom/T-Com	1 024	1 024	1 536	2 048	2 048	2 048	2 048	5 120	52.74	26.52	8.26	13.24	13.95	12.99	13.22	13.77	..	..	1 000	2 000	2 000	2 000	2 000	2 000	58.11	2.96
Slovenia	Telekom Slovenije	..	..	..	..	..	1 024	2 048	2 048	..	..	..	..	..	26	26.08	26.08	..	..	..	..	..	..	..	..	41.42	0.15
Spain	Telefonica	1 024	1 024	1 024	1 024	1 024	1 024	3 072	10 240	39.07	39.07	39.07	29.90	29.90	29.90	40.99	42.03	..	..	..	..	..	20 000	..	5 000	216.23	18.56
Sweden	TeliaSonera	24 576	24 576	24 576	24 576	24 576	24 576	30 720	30 720	419	399	379	359	359	344	447.11	475.61	..	..	..	..	..	..	..	..	11.80	17.58
Switzerland	Swisscom	2 400	3 584	3 584	5 120	5 120	5 120	10 000	10 000	99.00	69.00	49.00	49.00	49.00	49.00	74.35	74.35	..	..	..	..	..	..	..	..	39.75	23.18
Turkey	Turk Telecom	2 048	2 048	2 048	2 048	2 048	2 048	2 048	2 048	238.00	166.60	69.00	69.00	63.81	63.81	45.98	46.14	..	..	..	..	..	..	..	..	0.00	-14.97
United Kingdom	BT	2 200	8 192	8 192	8 192	20 480	20 480	20 480	40 960	24.99	26.99	24.99	24.99	24.46	24.99	25.46	28.69	15 000	40 000	..	..	..	..	..	..	41.42	7.15
United States	AT&T	3 072	3 072	3 072	3 072	3 072	3 072	3 072	3 072	36.99	24.99	24.99	30.00	30.00	35.00	24.24	44.09	..	..	..	..	..	..	..	..	0.00	12.24

Notes: The methodology used to collect all broadband offers is available at www.oecd.org/sti/ict/broadband. This data collection identified one DSL/fibre offer from each country (if available) in 2005. This offer was followed over time in terms of price, speed and bit cap. If the speeds on offer were no longer available the next highest available speed was used.

Source: OECD and Teligen.

StatLink http://dx.doi.org/10.1787/888932802534

Table 7.30. **Changes in cable offerings, September 2005 to 2012**

Cable		Speed (kbit/s)								Price (local currency)								Bitcap (MB)								CAGR (2010-12)	
		2005	2006	2007	2008	2009	2010	2011	2012	2005	2006	2007	2008	2009	2010	2011	2012	2005	2006	2007	2008	2009	2010	2011	2012	Speed	Price
Australia	Optus	2 880	10 240	10 240	20 480	20 480	20 480	20 480	20 480	74.95	79.95	109.99	109.99	79.99	59.99	64.95	59.99	12 000	20 000	30 000	30 000	30 000	120 000	120 000	120 000	0.00	0.00
Austria	UPC	16 384	16 384	16 384	16 384	20 480	20 480	20 480	35 840	89.00	89.00	69.00	26.90	29.90	29.90	24.15	26.42	..	..	..	..	..	..	..	..	32.29	-6.00
Belgium	Telenet	10 240	20 480	20 480	20 480	25 600	30 720	20 480	30 720	59.95	59.95	61.32	61.32	64.32	42.91	26.34	33.34	30 000	35 000	35 000	60 000	60 000	80 000	50 000	100 000	0.00	-11.85
Canada	Shaw	10 240	10 240	10 240	10 240	15 360	15 360	25 600	51 200	69.95	46.95	50.95	50.95	53.95	57.00	66.67	82.18	30 000	100 000	100 000	100 000	100 000	125 000	250 000	200 000	82.57	20.07
Chile	VTR	..	..	..	..	..	6 144	6 144	10 240	..	..	..	..	..	19900	19990	21407	..	..	..	..	..	..	..	..	29.10	3.72
Czech Republic	UPC	4 096	4 096	5 120	6 144	10 240	10 240	25 600	30 720	1 996	1 457	779	794	245	494	539	521	..	50 000	40 000	..	..	..	..	..	73.21	2.71
Denmark	Telia Stofa	4 096	4 096	4 096	4 096	4 096	8 192	10 240	15 000	499	459	339	239	159	171	171	187.28	-	-	-	-	-	150	-	-	35.32	4.65
Estonia	STV	..	..	..	..	..	2 048	5 120	5 120	..	..	..	..	..	16.96	12.72	12.72	..	..	..	..	..	..	..	..	58.11	-13.40
Finland	Welho	6 000	6 000	10 240	10 240	10 240	10 240	10 240	10 240	45.00	45.00	45.00	44.90	44.90	24.90	33.48	36.73	..	..	..	..	..	..	..	..	0.00	21.45
France	Noos/Numericable	10 240	20 000	30 720	30 720	102 400	102 400	102 400	102 400	34.90	34.90	19.90	21.90	21.90	39.90	44.01	41.29	..	..	..	..	..	..	..	..	0.00	1.73
Germany	Kabel Deutschland	6 200	2 200	6 144	6 144	6 144	6 144	6 144	8 000	29.89	29.90	19.90	19.90	19.90	19.90	19.78	18.68	..	..	..	..	..	..	..	..	14.11	-3.11
Hungary	UPC	5 120	6 144	5 120	10 240	15 360	15 360	25 000	25 000	29 990	28 790	5 990	7 500	4 750	3 000	3 235	3 285	60 000	..	..	..	..	..	..	..	27.58	4.64
Ireland	ntl / UPC Ireland	3 072	3 072	3 072	10 240	10 240	15 360	25 600	25 600	45.00	29.99	29.99	30.00	32.00	32.00	43.52	46.12	40 000	30 000	30 000	..	..	120 000	..	..	29.10	20.05
Israel	Hot	..	..	..	..	..	2 048	2 560	5 120	..	..	..	..	..	103.90	79.99	89.90	..	..	..	..	..	..	..	..	58.11	-6.98
Japan	J:COM	30 720	30 720	30 720	30 720	40 960	40 960	40 960	40 960	5 775	5 775	5 775	5 775	5 775	5 775	6 388	6 388	..	..	..	..	..	..	..	..	0.00	5.17
Korea	C&M	5 120	10 240	10 240	102 400	102 400	102 400	102 400	102 400	27 100	34 545	28 000	27 000	27 000	27 000	20 140	20 768	..	..	..	..	..	..	..	..	0.00	-12.30
Luxembourg	Coditel/Numericable	4 096	6 144	20 480	30 720	30 720	30 720	30 720	30 720	67.00	34.90	32.90	39.90	39.90	39.90	46.17	57.63	20 000	25 000	30 000	..	..	..	..	..	0.00	20.18
Mexico	Megacable	1 024	1 024	2 048	2 048	2 048	2 048	2 048	5 120	1093	345	299	299	299	399	379	399	..	..	..	..	..	..	..	..	58.11	0.00
Netherlands	UPC	20 480	20 480	20 480	24 576	25 600	25 600	25 600	25 600	79.95	59.95	59.95	60.00	25.00	25.00	43.02	42.57	..	..	..	..	..	..	..	..	0.00	30.49
New Zealand	TelstraClear	10 240	10 240	10 240	10 240	10 240	15 360	15 360	15 360	139.95	131.90	134.90	109.95	109.95	69.95	65.95	65.95	10 000	40 000	40 000	40 000	40 000	40 000	40 000	40 000	0.00	-2.90
Norway	Get	26 624	26 624	26 624	26 624	26 624	25 600	20 000	30 720	998	898	699	699	699	599	479	495	..	..	..	..	..	..	..	..	9.54	-9.09
Poland	UPC	12 288	12 288	20 480	20 480	30 720	51 200	51 200	76 440	299.00	299.00	249.00	149.00	90.00	100.00	80.03	76.67	..	..	..	..	..	..	..	..	22.19	-12.44
Portugal	Zon	8 192	8 192	12 288	18 432	20 480	20 480	30 720	102 400	61.00	49.50	35.59	35.30	35.59	46.25	43.44	49.99	8 000	30 000	30 000	..	..	..	..	..	123.61	3.96
Slovak Republic	UPC	3 072	4 096	4 096	10 240	10 240	10 240	25 000	30 720	79.63	47.40	36.48	21.58	16.00	16.00	9.93	10.98	..	..	..	..	..	..	..	..	73.21	-17.16
Slovenia	Amis	..	..	..	..	..	1 024	1 024	1 024	..	..	..	..	..	22.00	23.00	23.00	..	..	..	..	..	..	..	..	0.00	2.25
Spain	Ono	2 048	4 096	4 096	6 144	6 144	6 144	6 144	12 440	42.00	35.00	35.00	40.00	49.90	39.90	46.36	47.54	..	..	..	..	..	..	..	..	42.29	9.15
Sweden	Com Hem	8 192	8 192	8 192	10 240	10 240	10 240	10 240	10 240	389.00	299.00	319.00	299.00	279.00	279.00	255.78	255.78	..	..	..	..	..	..	..	..	0.00	-4.25
Switzerland	Cablecom	2 048	3 072	3 584	5 120	10 240	20 480	25 000	25 000	75.00	22.30	45.00	45.00	49.00	50.00	79.19	54.14	..	..	..	..	..	..	..	..	10.49	4.06
Turkey	Topaz / Turksat	2 048	2 048	2 048	2 048	2 048	5 120	5 120	5 120	220.00	289.00	209.00	59.00	54.56	69.00	69.77	69.77	..	..	..	..	..	..	..	..	0.00	0.56
United Kingdom	Telewest/Virgin	4 096	4 096	4 096	10 240	10 240	10 240	30 720	30 720	50.00	25.00	25.00	24.00	20.00	20.00	29.19	23.89	..	..	..	..	..	..	..	..	73.21	9.29
United States	Comcast	6 144	6 144	6 144	12 288	20 480	20 480	20 480	20 480	67.95	57.95	59.95	42.95	52.95	54.95	76.95	69.25	..	..	..	..	..	..	..	..	0.00	12.26

Notes: The methodology used to collect all broadband offers is available at www.oecd.org/sti/ict/broadband. This data collection identified one cable offer from each country (if available) in 2005. This offer was followed over time in terms of price, speed and bit cap. If the speeds on offer were no longer available the next highest available speed was used.

Source: OECD and Teligen.

StatLink http://dx.doi.org/10.1787/888932802553

Table 7.31. OECD 30 calls + 100 MB mobile basket, August 2012, VAT included

		Fixed		Voice		Messages		Data		Grand total		Contract type*
		USD	USD PPP	USD	USD PPP	USD	USD PPP	USD	USD PPP	USD	USD PPP	
Australia, Optus	Boost Pre-paid - Super Cap $25 , MyData 200 MB	26.35	16.78							26.35	16.78	PP
Austria, T-Mobile	Tele.ring - Masta Mini , Data 50 MB, SIM only	7.50	6.76					2.46	2.21	9.96	8.97	
Belgium, Proximus	Generation MTV €15 , Mobile Internet 100 MB, 24 months	15.01	12.83							15.01	12.83	
Canada, Bell Mobility	Virgin Prepaid $25 , Data 100 MB	28.70	22.60	1.01	1.01					29.71	23.61	
Chile, Entel Movil	Raimundo Plan , 140 SMS, Mobile Internet 150 MB, Calls to 1 sel.num.	13.25	18.15	10.37	14.21					23.62	32.35	PP / SD
Czech Republic, T-Mobile	Podnikatel Plus 250 HIT , Internet v Mobilu 100 MB, 24 months	21.36	26.70							21.36	26.70	
Denmark, TDC Mobil	Om Click Minut , Data 1000 MB	8.54	5.73	6.02	4.04	3.15	2.11			17.71	11.89	PP
Estonia, Tele2	Nuti 4.99 , Internet mobilis 5000 MB, SIM only	6.40	7.62	0.64	0.76					7.04	8.38	
Finland, Sonera	Sopiva Startti , 75 SMS, Data 1000 MB, SIM only	14.62	11.51							14.62	11.51	
France, SFR	Forfait Carre Bloques 1h 100MB , Internet 100 MB, SIM only, 24 months	14.74	12.82							14.74	12.82	SD
Germany, T-Mobile	Xtra Click , Unlimited data, SIM only	1.56	1.46	5.70	5.33	11.69	10.92			18.94	17.70	PP
Greece, Vodafone	Vodafone Prepaid , EUR10/month - any-net voice, Internet bundles 500 MB	18.60	18.60	0.29	0.29	2.67	2.67			21.56	21.56	PP / SD
Hungary, Telenor	Smart Tarifa 2 , 100 SMS, Smart Extra Data 250 MB	20.72	29.60							20.72	29.60	
Iceland, Vodafone	Risafrelsi Eitt verð , on-net calls & SMS, Net to the phone for RisaFrelsi 300 MB	13.10	11.10							13.10	11.10	PP
Ireland, O2	O2 Choices 300 , Data 0 MB, 24 months	18.33	14.66					3.12	2.50	21.45	17.16	
Israel, Cellcom	Unlimited 119 , Unlimited data	29.67	25.36							29.67	25.36	
Italy, Vodafone	Vodafone Al Secondo , Vodafone Smart Square, Data 1024 MB	18.42	18.61	10.35	10.45					28.77	29.07	PP / SD
Japan, KDDI au	Plan S Simple , Everybody Discount with 24M, Unlimited data, 24 months	91.70	61.54	1.46	0.98	0.24	0.16			93.40	62.68	
Korea, KT	EvergreenMobile Smart Zerozone 1(100M) , Data 100 MB, SIM only	5.45	6.82	5.23	6.54	1.95	2.43			12.63	15.79	
Luxembourg, Tango	Knock-out , + Tango Family (disc calcs for on-net only), Data 300 MB	6.14	4.95	3.85	3.11	8.50	6.85			18.49	14.91	SD
Mexico, Telcel	Amigo Plus , $300 Top-up, $150 bonus, Data 150 MB			6.36	9.94	4.95	7.74			11.31	17.68	PP / SD
Netherlands, Vodafone	Bel+SMS+Web Vaak 75 , Internet 500 MB, SIM only, 24 months	21.79	19.63							21.79	19.63	
New Zealand, Vodafone	Smart Prepay 25 , Data 300 MB	20.92	16.61							20.92	16.61	PP
Norway, Telenor	Prat , Data 500 MB, SIM only, Frifamilie Calls and SMS to 7 sel.num.	8.12	10.25	8.56	2.51	7.84	2.57			24.52	15.33	SD
Poland, Polkomtel	Na_Karte , Unlimited SMS 3000 SMS, Data 100 MB, SIM only	2.13	3.62	4.48	7.60	..	..			6.62	11.21	PP
Portugal, TMN	UZO 8 Cents , Minimum usage, Mobile Internet 135 MB	3.29	3.33	9.26	6.10	9.16	12.51			21.72	21.94	PP
Slovak Republic, T-Mobile	Easy Time , Mobile Internet 200 MB	3.17	4.18	5.14	6.77	10.56	13.90			18.88	24.84	PP
Slovenia, Si.mobil	SIMPL , 100 SMS, Mobile Internet Options 100 MB	12.69	14.10	6.42	7.13	..	..			19.11	21.23	PP
Spain, Vodafone	Tarjeta Internet y SMS Gatuitas , Top-up €10, TP Smartphone 100 MB	16.10	15.63	..	..	..	..			16.10	15.63	PP / SD
Sweden, Tele 2 Comviq	Comviq Kontant Smart , SEK 69 Top-up, Mobil Data 100 MB	0.88	0.69	6.18	4.83	7.21	5.64			14.27	11.15	PP
Switzerland, Sunrise	Sunrise Free , Unlimited data	1.59	0.91	19.30	11.09	12.27	7.05			33.16	19.06	PP
Turkey, Turkcell	Hepsi 300 , Mobile Internet Packs 300 MB, 12 months	16.15	28.83							16.15	28.83	
United Kingdom, T-Mobile	Pay Monthly 7 , Data 250 MB, 24 months	10.99	8.32							10.99	8.32	
United States, AT&T	Pay As You Go - 10c / min , 200 SMS, Data 200 MB	21.99	21.99	7.10	7.10					29.09	29.09	PP
OECD average		15.29	14.18	3.46	3.23	2.36	2.19	0.16	0.14	21.28	19.74	

Notes: The OECD basket of mobile telephone charges includes fixed and usage charges for 30 voice calls and 100 MB of data per month. USD purchasing power parities (PPP) are used to aid international comparisons. (*) Contact type: PP = Pre-paid plan ; SD = Including selective discounts.

Source: OECD and Teligen.

StatLink http://dx.doi.org/10.1787/888932802572

Table 7.32. **OECD 100 calls + 500 MB mobile basket, August 2012, VAT included**

		Fixed		Voice		Messages		Data		Grand total		Contract type*
		USD	USD PPP	USD	USD PPP	USD	USD PPP	USD	USD PPP	USD	USD PPP	
Australia, Optus	Boost Pre-paid - Super Cap $30 , MyData 500 MB	31.60	20.13							31.60	20.13	PP
Austria, T-Mobile	Tele.ring - Masta Max , Unlimited data, SIM only	11.19	10.08							11.19	10.08	
Belgium, Mobistar	Panthere Pro 45 , Mobile Mail&Surf 500 MB, 12 months	44.22	37.79							44.22	37.79	
Canada, Bell Mobility	Virgin Choice $25 , 5000 SMS, Data 0 MB, Voicemail option	42.48	33.45					5.74	4.52	48.22	37.97	
Chile, Entel Movil	Pablo Plan , Top-up $10000, 140 SMS, Mobile Internet 150 MB, Calls to 1 sel.num.	13.25	18.15	32.67	44.75					45.91	62.90	PP / SD
Czech Rep., T-Mobile	S námi 790 + HIT , Internet v Mobilu 600 MB, 24 months	42.37	52.97	1.11	1.39					43.48	54.35	
Denmark, Telenor	Tale 0 Timer , Surf 3000 MB, SIM only, 6 months, Mine nærmeste Calls to 10 sel.num.	11.85	7.95	12.33	8.27	5.78	3.88			29.96	20.11	SD
Estonia, Tele2	Nuti 4.99 , Internet mobilis 5000 MB, SIM only	6.40	7.62	2.39	2.84	0.49	0.58			9.28	11.05	
Finland, Sonera	Sopiva Nappara , 150 SMS, Data 1000 MB, SIM only	20.76	16.35							20.76	16.35	
France, Orange	Sosh - 24/7 2GB , Unlimited data, SIM only	24.44	21.25							24.44	21.25	
Germany, Vodafone	CallYa Smartphone Fun , Top up EUR 15, Unlimited data	18.74	17.51	20.43	19.09					39.17	36.61	PP
Greece, Vodafone	Vodafone Prepaid , EUR20/month - any-net voice, Internet bundles 500 MB, Calls to 2 sel.num.	30.88	30.88	0.73	0.73					31.61	31.61	PP / SD
Hungary, Telenor	Red Bull Mobile All-in , Data traffic 500 MB, SIM only, 24 months	24.16	34.51	1.77	2.53	5.98	8.55			31.91	45.59	
Iceland, Vodafone	Risafrelsi Eitt verð , on-net calls & SMS, Net to the phone for RisaFrelsi 1000 MB	17.57	14.89	9.27	7.85					26.84	22.74	PP
Ireland, O2	O2 Choices 300 , Data 350 MB, SIM only, 1 months	27.67	22.14					4.68	3.74	32.35	25.88	
Israel, Cellcom	Unlimited 119 , Unlimited data	29.67	25.36							29.67	25.36	
Italy, Vodafone	Smart 300 , Data 500 MB, 12 months, Calls to 1 sel.num.	43.19	43.62							43.19	43.62	SD
Japan, KDDI au	Plan M Simple , Everybody Discount with 24M, Unlimited data	109.45	73.46	1.40	0.94					110.85	74.40	SD
Korea, KT	CJ HelloMobile Hello Smart 37 , Data 500 MB	23.93	29.92							23.93	29.92	
Luxembourg, Tango	Knock-out , + Tango Family (disc calcs for on-net only), Data 300 MB	6.14	4.95	15.78	12.72	12.37	9.98	6.14	4.95	40.43	32.61	SD
Mexico, MoviStar	Prepago Lite , Top-Up $150, 100 SMS, Data 100 MB, Calls to 1 sel.num.	19.31	30.17	4.35	6.80	1.06	1.66	0.60	0.94	25.33	39.57	PP / SD
Netherlands, Vodafone	Bel+SMS+Web Vaak 160 , Internet 500 MB, SIM only, 24 months	29.21	26.31							29.21	26.31	
New Zealand, Vodafone	Smart $65 , Data 750 MB, 12 months, 1 x NZBestmates Calls and SMS to 1 sel.num.	54.26	43.07							54.26	43.07	SD
Norway, Telenor	djuice All-in-One M , All in 1, 500 MB, SIM only	32.97	20.61							32.97	20.61	
Poland, Polkomtel	Na_Karte , Top-Up ZL 150, Unlimited SMS 3000 SMS, Data 100 MB	12.69	21.52					2.39	4.05	15.09	25.57	PP / SD
Portugal, TMN	Pack Negocios Rede 12 , 200 SMS, Mobile Internet 500 MB	35.07	35.42	0.12	0.13					35.19	35.55	
Slovak Rep., T-Mobile	Happy M , Mobile Internet 500 MB, 24 months	29.81	39.22							29.81	39.22	
Slovenia, Mobitel	Paketi 2012 , Mobile Internet 503 MB, Calls to 3 sel.num.	25.12	27.91							25.12	27.91	SD
Spain, Vodafone	Tarifa @S , TP Smartphone 500 MB, Calls to 1 sel.num.	54.99	53.39							54.99	53.39	SD
Sweden, Tele 2 Comviq	Volym Låg , Surf 1000 MB	28.23	22.05							28.23	22.05	
Switzerland, Swisscom	Natel Infinity XS , Unlimited data	61.47	35.33							61.47	35.33	
Turkey, Vodafone	Hesabini Bilen Hepsi 500 , Mobile Internet 500 MB	19.49	34.80							19.49	34.80	
United Kingdom, T-Mobile	Business Pay Monthly 500 Mins - Internet Booster , Internet Booster 500 MB, SIM only, 12 months	14.65	11.10							14.65	11.10	
United States, AT&T	Pay As You Go , $25 Monthly Plan, Data 1024 MB	55.00	55.00							55.00	55.00	PP
OECD average		30.93	29.05	3.10	3.27	0.78	0.75	0.59	0.55	35.40	33.63	

Notes: The OECD basket of mobile telephone charges includes fixed and usage charges for 100 voice calls and 500 MB of data per month. USD purchasing power parities (PPP) are used to aid international comparisons. (*) Contact type : PP = Pre-paid plan ; SD = Including selective discounts.

Source: OECD and Teligen.

StatLink http://dx.doi.org/10.1787/888932802591

Table 7.33. **OECD 300 calls + 1 GB mobile basket, August 2012, VAT included**

		Fixed		Voice		Messages		Data		Grand total		Contract type*
		USD	USD PPP	USD	USD PPP	USD	USD PPP	USD	USD PPP	USD	USD PPP	
Australia, Optus	Boost Pre-paid - Super Cap $40 , MyData 3072 MB	42.12	26.83	..	..	..	..	..	..	42.12	26.83	PP
Austria, T-Mobile	Tele.ring - Masta Max , Unlimited data, SIM only	11.19	10.08	..	..	..	..	..	..	11.19	10.08	
Belgium, Mobistar	Panthere Pro 60 , Mobile Mail&Surf 1024 MB, 12 months	58.96	50.39	13.71	11.72	..	..	..	..	72.67	62.11	
Canada, Bell Mobility	Virgin Choice $40 A , 5000 SMS, Data 100 MB, Calls to 5 sel.num., Voicemail option	59.70	47.01	6.77	5.33	..	..	10.33	8.14	76.81	60.48	
Chile, Entel Movil	Planes Red Smartphone 1500 , 140 SMS, Mobile Internet 1200 MB	92.00	126.03	8.26	11.31	..	..	..	..	100.26	137.34	
Czech Rep., T-Mobile	Tarify Grand v Siti , Internet v Mobilu 1000 MB, 24 months	72.50	90.63	..	..	..	..	..	..	72.50	90.63	
Denmark, TDC Mobil	Mobil 6hrs , Fri SMS, Data 1000 MB, SIM only, 6 months	41.56	27.90	..	..	..	..	..	..	41.56	27.90	
Estonia, Tele2	Nuti 7.99 , Internet mobilis 5000 MB, SIM only	10.09	12.01	..	..	2.76	3.29	..	..	12.85	15.30	
Finland, Sonera	Sopiva Fiksu , Data 10000 MB, SIM only	24.44	19.25	..	..	..	..	..	..	24.44	19.25	
France, Orange	Sosh - 24/7 2GB , Unlimited data, SIM only	24.44	21.25	..	..	..	..	..	..	24.44	21.25	
Germany, Vodafone	SuperFlat Internet Allnet Spezial , Unlimited data, SIM only, 24 months	99.21	92.72	..	..	..	..	..	..	99.21	92.72	
Greece, Vodafone	Vodafone Unlimited 60 , Mobile internet 1500 MB, SIM only	66.33	66.33	..	..	..	..	..	..	66.33	66.33	
Hungary, Telenor	Red Bull Mobile All-in + , Data traffic 1000 MB, SIM only, 24 months	43.96	62.79	28.26	40.37	3.19	4.56	..	..	75.40	107.72	
Iceland, Vodafone	Risafrelsi Eitt verð , 5 numbers + on-net calls & SMS, Net to the phone for RisaFrelsi 1000 MB	34.14	28.93	12.63	10.70	..	..	..	..	46.77	39.63	PP / SD
Ireland, O2	O2 Choices 800 (600 mins & 200 SMS) , Data 1024 MB, SIM only, 12 months	45.45	36.36	..	..	..	..	..	..	45.45	36.36	
Israel, Cellcom	Unlimited 119 , Unlimited data	29.67	25.36	..	..	..	..	..	..	29.67	25.36	
Italy, Vodafone	Smart 600 , Data 1024 MB, 12 months, Calls to 1 sel.num.	55.47	56.03	..	..	..	..	..	..	55.47	56.03	SD
Japan, KDDI au	Plan LL Simple , Everybody Discount with 24M, Unlimited data, 24 months	160.87	107.97	2.70	1.82	..	..	..	..	163.58	109.78	
Korea, KT	Control 540 w/o SMS pack (WIBRO) , KT 50% Discount, WIBRO 10240 MB, 24 months	41.20	51.50	..	..	..	..	..	..	41.20	51.50	
Luxembourg, Tango	Easy Total , Data 1536 MB	55.27	44.58	..	..	..	..	..	..	55.27	44.58	
Mexico, Telcel	Telcel Plus 300 , Data 1424 MB, Calls to 2 sel.num.	45.11	70.48	0.81	1.26	9.96	15.56	..	..	55.87	87.30	SD
Netherlands, Vodafone	Bel+SMS+Web Altijd 350 , Internet 1000 MB, SIM only, 24 months	46.52	41.91	..	..	..	..	..	..	46.52	41.91	
New Zealand, Vodafone	Smart $100 , Data 1000 MB, 12 months, 3 x NZBestmates Calls and SMS to 3 sel.num.	74.51	59.14	..	..	..	..	..	..	74.51	59.14	SD
Norway, Telenor	djuice All-in-One L , All in 1, 1000 MB, SIM only	49.54	30.97	..	..	..	..	..	..	49.54	30.97	
Poland, Orange	Rozmowy bez limitu 88.88 , Internet package 2000 MB, SIM only, 24 months	34.72	58.86	..	..	..	..	..	..	34.72	58.86	
Portugal, Vodafone	Plano Pro 91 Net 240 , Mobile Internet 1024 MB	73.98	74.73	..	..	..	..	..	..	73.98	74.73	
Slovak Rep., T-Mobile	Happy L , Mobile Internet 2048 MB, 24 months	37.18	48.92	2.49	3.28	..	..	..	..	39.67	52.20	
Slovenia, Mobitel	Paketi 2012 , Mobile Internet 1003 MB, Calls to 3 sel.num.	36.18	40.20	..	..	..	..	..	..	36.18	40.20	SD
Spain, Vodafone	Tarifa @M , TP Smartphone 1000 MB, Calls to 2 sel.num.	71.34	69.26	16.19	15.72	..	..	..	..	87.53	84.98	SD
Sweden, Tele 2 Comviq	Comviq fastpris , Mobil Data 1000 MB	36.06	28.18	..	..	..	..	..	..	36.06	28.18	
Switzerland, Swisscom	Natel Infinity XS , Unlimited data	61.47	35.33	..	..	..	..	..	..	61.47	35.33	
Turkey, Vodafone	Hesabini Bilen Hepsi 1000 , Mobile Internet 1000 MB	30.62	54.69	..	..	..	..	..	..	30.62	54.69	
United Kingdom, T-Mobile	Pay Monthly 8 , Data 1536 MB, SIM only, 12 months	30.87	23.38	..	..	..	..	..	..	30.87	23.38	
United States, AT&T	Pay As You Go , Monthly Unltd Plan, Unlimited data	55.00	55.00	..	..	..	..	..	..	55.00	55.00	PP
OECD		51.81	50.55	2.78	3.08	0.48	0.71	0.31	0.25	55.38	54.58	

Notes: The OECD basket of mobile telephone charges includes fixed and usage charges for 300 voice calls and 1 GB of data per month. USD purchasing power parities (PPP) are used to aid international comparisons. (*) Contact type : PP = Pre-paid plan ; SD = Including selective discounts.

Source: OECD and Teligen.

StatLink http://dx.doi.org/10.1787/888932802610

Table 7.34. **OECD 900 calls + 2 GB mobile basket, August 2012, VAT included**

		Fixed		Voice		Messages		Data		Grand total		Contract type*
		USD	USD PPP	USD	USD PPP	USD	USD PPP	USD	USD PPP	USD	USD PPP	
Australia, Optus	Boost Pre-paid - Super Cap $40 , MyData 3072 MB	42.12	26.83	..	..	..	..	..	..	42.12	26.83	PP
Austria, T-Mobile	Tele.ring - Masta Mega , Unlimited data, SIM only	19.79	17.83	..	..	..	..	..	..	19.79	17.83	
Belgium, Mobistar	Panthere Unlimited , Mail&Surf 2048 MB	110.55	94.48	..	..	..	..	..	..	110.55	94.48	
Canada, Bell Mobility	Virgin Choice $50 , 5000 SMS, Data 0 MB, Voicemail option	71.19	56.05	..	..	..	..	22.96	18.08	94.15	74.13	
Chile, Entel Movil	Planes Red 1500 , 250 SMS, Mobile Internet 3072 MB	104.85	143.63	54.53	74.70	12.44	17.04	..	..	171.82	235.37	
Czech Rep., O2	Smart Podnikani XXL , 200 SMS, Unlimited data, SIM only, 24 months	103.77	129.72	2.46	3.07	..	..	..	..	106.23	132.79	
Denmark, TDC Mobil	Mobilpakke 12hrs talktime - 10GB , Data 10000 MB, SIM only, 6 months	58.07	38.98	3.84	2.58	..	..	..	..	61.92	41.55	
Estonia, Tele2	Nuti 7.99 , Internet mobilis 5000 MB, SIM only	10.09	12.01	10.03	11.93	4.30	5.12	..	..	24.41	29.06	
Finland, Elisa	Saunalahti Nopsa Paketti + , 500 SMS, Unlimited data, 12 months	47.79	37.63	..	..	..	..	..	..	47.79	37.63	
France, Orange	Sosh - 24/7 2GB , Unlimited data, SIM only	24.44	21.25	..	..	..	..	..	..	24.44	21.25	
Germany, T-Mobile	Complete Mobile XL , Unlimited data, SIM only	111.37	104.09	..	..	..	..	..	..	111.37	104.09	
Greece, Cosmote	Cosmote Unlimited 100 , Internet data 2524 MB, SIM only	128.97	128.97	6.60	6.60	..	..	..	..	135.57	135.57	
Hungary, T-Mobile	Mozaik XXL , Extra SMS 500 SMS, Unlimited data	94.07	134.38	96.66	138.09	..	..	..	..	190.73	272.47	SD
Iceland, Vodafone	Risafrelsi Eitt verð , 5 numbers + on-net calls & SMS	41.59	35.25	115.75	98.09	..	..	..	..	157.34	133.34	PP / SD
Ireland, O2	O2 Choices 1200 (800 mins & 400 SMS) , Data 2048 MB	61.51	49.21	..	..	..	..	..	..	61.51	49.21	
Israel, Cellcom	Unlimited 129 , Unlimited data	32.16	27.49	..	..	..	..	..	..	32.16	27.49	
Italy, TIM	Tutto Compreso 1500 , Opzione Full Maxi, Unlimited data	114.67	115.83	..	..	..	..	..	..	114.67	115.83	SD
Japan, KDDI au	Plan LL Simple , Everybody Discount with 24M	165.86	111.32	148.79	99.86	4.66	3.13	..	..	319.31	214.30	SD
Korea, KT	Style 550 (WIBRO) , Style SMS 200 SMS	44.60	55.75	12.42	15.52	1.95	2.43	..	..	58.96	73.71	
Luxembourg, Tango	Smart 45 , Unlimited data, 24 months	55.27	44.58	..	..	..	..	..	..	55.27	44.58	
Mexico, Telcel	Telcel Plus 1000 , Data 2748 MB, Calls to 7 sel.num.	88.10	137.66	2.42	3.79	11.62	18.15	..	..	102.14	159.60	SD
Netherlands, KPN	Mobiel Instap - Call-SMS Unlimited , Data 2500 MB	80.86	72.85	..	..	..	..	..	..	80.86	72.85	
New Zealand, Vodafone	Business Smart $200 , Data 2048 MB, 12 months, FCN Calls to 1 sel.num.	145.78	115.70	..	..	..	..	..	..	145.78	115.70	SD
Norway, Telenor	Komplett L , Data 2000 MB, SIM only, Frifamilie Calls and SMS to 7 sel.num.	79.20	49.50	..	..	..	..	..	..	79.20	49.50	SD
Poland, Orange	Rozmowy bez limitu 88.88 , Internet package 2000 MB	34.72	58.86	..	..	..	..	..	..	34.72	58.86	
Portugal, Vodafone	Plano Pro Premium , Mobile Internet 2048 MB	126.64	127.92	..	..	..	..	..	..	126.64	127.92	
Slovak Rep., T-Mobile	Happy XL , Mobile Internet 2048 MB, 24 months	55.60	73.16	..	..	..	..	..	..	55.60	73.16	
Slovenia, Si.mobil	L Zate , Mobile Internet Plus 3801 MB	65.39	72.65	..	..	..	..	..	..	65.39	72.65	
Spain, Vodafone	Tarifa @XL , TP Smartphone 2048 MB, Calls to 4 sel.num.	136.73	132.75	49.13	47.70	..	..	..	..	185.87	180.46	SD
Sweden, Tele 2 Comviq	Comviq fastpris , Mobil Data 4000 MB	48.58	37.95	..	..	..	..	..	..	48.58	37.95	
Switzerland, Swisscom	Natel Infinity XS , Unlimited data	61.47	35.33	..	..	..	..	..	..	61.47	35.33	
Turkey, Vodafone	Red Mini , Mobile Internet Packs 2000 MB	38.98	69.60	..	..	..	..	..	..	38.98	69.60	
United Kingdom, O2	On & On SIMplicity , Data 2048 MB, SIM only, 12 months	39.24	29.73	..	..	..	..	..	..	39.24	29.73	
United States, AT&T	Pay As You Go , Monthly Unltd Plan, Unlimited data	55.00	55.00	..	..	..	..	..	..	55.00	55.00	PP
OECD average		74.45	73.55	15.23	15.21	1.06	1.39	0.70	0.55	91.44	90.70	

Notes: The OECD basket of mobile telephone charges includes fixed and usage charges for 900 voice calls and 100 MB of data per month. USD purchasing power parities (PPP) are used to aid international comparisons. (*) Contact type : PP = Pre-paid plan ; SD = Including selective discounts.

Source: OECD and Teligen.

StatLink http://dx.doi.org/10.1787/888932802629

Table 7.35. **OECD 100 calls + 2 GB mobile basket, August 2012, VAT included**

		Fixed		Voice		Messages		Data		Grand total		Contract type*
		USD	USD PPP	USD	USD PPP	USD	USD PPP	USD	USD PPP	USD	USD PPP	
Australia, Optus	Boost Pre-paid - Super Cap $40 , MyData 3072 MB	42.12	26.83	..	..	..	..	..	..	42.12	26.83	PP
Austria, T-Mobile	Tele.ring - Masta Max , Unlimited data, SIM only	11.19	10.08	..	..	..	..	..	..	11.19	10.08	
Belgium, Mobistar	Panthere 30 , Internet Everywhere Multi 2248 MB, 12 months	66.33	56.69	..	..	..	..	..	..	66.33	56.69	
Canada, Bell Mobility	Virgin Choice $25 , 5000 SMS, Data 0 MB, Voicemail option	42.48	33.45	..	..	..	..	22.96	18.08	65.45	51.53	
Chile, Entel Movil	Pablo Plan , Top-up $10000, 140 SMS, Mobile Internet 150 MB	13.25	18.15	31.53	43.20	..	..	..	..	44.78	61.34	PP / SD
Czech Rep., T-Mobile	S námi 790 + HIT , Internet v Mobilu 2300 MB, 24 months	55.03	68.79	0.22	0.27	..	..	..	..	55.25	69.06	
Denmark, Telenor	Tale 0 Timer , Surf 3000 MB, SIM only, 6 months	11.85	7.95	11.44	7.68	5.78	3.88	..	..	29.07	19.51	Sel
Estonia, Tele2	Nuti 4.99 , Internet mobilis 5000 MB, SIM only	6.40	7.62	2.31	2.75	0.49	0.58	..	..	9.20	10.96	
Finland, Sonera	Sopiva Nappara , 150 SMS, Data 3000 MB, SIM only	23.21	18.28	..	..	..	..	..	..	23.21	18.28	
France, Orange	Sosh - 24/7 2GB , Unlimited data, SIM only	24.44	21.25	..	..	..	..	..	..	24.44	21.25	
Germany, T-Mobile	Complete Mobile L , SMS Flat 3000 SMS, Unlimited data, SIM only	86.68	81.01	4.13	3.86	..	..	..	..	90.81	84.87	
Greece, Cosmote	Cosmote Economy 30 , 150 SMS, Internet data 20 MB, SIM only	46.68	46.68	0.73	0.73	..	..	..	..	47.41	47.41	
Hungary, T-Mobile	ikon 200 , 25 SMS, Unlimited data	33.83	48.33	0.25	0.36	10.87	15.52	..	..	44.95	64.21	
Iceland, Vodafone	Risafrelsi Eitt verð , on-net calls & SMS	25.03	21.21	8.01	6.79	..	..	..	..	33.04	28.00	PP
Ireland, Vodafone	Simply 100 , Unlimited on-net calls & SMS, Data 10240 MB	49.53	39.62	..	..	..	..	..	..	49.53	39.62	
Israel, Cellcom	Unlimited 129 , Unlimited data	32.16	27.49	..	..	..	..	..	..	32.16	27.49	
Italy, TIM	Tutto Compreso 250 , Opzione Full Maxi, Unlimited data	53.25	53.79	..	..	..	..	..	..	53.25	53.79	SD
Japan, KDDI au	Plan M Simple , Everybody Discount with 24M, Unlimited data	109.45	73.46	1.40	0.94	..	..	..	..	110.85	74.40	SD
Korea, KT	Control 340 (WIBRO) , WIBRO 10240 MB, 24 months	25.15	31.44	..	..	..	..	..	..	25.15	31.44	
Luxembourg, Tango	iPhone 30 , Data 2048 MB, 24 months	36.85	29.72	11.39	9.19	4.42	3.57	..	..	52.66	42.47	
Mexico, MoviStar	Prepago Lite , Top-Up $150, 100 SMS, Data 100 MB	19.31	30.17	3.96	6.18	1.00	1.57	2.87	4.48	27.14	42.40	PP / SD
Netherlands, Vodafone	Bel+SMS+Web Altijd 160 , Internet 2000 MB, SIM only, 24 months	48.91	44.06	..	..	..	..	..	..	48.91	44.06	
New Zealand, Telecom	Smartphone 300 , Data 2048 MB, Favourites Calls to 1 sel.num.	85.85	68.13	0.49	0.39	..	..	..	..	86.34	68.52	SD
Norway, Telenor	djuice All-in-One M , All in 1, 2500 MB, SIM only	59.15	36.97	..	..	..	..	..	..	59.15	36.97	
Poland, Polkomtel	Na_Karte , Top-Up ZL 100, Unlimited SMS 3000 SMS, Data 100 MB	10.17	17.24	2.24	3.80	..	..	11.35	19.24	23.76	40.28	PP / SD
Portugal, TMN	Pack Negocios Rede 12 , 200 SMS, Mobile Internet 4096 MB	55.07	55.63	0.12	0.13	..	..	..	..	55.19	55.75	
Slovak Rep., T-Mobile	Podľa seba 1 , SMS + Redu Chrg, Mobile Internet 2048 MB	19.97	26.28	13.16	17.32	..	..	..	..	33.13	43.60	
Slovenia, Mobitel	Itak Dzabest , Mobile Internet 5420 MB, Calls to 3 sel.num.	41.76	46.40	..	..	..	..	..	..	41.76	46.40	SD
Spain, Vodafone	Tarifa XS , TP Smartphone 2048 MB, Qtal Calls and SMS	60.69	58.92	..	..	..	..	..	..	60.69	58.92	SD
Sweden, Tele 2 Comviq	Volym Låg , Surf 3000 MB	34.11	26.65	..	..	..	..	..	..	34.11	26.65	
Switzerland, Swisscom	Natel Infinity XS , Unlimited data	61.47	35.33	..	..	..	..	..	..	61.47	35.33	
Turkey, Vodafone	Hesabini Bilen Mini , 500 SMS, Mobile Internet 4020 MB	34.47	61.55	..	..	..	..	..	..	34.47	61.55	
UK, O2	On & On SIMplicity , Data 2048 MB, SIM only, 12 months	39.24	29.73	..	..	..	..	..	..	39.24	29.73	
USA, AT&T	Pay As You Go , Monthly Unltd Plan, Unlimited data	55.00	55.00	..	..	..	..	..	..	55.00	55.00	PP
OECD		41.76	39.00	2.77	3.14	0.68	0.76	1.13	1.27	46.34	44.17	

Notes: The OECD basket of mobile telephone charges includes fixed and usage charges for 100 voice calls and 2 GB of data per month. USD purchasing power parities (PPP) are used to aid international comparisons. (*) Contact type : PP = Pre-paid plan ; SD = Including selective discounts.

Source: OECD and Teligen.

StatLink http://dx.doi.org/10.1787/888932802648

Table 7.36. **OECD Wireless Broadband basket, OECD Laptop 500M, September 2012**

	ISP	Package	Total		Speed (Mbit/s)*
			USD	USD PPP	
Australia	Vodafone	Mobile Broadband SIM only 2GB	19.60	12.02	6.0
Austria	Orange	Mobiles Internet 1GB Sim Only	7.10	6.51	2.0
Belgium	Mobistar	Internet Everywhere Multi Relax for PC	18.82	16.37	3.6
Canada	Telus Wireless	Mobile High Speed Data Line Flex 10+ (up to 500MB)	35.36	27.41	42.0
Chile	Claro	Connect 1GB +	17.18	22.91	4.0
Czech Republic	O2	Start (for O2 internet subscribers)	10.65	13.48	2.0
Denmark	Telia	4Life by Telia Micro 500MB	8.25	5.65	80.0
Estonia	Tele2	M2M Small	7.51	9.04	1.0
Finland	DNA	DNA Mobile Broadband S	6.25	5.04	0.5
France	Bouygues	Forfait Internet Occasionnel 3G+ Sans Engagement SIM Only	12.42	11.09	42.0
Germany	E-Plus	Internet Flat	14.29	13.61	7.2
Greece	Vodafone	500MB	11.85	12.22	42.2
Hungary	T-Mobile	Domino Web SIM package 250MB	8.84	12.63	7.2
Iceland	Vodafone	3G Net 1	8.96	7.53	7.2
Ireland	Meteor Mobile	Broadband To Go 1GB	10.03	8.15	14.4
Israel	Orange	Mobile Broadband Clear 1GB	21.57	18.92	7.2
Italy	TiM	Internet Pack Start	10.35	10.67	14.4
Japan	SoftBank	Unlimited Data Discount Ultra Speed Flat Rate	49.46	33.20	42.0
Korea	KT	4G Wibro 10G	9.95	12.29	40.3
Luxembourg	Tango	Special 2GB - 1 year mobile internet offer	10.35	8.48	7.2
Mexico	Nextel	BA 500 MB plan	7.44	10.95	2.2
Netherlands	T-Mobile	Internet Standaard: USB stick + Internet standaard (laptop)	12.55	11.62	3.6
New Zealand	2degrees	Broadband 1GB	17.11	13.16	7.2
Norway	Netcom	Mobilt Bredband Small (SIM only)	13.61	8.51	80.0
Poland	PlusGSM	iPlus prywatnie 15 SIM Only	4.88	8.28	42.0
Portugal	Optimus	Optimus Kanguru Up to 2Mbps	12.65	13.05	2.0
Slovak Republic	Orange	Mini +	6.31	8.52	3.6
Slovenia	Sl.mobil	Enkratni zakup 1GB	10.46	11.75	42.0
Spain	Vodafone	TP Diaria ilimitada	15.14	14.99	21.6
Sweden	Tele2	Mobilt Bredband Small	12.92	9.94	5.0
Switzerland	Swisscom	NATEL data start	19.88	11.70	7.2
Turkey	Avea	Internet Package 1GB Monthly	7.96	14.22	7.2
United Kingdom	Three	Broadband 0.5GB	8.37	6.34	5.6
United States	Sprint	Peel 3G Mobile Broadband 1GB	32.99	32.99	3.1
OECD average			14.15	13.04	

Notes: The OECD basket of wireless broadband includes total charges for 500 MB of data for laptop use per month. USD purchasing power parities (PPP) are used to aid international comparisons. (*) Wireless broadband speeds included in this basket represent a mix between theoretical maximum speeds and more realistic speed measurements, closer to what a customer will typically receive over the network.

Source: OECD and Teligen.

StatLink http://dx.doi.org/10.1787/888932802667

Table 7.37. **OECD Wireless Broadband basket, OECD Laptop 1GB, September 2012**

	ISP	Package	Total		Speed (Mbit/s)*
			USD	USD PPP	
Australia	Vodafone	Mobile Broadband SIM only 2GB	19.60	12.02	6.0
Austria	Orange	Mobiles Internet 1GB Sim Only	7.10	6.51	2.0
Belgium	Mobistar	Internet Everywhere Multi Relax for PC	18.82	16.37	3.6
Canada	Telus Wireless	Mobile High Speed 35	41.07	31.83	42.0
Chile	Claro	Connect 1GB +	17.18	22.91	4.0
Czech Republic	T-Mobile	Standard Internet on the Go (for T-Mobile voice customers)	13.98	17.70	10.8
Denmark	Telenor Denmark	Mobilt Bredband Medium (for Telenor subscribers)	15.29	10.47	16.0
Estonia	Tele2	M2M Small	7.51	9.04	1.0
Finland	DNA	DNA Mobile Broadband S	6.25	5.04	0.5
France	SFR	CARRE TABLETTE & CLE 1Go 24 months SIM	23.84	21.29	7.2
Germany	O2	Surf Flat M Business (no contract)	19.43	18.50	3.6
Greece	Cosmote	Internet On The Go 1GB	15.65	16.13	42.2
Hungary	T-Mobile	Domino Web SIM package 1GB	8.84	12.63	7.2
Iceland	Vodafone	3G Net 1	8.96	7.53	7.2
Ireland	Meteor Mobile	Broadband To Go 1GB	10.03	8.15	14.4
Israel	Orange	Mobile Broadband Clear 1GB	21.57	18.92	7.2
Italy	TiM	Internet Pack Start	10.35	10.67	14.4
Japan	SoftBank	Unlimited Data Discount Ultra Speed Flat Rate	49.46	33.20	42.0
Korea	KT	4G Wibro 10G	9.95	12.29	40.3
Luxembourg	Tango	Special 2GB - 1 year mobile internet offer	10.35	8.48	7.2
Mexico	Movistar	Plan Navega 1GB	17.22	25.32	3.6
Netherlands	KPN	Mobiel Internet SIM Only Extra (for PC)	19.17	17.75	7.2
New Zealand	2degrees	Broadband 1GB	17.11	13.16	7.2
Norway	Netcom	Mobilt Bredband Small (SIM only)	13.61	8.51	80.0
Poland	T-Mobile	blueconnect 19 SIM Only	6.98	11.83	42.0
Portugal	Optimus	Optimus Kanguru Up to 2Mbps	12.65	13.05	2.0
Slovak Republic	O2	Internet for a laptop	13.80	18.65	14.4
Slovenia	SI.mobil	Enkratni zakup 1GB	10.46	11.75	42.0
Spain	Vodafone	TP Diaria ilimitada	15.14	14.99	21.6
Sweden	Tele2	Mobilt Bredband Small	12.92	9.94	5.0
Switzerland	Orange	Internet Everywhere One	30.33	17.84	7.2
Turkey	Avea	Internet Package 1GB Monthly	7.96	14.22	7.2
United Kingdom	T-Mobile	Mobile Broadband 1GB (18 months)	11.87	8.99	4.5
United States	Sprint	Peel 3G Mobile Broadband 1GB	32.99	32.99	3.1
OECD average			16.40	15.26	

Notes: The OECD basket of wireless broadband includes total charges for 1 GB of data for laptop use per month. USD purchasing power parities (PPP) are used to aid international comparisons. (*) Wireless broadband speeds included in this basket represent a mix between theoretical maximum speeds and more realistic speed measurements, closer to what a customer will typically receive over the network.

Source: OECD and Teligen.

StatLink http://dx.doi.org/10.1787/888932802686

Table 7.38. **OECD Wireless Broadband basket, OECD Laptop 2GB, September 2012**

	ISP	Package	Total		Speed (Mbit/s)*
			USD	USD PPP	
Australia	Vodafone	Mobile Broadband SIM only 2GB	19.60	12.02	6.0
Austria	Orange	Mobiles Internet 6GB Sim Only	12.12	11.12	21.0
Belgium	Mobistar	Internet Everywhere Multi Max for PC	37.64	32.73	3.6
Canada	Telus Wireless	Mobile High Speed 35	41.07	31.83	42.0
Chile	Claro	Browse 2GB +	22.39	29.85	12.0
Czech Republic	T-Mobile	Standard Internet on the Go (for T-Mobile voice customers)	13.98	17.70	10.8
Denmark	Telenor Denmark	Mobilt Bredband Medium (for Telenor subscribers)	15.29	10.47	16.0
Estonia	Tele2	M2M Small	7.51	9.04	1.0
Finland	DNA	DNA Mobile Broadband S	8.82	7.11	0.5
France	Bouygues	Forfait Internet illimité 3G+ 24/24 (24M) SIM Only	24.97	22.29	42.0
Germany	Telekom	CombiCard Connect M	26.11	24.87	21.6
Greece	Cosmote	Internet Any Way 2GB SIM Only	21.96	22.64	42.2
Hungary	T-Mobile	Gold Partner Card RelaxNet M (speed reduction)	8.84	12.63	7.2
Iceland	Simin	3G Net 3 GB	14.35	12.06	7.2
Ireland	O2	Broadband Plus - For Light Users 2GB	15.43	12.55	7.2
Israel	Orange	Mobile Broadband Clear 2GB	26.55	23.29	7.2
Italy	Vodafone	Internet Ready	12.55	12.94	14.4
Japan	SoftBank	Unlimited Data Discount Ultra Speed Flat Rate	49.46	33.20	42.0
Korea	KT	4G Wibro 10G	9.95	12.29	40.3
Luxembourg	Tango	Special 2GB - 1 year mobile internet offer	10.35	8.48	7.2
Mexico	Telcel	Plan de Renta Internet Banda Ancha Movil 3GB	21.73	31.96	7.2
Netherlands	KPN	Mobiel Internet SIM Only Extra (for PC)	19.17	17.75	7.2
New Zealand	2degrees	Broadband 1GB	33.13	25.48	7.2
Norway	Tele2	Mobile Broadband Leisure	23.66	14.79	3.6
Poland	T-Mobile	blueconnect 19 SIM Only	6.98	11.83	42.0
Portugal	Optimus	Optimus Kanguru Up to 2Mbps	12.65	13.05	2.0
Slovak Republic	O2	Internet for a laptop	13.80	18.65	14.4
Slovenia	Mobitel	Internet Mini	14.22	15.97	21.6
Spain	Vodafone	Tarifa Internet Movil 2GB	34.79	34.45	21.6
Sweden	Tele2	Mobilt Bredband Small	12.92	9.94	5.0
Switzerland	Orange	Internet Everywhere Young	40.78	23.99	7.2
Turkey	Turkcell	Kamu Internet Paketi 5GB	15.54	27.75	21.6
United Kingdom	T-Mobile	Mobile Broadband 10GB	15.82	11.99	4.5
United States	Verizon	4GB Shared Data 4G	33.61	33.61	12.0
OECD average			20.52	19.13	

Notes: The OECD basket of wireless broadband includes total charges for 2 GB of data for laptop use per month. USD purchasing power parities (PPP) are used to aid international comparisons. (*) Wireless broadband speeds included in this basket represent a mix between theoretical maximum speeds and more realistic speed measurements, closer to what a customer will typically receive over the network.

Source: OECD and Teligen.

StatLink http://dx.doi.org/10.1787/888932802705

Table 7.39. **OECD Wireless Broadband basket, OECD Laptop 5 GB, September 2012**

	ISP	Package	Total		Speed (Mbit/s)*
			USD	USD PPP	
Australia	Optus	BMB Plan 5GB	26.59	16.31	3.0
Austria	Orange	Mobiles Internet 6GB Sim Only	12.12	11.12	21.0
Belgium	Mobistar	Internet Everywhere Multi Max for PC	69.01	60.01	3.6
Canada	Rogers Wireless	LTE Promotional Plan	62.06	48.11	100.0
Chile	Claro	Download 6GB +	34.54	46.05	12.0
Czech Republic	O2	LTE Mobile Internet XL	25.39	32.14	10.0
Denmark	Telenor Denmark	Mobilt Bredband Medium (for Telenor subscribers)	15.29	10.47	16.0
Estonia	Tele2	M2M Small	7.51	9.04	1.0
Finland	Elisa - saunalahti	Saunalahti Prepaid Mobile Broadband - 1 month	9.89	7.98	0.4
France	Bouygues	Forfait Internet illimité 3G+ 24/24 (24M) SIM Only	24.97	22.29	42.0
Germany	E-Plus	Internet Flat XL	26.83	25.56	7.2
Greece	Cosmote	Internet Any Way 5GB SIM Only	33.46	34.49	42.2
Hungary	T-Mobile	Gold Partner Card RelaxNet L (speed reduction)	21.33	30.48	7.2
Iceland	Vodafone	3G Net 5	16.36	13.75	7.2
Ireland	Meteor Mobile	Broadband To Go 5GB	21.67	17.62	14.4
Israel	Pelephone (Bezeq)	High Speed 99	27.08	23.76	3.6
Italy	WIND	Super Internet Unlimited	15.06	15.52	14.4
Japan	au KDDI	WiMAX Flat Rate (discounted price)	51.74	34.72	40.0
Korea	KT	4G Wibro 10G	9.95	12.29	40.3
Luxembourg	Tango	Unlimited Tango Mobile ADSL	25.42	20.84	7.2
Mexico	Movistar	Plan Descarga 5GB	30.00	44.11	3.6
Netherlands	KPN	Mobiel Internet SIM Only Premium (for PC)	28.58	26.46	7.2
New Zealand	2degrees	Broadband 12GB	80.39	61.84	7.2
Norway	Netcom	Mobilt Bredband Medium (SIM only)	25.67	16.04	80.0
Poland	PlusGSM	iPlus prywatnie 50	15.13	25.64	42.0
Portugal	Optimus	Optimus Kanguru Up to 5Mbps	25.31	26.09	5.0
Slovak Republic	Orange	Klasik +	17.87	24.14	7.2
Slovenia	SI.mobil	Enkratni zakup 5GB	16.73	18.80	42.0
Spain	Orange	Internet Everywhere Pro 29	44.03	43.59	7.2
Sweden	Tele2	Mobilt Bredband Medium	26.45	20.34	15.0
Switzerland	Orange	Internet Everywhere Young	40.78	23.99	7.2
Turkey	Turkcell	Kamu Internet Paketi 5GB	15.54	27.75	21.6
United Kingdom	T-Mobile	Mobile Broadband 10GB	15.82	11.99	4.5
United States	Verizon	6GB Shared Data 4G	44.61	44.61	12.0
OECD average			28.33	26.70	

Notes: The OECD basket of wireless broadband includes total charges for 5 GB of data for laptop use per month. USD purchasing power parities (PPP) are used to aid international comparisons. (*) Wireless broadband speeds included in this basket represent a mix between theoretical maximum speeds and more realistic speed measurements, closer to what a customer will typically receive over the network.

Source: OECD and Teligen.

StatLink http://dx.doi.org/10.1787/888932802724

Table 7.40. **OECD Wireless Broadband basket, OECD Laptop 10 GB, September 2012**

	ISP	Package	Total		Speed (Mbit/s)*
			USD	USD PPP	
Australia	Optus	BMB Plan 10GB	36.91	22.64	3.0
Austria	T-Mobile	All inclusive Internet 3G	13.87	12.72	4.0
Belgium	Mobistar	Internet Everywhere Multi Max for PC	225.85	196.39	3.6
Canada	Rogers Wireless	LTE Promotional Plan	62.06	48.11	100.0
Chile	Claro	Download 10GB +	43.22	57.62	12.0
Czech Republic	O2	LTE Mobile Internet XL	25.39	32.14	10.0
Denmark	Telenor Denmark	Mobilt Bredband Large (for Telenor subscribers)	30.72	21.04	32.0
Estonia	Tele2	M2M Small	7.51	9.04	1.0
Finland	Elisa - saunalahti.fi	Saunalahti Prepaid Mobile Broadband - 1 month	9.89	7.98	0.4
France	Bouygues	Forfait Internet illimité PRO 3G+ 24/24 (24M) SIM Only	37.52	33.50	42.0
Germany	Telekom	CombiCard Connect L	45.07	42.92	42.2
Greece	Vodafone	10GB	46.01	47.43	42.2
Hungary	T-Mobile	BusinessNet L	25.37	36.24	21.0
Iceland	Nova	3G Internet subscription 10GB	26.18	22.00	5.0
Ireland	Vodafone	Pay Monthly Mobile Broadband 10GB	25.08	20.39	7.2
Israel	Cellcom	Unlimited 2.8Mbps	31.06	27.25	2.8
Italy	WIND	Super Internet Unlimited	15.06	15.52	14.4
Japan	au KDDI	WiMAX Flat Rate (discounted price)	51.74	34.72	40.0
Korea	KT	4G Wibro 10G	9.95	12.29	40.3
Luxembourg	Tango	Unlimited Tango Mobile ADSL	25.42	20.84	7.2
Mexico	Movistar	Plan Comparte 10GB	42.03	61.80	3.6
Netherlands	T-Mobile	Internet Maximaal	50.19	46.47	14.4
New Zealand	2degrees	Broadband 12GB	80.39	61.84	7.2
Norway	Netcom	Mobilt Bredband Large (SIM only)	39.45	24.66	80.0
Poland	Orange	Business Everywhere Standard 12 SIM Only	17.77	30.11	42.0
Portugal	Optimus	Optimus Kanguru Up to 5Mbps	25.31	26.09	5.0
Slovak Republic	O2	Superfast LTE internet	18.82	25.43	100.0
Slovenia	Sl.mobil	Mobilni internet XL	22.62	25.42	42.0
Spain	Orange	Internet Everywhere Pro 39	59.21	58.62	7.2
Sweden	Tele2	Mobilt Bredband Medium	26.45	20.34	15.0
Switzerland	Sunrise	Take Away max	52.39	30.82	7.2
Turkey	Turkcell	Super internet Paketi 10GB	29.21	52.17	21.6
United Kingdom	T-Mobile	Mobile Broadband 10GB	15.82	11.99	4.5
United States	Verizon	10GB Shared Data 4G	66.61	66.61	12.0
OECD average			39.42	37.15	

Notes: The OECD basket of wireless broadband includes total charges for 10 GB of data for laptop use per month. USD purchasing power parities (PPP) are used to aid international comparisons. (*) Wireless broadband speeds included in this basket represent a mix between theoretical maximum speeds and more realistic speed measurements, closer to what a customer will typically receive over the network.

Source: OECD and Teligen.

StatLink http://dx.doi.org/10.1787/888932802743

Table 7.41. **OECD Wireless Broadband basket, OECD Tablet 250M, September 2012**

	ISP	Package	Total		Speed (Mbit/s)*
			USD	USD PPP	
Australia	Vodafone	Mobile Broadband SIM only 2GB	19.60	12.02	6.0
Austria	Orange	Mobiles Internet 1GB Sim Only	7.10	6.51	2.0
Belgium	Proximus	Mobile Internet Pay&Surf for iPad (€10 top-up)	12.55	10.91	3.6
Canada	Telus Wireless	TELUS Data Plan for iPad 250MB	17.12	13.27	7.2
Chile	Entel	Planes Controlado 500MB - Tablet	15.45	20.60	2.0
Czech Republic	Vodafone	Super for tablets (12 months contract)	10.05	12.72	21.6
Denmark	Telia	4Life by Telia Micro 500MB	8.25	5.65	80.0
Estonia	Tele2	M2M Large	3.99	4.81	1.0
Finland	DNA	DNA Mobile Broadband S	6.25	5.04	0.5
France	SFR	CARRE TABLETTE & CLE 250Mo 12 months SIM	8.78	7.84	7.2
Germany	E-Plus	Internet Flat SIM Only for tablet	12.55	11.95	7.2
Greece	Vodafone	500MB for iPad	11.50	11.86	42.2
Hungary	T-Mobile	Domino Web SIM package 250MB	4.42	6.31	7.2
Iceland	Nova	3G Internet subscription 1GB (with tablet)	8.14	6.84	5.0
Ireland	Meteor Mobile	Broadband To Go 1GB	10.03	8.15	14.4
Israel	Cellcom	SIM only 2GB for tablets	17.16	15.05	2.8
Italy	TiM	Internet Pack Start	10.35	10.67	14.4
Japan	SoftBank	Unlimited Data Discount Ultra Speed Flat Rate	49.46	33.20	42.0
Korea	KT	4G Wibro 10G	9.95	12.29	40.3
Luxembourg	Tango	Tango Mobile ADSL 500MB	11.75	9.63	7.2
Mexico	Nextel	BA 500 MB plan	7.44	10.95	2.2
Netherlands	T-Mobile	Internet Standaard: USB stick + Internet standaard	12.55	11.62	3.6
New Zealand	2degrees	Broadband 1GB	17.11	13.16	7.2
Norway	Netcom	Mobilt Bredbard Small (SIM only)	13.61	8.51	80.0
Poland	PlusGSM	iPlus prywatnie 15 SIM Only	4.88	8.28	42.0
Portugal	Optimus	Optimus Kanguru Up to 2Mbps	12.65	13.05	2.0
Slovak Republic	Orange	Mini +	6.31	8.52	3.6
Slovenia	Sl.mobil	Enkratni zakup 1GB	10.46	11.75	42.0
Spain	Vodafone	TP Diaria ilimitada	15.14	14.99	21.6
Sweden	Tele2	Surf for tablet SIM Only	7.37	5.67	5.0
Switzerland	Swisscom	NATEL data start Tab	19.85	11.68	7.2
Turkey	Avea	iPad 1GB	4.94	8.82	7.2
United Kingdom	Vodafone	Top Up £5	7.91	5.99	3.6
United States	AT&T	DataConnect 250MB for Tablets	16.49	16.49	7.2
OECD average			12.09	11.02	

Notes: The OECD basket of wireless broadband includes total charges for 250 MB of data for tablet use per month. USD purchasing power parities (PPP) are used to aid international comparisons. (*) Wireless broadband speeds included in this basket represent a mix between theoretical maximum speeds and more realistic speed measurements, closer to what a customer will typically receive over the network.

Source: OECD and Teligen.

StatLink http://dx.doi.org/10.1787/888932802762

Table 7.42. **OECD Wireless Broadband basket, OECD Tablet 500M, September 2012**

	ISP	Package	Total		Speed (Mbit/s)*
			USD	USD PPP	
Australia	Vodafone	Mobile Broadband SIM only 2GB	19.60	12.02	6.0
Austria	Orange	Mobiles Internet 1GB Sim Only	7.10	6.51	2.0
Belgium	Proximus	Mobile Internet Pay&Surf for iPad (€10 top-up)	12.55	10.91	3.6
Canada	Telus Wireless	TELUS Data Plan for iPad 500MB	22.83	17.70	7.2
Chile	Entel	Planes Controlado 500MB - Tablet	15.45	20.60	2.0
Czech Republic	Vodafone	Super for tablets (12 months contract)	10.05	12.72	21.6
Denmark	Telia	4Life by Telia Micro 500MB	8.25	5.65	80.0
Estonia	Tele2	M2M Small	7.51	9.04	1.0
Finland	DNA	DNA Mobile Broadband S	6.25	5.04	0.5
France	Bouygues	Forfait Internet Occasionnel 3G+ Sans Engagement SIM Only	12.42	11.09	42.0
Germany	E-Plus	Internet Flat SIM Only for tablet	12.55	11.95	7.2
Greece	Vodafone	500MB for iPad	11.50	11.86	42.2
Hungary	Telenor	Online iDea 500MB (for iPad)	6.58	9.41	7.2
Iceland	Nova	3G Internet subscription 1GB (with tablet)	8.14	6.84	5.0
Ireland	Meteor Mobile	Broadband To Go 1GB	10.03	8.15	14.4
Israel	Cellcom	SIM only 2GB for tablets	17.16	15.05	2.8
Italy	TiM	Internet Pack Start	10.35	10.67	14.4
Japan	SoftBank	Unlimited Data Discount Ultra Speed Flat Rate	49.46	33.20	42.0
Korea	KT	4G Wibro 10G	9.95	12.29	40.3
Luxembourg	Tango	Tango Mobile ADSL 500MB	11.75	9.63	7.2
Mexico	Nextel	BA 500 MB plan	7.44	10.95	2.2
Netherlands	T-Mobile	Internet Standaard: USB stick + Internet standaard	12.55	11.62	3.6
New Zealand	2degrees	Broadband 1GB	17.11	13.16	7.2
Norway	Netcom	Mobilt Bredband Small (SIM only)	13.61	8.51	80.0
Poland	PlusGSM	iPlus prywatnie 15 SIM Only	4.88	8.28	42.0
Portugal	Optimus	Optimus Kanguru Up to 2Mbps	12.65	13.05	2.0
Slovak Republic	Orange	Mini +	6.31	8.52	3.6
Slovenia	Sl.mobil	Enkratni zakup 1GB	10.46	11.75	42.0
Spain	Vodafone	TP Diaria ilimitada	15.14	14.99	21.6
Sweden	Tele2	Surf for tablet SIM Only	7.37	5.67	5.0
Switzerland	Swisscom	NATEL data start Tab	19.85	11.68	7.2
Turkey	Avea	iPad 1GB	4.94	8.82	7.2
United Kingdom	Three	Broadband 0.5GB	8.37	6.34	5.6
United States	Sprint	Peel 3G Mobile Broadband 1GB	32.99	32.99	3.1
OECD average			13.03	11.96	

Notes: The OECD basket of wireless broadband includes total charges for 500 MB of data for tablet use per month. USD purchasing power parities (PPP) are used to aid international comparisons. (*) Wireless broadband speeds included in this basket represent a mix between theoretical maximum speeds and more realistic speed measurements, closer to what a customer will typically receive over the network.

Source: OECD and Teligen.

StatLink http://dx.doi.org/10.1787/888932802781

Table 7.43. **OECD Wireless Broadband basket, OECD Tablet 1GB, September 2012**

	ISP	Package	Total		Speed (Mbit/s)*
			USD	USD PPP	
Australia	Vodafone	Mobile Broadband SIM only 2GB	19.60	12.02	6.0
Austria	Orange	Mobiles Internet 1GB Sim Only	7.10	6.51	2.0
Belgium	Mobistar	Internet Everywhere Multi Relax for tablet	18.82	16.37	3.6
Canada	Rogers Wireless	Rogers Plan 250MB for iPad	35.77	27.73	21.0
Chile	Entel	Planes Controlado 500MB - Tablet	15.45	20.60	2.0
Czech Republic	T-Mobile	Standard Internet on the Go	13.98	17.70	10.8
Denmark	Telenor Denmark	Mobilt Bredband Medium	15.29	10.47	16.0
Estonia	Tele2	M2M Small	7.51	9.04	1.0
Finland	DNA	DNA Mobile Broadband S	6.25	5.04	0.5
France	SFR	CARRE TABLETTE & CLE 1Go 24 months SIM	23.84	21.29	7.2
Germany	O2	Surf Flat M Business (no contract) + Samsung Galaxy Tab 2. 10.1	18.35	17.48	3.6
Greece	Cosmote	Cosmote i-Data 500MB for iPad	12.55	12.94	42.2
Hungary	T-Mobile	Domino Web SIM package 1GB	8.84	12.63	7.2
Iceland	Nova	3G Internet subscription 1GB (with tablet)	8.14	6.84	5.0
Ireland	Meteor Mobile	Broadband To Go 1GB	10.03	8.15	14.4
Israel	Cellcom	SIM only 2GB for tablets	17.16	15.05	2.8
Italy	TiM	Internet Pack Start	10.35	10.67	14.4
Japan	SoftBank	Unlimited Data Discount U tra Speed Flat Rate	49.46	33.20	42.0
Korea	KT	4G Wibro 10G	9.95	12.29	40.3
Luxembourg	LuxGSM	iPad S	12.55	10.28	7.2
Mexico	Movistar	Plan iPad 1GB	17.22	25.32	3.6
Netherlands	KPN	Mobiel Internet SIM Only Extra (for tablets)	19.17	17.75	7.2
New Zealand	2degrees	Broadband 1GB	17.11	13.16	7.2
Norway	Netcom	Mobilt Bredband Small (SIM only)	13.61	8.51	80.0
Poland	T-Mobile	blueconnect 19 SIM Only	6.98	11.83	42.0
Portugal	Optimus	Optimus Kanguru Up to 2Mbps	12.65	13.05	2.0
Slovak Republic	O2	Internet for a laptop	13.80	18.65	14.4
Slovenia	Sl.mobil	Enkratni zakup 1GB	10.46	11.75	42.0
Spain	Vodafone	TP Diaria ilimitada	15.14	14.99	21.6
Sweden	Tele2	Surf for tablet SIM Only	7.37	5.67	5.0
Switzerland	Orange	Internet Everywhere One	30.33	17.84	7.2
Turkey	Avea	iPad 1GB	4.94	8.82	7.2
United Kingdom	T-Mobile	Mobile Broadband 1GB (13 months)	11.87	8.99	4.5
United States	Sprint	Peel 3G Mobile Broadband 1GB	32.99	32.99	3.1
OECD average			15.72	14.58	

Notes: The OECD basket of wireless broadband includes total charges for 1 GB of data for tablet use per month. USD purchasing power parities (PPP) are used to aid international comparisons. (*) Wireless broadband speeds included in this basket represent a mix between theoretical maximum speeds and more realistic speed measurements, closer to what a customer will typically receive over the network.

Source: OECD and Teligen.

StatLink http://dx.doi.org/10.1787/888932802800

Table 7.44. **OECD Wireless Broadband basket, OECD Tablet 2 GB, September 2012**

	ISP	Package	Total		Speed (Mbit/s)*
			USD	USD PPP	
Australia	Vodafone	Mobile Broadband SIM only 2GB	19.60	12.02	6.0
Austria	Orange	Mobiles Internet 6GB Sim Only	12.12	11.12	21.0
Belgium	Mobistar	Internet Everywhere Multi Max for tablet	37.64	32.73	3.6
Canada	Telus Wireless	TELUS Data Plan for iPad 5GB	39.96	30.97	7.2
Chile	Entel	Planes Controlado 500MB - Tablet	15.45	20.60	2.0
Czech Republic	T-Mobile	Standard Internet on the Go	13.98	17.70	10.8
Denmark	Telenor Denmark	Mobilt Bredband Medium	15.29	10.47	16.0
Estonia	Tele2	M2M Small	7.51	9.04	1.0
Finland	DNA	DNA Mobile Broadband S	8.82	7.11	0.5
France	Bouygues	Forfait Internet illimité 3G+ 24/24 (24M) SIM Only	24.97	22.29	42.0
Germany	E-Plus	Internet Flat XL SIM Only for tablet	25.09	23.90	7.2
Greece	Cosmote	Internet Any Way 2GB SIM Only	21.96	22.64	42.2
Hungary	T-Mobile	Gold Partner Card RelaxNet M (speed reduction)	8.84	12.63	7.2
Iceland	Simin	3G Net 3 GB	14.35	12.06	7.2
Ireland	O2	Broadband Plus - For Light Users 2GB	15.43	12.55	7.2
Israel	Cellcom	SIM only 2GB for tablets	17.16	15.05	2.8
Italy	Vodafone	Internet Ready	12.55	12.94	14.4
Japan	SoftBank	Unlimited Data Discount Ultra Speed Flat Rate	49.46	33.20	42.0
Korea	KT	4G Wibro 10G	9.95	12.29	40.3
Luxembourg	LuxGSM	iPad S	12.55	10.28	7.2
Mexico	Telcel	Plan de Renta Internet Banda Ancha Movil 3GB	21.73	31.96	7.2
Netherlands	KPN	Mobiel Internet SIM Only Extra (for tablets)	19.17	17.75	7.2
New Zealand	2degrees	Broadband 1GB	33.13	25.48	7.2
Norway	Tele2	Mobile Broadband Leisure	23.66	14.79	3.6
Poland	T-Mobile	blueconnect 19 SIM Only	6.98	11.83	42.0
Portugal	Optimus	Optimus Kanguru Up to 2Mbps	12.65	13.05	2.0
Slovak Republic	O2	Internet for a laptop	13.80	18.65	14.4
Slovenia	Mobitel	Internet Mini with tablet	13.80	15.51	21.6
Spain	Vodafone	Tarifa Internet Movil 2GB	34.79	34.45	21.6
Sweden	Tele2	Mobilt Bredband Small	12.92	9.94	5.0
Switzerland	Orange	Internet Everywhere Young	40.78	23.99	7.2
Turkey	Turkcell	Kamu Internet Paketi 5GB	15.54	27.75	21.6
United Kingdom	T-Mobile	Mobile Broadband 10GB	15.82	11.99	4.5
United States	AT&T	DataConnect 3GB for Tablets	33.00	33.00	7.2
OECD average			20.01	18.58	

Notes: The OECD basket of wireless broadband includes total charges for 2 GB of data for tablet use per month. USD purchasing power parities (PPP) are used to aid international comparisons. (*) Wireless broadband speeds included in this basket represent a mix between theoretical maximum speeds and more realistic speed measurements, closer to what a customer will typically receive over the network.

Source: OECD and Teligen.

StatLink http://dx.doi.org/10.1787/888932802819

Table 7.45. **OECD Wireless Broadband basket, OECD Tablet 5 GB, September 2012**

	ISP	Package	Total		Speed (Mbit/s)*
			USD	USD PPP	
Australia	Optus	Optus Data Plan $24.95 + Acer Iconia A501	25.73	15.79	3.0
Austria	Orange	Mobiles Internet 6GB Sim Only	12.12	11.12	21.0
Belgium	Mobistar	Internet Everywhere Multi Max for tablet	69.01	60.01	3.6
Canada	Telus Wireless	TELUS Data Plan for iPad 5GB	39.96	30.97	7.2
Chile	Entel	Planes Controlado 500MB - Tablet	15.45	20.60	2.0
Czech Republic	O2	LTE Mobile Internet XL	25.39	32.14	10.0
Denmark	Telenor Denmark	Mobilt Bredband Medium	15.29	10.47	16.0
Estonia	Tele2	M2M Small	7.51	9.04	1.0
Finland	Elisa - saunalahti.fi	Saunalahti Prepaid Mobile Broadband - 1 month	9.89	7.98	0.4
France	Bouygues	Forfait Internet illimité 3G+ 24/24 (24M) SIM Only	24.97	22.29	42.0
Germany	E-Plus	Internet Flat XL SIM Only for tablet	25.09	23.90	7.2
Greece	Cosmote	Internet Any Way 5GB SIM Only	33.46	34.49	42.2
Hungary	T-Mobile	Net icon 2GB	13.04	18.62	7.2
Iceland	Vodafone	NetFrelsi 5GB (with tablet)	16.36	13.75	7.2
Ireland	Meteor Mobile	Broadband To Go 5GB	21.67	17.62	14.4
Israel	Pelephone (Bezeq)	High Speed 99	27.08	23.76	3.6
Italy	WIND	Super Internet Unlimited	15.06	15.52	14.4
Japan	au KDDI	WiMAX Flat Rate (discounted price)	51.74	34.72	40.0
Korea	KT	4G Wibro 10G	9.95	12.29	40.3
Luxembourg	Tango	Unlimited Data Plan + WiFI for iPad	23.71	19.44	7.2
Mexico	Movistar	Plan iPad 5GB	29.25	43.01	3.6
Netherlands	KPN	Mobiel Internet SIM Only Premium (for tablets)	28.58	26.46	7.2
New Zealand	2degrees	Broadband 12GB	80.39	61.84	7.2
Norway	Netcom	Mobilt Bredband Medium (SIM only)	25.67	16.04	80.0
Poland	Orange	Business Everywhere Standard 6 SIM Only	14.38	24.38	42.0
Portugal	Optimus	Optimus Kanguru Up to 5Mbps	25.31	26.09	5.0
Slovak Republic	Orange	Klasik +	17.87	24.14	7.2
Slovenia	Si.mobil	Enkratni zakup 5GB	16.73	18.80	42.0
Spain	Orange	Internet Everywhere Pro 29	44.03	43.59	7.2
Sweden	Tele2	Surf for iPad Medium SIM Only	25.25	19.43	6.0
Switzerland	Orange	Internet Everywhere Young	40.78	23.99	7.2
Turkey	Turkcell	Kamu Internet Paketi 5GB	15.54	27.75	21.6
United Kingdom	T-Mobile	Mobile Broadband 10GB	15.82	11.99	4.5
United States	Verizon	6GB Shared Data 4G with tablet	44.00	44.00	12.0
OECD			26.65	24.88	

Notes: The OECD basket of wireless broadband includes total charges for 500 MB of data for laptop use per month. USD purchasing power parities (PPP) are used to aid international comparisons. (*) Wireless broadband speeds included in this basket represent a mix between theoretical maximum speeds and more realistic speed measurements, closer to what a customer will typically receive over the network.

Source: OECD and Teligen.

StatLink http://dx.doi.org/10.1787/888932802838

Chapter 8

Recent developments in household and individual communication expenditures and use

Information and communication technologies (ICTs) are an increasing part of everyday life. Communication and computing devices along with the Internet and broadband are now widely available in most OECD countries. Types of devices and services evolve continuously, so penetration rates can vary when benchmarked against leaders. ICT expenditures comprised the most dynamic component of household expenditure in recent years, but growth began to slow after 2005.

The statistical data for Israel are supplied by and under the responsibility of the relevant Israeli authorities. The use of such data by the OECD is without prejudice to the status of the Golan Heights, East Jerusalem and Israeli settlements in the West Bank under the terms of international law.

This chapter explores the evolution of household expenditure in ICT goods and services, focusing specifically on the rise of mobile communications, the most dynamic and significant component of telecommunication services expenditure. This expenditure mirrors the massive diffusion of mobile devices and the uptake of a growing number of mobile wireless services. The growth in use of data services, driven by the deployment of mobile broadband networks and the extraordinarily rapid uptake of smartphones, is now replicating that of SMS, the earliest addition to mobile telephony. The chapter also examines mobile device and smartphone activities with a particular focus on mobility, such as Internet access on the move, mobile broadband access and m-commerce. The development of new technologies, services and lower prices driven by innovation, scale economies and competition, are a clear influence on the increasing use of communication services by households.

ICT household expenditure in OECD countries

In 2011, the amount of household final consumption expenditures devoted to ICT goods and services in 29 OECD countries reached USD 1 103 billion PPP, around 4.9% of their total final consumption expenditures (Figure 8.1). The share of communication within ICT expenditures was dominant in most countries, reaching 75% and more in Greece, Hungary, Ireland, Israel, Mexico and Slovenia. The only exceptions to this general pattern were Australia, Denmark and the United States, where expenditures in information technology (IT) goods and services exceeded those for communication.

The share of communication in total household expenditures increased steeply across the OECD area during the second half of the 1990s. This trend reflects the development of mobile telephony, the Internet and broadband. After the "Internet bubble" burst in 2001 the trend slowed, but consumer demand for an increasing array of communication products and services continued to grow steadily, stimulated by lower prices for both equipment and services. Overall, the effect of this increase compensated for declining prices both for communication goods and services in total expenditure.

Beyond communication prices, additional underlying factors may affect this ratio. On one hand, total expenditures are impacted by the income level. On the other, communication expenditures are influenced by the amount and the quality of communication services actually used by the consumers.

Between 1995 and 2000, communication expenditure grew at a rapid rate, exceeding every other type of expenditure in pace (OECD, 2011). In 2000, total household communication expenditures in the OECD area were USD 422 billion PPP or 2.6% of final consumption expenditure, against 2.2% in 1995 (Table 8.1). The turn of the century can be considered a high point in terms of relative growth. In contrast, from 2001-10 communication expenditures grew at almost the same pace as total final consumption expenditures, and were outpaced by health, education, and housing, water and energy. This pattern mirrors the relative rise in energy prices and to a lesser extent services with

Figure 8.1. **Share of households' ICT expenditures[1] in OECD countries, 2011**

% of final consumption expenditure of households on the territory

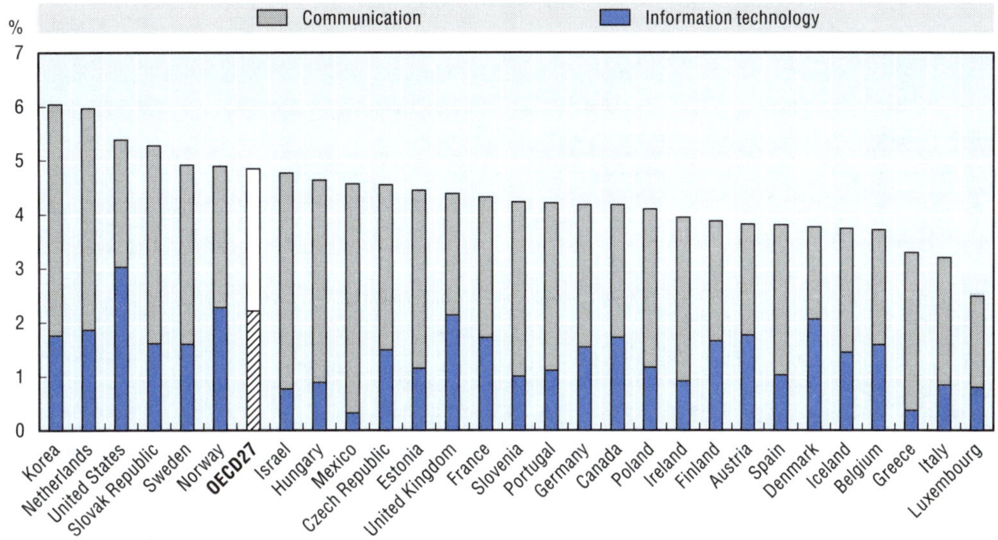

1. Based on National Accounts using the COICOP classification. Information technology includes audiovisual, photographic and information-processing equipment. Communication includes telecommunication equipment and services and postal services. The detailed definition of information technology and communication expenditures is provided in Annex 1.
2. OECD 29: Data for Chile, Japan, New Zealand, Switzerland and Turkey are not included.
3. Estimates for Information technology data were used for Australia, Austria, Canada and Norway.

Source: OECD, based on data from the National Accounts database, March 2013.

StatLink http://dx.doi.org/10.1787/888932800197

higher labour content, as well as the growing importance accorded to health and investment in education. In 2011, communication expenditures amounted (in current terms) to USD 684 billion PPP or 2.7% of final consumption expenditures, slightly above the relative level of 2000 (Figure 8.2).

Total consumption expenditures including communication decreased in households across more than half of OECD countries in 2009 (see also Finland and France below). However, expressed in terms of budgetary coefficient, the relative propensity of households to consume communication in spite of the global economic crisis remained unchanged at the OECD level. It declined marginally in some countries and more severely in Estonia. In contrast, budgetary coefficients increased notably from 2009 to 2011 in Greece, Japan and the United Kingdom (see Table 8.1).

Shares of communication and IT have not always been constant (Figure 8.3) and have evolved in different ways across OECD countries for which data are available.[1] Between 1990 and 2011, communication expenditures surpassed IT expenditures in a majority of countries. This change began during the mid-1990s and was more pronounced in some countries (i.e. Korea, Iceland, Italy and the Netherlands). After the turn of the century, stabilisation of shares occurred in most OECD countries.

Data on ICT expenditures in Finland and France are available over a longer period and thus provide a historical perspective to this change (Figure 8.4). In France, IT expenditures remained higher than communication from 1956 to 1979. The two items fluctuated around the same share between 1979 and 1999 when communication started to exceed IT expenditures. Then both shares declined slightly from 2010 onwards under the effect of the

Figure 8.2. **Changes in the proportion of households' expenditure by category in the OECD,[1] 2000-11**

Base 100 in 2000

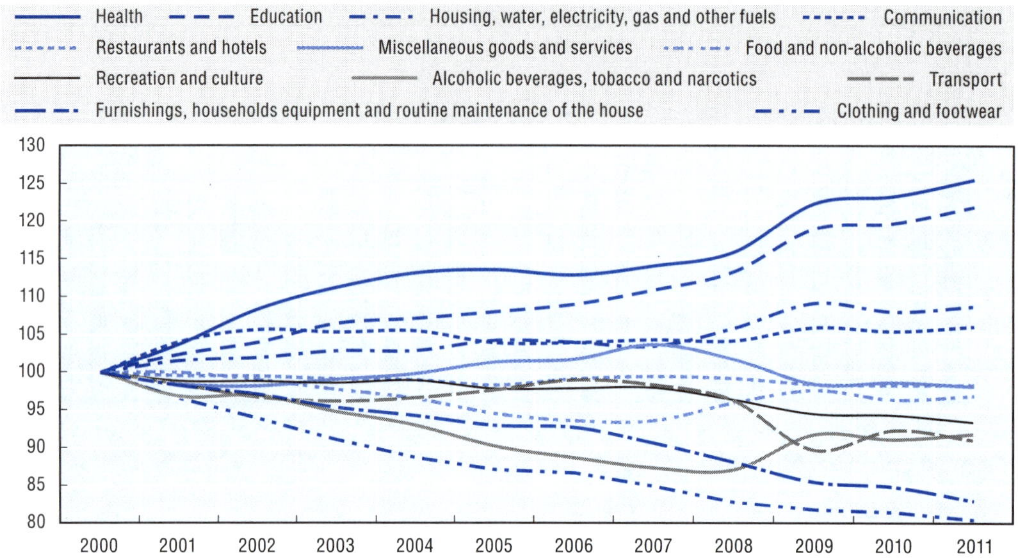

1. Chile, Greece, New Zealand and Turkey are not included in the calculations.

Source: OECD, based on data from the National Accounts Database, March 2013.

StatLink http://dx.doi.org/10.1787/888932800216

Figure 8.3. **Difference between communication and information technology expenditures[1] in selected OECD countries, 1990-2011**

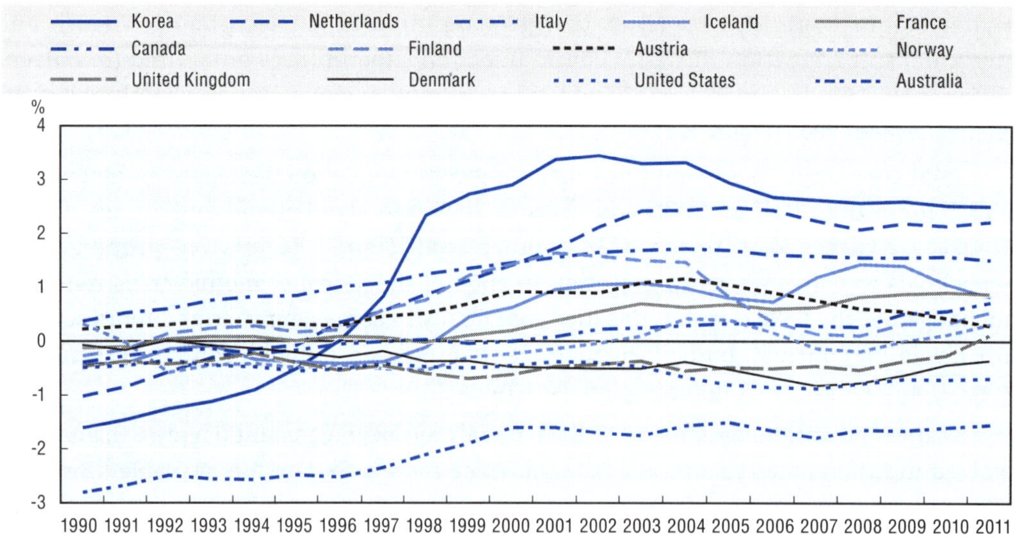

1. See note 1 of Figure 8.1.

Source: OECD, based on data from the National Accounts Database, March 2013.

StatLink http://dx.doi.org/10.1787/888932800235

2009 global economic crisis. Despite the continuous price decline communication service expenditures stopped growing in volume in 2009 for the first time since the development of mobile wireless and the Internet, and in 2011 they decreased in connection with the tax increase on ISP triple-play proposals (INSEE, 2010, 2012).

Figure 8.4. **Share of ICT expenditures in the total consumption expenditures of households in Finland and France, 1959-2011**

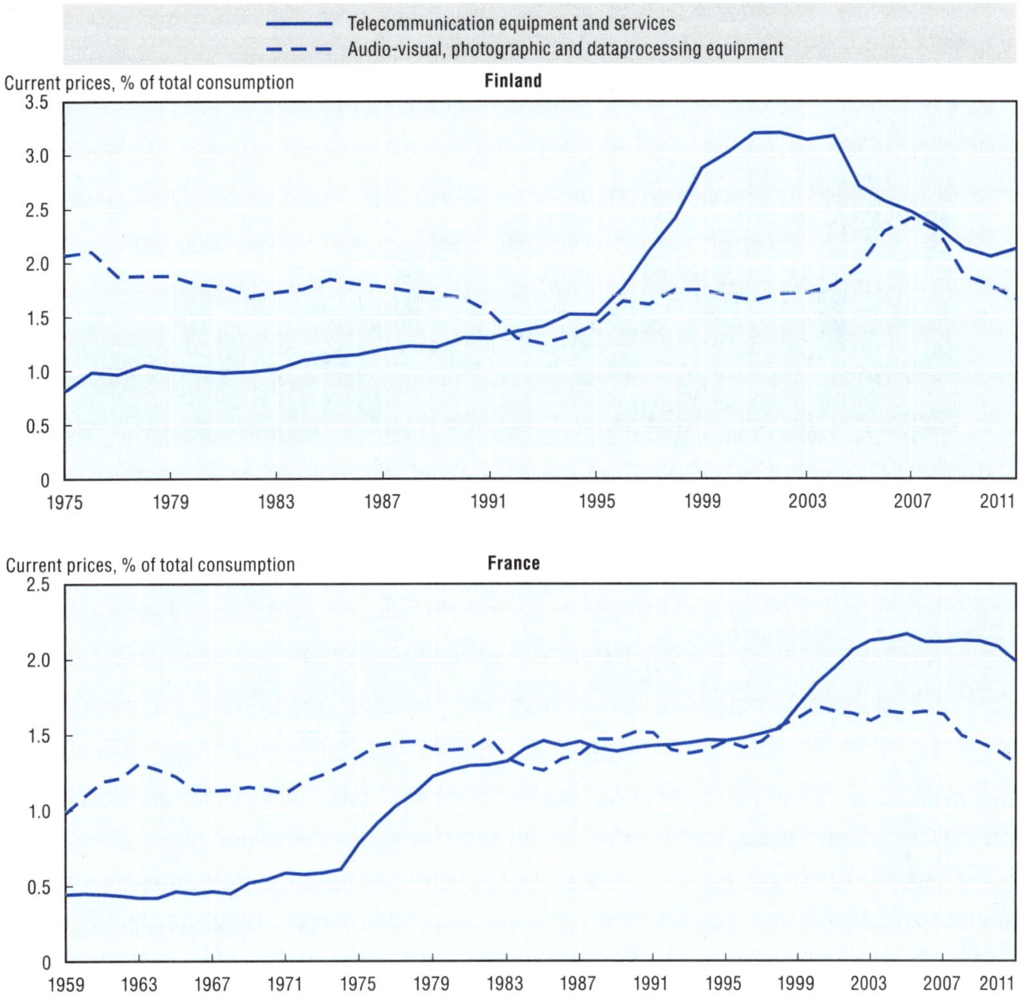

Source: OECD, based on data from National Accounts, March 2013.

StatLink http://dx.doi.org/10.1787/888932800254

In Finland, the prevalence of IT expenditures over communication dates back to 1975 and continued until 1992, when early adoption of mobile telecommunication services and, thereafter, Internet access increased the share of communication. From 2005 onwards the gap began to close. The share of communication started to decrease in 2004 and appeared to stabilise in 2011. In contrast, the decrease in IT expenditures started as recently as 2008 and appeared significant in 2011.

Using national surveys it is possible to compare monthly household spending on communication equipment and services expenditures in USD PPP, and provide a breakdown of Internet, mobile and fixed-related communication expenditures, where available (Figure 8.5). Spending patterns vary widely across OECD countries in terms of global amount spent and among the three categories. In addition, comparison is less than straightforward as national surveys are sometimes based on different definitions and methodologies. In 2011, average expenditure on communication was USD 160 PPP per

Figure 8.5. **Monthly household expenditures on communications in the OECD, 2011**

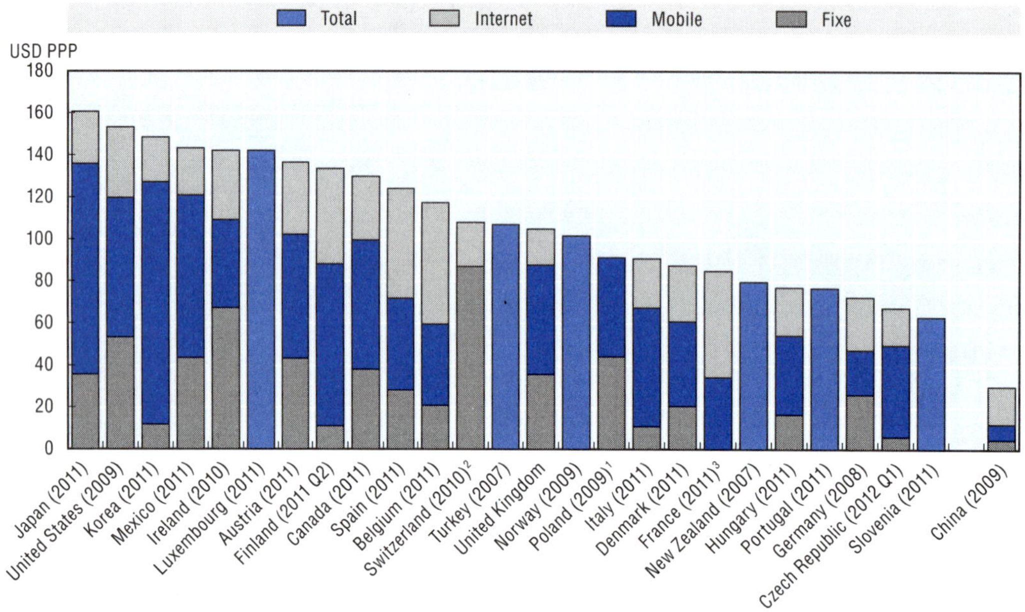

1. Internet expenditure are not included.
2. Cellular mobile expenditures are included in fixed line expenditures.
3. Fixed expenditures are included in Internet expenditures.

Source: OECD, Telecommunication Database, March 2013.

StatLink http://dx.doi.org/10.1787/888932800273

month in Japan and USD 30 PPP per month in China (where 60% was devoted to the Internet, representing the highest share of all countries considered here).

A number of factors influence household expenditure and those countries with high smartphone penetration are likely to have experienced changes alongside the life cycles for these devices compared to "feature phones". Koreans, for example, are among the leaders in the OECD area relative to the total amount of household expenditure that they devote to communication devices and services. This reflects the rapid adoption and high take-up of smartphones and mobile broadband (see also Chapter 4, Figure 4.10). Korean consumers also make among the greatest use of their smartphones, with Cisco data suggesting average use of 1.2GB per month. This monthly usage is significantly higher than Cisco data for other countries and generally larger than for similar data collected by official statistical sources in those countries.

Mobile telephones as a share of telecommunication services expenditure

The first decade of the 21st century witnessed a significant shift in telephony expenditures toward mobile telephony. The structure of telecommunication services expenditure has evolved significantly over the last decade, reflecting the spread of mobile telephone use within society. In the United States, from 2001 to 2011, the increase in telephone service expenditures was due almost exclusively to higher expenditure on cellular phone services at all income levels (Figure 8.6). Spending on cellular phone services rose from USD 210 on average in 2001 to USD 826 in 2011, but declined for residential phone services from USD 686 on average in 2001 to USD 380 in 2011 (United States Bureau of Labor Statistics, 2012a, 2012b).

Figure 8.6. **Household's mobile telephone expenditures by income level and prices in the United States, 2001-11**

% of total telephone services expenditures by income quintile

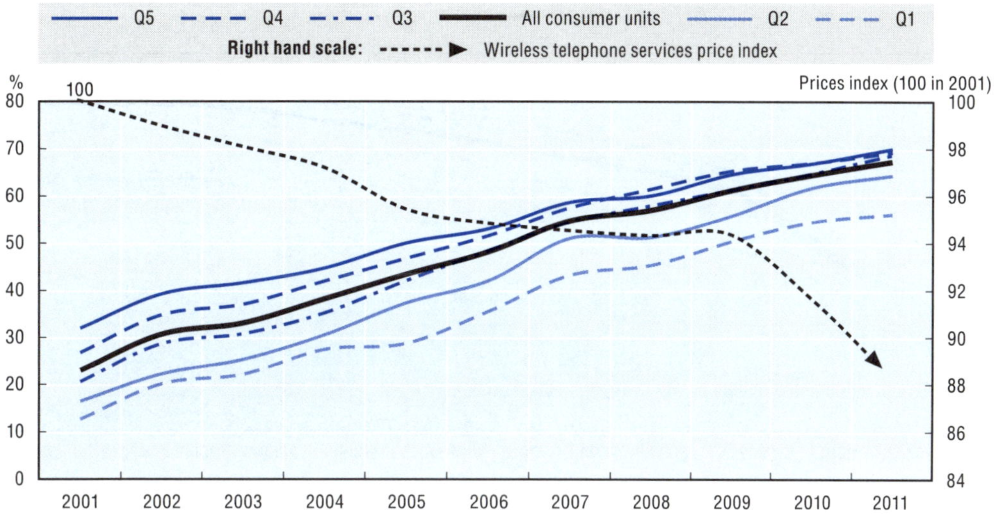

Source: United States Bureau of Labor Statistics.

StatLink http://dx.doi.org/10.1787/888932800292

The share of mobile services in the United States rose correspondingly on average from 23% in 2001 to more than 67% of total telecommunication expenditures in 2011. People with the highest incomes were the first adopters, devoting more than 50% of their telecommunication expenditures to mobile services in 2005. People with lower incomes followed, reaching the 50% threshold in 2009. The gap between highest and lowest income quintiles fell from 19.3 percentage points in 2001 to 13.8 in 2011. In parallel, wireless telephone service prices experienced a significant decrease over the decade.

The increasing share of mobile phone expenditure is also observable in Japan where the 50% threshold was reached in 2002, climbing to 70% share in 2011, comparable to the United States (67%). People in Japan with lower incomes followed the general trend, reaching the 55% threshold in 2011 (Figure 8.7). The share of mobile phones is also growing in Canada, but reached only 46% in 2011 (Figure 8.8).

The trend in the United Kingdom was similar, though less pronounced. Household expenditure fell in 2009 for the first time in 10 years and the share of total telecommunication services attributed to mobile services paused relative to recent growth. Growth restarted in 2010 due to significant changes in the volume and price of fixed and mobile communication services. Average call charges per minute have declined since 2006, more rapidly for mobile than for fixed lines. This parallels a fall in volume of fixed voice calls and a dramatic increase in the volume of mobile voice calls. The use of mobile data and mobile broadband services is also starting to increase significantly, and the share of mobile telephone services in telecommunication services expenditures crossed the 50% threshold in 2011.

In the Czech Republic households devoted around half of their telecommunication service expenditures to mobile services in 2003, a share that increased regularly to 88% in 2011. The percentage of European households with a fixed telephone line was relatively low in 2003 (72.4%) and the cost of a fixed line was relatively high compared to mobile telephony. The rapid adoption of mobile telephony in the first decade of the century saw

Figure 8.7. **Mobile wireless charges expenditures as a % of total phone charges expenditures[1] in Japan, 2000-11**

By income quintile

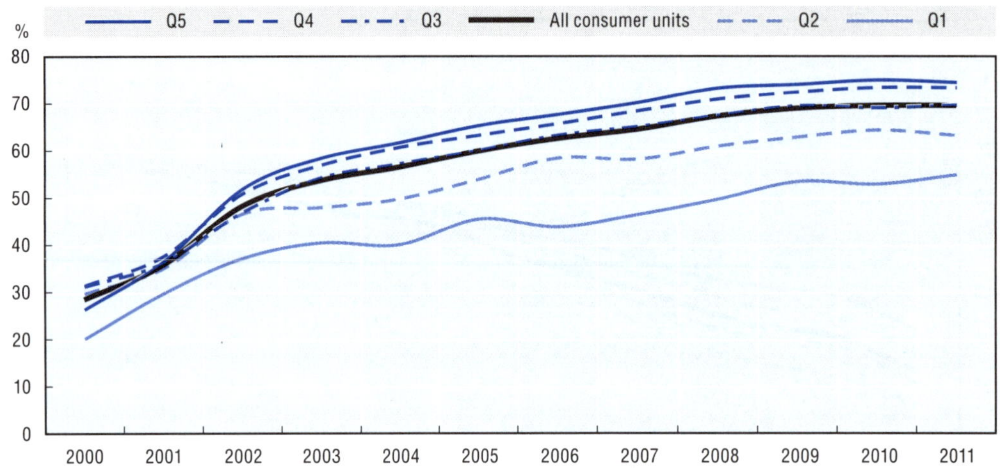

1. Total phone charges expenditures include telephone charges, mobile telephone charges and forwarding charges.

Source: Family income and Expenditure Survey, Statistics Bureau, Ministry of Internal Affairs and Communications, Japan.

StatLink http://dx.doi.org/10.1787/888932800311

Figure 8.8. **Share of mobile phone expenditures in telecommunication services expenditures in Canada, level of income**

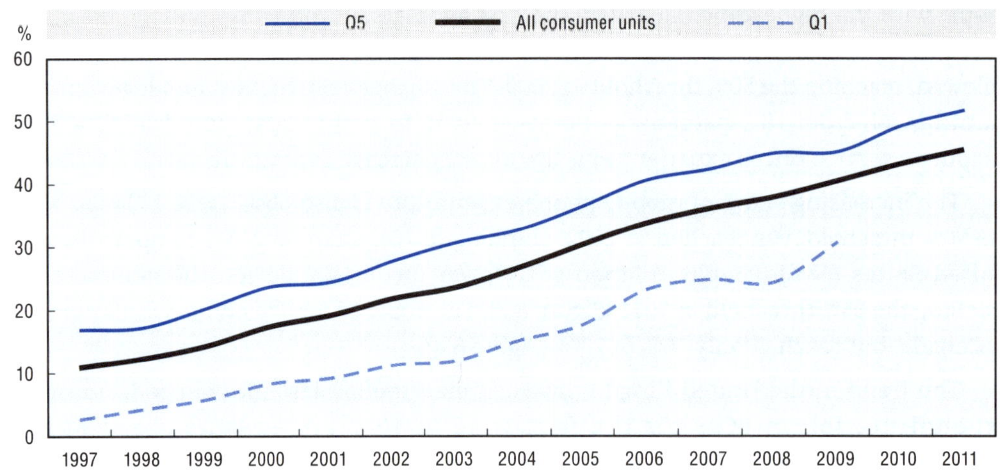

Note: First quintile not available from 2010 onwards.

Source: Statistics Canada, *Survey of Household Spending*, various years.

StatLink http://dx.doi.org/10.1787/888932800330

many households give up or not acquire a fixed line, and by 2011 only 23.4% of households in the Czech Republic were equipped with a fixed telephone line (Figure 8.9).

A generational effect

Younger generations have been clear leaders in the shift towards mobile telephony within telecommunication expenditures. In 2011, people in the United States aged below 25 were devoting more than 87% of their phone expenditures to mobile, compared to 30% for those aged above 75. The time lag for the latter appears to be about a decade. In the

Figure 8.9. **Share of mobile telephone services expenditures in total telecommunication services expenditures in the Czech Republic and the United Kingdom, 2001-11**[1]

% of total phone service expenditures

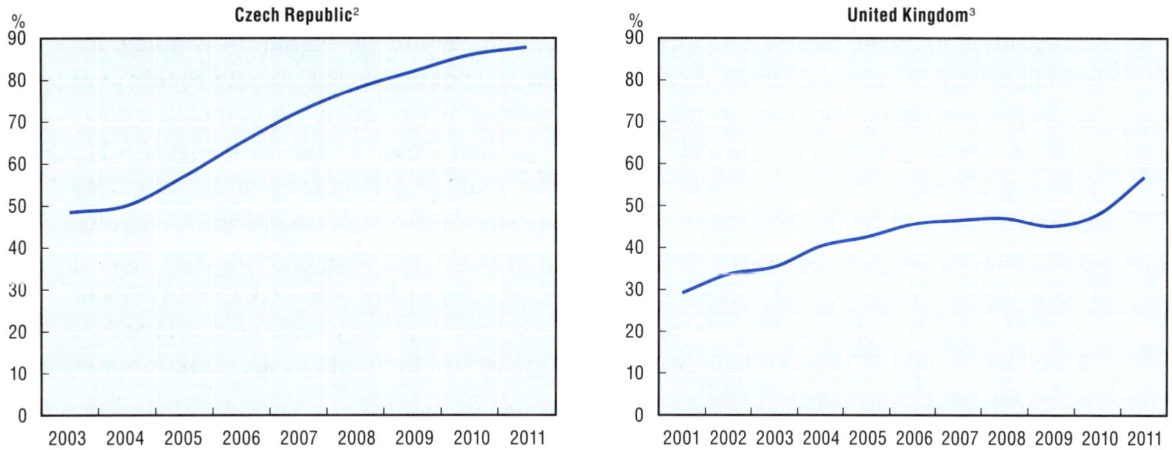

1. For the United Kingdom, financial year April 2000 to March 2001 instead 2001, and similarly until 2006 included.
2. Based on the average yearly expenditures of the households.
3. Based on average weekly expenditures of the households.

Source: OECD, based on Family Spending and Family Expenditure Survey, Office for National Statistics, various issues, and data provided by the Czech Statistical Office (2012).

StatLink http://dx.doi.org/10.1787/888932800349

Czech Republic, the 50% level was only reached in 2006 for people aged 60 to 69, and in 2008 for those aged above 69. In 2011, the latter were devoting around 68% of their telephone expenditures to mobile services, a threshold reached five years earlier for the population on average (Figure 8.10).

Figure 8.10. **Recent evolution of cell phone expenditures as a percentage of total telephone expenditures by age group**[1] **in the United States and the Czech Republic**

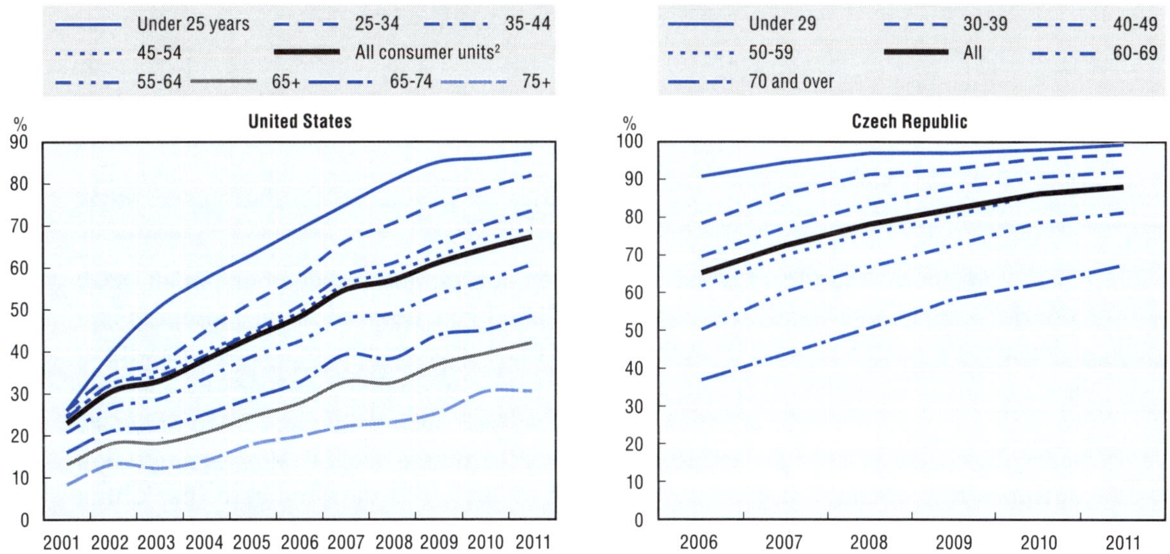

1. Age group of the reference person of the household.
2. Consumer unit refers to household.

Source: United States Bureau of Labor Statistics (2012b) and Czech Statistical Office (2012).

StatLink http://dx.doi.org/10.1787/888932800368

The pervasiveness of mobile telephones

Rapid diffusion in OECD countries

Diffusion of mobile communication has been extremely rapid in OECD countries – comparable with the earlier take up of television or colour television. In Norway, for example, only 1 person in 3 made a private mobile call per day in 2000, but 8 people out of 10 made one in 2011. Even among the older generation the pervasiveness of mobile calls is striking, reaching 6 persons out of 10, against less than 1 out of 10 in 2000 (Figure 8.11). Between 2009-11 there was a slowdown in use by younger generations, which might be explained by increasing engagement with new patterns of communication, with less voice calls and more Internet-oriented activities (fixed or mobile including voice or data). During this period, time spent on the Internet in an average day in Norway increased by 10% to reach 80 minutes, but the highest increase was seen among young people aged 9-15 at 25% reaching 83 minutes. One third of the 13-15 age group also increasingly accessed Internet via mobile phones in 2011.

Figure 8.11. **Individuals giving/receiving private mobile calls on an average day in Norway**

% of individuals in each age category

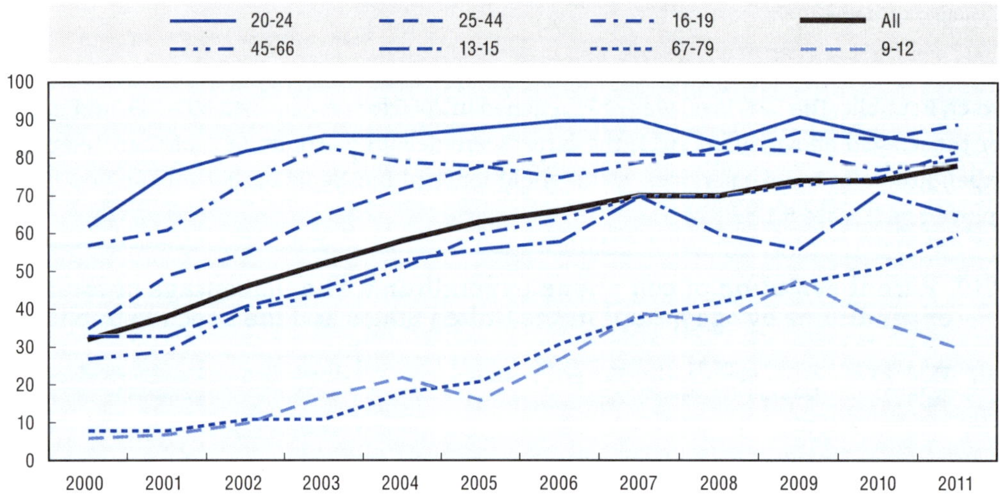

Source: Statistics Norway, Norvegian Mediabarometer.

StatLink http://dx.doi.org/10.1787/888932800387

In 2010, more than 80% of households owned at least one mobile phone in all OECD countries except Canada and Mexico (Figure 8.12). In India, this device was present in half of households in 2010. In the EU27 at the end of 2011, the proportion of households with at least one mobile telephone was about 90%. In terms of single households by age, this proportion ranged from above 92% among those aged between 30 and 59 to only 63% among those aged 60 and older according to the Eurobarometer (2012a). Mobile penetration continues to increase worldwide. At the end of 2011, ITU data indicate that China accounted for more than 1 billion subscriptions out of 6 billion worldwide, and India is expected to reach the billion mark by 2012.

Figure 8.12. **Households with a mobile telephone in selected countries, 2010**

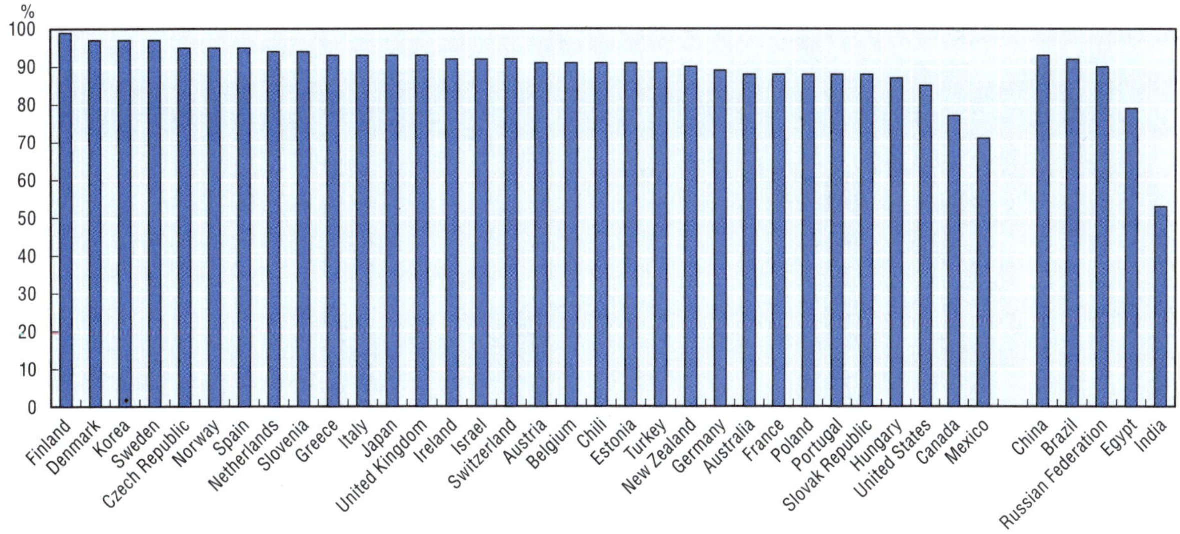

Source: ITU data, as reported in World Bank, 2012.

StatLink http://dx.doi.org/10.1787/888932800406

Significant growth of the mobile-only population

At the end of 2006, more than one household out of six (18%) on average was a mobile-only user at the EU27 level. At the end of 2011, Eurobarometer (2012a) records that this share increased to 27%. The proportion was significantly higher in new member states (52%) of the European Union than in the EU15 (21%), with the exception of Finland (80%), Austria (52%), and Denmark (48%). Recent data also show that the proportion of single households or people without a fixed line, but with mobile phones, has increased at the European level among both the youngest and older generations (Figure 8.13). The growth of smartphones, including Internet and video-access capabilities, may contribute to an increase in the community of mobile-only users, especially among younger generations. Currently, the two main factors of influence are income and age. In France, for example, mobile-only possession is most frequent among two groups: single household, low-income level people and blue collar workers. In this country, more than 80% of those under 60 were equipped with both a fixed and a mobile phone in 2011.

Surveys indicate that the mobile-only population grew significantly in some countries (Czech Republic, United Kingdom and United States) during the current decade, especially among young people (Figures 8.14 and 8.15). However, many respondents, while not having a traditional PSTN line, may have an alternative fixed line connection (e.g. via an unbundled DSL line or a cable television connection through which they access a VoIP service).

In the Czech Republic, the mobile-only population was practically non-existent in 1999 but reached three quarters of households by 2010 (Figure 8.15). As in the United Kingdom and the United States, younger generations led the trend, but more than half of households with a head over 70 are also embracing this approach. The move towards mobile use has been rapid, massive and is due to a number of factors including users surrendering traditional telephone lines for a combination of mobile services and fixed broadband Internet access connections.

Figure 8.13. **Mobile telephone access but no fixed telephone access in EU, 2009-11**

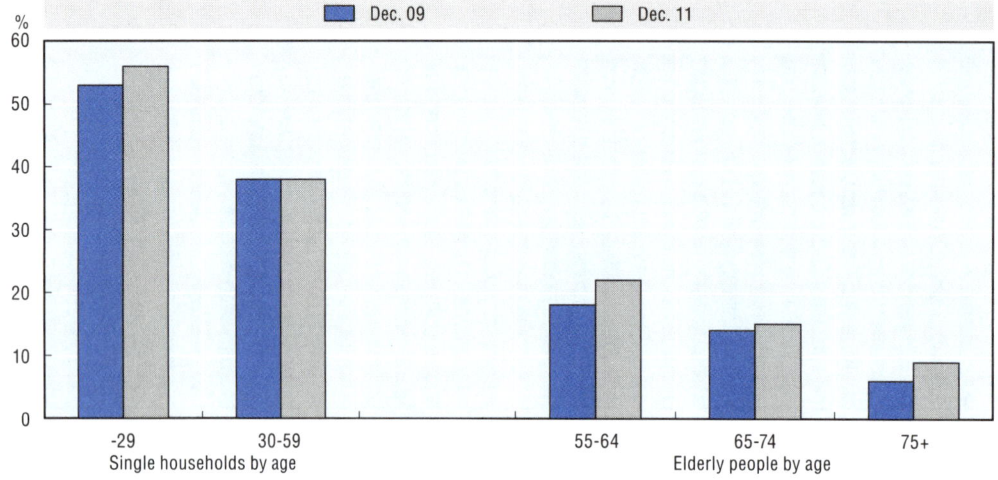

Source: Eurobarometer, E-Communications Household Survey, various years.

StatLink http://dx.doi.org/10.1787/888932800425

Figure 8.14. **The mobile-only population in United Kingdom and United States, 2002-12**

% of adults[1]

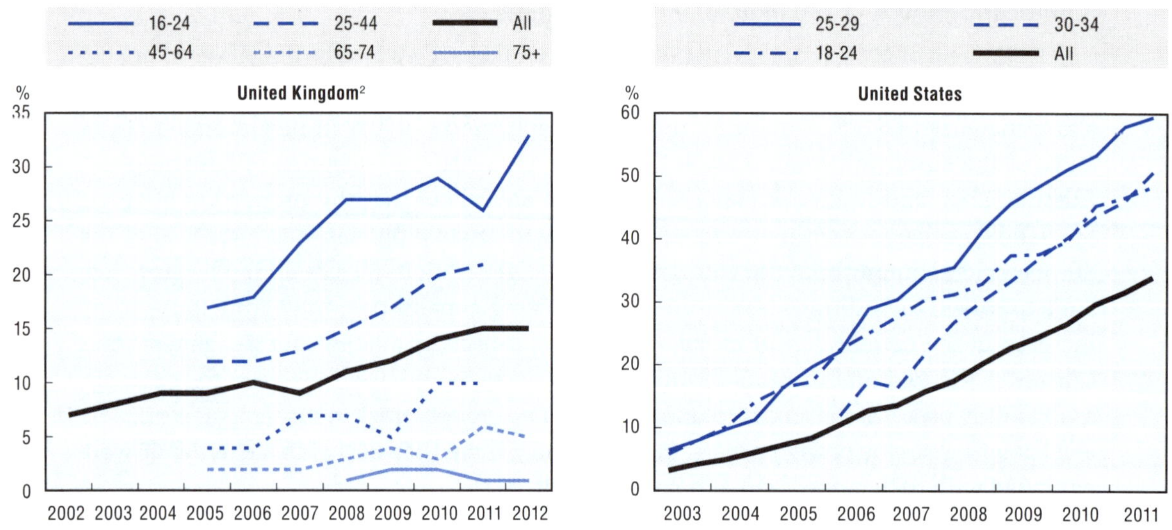

1. 16+ in the United Kingdom and 18+ in the United States. In the United States, mobile only refers to a person living in a family having a working cellular telephone and no working landline inside the household. In the United Kingdom, mobile only refers to the share of the population that relies solely on a mobile handset for voice telephony.
2. In the United Kingdom, for the total population ("All"), Q4 of each year until 2004, Q1 from 2005 to 2007, and Q2 from 2008 onwards. For detailed age categories, Q2 of each year.

Sources: OFCOM Research and US National Health Interview Survey.

StatLink http://dx.doi.org/10.1787/888932800444

The mobile-only population in France declined slightly during the same period. This trend is more evident for younger cohorts, more sensitive to the mobile-only dimension. The need in some locations to use a fixed line to connect to Internet via xDSL is one explanatory factor, and the popularity of "bundled services" (i.e. including telephony, video and Internet access) has undoubtedly contributed to these results. In 2011, for example, around 58% of individuals in France were equipped with a "multiservice box" enabling telephony over the Internet, against only 7% in 2005. In addition, quadruple offers bundled

Figure 8.15. **The mobile-only population in Czech Republic and France, 1999-2012**

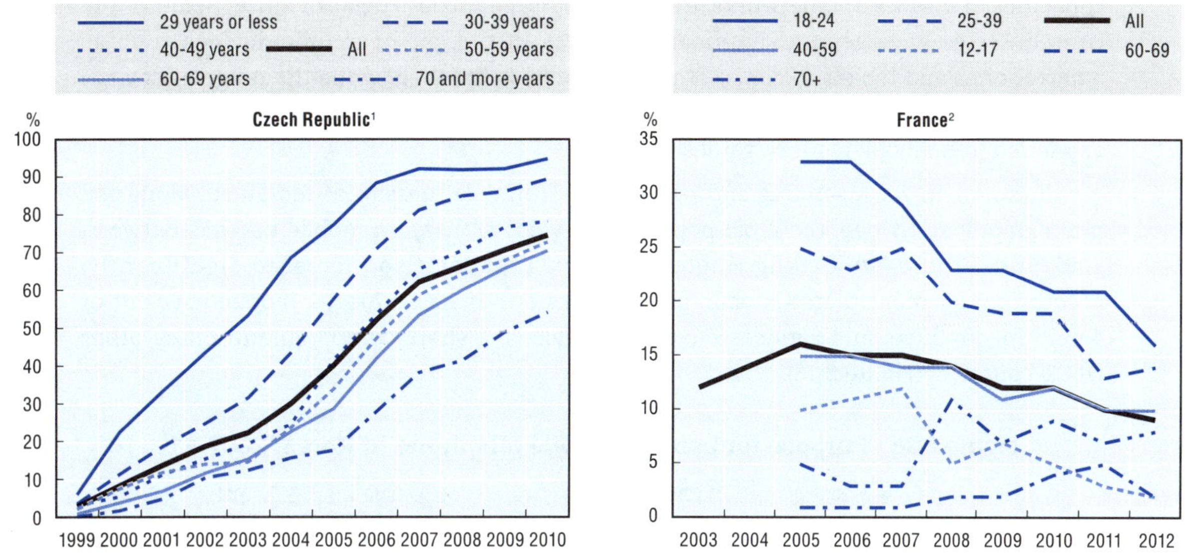

1. Percentage of households, by age of the household's head.
2. Percentage of individuals, by age.
Source: Czech Statistical Office (2012) and CREDOC (2012).

StatLink http://dx.doi.org/10.1787/888932800463

with fixed services were first launched in 2010 by fixed and mobile operators. In 2011, among the population equipped both with home Internet access and a mobile phone, 43% of individuals declared that their ISP was also providing mobile phone services. In 2012, this share reached 54% (CREDOC, 2012). This percentage is expected to continue to increase, given that at least one fixed broadband operator now includes a mobile telephone line at no additional charge to fixed broadband subscribers, together with substantial discounts for subscriptions that include bundled mobile data and additional subscriptions.

Mobile telephone activities: From voice to multimedia

The variety of activities on mobile devices has increased with the evolution of both terminals and networks. Smartphones and the tariff plans that support them offer many options for triple-play service (voice, data and video), while capabilities such as Bluetooth, Wi-Fi and Near Field Communications (NFC) add options for connectivity and data exchange.

Surveys reveal that traditional uses for mobile telephones, including SMS, downloading ringtones, games or music, are still widely used and are in some cases increasing. In many countries, texting or SMS (now becoming MMS) remain one of the main activities on mobile telephones, generally followed by photography or video. Technological improvements in the quality of photographs and video, together with the ability to share, store and display output across "home networks" and the Internet via broadband connectivity, have all contributed to this development. The use of Internet-connected mobile devices is increasing together with the associated features, which allow direct connection to social networks or new applications. Finally, but not exhaustively, GPS is increasingly incorporated into mobile devices and can be used by applications from maps to simply recording where a photograph was taken.

The following figures illustrate the recent evolution of mobile activities in some OECD countries (Figures 8.16 to 8.20). The increasing use of mobile telephones for GPS-based applications, video calls, and watching video and television, mirrors the increasing capacities of devices and networks. In recent years, countries such as France, Spain or the United States have seen a significant increase in the use of mobile devices such as smartphones and tablets to consult emails or browse the Web, while the range of devices that can access the Internet and perform other forms of communication has similarly expanded (e.g. e-book readers, games consoles).

In Korea, in 2011, sending and receiving multimedia messages[2] remained, alongside photography and video, the most popular non-voice activity on mobile devices. Between 2007 and 2010, downloading ringtones and wallpaper also increased significantly, accounting for more than 8 mobile phone users out of 10. In 2011, 1 mobile user out of 4 paid for products and services on mobile devices, a share comparable to many other OECD countries (Figure 8.16).

Figure 8.16. **Purpose for using a mobile telephone in Korea, 2007-11**

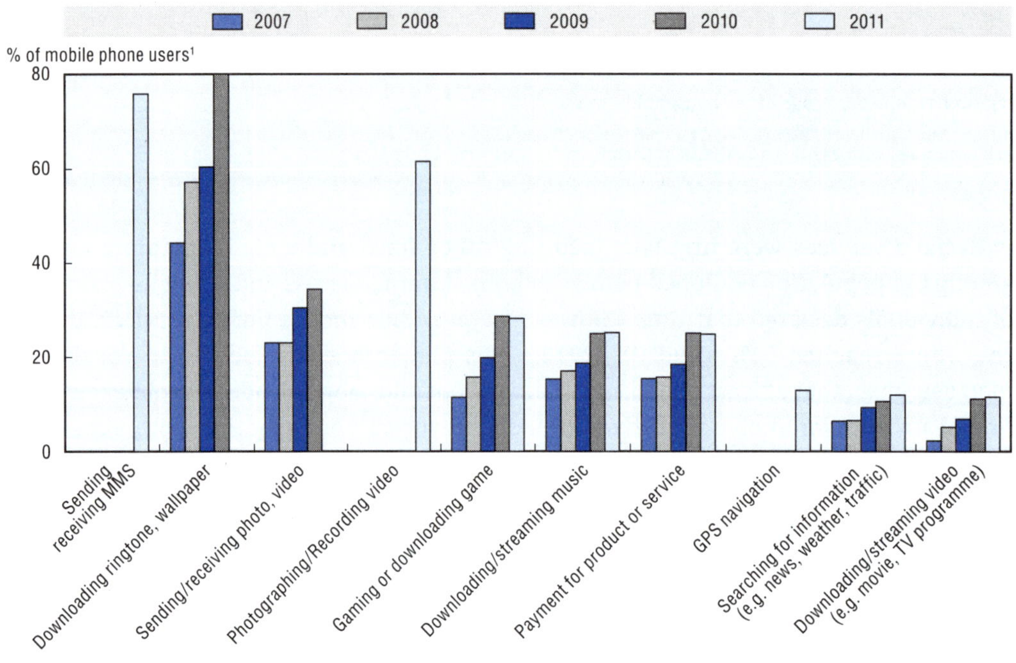

1. Aged 3 and over.
Source: KISA, *Survey on the Internet Usage*, 2008-11 issues.

StatLink http://dx.doi.org/10.1787/888932800482

In the United States all mobile telephone device activities increased among telephone users between 2007 and 2011. Taking photographs and sending/receiving text messages remained the most popular activities. Accessing the Internet and sending/receiving emails have gained the highest share of mobile users over the five-year period. Downloading apps and recording video also became significant activities (Figure 8.17).

In Spain, as in other countries, use of mobile telephones is led by telephony (calls made and received). In 2011, most mobile telephone users (around 90%) undertook this activity, and more than 80% did so daily or weekly (as opposed to monthly or sporadically). The percentage of mobile users receiving or sending SMS increased from 75% to 85% in seven years, around 60% of which undertook this activity daily or weekly (Figure 8.18).

Figure 8.17. **Mobile phone activities in the United States, 2007-12**

% of cell phone or PDA owners

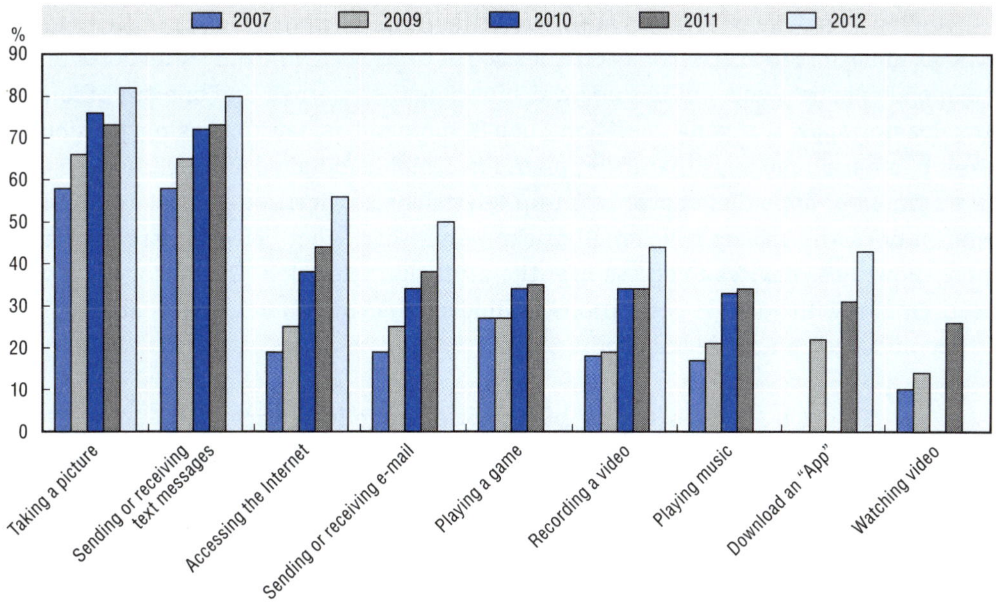

Source: Compiled from PEW Internet Research, various issues.

StatLink http://dx.doi.org/10.1787/888932800501

Figure 8.18. **Main use of mobile telephony in Spain, 2005-11**

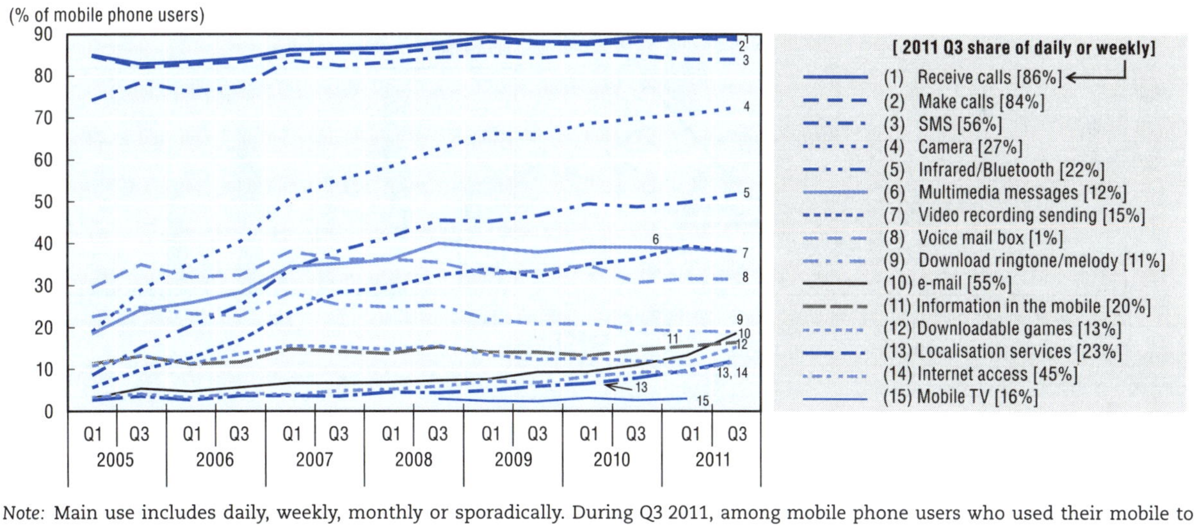

Note: Main use includes daily, weekly, monthly or sporadically. During Q3 2011, among mobile phone users who used their mobile to make calls, 86% did so daily or weekly.

Source: ICTs in Households Survey Panel, 2005-2011 issues, ONTSI (Spanish Observatory for Telecommunications and the Information Society).

StatLink http://dx.doi.org/10.1787/888932800520

In Spain, several mobile activities have increased dramatically since 2005. More than 7 out of 10 mobile users now use their mobile devices as a camera, against only 1 out of 5 at the beginning of 2005. More than one out of five does so daily or weekly. Mobile telephones are also increasingly used within an ICT environment. More than half of mobile users now use infrared bluetooth capabilities, against less than 1 out of 10 in 2005. Sending video recordings or multimedia messages has also spread significantly among mobile users.

After 2009, Internet access on mobiles started to increase, accompanied by emails or localization services. Mobile television remained little used, however.

In France, mobile telephones are used (aside from telephone calls) primarily to send SMS. As in Spain, Internet-related activities began to grow significantly after 2009. In 2012, one mobile phone owner out of three was browsing the Internet, and one out of four was downloading apps. Watching television, though marginal, is now increasing, possibly as a result of smartphone apps that incorporate fixed/mobile bundled services, as well as the development of apps that enable catch-up televisions services. In addition, a number of smartphones and tablets have small attachments that, when used in association with apps, allow these devices to access over-the-air digital television. Overall, age is a strong discriminator with younger generations leading significantly in Internet-related activities with a mobile telephone device (Figure 8.19).

Figure 8.19. **Selected mobile phone usage in France, 2003-11**

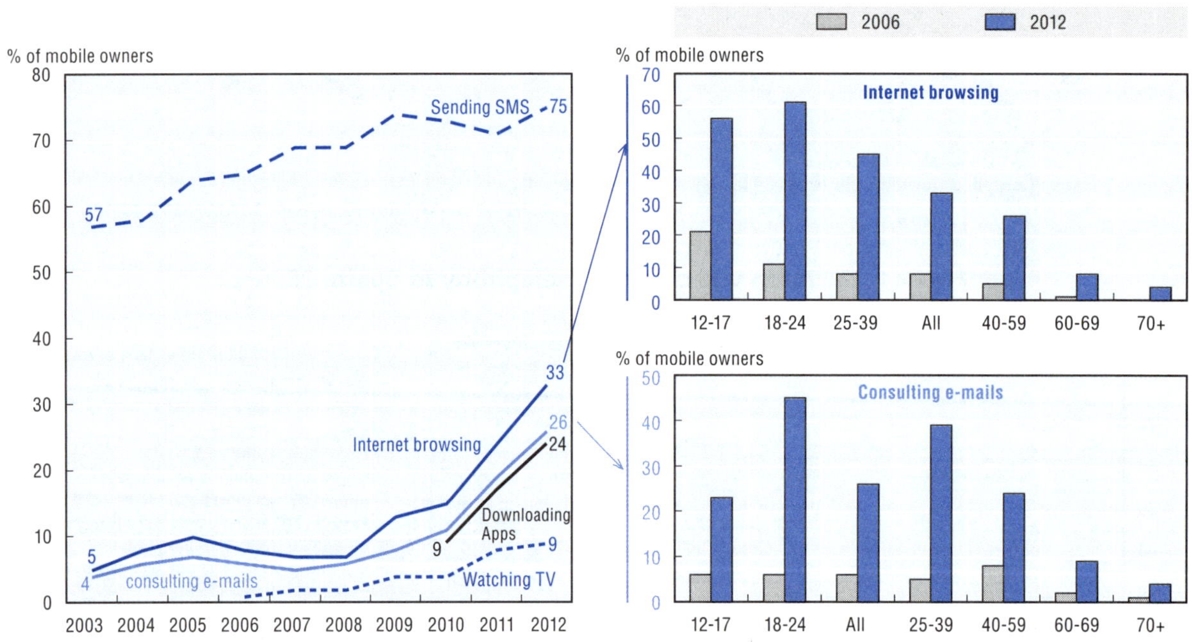

Source: OECD, compiled from CREDOC, various issues.

StatLink http://dx.doi.org/10.1787/888932800539

Australia shows similar patterns for non-voice activities undertaken by mobile telephone users. SMS comes first followed by photos or videos. Bluetooth applications are already used by 40% of users. More than one quarter of mobile device users now use GPS navigation, and video calls are starting to emerge (Figure 8.20).

SMS

In the United Kingdom, OFCOM has noted the increasing use of non-voice forms of communication in recent years, a major trend valid in many other countries (Figures 8.16 to 8.20). The most prevalent form is mobile messaging. As with most mobile-related activities age is a primary factor influencing uptake, and a significant age gap remains even after a decade of usage. In France as in other countries SMS was adopted first by younger people accounting for a high proportion of users by 2003. Older generations progressively

Figure 8.20. **Mobile telephone activities in Australia, April 2011**[1]

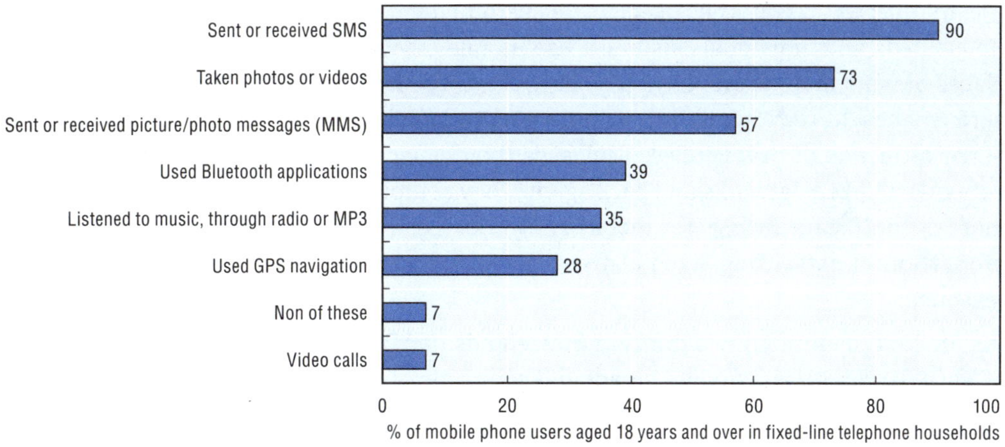

1. In the six months to April 2011.
Source: ACMA (2011).

StatLink http://dx.doi.org/10.1787/888932800558

embraced SMS though with less enthusiasm proportionate to age (Figure 8.21). The availability of packages that bundled unlimited SMS encouraged this trend, especially offers with a charge for voice. An increasing number of offers in France now bundle voice, data and SMS, leaving little to discriminate between services in terms of price incentive, an aspect younger generations are often more sensitive to than older generations. For the future, the ability of smartphones to voice-record SMS will blur these distinctions and it may become less clear whether texts are sent via SMS or from within an application. In the future, texts will be recorded as data rather than as a specific service, especially if incentives exist to send them via SMS.

Figure 8.21. **Mobile owners sending SMS and weekly numbers sent by age in France, 2003-12**[1]

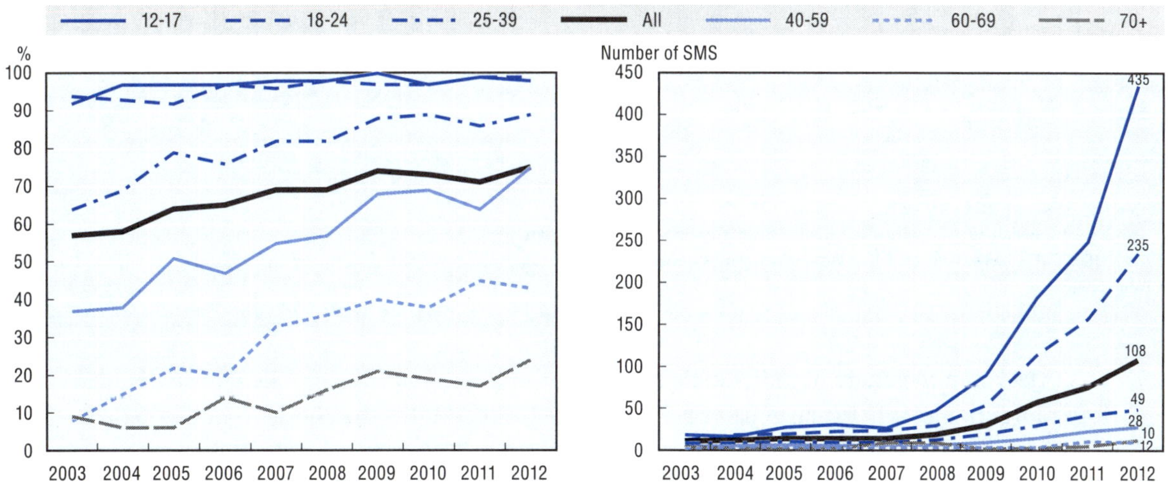

1. June of each year.
Source: OECD, compiled from CREDOC, various issues.

StatLink http://dx.doi.org/10.1787/888932800577

Mobile telephones: An entry door to the information society

In most OECD countries users are accustomed to using fixed broadband connections. Households and individuals also now use mobile devices to access the Internet at home, places of work and leisure, as well as between these locations. The deployment of higher performance 3G/UMTS and, increasingly, 4G/LTE mobile networks support this trend. Part of the attraction of new wireless devices is their capacity to shift seamlessly from mobile networks to fixed networks and back again. They complement fixed networks and perform most of the "heavy lifting" for mobile networks (i.e. mobile users download the greatest proportion of data using Wi-Fi at fixed network locations for financial and bandwidth reasons).

In 2011, more than two thirds of households had broadband Internet access in 20 out of 34 OECD countries. Wireless broadband connections are spreading significantly in some countries. In Austria and the Nordic European countries (except Denmark), more than one third of households have mobile high-speed access to the Internet, and more than one quarter in Portugal (Figure 8.22).

Figure 8.22. **Households with broadband access, 2011 or latest available year**[1]

% of all households

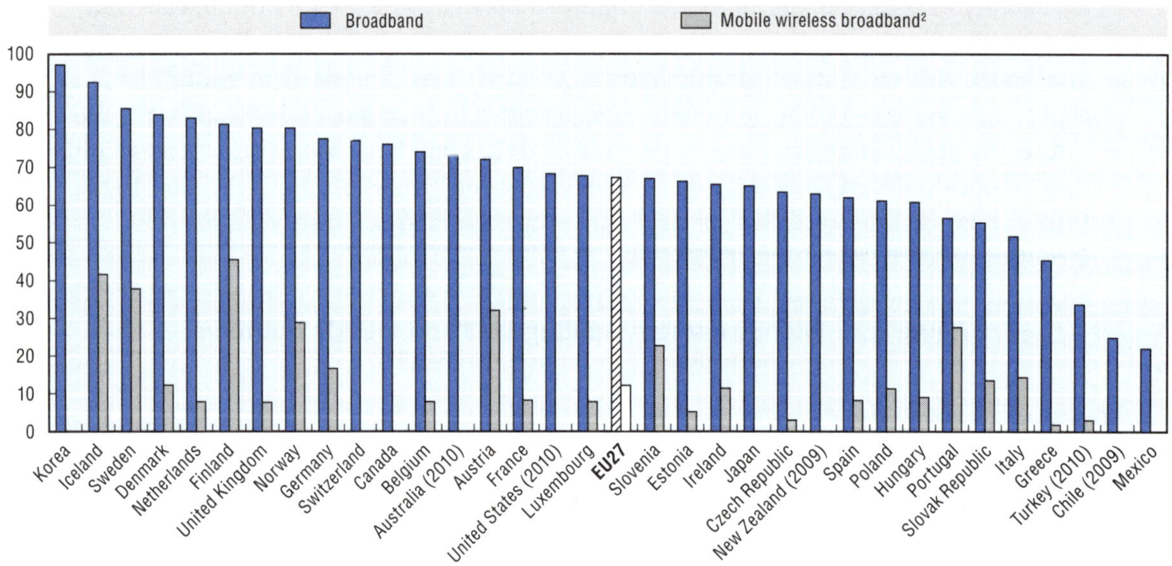

1. See the notes of Table 8.2.
2. Household uses a mobile wireless broadband connection (3G/UMTS, etc.).

Source: OECD, ICT Database and Eurostat, Community Survey on ICT usage in households and by individuals, and Canadian CRTC, October 2012.

StatLink http://dx.doi.org/10.1787/888932800596

In Japan the share of individuals accessing the Internet with a mobile device doubled to more than 80% between 2002 and 2011. In the United States the share of mobile cells in which users access the Internet reached 55% in April 2012, against only 31% three years before (Figure 8.23). In Canada one third (33%) of Internet users went online with a wireless handheld device in 2010. In Europe an increasing number of people now access the Internet with their mobile device – more than one third of individuals in seven countries with greatest use found in the Netherlands and the United Kingdom (Figure 8.24).

Figure 8.23. **Individuals accessing the Internet via mobile devices in Japan and United States**

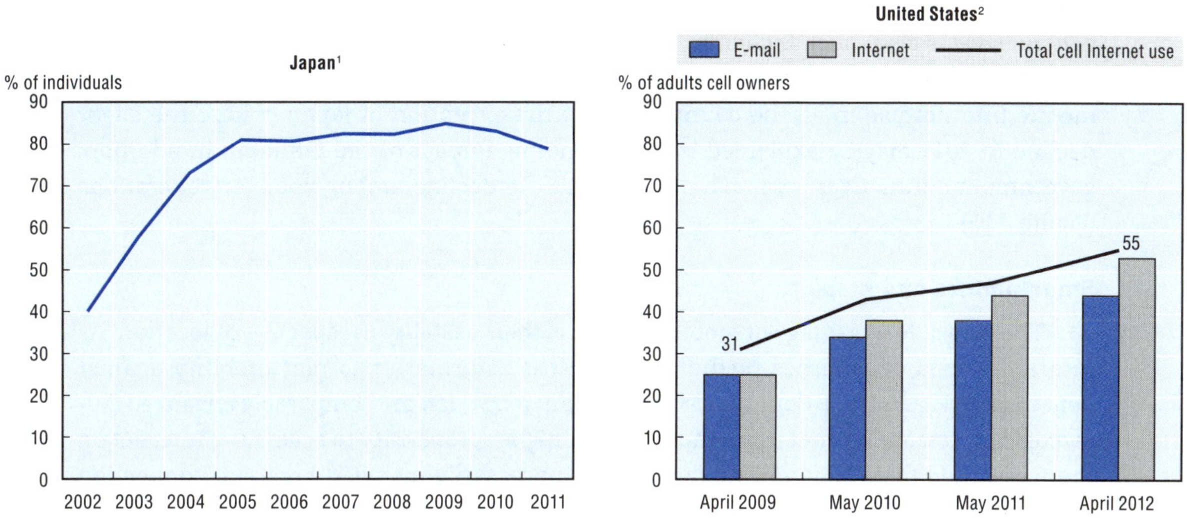

1. Via mobile devices.
2. Via mobile phone.

Source: Based on MIC (2011a and b) and PEW (2012).

StatLink http://dx.doi.org/10.1787/888932800615

Figure 8.24. **Individuals using their mobile phone (or smartphone) to access the Internet[1] in selected OECD countries, 2011**

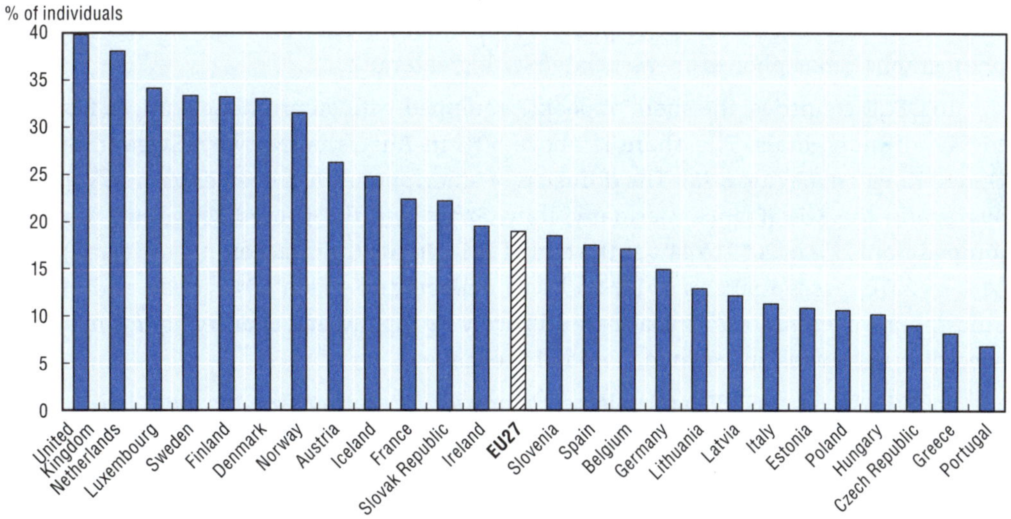

1. Away from home or work.

Source: EUROSTAT.

StatLink http://dx.doi.org/10.1787/888932800634

As might be expected, younger generations are making greater use of mobile Internet access. The European Union's Kids Online survey reported that by 2010 an average of 22% of children (aged 9-16) using the Internet in EU25 accessed it via mobile telephones (in addition to fixed networks). The share for people using the Internet in this age group was 66% in Greece, 37% in Germany and 33% in the Czech Republic and the United Kingdom. In 2010, in Canada, Internet users going online with a wireless handheld device tended to be younger and more experienced Internet users. The majority (59%) were under the age of 35.

When looking at selected OECD countries, it is clear that despite the varying level of mobile Internet use among countries, the generation gap is a global phenomenon. The general level is also much higher in Japan and Korea as compared to European countries, and gaps across European countries are much smaller. Recent trends shows increasing mobile Internet use in all the countries, with the exception of Japan, where the slight decline in 2011 may be explained by the economic impact of the tsunami. In addition, governmental initiatives[3] may have influenced the decline among younger generations (Figure 8.25).

Smartphones and mobility

The principle advantage of mobile devices is their usability irrespective of location. As Internet access capabilities further develop it is informative to plot mobility against penetration (Figure 8.26). Countries such as Japan or Sweden are comparably advanced, but Swedish or Korean Internet users show a much higher propensity to access the Internet on the move. Additional influencing factors, not necessarily included here, are connection prices and quality of the network (in terms of speed and coverage). For example, in Sweden more than 80% of people accessing Internet via a mobile phone used a 3G+ or 4G network in 2011. Japan and Sweden were among the first countries to introduce 4G networks, which were accompanied in both countries by new tariff plans and higher speeds.

In France, Japan and the United States usage of smartphones and feature phones differs (Figure 8.27), particularly for Internet-related activities, such as Internet access, web browsing, reading/sending email, downloading apps, using social media or viewing online videos. For established activities pioneered on feature phones, such as SMS or taking photographs, smartphones are associated with greater use.

In OECD countries, the share of people equipped with smartphones ranges from 20% to 54% of individuals. The share is above 50% in Australia, Norway, Sweden and the United Kingdom (Figure 8.28). The diffusion of smartphones has been extremely rapid. On average in the EU5 (France, Germany, Italy, Spain and the United Kingdom) the share jumped from 22% in 2009 to 44% at the end of 2011. In Australia (according to Telstra) it was 31% in 2010, passed 50% in 2011, and was expected to reach 60% at the end of 2012. Smartphone penetration and usage are still evolving rapidly, and country figures may differ according to sources and related period of the survey.

In 2012, in most OECD and selected countries, operating systems were led by Apple (iOS) and Google (Android). In some countries, a third operating system holds a significant or even the largest market share. Symbian (Nokia) has a share in Finland (43%), China (32%), Italy (25%) and Spain (21%). Meanwhile, Blackberry (RIM), has markets shares in Mexico (25%) and Canada and the United Kingdom (23%).

People are spending more time using their mobile devices. This is unsurprising given the increased capabilities of smartphones. In the United Kingdom, Ofcom noted an increase in the average monthly time devoted to mobile Internet from 1.7 to 2.1 hours between 2010 and 2011. When smartphones users are asked where they use their devices, the most frequent answer is home followed by mobility ("on the go"). Smartphones are also frequently used at work and while shopping in stores.

In 2012, Google analysed the ways in which smartphone users most connect to the Internet and the frequency of connections, specifically on computers or smartphones (Figure 8.29). In all selected countries (except the United Kingdom), smartphone users still

Figure 8.25. **Individuals using mobile telephone to access the Internet by age in recent years, selected OECD countries**

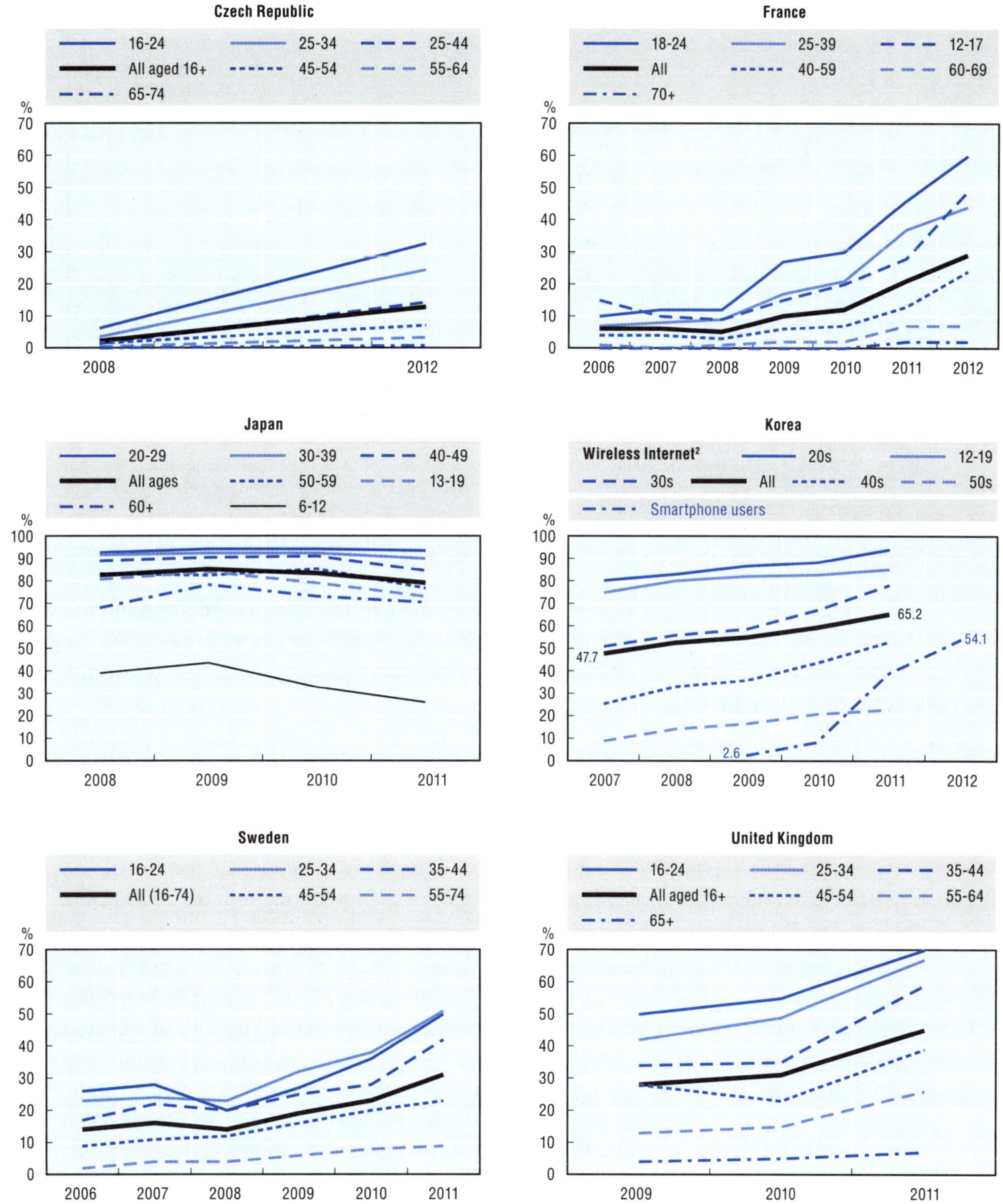

1. Mobile telephone or similar handheld device (excluding tablet computer).
2. Wireless Internet: Internet services such as mobile network (2G/3G), Wireless LAN (WiFi), WiBro, etc., using Wireless devices such as mobile phone (incl. smartphone) and notebook computer, etc. Age of individuals varies across countries: 12+ in France, 16-74 in Sweden, 6+ in Japan, 12-59 in Korea, and 16+ in the Czech Republic and the United Kingdom.

Source: OECD, compiled from Czech Statistical Office (2012), CREDOC (2011), MIC (2012b), KISA (2012a), Statistics Sweden (2012) and OFCOM (2012 and 2013).

StatLink http://dx.doi.org/10.1787/888932800653

Figure 8.26. **Internet mobility, smartphones and tablet computers in selected OECD countries, March 2012**

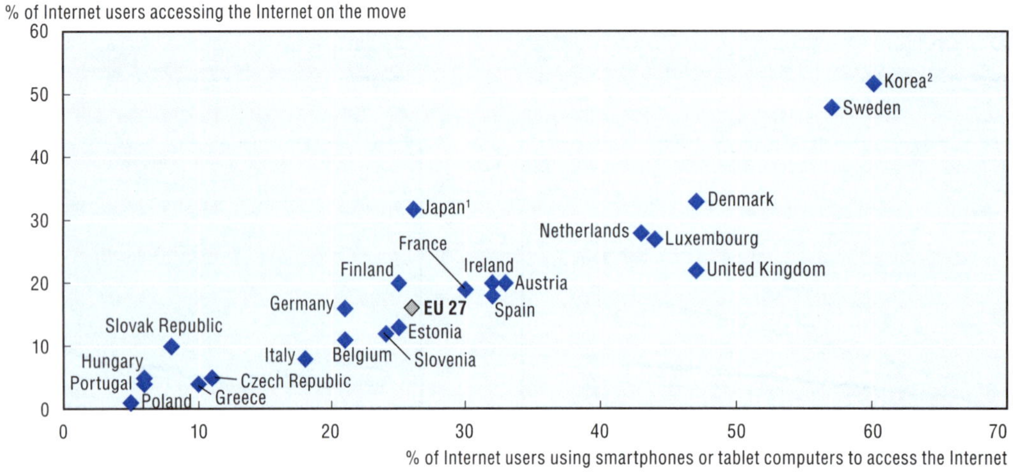

1. End of 2011 for Japan. In this country, accessing the Internet on the move refers to the percentage of Internet users who access the Internet outside home, and where it is neither office nor school.
2. 2011 for Korea. "on the move" refers to" irrespective of the location". "% of Internet users using smartphones or tablet computers to access the Internet" refers to wireless Internet users accessing wireless Internet by smartphones, tablet computers excluded.

Source: Based on data from Eurobarometer (2012b), KISA (2012a and b), and MIC (2012b).

StatLink http://dx.doi.org/10.1787/888932800672

opted to use a computer for frequent daily connections to the Internet. Smartphone use comes second. Likewise, the group of smartphone users that have never connected to the Internet with their computer is smaller than those that have never connected to the Internet with their smartphone.

In many countries, above 60% of smartphones users are likely to use computers on a daily basis, a figure that reaches 80% in many countries. In most countries there is at least a 10-point higher propensity to connect using computers on a daily basis rather than a smartphone, except in Japan and the United Kingdom, where use is fairly equal (Figure 8.30).

Smartphones are fast becoming "tool boxes", not only because of their intrinsic functionalities and capabilities, but also as a result of the increasing number of available applications (Figure 8.31).

A survey undertaken by Google found that people increasingly perform tasks on other devices while using their smartphone, tablet or other device, for example, watching television at home or using computers at the office. In 2012, in a majority of selected countries, more than 70% of smartphone users were multi-tasking. The share was particularly high in Brazil, Ireland, Mexico and Norway, as well as in Egypt. In Italy, smartphone users seem to multi-task less frequently compared to the other countries. The most frequently performed task while using a smartphone is watching television or listening to music of any type. Use of the Internet on another device follows, generally in third position. In Brazil, a different pattern of usage is visible: the most frequent multi-tasking activity is listening to music, followed by use of the Internet on another device, and watching television follows. In all countries, except China, reading a book is the least likely activity to be multi-tasked. (Figure 8.32).

In relation to mobility, Google also found that smartphone users who use their devices daily are generally engaged in short online sessions, usually several per day, and less

Figure 8.27. **Differences between mobile phone and smartphone activities in France, Japan and the United States**

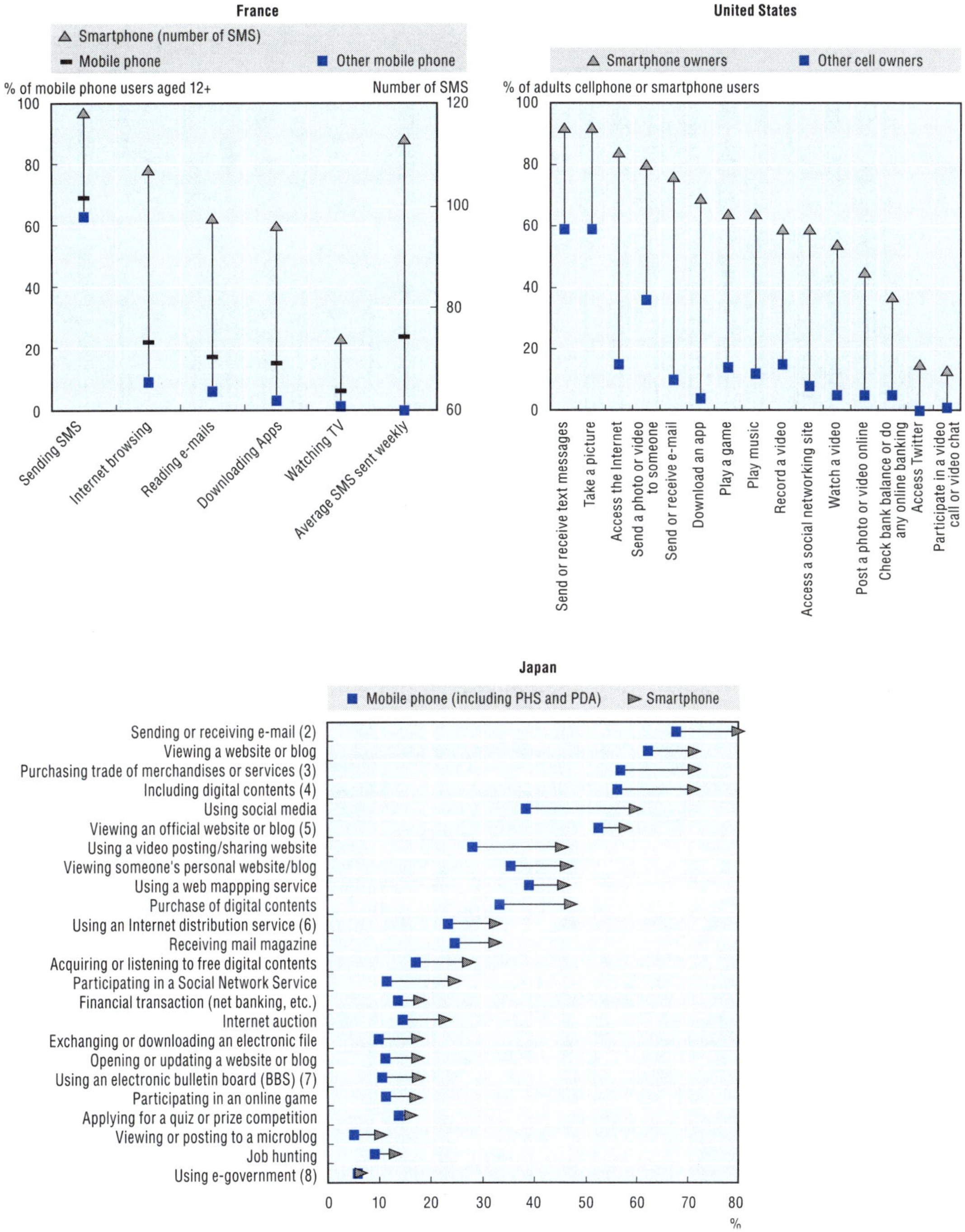

1. Purpose of Internet usage for selected devices.
2. Except magazines.
3. Including purchase of digital content; except financial trading.
4. Except financial transaction.
5. Company, government, public office website or blog (weblog).
6. Internet distribution service of radio, TV or video.
7. Or a Chat Room.
8. National or local government (e-application, e-submission, e-report).

Source: Based on CREDOC (2011), PEW (2011) and MIC (2012b).

StatLink http://dx.doi.org/10.1787/888932800691

Figure 8.28. **Smartphone and operating systems penetration in selected countries, 1Q 2012**

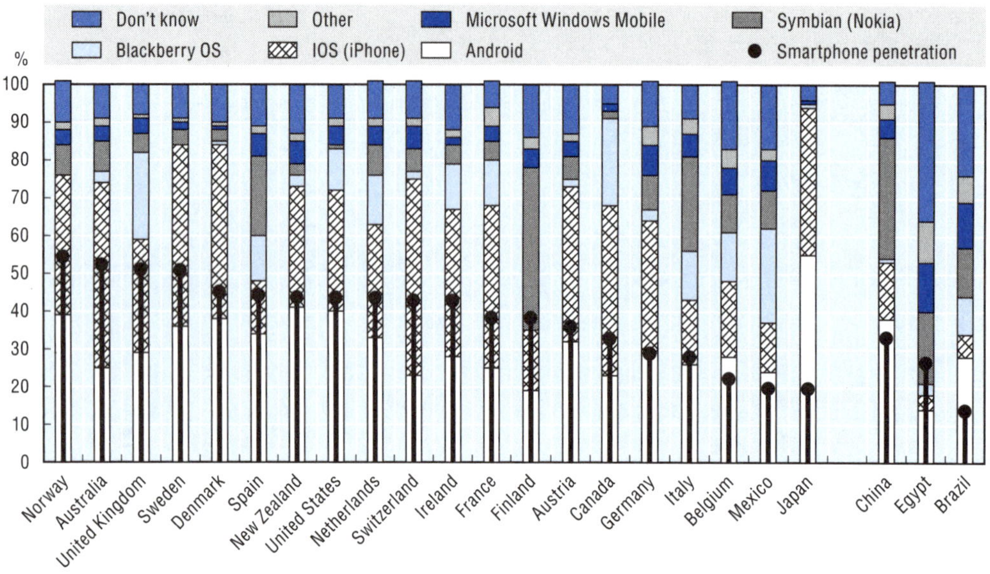

Note: Smartphone penetration as a percentage of all individuals. Operating system penetration as a percentage of all smartphone users. Smartphone users were asked "Which operating system do you have on your phone? With operating system we mean the software which is pre-installed on smartphones."
Source: Google (2012).

StatLink http://dx.doi.org/10.1787/888932800710

Figure 8.29. **Intra-day frequency of smartphone and computer online use in selected countries, 1Q 2012**

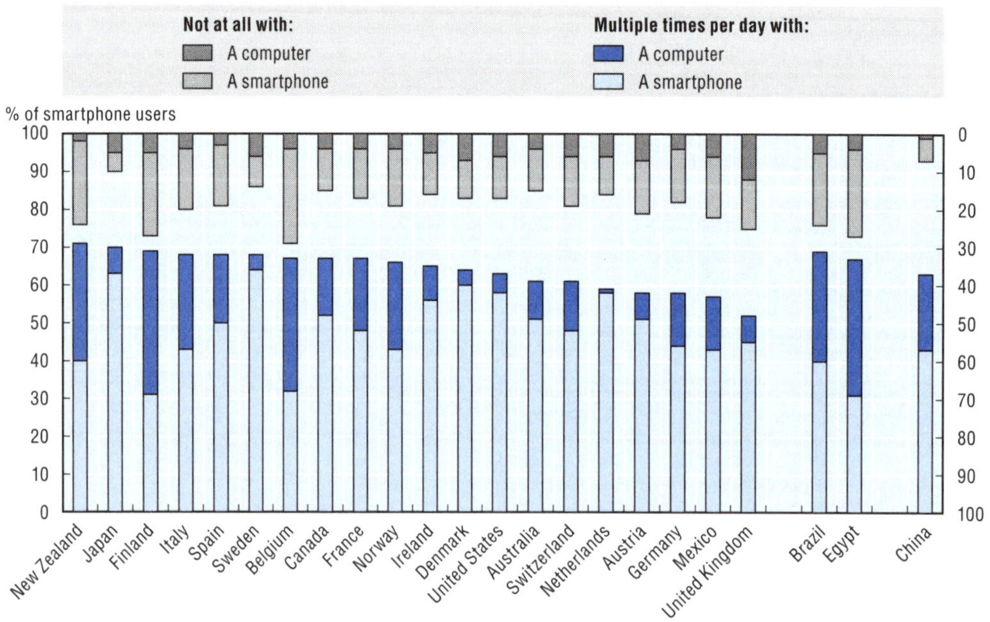

Note: The question raised was: "Now thinking about yesterday, how often were you online with your Smartphone?/ Computer?" Options proposed were: "Don't know", "Not at all"; "Once", "2-3 times a day" and "Multiple times a day".
Source: Google (2012).

StatLink http://dx.doi.org/10.1787/888932800729

Figure 8.30. **Daily Internet connection among smartphone users in selected countries, 1Q 2012**

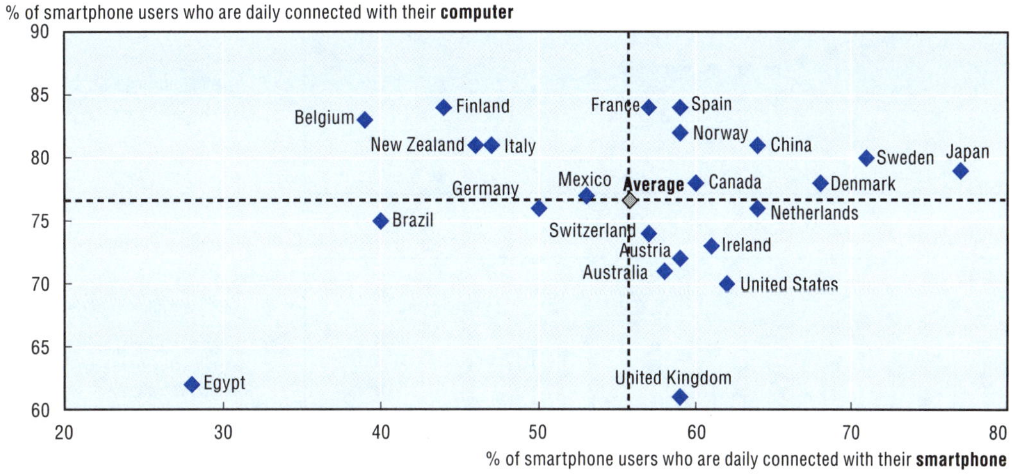

Note: The question raised was: "Thinking about the last seven days, on how many days were you online with your Smartphone?/Computer? (Please select one answer per device)." Options proposed were: "0 days"; "1 out of 7"; "2 out of 7"; . . . ;"7 out of 7". Those having selected the latter option are shown in the Figure.
Source: Google (2012).

StatLink http://dx.doi.org/10.1787/888932800748

Figure 8.31. **Number of applications, Apple and Android universes**

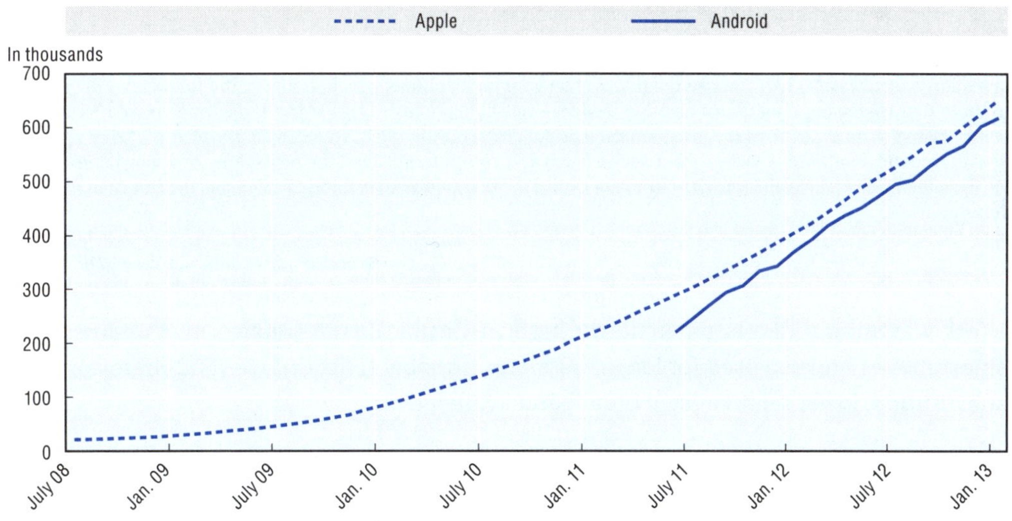

Source: Based on data from 148Apps.biz and Appbrain, February 2013.

StatLink http://dx.doi.org/10.1787/888932800767

frequently one session. In some countries, however (China, Egypt, Japan, Mexico, Spain), smartphones are also used for longer sessions, possibly suggesting substitution for computers (Figure 8.33).

Smartphone users are increasingly watching online videos via websites or apps, but computers seemingly remain the preferred device for this activity in all countries (Figure 8.34). However, this trend reflects a preference among users for watching video when connected to a fixed network including via Wi-Fi. In other words, the devices is not as significant as the available bandwidth and cost of access. The share of smartphone

Figure 8.32. **Smartphone use and multi-tasking in selected countries, 2012**

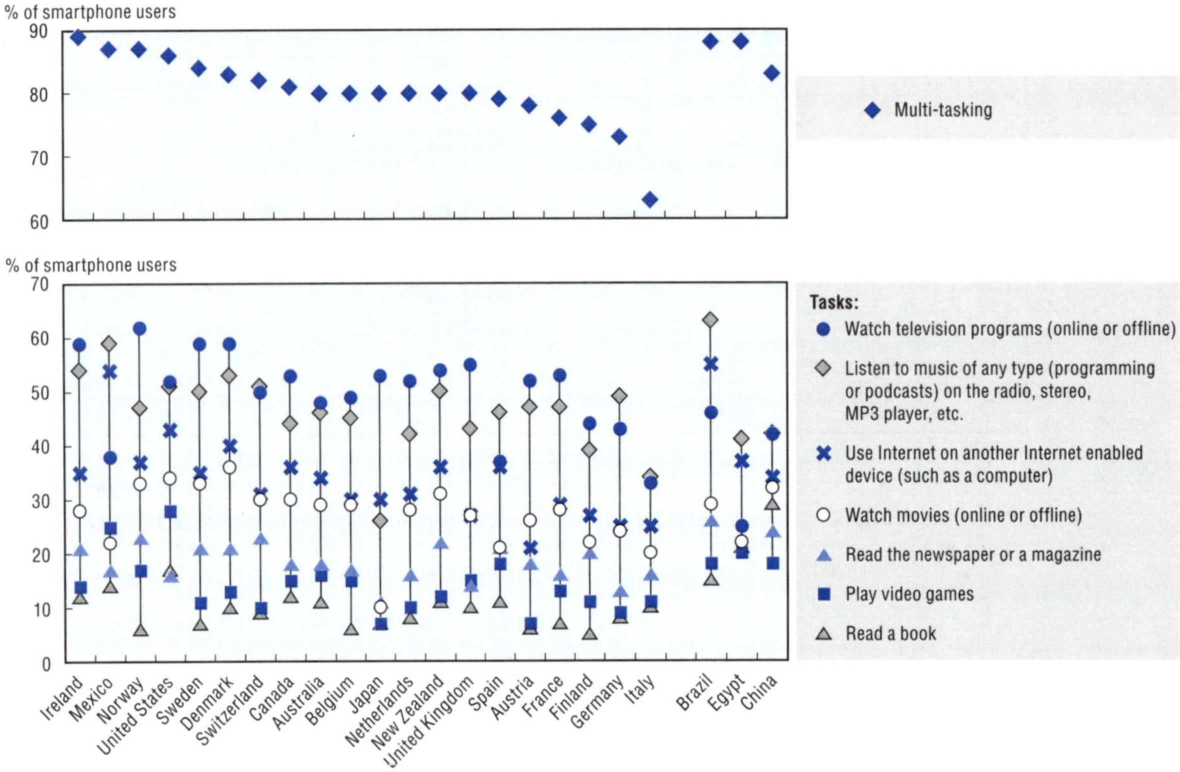

Note: The question raised was: "When you use the Internet on your smartphone, which, if any of the following – do you do at the same time? Please note that we mean doing things on other devices – not doing many things at the same time on your smartphone." Options proposed were: "Listen to music of any type (programming or podcasts) on the radio, stereo, MP3 player, etc."; "Play video games"; "Read a book"; "Use Internet on another Internet enabled device (such as a computer)"; "Read the newspaper or a magazine"; "Watch television programmes (online or offline)"; "Watch movies (online or offline)"; "Do not do other things while I use the Internet on my smartphone"; "None of these".

Source: Google (2012).

StatLink http://dx.doi.org/10.1787/888932800786

users watching video is particularly high in China, Egypt, Mexico and Spain, where smartphones are also used for longer sessions, possibly reflecting, beyond mere computer substitution, lower relative fixed broadband availability.

Applications and m-commerce

The primary use of smartphones in all countries, with the possible exception of the Netherlands, was found to be browsing the Web (Google, 2012), possibly because users can download alternative Web browsers as apps. Somewhat surprisingly, not all smartphone users make use of apps that do come bundled with their devices. Use of applications is widespread for around three quarters of all smartphone users in Denmark, Sweden and Switzerland, but only half in Finland, Italy, Mexico or Spain (Figure 8.35). Within the group of selected countries, smartphones contained an average of 25 apps of which 9 were used during the last 30 days and 10 were purchased in apps stores (Google, 2012).

M-commerce is an area where smartphones are expected to have increasing utility. A variety of applications allow users to undertake actions such as comparing prices or completing purchases. Mobile telecommunication companies, manufacturers of operating systems and producers of "over-the-top" applications are pursuing a range of approaches,

Figure 8.33. **Online sessions patterns of intensive smartphone users[1] in selected countries, 1Q 2012**

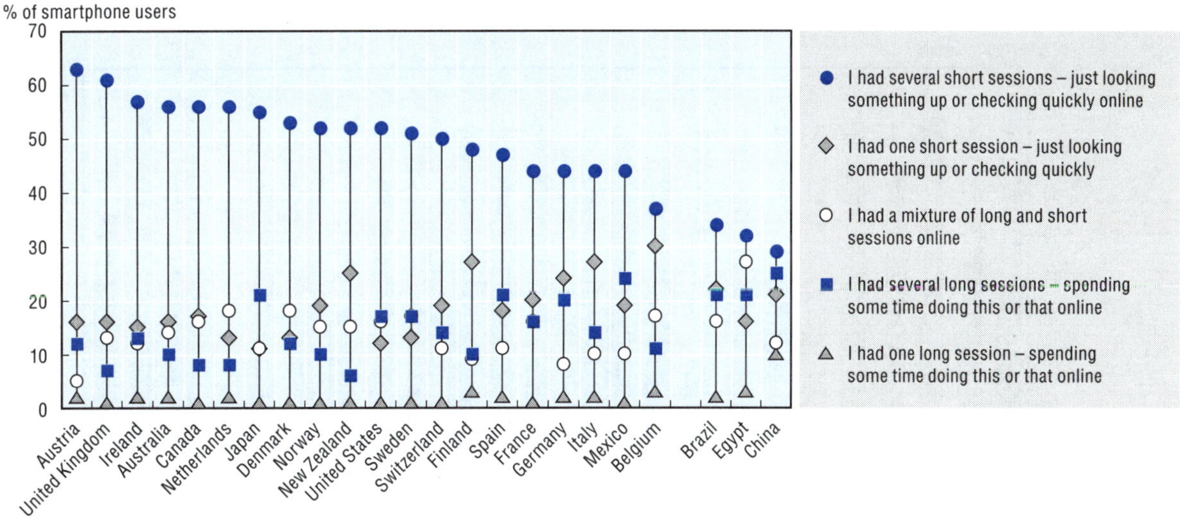

1. The question addressed to smartphone users was the following: (Among smartphone users who were online yesterday): "And how long were you online for each of these sessions yesterday?"

Source: Google (2012).

StatLink http://dx.doi.org/10.1787/888932800805

Figure 8.34. **Daily or weekly frequency of video watching via websites or apps in selected countries, 1Q 2012**

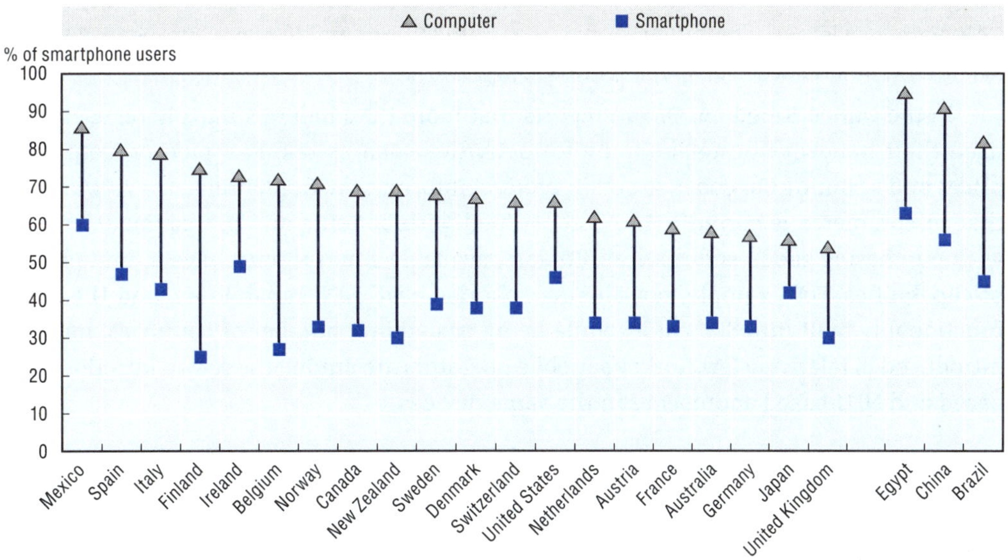

Source: Google (2012).

StatLink http://dx.doi.org/10.1787/888932800824

including trialing or incorporating functions into devices such as the use of SIM cards, or adding capabilities such as NFC.

These developments suggest that mobile devices may enhance the ability of point-of-purchase retailers to compete with the growing use of online shopping, including over mobile devices. Recent changes in eBay exemplify the growth in online use of mobile

Figure 8.35. **Internet browsing, apps and m-commerce with smartphones, 1Q 2012**

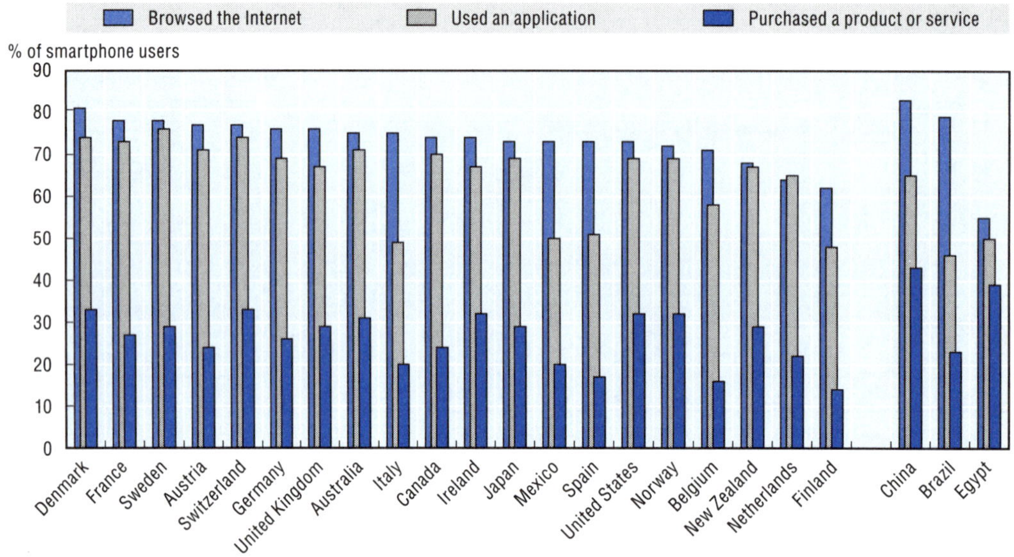

Source: Google (2012).

StatLink http://dx.doi.org/10.1787/888932800843

devices. eBay's m-commerce revenue went from USD 2 billion in 2010 to USD 5 billion in 2011. At this stage, m-commerce accounted for 7% of eBay's total revenue for that year. Google has found that smartphone usage for m-commerce is already significantly high in China and to a lesser extent in Australia, the United States and in some European countries (Denmark, Ireland and Norway) (Figure 8.35).

In the United Kingdom, Ofcom reports that more than half of smartphone users have used their handset during shopping for activities such as taking a picture, comparing prices and scanning bar codes to obtain product information. In Australia, the ACMA states that m-commerce is currently more concerned with banking and bill payment and less with purchasing goods and services. To date, available smartphones have a limited range of capabilities. For example, not all smartphones have NFC capabilities, and the use of functionality built into SIM cards, while being trialed in a number of countries, has been introduced in relatively few. Turkey's mobile operators are among the few to introduce SIM-based and NFC-based approaches in the same devices.

Recent consumer price trends in communication services

In Europe, the all-items harmonised index of consumer prices increased by 35% from 1999 to 2011. The indices for communications, instead, declined by 16%,[4] with a fall of 77.4% for the case of equipment and 20.2% for services[5] (Figure 8.36). Important scale economies, due to the explosion of volumes produced, is one of the main factors having led to significant price decline. Overall, telecommunication equipment and services tend to become more affordable for consumers. In the most recent period, between 2009 and 2011, communication equipment prices continued to fall strongly (17%), while communication services prices declined only slightly (1.6%).

During the same period in the United States, the general index for all items increased by 35%, while the communication index declined by 13.2% (Figure 8.37). As for Europe,

Figure 8.36. **Trend of harmonised indices of consumer prices (HICP) for communication, EU27**

Index 1999 = 100

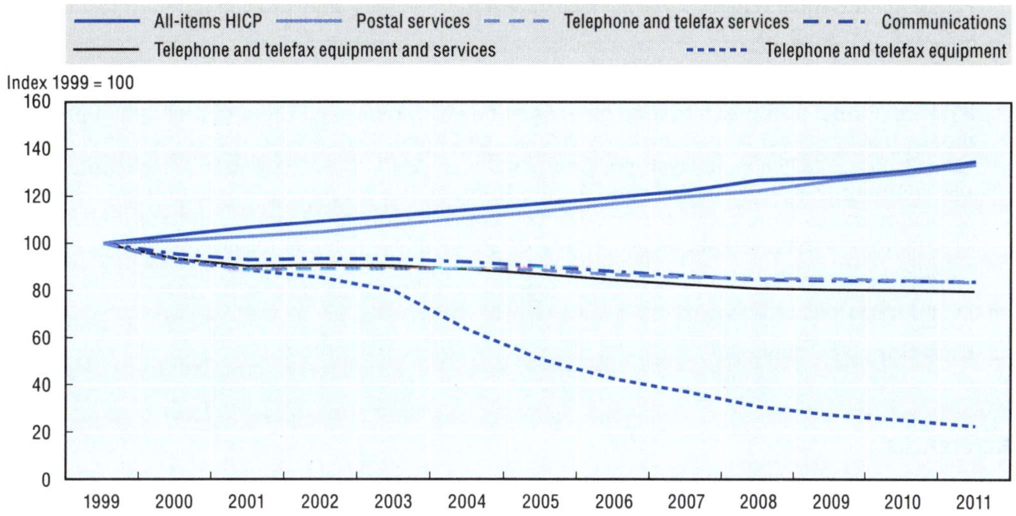

Note: Communications includes telephone and telefax equipment and services, telephone and telefax equipment and postal services.

Source: Eurostat.

StatLink http://dx.doi.org/10.1787/888932800862

Figure 8.37. **Trend of indices of consumer prices for communication, United States**

Index 1999 = 100

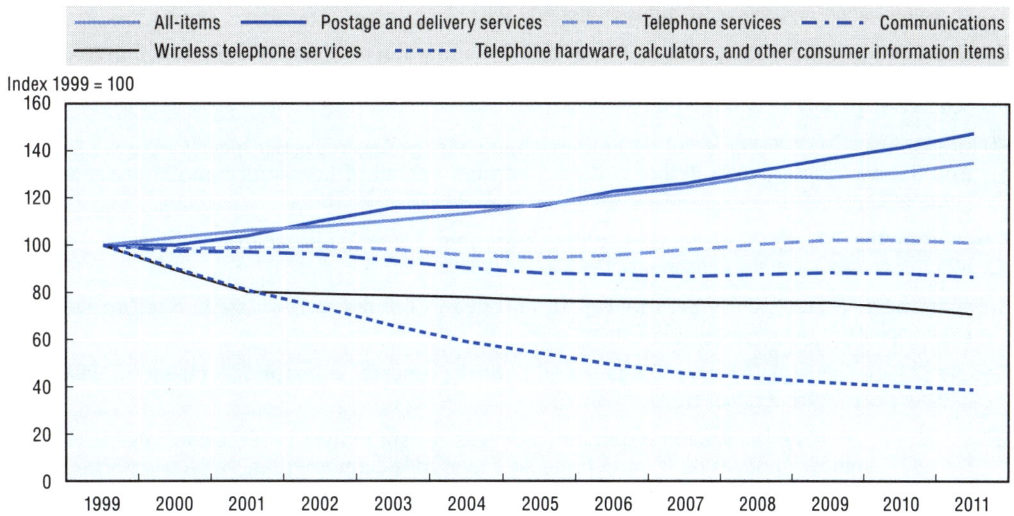

Notes: Communications includes: "Postage and delivery services" and "Information and information processing". Telephone services includes wireless telephone services and landline telephone services.

Source: US Bureau of Labor Statistics.

StatLink http://dx.doi.org/10.1787/888932800881

communication equipment prices diminished far more (61%) than those for communication services (13%). This latter decline is entirely due to wireless telephone services (29.1%), as telephone services have shown a slight increase (1.1%).

Notes

1. It should be noted that those shares have been calculated based on national accounts data using the COICOP classification (see Annex 1 for details). Using other national sources, such as the National Household Budget Surveys, to calculate those shares, may provide different results.
2. Text messages before 2011.
3. In February 2009, the Ministry of Education, Culture, Sports, Science and Technology (MEXT) issued a notice (formal document) to local governments which requested them to prohibit, in principle, mobile use by students at elementary schools and junior high schools (*www.mext.go.jp/b_menu/hakusho/nc/1234695.htm, in Japanese*). Additionally, in April 2009, the Act on Development of an Environment that Provides Safe and Secure Internet Use for Young People entered into force. Mobile operators were obliged to provide young people (under 18 years old) with their filtering service, unless their parents request not to do so, by this legislation (*www8.cao.go.jp/youth/youth-harm/law/pdf/english.pdf*).
4. Including postal services.
5. Excluding postal services.

References

ACMA (2011), "Converging Communication Channels: Preferences and Behaviours of Australian Communication Consumers", *2010-11 Communications Report Series*, Report 2, Australian Communications and Media Authority, *www.acma.gov.au/webwr/_assets/main/lib410148/report2-convergent_comms.pdf*.

CREDOC (2012), "La diffusion des technologies de l'information dans la société française", Centre de recherche pour l'étude et l'observation des conditions de vie, Paris, *www.arcep.fr/uploads/tx_gspublication/rapport-credoc-diffusion-tic-2012.pdf*.

CREDOC (2011), "La diffusion des technologies de l'information dans la société française", Centre de recherche pour l'étude et l'observation des conditions de vie, Paris, *www.arcep.fr/uploads/tx_gspublication/rapport-credoc-diffusion-tic-2011.pdf*.

Czech Statistical Office (2012), "Expenditures and Consumption of Households included in Household Budget Survey", 2012, Czech Statistical Office, Praha, *www.czso.cz/csu/2012edicniplan.nsf/engpubl/3001-12-eng_r_2012*.

EU Kids Online (2011), "EU Kids Online: Enhancing Knowledge Regarding European Children's Use, Risk And Safety Online", 2010, *Technical Report*, London School of Economics and Political Science, London, *www.esds.ac.uk/doc/6885%5Cmrdoc%5Cpdf%5C6885_reports.pdf*.

Eurobarometer (2012a), "E-Communications Household Survey", European Commission, Brussels, *http://ec.europa.eu/public_opinion/archives/ebs/ebs_381_en.pdf*.

Eurobarometer (2012b), *Cyber Security Report*, European Commission, Brussels, *http://ec.europa.eu/public_opinion/archives/ebs/ebs_390_en.pdf*.

Google (2012), *Our Mobile Planet*, dataset and country reports, smartphone research, *www.thinkwithgoogle.com/mobileplanet/en/downloads/*.

INSEE (2010), "En 2009, la consommation des ménages résiste malgré la récession", INSEE Première No. 1301, Institut national de la statistique de des études économiques, Paris, *www.insee.fr/fr/ffc/ipweb/ip1301/ip1301.pdf*.

INSEE (2012), "En 2011, la consommation des ménages marque le pas", INSEE Première No. 1402, Institut national de la statistique de des études économiques, Paris, *www.insee.fr/fr/ffc/ipweb/ip1402/ip1402.pdfhttp://www.insee.fr/fr/ffc/ipweb/ip1301/ip1301.pdf*.

ITU (2012), "Measuring the Information Society 2012", *Executive Summary*, International Telecommunication Union, Geneva, *www.itu.int/dms_pub/itu-d/opb/ind/D-IND-ICTOI-2012-SUM-PDF-E.pdf*.

KISA (2012a), "2011 Survey on the Internet Usage", *Executive Summary*, Korea Internet and Security Agency, Seoul, *http://isis.kisa.or.kr/eng/board/index.jsp?pageId=040100&bbsId=10&itemId=317&pageIndex=1*.

KISA (2012b), "2011 Survey on the Wireless Internet Usage", Korea Internet and Security Agency, Seoul.

MIC (2012a), *Information and Communications in Japan 2011*, Ministry of Internal Affairs and Communications, Tokyo.

MIC (2012b), *Communication Usage Trend Survey 2011* (in Japanese), Ministry of Information and Communications, Tokyo.

OECD (2011), *OECD Communications Outlook 2011*, OECD Publishing; doi: *10.1787/comms_outlook-2011-en*.

OFCOM (2012), *Communications Market Report 2012*, Research Document, Office of Communications, London, *http://stakeholders.ofcom.org.uk/market-data-research/market-data/communications-market-reports/*.

OFCOM (2013), Adults'media use and attitudes report, Research Document, Office of Communications, London, April 2013, *http://stakeholders.ofcom.org.uk/market-data-research/media-literacy-pubs/*.

PEW (2012), *Cell Phone Internet Use*, 26 June, PEW, Washington, DC, *http://pewinternet.org/~/media//Files/Reports/2012/PIP_Cell_Phone_Internet_Access.pdf*.

PEW (2011), *Americans and their cell phones*, 15 August, PEW, Washington, DC, *www.pewinternet.org/~/media//Files/Reports/2011/Cell%20Phones%202011.pdf*.

Statistics Canada (2011), "Individual Internet use and e-commerce, 2010", *The Daily*, 12 October 2011, *www.statcan.gc.ca/daily-quotidien/111012/dq111012-eng.pdf*.

US Bureau of Labor Statistics (2012a), "Consumer Spending in 2010, Focus on Prices and Spending", December, Vol. 2, No. 12, US Government, Washington, DC, *www.bls.gov/opub/focus/volume2_number12/cex_2_12.pdf*.

US Bureau of Labor Statistics (2012b), *Consumer Expenditure Survey 2011*, and previous years, US Government, Washington, DC, *www.bls.gov/opub/focus/volume2_number12/cex_2_12.pdf*.

World Bank (2012), *Information and Communications for Development 2012: Maximizing Mobile*, World Bank, Washington, DC, *www.worldbank.org/ict/IC4D2012*.

Table 8.1. **Communication expenditures[1] as a share of disposable income in OECD countries, 1990-2011**

	1990	1991	1992	1993	1994	1995	1996	1997	1998	1999	2000	2001	2002	2003	2004	2005	2006	2007	2008	2009	2010	2011
Australia	1.4	1.5	1.6	1.8	1.8	2.0	2.1	2.2	2.3	2.3	2.6	2.6	2.7	2.7	2.8	2.9	2.8	2.7	2.7	2.6	2.5	2.3
Austria	1.9	1.9	1.9	1.9	2.0	1.9	2.0	2.2	2.3	2.5	2.8	2.8	2.8	2.8	2.8	2.7	2.6	2.5	2.4	2.3	2.2	2.1
Belgium	..	..	..	..	..	1.6	1.7	1.9	2.0	2.2	2.3	2.5	2.7	2.9	2.9	2.9	2.6	2.5	2.3	2.3	2.2	2.1
Canada	1.8	1.8	1.8	1.8	1.8	1.8	1.9	2.0	2.1	2.0	2.1	2.2	2.3	2.3	2.3	2.4	2.4	2.3	2.3	2.4	2.4	2.5
Chile	..	..	..	..	..	..	..	..	..	..	..	..	..	..	..	..	..	..	4.0	4.0	4.1	4.1
Czech Republic	..	..	..	1.9	1.8	1.8	1.9	1.8	1.8	1.9	2.2	2.6	2.8	3.2	3.3	3.3	3.5	3.4	3.3	3.3	3.2	3.1
Denmark	1.6	1.6	1.8	1.8	1.8	1.8	1.7	1.9	1.8	1.9	2.0	2.0	1.9	2.1	2.1	2.0	2.0	1.9	1.9	1.9	1.7	1.7
Estonia	..	..	..	..	..	1.8	2.0	2.0	2.2	2.7	3.0	3.1	3.0	3.2	3.2	3.0	3.0	2.8	3.8	3.7	3.6	3.3
Finland	1.4	1.4	1.5	1.5	1.6	1.6	1.9	2.2	2.5	3.0	3.1	3.3	3.3	3.2	3.3	2.8	2.7	2.6	2.4	2.2	2.1	2.2
France	1.8	1.8	1.8	1.8	1.9	1.9	1.9	1.9	2.0	2.2	2.4	2.5	2.7	2.8	2.8	2.8	2.7	2.7	2.8	2.8	2.7	2.6
Germany	..	1.7	1.8	1.9	1.9	2.0	2.0	2.2	2.2	2.3	2.5	2.8	2.8	2.9	2.9	2.9	2.8	2.8	2.7	2.7	2.7	2.6
Greece	..	..	..	..	..	..	..	..	..	..	..	..	..	..	..	3.1	3.7	3.6	3.5	3.4	3.1	2.9
Hungary	..	..	..	..	..	1.8	2.4	2.7	3.2	3.7	3.9	4.2	4.5	4.2	4.0	4.2	4.0	3.7	3.8	3.6	3.8	3.7
Iceland	1.4	1.3	1.3	1.3	1.2	1.1	1.1	1.0	1.4	1.9	2.1	2.4	2.7	2.8	2.7	2.6	2.5	2.7	2.7	2.6	2.5	2.3
Ireland	..	..	..	..	..	1.8	2.0	2.0	1.9	2.0	2.2	2.4	2.6	2.9	3.3	3.4	3.5	3.3	3.2	3.4	3.1	3.0
Israel	..	..	..	..	..	2.9	3.0	3.2	3.4	4.0	4.0	4.1	4.5	4.4	4.6	4.4	4.3	4.8	4.3	4.4	4.1	4.0
Italy	1.6	1.7	1.7	1.7	1.7	1.8	1.9	2.1	2.2	2.4	2.5	2.6	2.7	2.7	2.7	2.7	2.6	2.5	2.5	2.5	2.5	2.4
Japan	..	..	..	..	..	..	..	..	..	..	..	2.5	2.5	2.6	2.6	2.6	2.6	2.7	2.7	2.9	3.0	3.1
Korea	1.5	1.5	1.6	1.7	1.9	2.2	2.6	3.2	4.0	4.8	5.4	5.6	5.6	5.3	5.3	5.1	4.8	4.6	4.5	4.4	4.3	4.3
Luxembourg	..	..	..	..	..	1.3	1.5	1.6	1.9	1.8	1.7	1.8	1.7	1.9	1.9	1.8	1.8	1.8	1.7	1.7	1.7	1.7
Mexico	2.2	2.6	2.9	3.1	3.3	3.2	3.1	3.0	2.9	3.1	3.1	3.3	3.3	3.3	3.6	3.8	4.1	4.4	4.2	4.6	4.5	4.2
Netherlands	1.9	1.9	1.9	2.1	2.1	2.2	2.6	2.8	3.2	3.5	3.9	4.2	4.5	4.7	4.6	4.6	4.7	4.5	4.3	4.2	4.1	4.1
New Zealand	2.3	2.5	2.5	2.5	2.4	2.6	2.6	2.9	2.8	2.7	2.7	2.7	2.7	2.7	2.7	2.7	2.8	2.9	3.1	3.1	3.1	3.1
Norway	2.1	1.9	1.6	1.8	1.7	1.8	1.9	1.9	2.1	2.5	2.6	2.7	2.8	2.8	3.2	3.2	3.1	2.9	2.8	2.8	2.7	2.6
Poland	..	..	..	..	..	2.2	2.1	1.6	2.1	2.6	2.7	2.9	3.0	3.1	3.1	3.4	3.3	3.2	3.2	3.0	3.1	2.9
Portugal	..	..	..	..	..	2.2	2.3	2.4	2.4	2.5	2.6	3.1	3.2	3.3	3.3	3.2	3.2	3.1	3.0	3.0	3.0	3.1
Slovak Republic	..	..	..	..	..	2.1	2.1	2.1	2.5	3.0	3.2	3.8	3.7	3.7	3.6	3.6	3.7	3.7	3.6	3.8	3.8	3.7
Slovenia	..	..	..	..	..	1.8	1.9	1.9	1.9	2.1	2.2	2.5	3.0	3.0	3.2	3.5	3.6	3.2	3.1	3.3	3.2	3.2
Spain	..	..	..	..	..	1.8	1.9	2.0	2.1	2.3	2.4	2.6	2.7	2.5	2.6	2.6	2.6	2.6	2.7	2.8	2.8	2.8
Sweden	..	..	..	1.9	2.0	2.2	2.5	2.7	3.0	3.2	3.1	3.4	3.6	3.7	3.7	3.6	3.4	3.3	3.3	3.3	3.3	3.3
Switzerland	1.8	1.9	1.9	2.0	2.0	2.0	2.0	2.0	2.0	2.0	2.0	2.2	2.3	2.4	2.5	2.6	2.6	2.5	2.4	2.5	2.4	2.4
Turkey	..	..	..	..	..	..	..	..	..	..	..	..	..	..	..	..	..	..	..	..	..	..
United Kingdom	1.9	1.9	1.9	2.0	2.0	2.0	1.9	1.9	2.0	2.1	2.2	2.2	2.2	2.2	2.2	2.2	2.1	2.1	2.0	2.0	2.1	2.2
United States	1.9	1.9	1.9	1.9	2.0	2.0	2.1	2.2	2.3	2.3	2.4	2.5	2.5	2.4	2.3	2.3	2.3	2.3	2.4	2.4	2.4	2.4
OECD 31[2]						2.2	2.2	2.3	2.3	2.5	2.6	2.7	2.7	2.7	2.7	2.7	2.7	2.7	2.7	2.7	2.7	2.7

Notes: 1. See detailed definition of communication expenditures in Annex 1. As indicated at the beginning of this chapter, these shares have been calculated based on national accounts data using the COICOP classification (see Annex 1 for details). Using other national sources, such as the National Households Budget Surveys, to calculate those shares may provide different results. 2. OECD 31: Chile, Greece and Turkey are not included. Japan is not included before year 2001.

Source: OECD, based on SNA database, March 2013.

StatLink http://dx.doi.org/10.1787/888932802857

Table 8.2. **Households with broadband access, 2000-11**

% of all households

	2000	2001	2002	2003	2004	2005	2006	2007	2008	2009	2010	2011
Australia (*)	..	..	..	..	16.3	28.3	43.0	52.0	62.0	..	72.6	..
Austria	..	..	..	10.3	15.9	23.1	33.1	46.1	54.5	57.8	63.7	72.0
Belgium	..	..	..	..	..	40.6	48.0	56.4	60.3	63.4	70.0	74.0
Canada (*)	..	21.6	29.3	35.5	44.1	50.1	57.9	64.2	66.9	72.2	72.7	..
Chile	..	..	..	6.4	..	..	15.1	..	..	24.9	..	..
Czech Republic (*)	..	..	..	1.5	4.5	5.1	16.6	28.1	36.4	48.9	53.6	63.4
Denmark	..	..	..	25.1	35.8	51.2	63.3	69.5	74.1	76.0	80.1	83.9
Estonia	..	..	..	..	20.3	29.8	36.6	47.6	54.4	62.0	64.5	66.2
Finland	..	..	..	12.4	21.3	36.1	52.9	62.9	66.1	73.7	75.8	81.3
France	..	..	..	..	..	..	30.3	48.9	57.1	62.7	66.4	70.0
Germany	..	..	..	9.3	18.0	23.2	33.5	49.6	54.9	64.6	75.2	77.5
Greece	..	..	..	0.6	0.2	0.6	3.8	7.5	22.5	33.1	41.2	45.4
Hungary	..	..	..	..	5.8	10.9	22.0	33.0	42.3	50.9	52.2	60.8
Iceland	..	..	..	..	45.4	63.5	72.1	76.1	83.2	86.7	87.0	92.4
Ireland	..	..	..	0.6	2.9	7.4	13.1	31.2	42.9	53.7	57.5	65.4
Israel (*)	19.8	22.5	25.4	30.8	40.7	48.9	54.6	59.3	61.8	66.3	68.1	..
Italy	..	..	..	..	..	12.9	16.2	25.3	30.8	39.0	48.9	51.7
Japan (*)	..	..	16.6	31.1	40.4	42.1	40.7	51.7	58.5	63.4	63.4	65.1
Korea (*)	30.3	56.4	68.0	66.0	85.7	90.8	94.0	94.1	94.3	95.9	96.8	97.2
Luxembourg (*)	..	..	..	7.4	16.3	33.4	44.1	57.8	61.0	71.1	70.3	67.7
Mexico (*)	..	0.3	0.4	..	1.8	2.3	4.1	6.1	9.6	13.8	21.1	22.1
Netherlands	..	..	..	20.0	31.3	53.9	66.2	73.8	74.0	77.0	79.5	82.9
New Zealand (*)	..	..	..	..	..	..	33.0	..	..	63.0	..	..
Norway (*)	..	..	..	22.9	30.0	41.4	57.1	66.7	73.0	77.8	82.6	80.4
Poland	..	..	..	..	8.3	15.6	21.6	29.6	37.9	51.1	56.8	61.1
Portugal	..	..	..	7.9	12.3	19.7	24.0	30.4	39.3	46.2	50.3	56.6
Slovak Republic	..	..	..	..	3.6	7.1	11.4	26.5	35.3	41.7	49.4	55.3
Slovenia	..	..	..	..	10.2	19.4	33.6	43.6	49.7	56.1	62.0	67.0
Spain	..	..	..	..	15.0	20.8	29.3	39.2	44.6	51.3	57.4	61.9
Sweden	..	..	..	..	..	40.2	51.0	66.6	70.7	79.5	82.6	85.5
Switzerland	..	..	..	..	..	..	52.8	63.0	70.8	73.9	77.0	..
Turkey	..	..	..	..	0.2	1.7	..	16.5	22.0	26.2	33.7	..
United Kingdom	..	..	..	10.7	15.8	31.5	43.9	56.7	61.5	69.5	..	80.4
United States	4.4	9.1	..	19.9	..	..	..	50.8	..	63.5	68.2	..
EU27	..	..	..	..	14.9	23.0	30.4	42.4	48.6	56.7	61.3	67.3

Notes: (*) See coutries' notes below. Generally, data from the EU Community Survey on household use of ICT, which covers EU countries plus Iceland, Norway and Turkey, relate to the first quarter of the reference year.

For Australia: data was based on a multi-staged area sample of private and non-private dwellings, and covers the civilian population only. Households in remote and sparsely settled parts of Australia are excluded from the survey. Data for 2000, 2002 and 2003 are based on a calendar year; data provided relate to the reference year. Data for 2001 and for 2004 onwards are based on a financial year; data provided relate to the second half of the reference year and the first half of the following year. Data for 2005-06 to 2010-11 include persons aged 15 years and over except for members of the permanent defence forces, certain diplomatic personnel of overseas governments customarily excluded from census and estimated population counts, overseas residents in Australia, and members of non-Australian defence forces (and their dependants) stationed in Australia. Data for 2000 to 2004-05 include persons aged 18 years and over except members of the permanent defence forces, certain diplomatic personnel of overseas governments customarily excluded from census and estimated population counts, overseas residents in Australia, and members of non-Australian defence forces (and their dependants) stationed in Australia.

For Canada: Estimates for 2001 and 2003 include the territories (Northwest Territories, Yukon Territory and Nunavut). All other estimates include data from the ten provinces only. Due to a change in survey methodology caution should be used in comparing the 2010 data with previous years.

For the Czech Republic, data relate to Q4 of the reference year.

For Israel: The Internet question does not distinguish between broadband and other bandwidth Internet channels.

For Japan: Households with Internet access via FTTx, ADSL, cable and fixed wireless broadband.

For Korea: For 2000 to 2003, data include broadband access modes such as xDSL, cable and other fixed and wireless broadband via computers. As of 2004, data also include mobile phone access.

For Luxembourg: For 2004, data include wireless access.

For Mexico: For 2001 and 2002, households with Internet access via cable. From 2004, households with Internet access via cable, ADSL or fixed wireless.

For New Zealand: The information is based on households in private occupied dwellings. Visitor-only dwellings, such as hotels, are excluded.

For Norway: For 2003, data include LAN (wireless or cable).

For Switzerland: data for 2004 to 2009 include all private households with at least one person aged 15 years and over; data for 2010 include all private households with at least one member aged 16 to 74 years old.

Source: OECD, ICT database and Eurostat, Community Survey on ICT usage in households and by individuals, February 2013.

StatLink http://dx.doi.org/10.1787/888932801280

ANNEX 8.A1

Definitions

Definition of communication expenditures

Definitions COICOP

08.1.0 Postal services:

- Payments for the delivery of letters, postcards and parcels.
- Private mail and parcel delivery.

Includes: All purchases of new postage stamps, pre-franked postcards and aerogrammes.

Excludes: Purchase of used or cancelled postage stamps (09.3.1); financial services of post offices (12.6.2).

08.2.0 Telephone and fax equipment:

- Purchases of telephones, radio-telephones, telefax machines, telephone-answering machines and telephone loudspeakers.
- Repair of such equipment.

Excludes: Telefax and telephone-answering facilities provided by personal computers (09.1.3).

08.3.0 Telephone and telefax services:

- Installation and subscription costs of personal telephone equipment.
- Telephone calls from a private line or from a public line (public telephone box, post office cabin, etc.); telephone calls from hotels, cafés, restaurants and similar.
- Telegraphy, telex and telefax services.
- Information transmission services; Internet connection services.
- Hire of telephones, telefax machines, telephone-answering machines and telephone loudspeakers.

Includes: Radio-telephony, radio-telegraphy and radiotelex services.

Excludes: Telefax and telephone answering facilities provided by personal computers (09.1.3).

Definition of audiovisual, photographic and information processing equipment

Definitions COICOP

09.1 Audiovisual, photographic and information-processing equipment.

09.1.1 Equipment for the reception, recording and reproduction of sound and pictures.

09.1.2 Photographic and cinematographic equipment and optical instruments.

09.1.3 Information-processing equipment.

09.1.4 Recording media.

09.1.5 Repair of audiovisual, photographic and information-processing equipment.

Annex Table A.1. **Average annual exchange rates**

In national currency units per USD

	1996	1997	1998	1999	2000	2001	2002	2003	2004	2005	2006	2007	2008	2009	2010	2011
Australia	1.28	1.35	1.59	1.55	1.72	1.93	1.84	1.54	1.36	1.31	1.33	1.20	1.19	1.28	1.09	0.97
Austria	0.77	0.89	0.90	0.94	1.09	1.12	1.06	0.89	0.81	0.80	0.80	0.73	0.68	0.72	0.76	0.72
Belgium	0.77	0.89	0.90	0.94	1.09	1.12	1.06	0.89	0.81	0.80	0.80	0.73	0.68	0.72	0.76	0.72
Canada	1.36	1.38	1.48	1.49	1.49	1.55	1.57	1.40	1.30	1.21	1.13	1.07	1.07	1.14	1.03	0.99
Chile	412	419	460	509	540	635	689	691	610	560	530	522	522	561	510	484
Czech Republic	27.14	31.70	32.28	34.57	38.60	38.04	32.74	28.21	25.70	23.96	22.60	20.29	17.07	19.06	19.1	17.70
Denmark	5.80	6.60	6.70	6.98	8.08	8.32	7.89	6.59	5.99	6.00	5.95	5.44	5.10	5.36	5.62	5.37
Estonia	0.77	0.89	0.90	0.94	1.08	1.12	1.06	0.89	0.81	0.80	0.80	0.73	0.68	0.72	0.75	0.72
Finland	0.77	0.87	0.90	0.94	1.09	1.12	1.06	0.89	0.81	0.80	0.80	0.73	0.68	0.72	0.76	0.72
France	0.78	0.89	0.90	0.94	1.09	1.12	1.06	0.89	0.81	0.80	0.80	0.73	0.68	0.72	0.76	0.72
Germany	0.77	0.89	0.90	0.94	1.09	1.12	1.06	0.89	0.81	0.80	0.80	0.73	0.68	0.72	0.76	0.72
Greece	0.71	0.80	0.87	0.90	1.07	1.12	1.06	0.89	0.81	0.80	0.80	0.73	0.68	0.72	0.76	0.72
Hungary	153	187	214	237	282	286	258	224	203	200	210	184	172	202	208	201
Iceland	67	71	71	72	79	97	92	77	70	63	70	64	88	124	122	116
Ireland	0.79	0.84	0.89	0.94	1.09	1.12	1.06	0.89	0.81	0.80	0.80	0.73	0.68	0.72	0.76	0.72
Israel	3.19	3.45	3.80	4.14	4.08	4.21	4.74	4.55	4.48	4.49	4.46	4.11	3.59	3.93	3.74	3.58
Italy	0.80	0.88	0.90	0.94	1.09	1.12	1.06	0.89	0.81	0.80	0.80	0.73	0.68	0.72	0.76	0.72
Japan	109	121	131	114	108	122	125	116	108	110	116	118	103	94	88	80
Korea	804	951	1401	1189	1131	1291	1251	1192	1145	1024	955	929	1 102	1 277	1 156	1 108
Luxembourg	0.77	0.89	0.90	0.94	1.09	1.12	1.06	0.89	0.81	0.80	0.80	0.73	0.68	0.72	0.76	0.72
Mexico	7.60	7.92	9.14	9.56	9.46	9.34	9.66	10.79	11.29	10.90	10.90	10.93	11.13	13.51	12.64	12.42
Netherlands	0.77	0.89	0.90	0.94	1.09	1.12	1.06	0.89	0.81	0.80	0.80	0.73	0.68	0.72	0.76	0.72
New Zealand	1.45	1.51	1.87	1.89	2.20	2.38	2.16	1.72	1.51	1.42	1.54	1.36	1.42	1.60	1.39	1.27
Norway	6.45	7.07	7.55	7.80	8.80	8.99	7.98	7.08	6.74	6.44	6.41	5.86	5.64	6.29	6.04	5.60
Poland	2.70	3.28	3.48	3.97	4.35	4.09	4.08	3.89	3.66	3.24	3.10	2.77	2.41	3.12	3.02	2.96
Portugal	0.77	0.87	0.90	0.94	1.09	1.12	1.06	0.89	0.81	0.80	0.80	0.73	0.68	0.72	0.76	0.72
Slovak Republic	1.02	1.12	1.17	1.37	1.53	1.61	1.50	1.22	1.07	1.03	0.99	0.82	0.71	0.72	0.76	0.72
Slovenia	0.56	0.67	0.69	0.76	0.93	1.01	1.00	0.86	0.80	0.80	0.80	0.73	0.68	0.72	0.76	0.72
Spain	0.76	0.88	0.90	0.94	1.09	1.12	1.06	0.89	0.81	0.80	0.80	0.73	0.68	0.72	0.76	0.72
Sweden	6.71	7.63	7.95	8.26	9.16	10.33	9.74	8.09	7.35	7.47	7.38	6.76	6.59	7.65	7.21	6.49
Switzerland	1.24	1.45	1.45	1.50	1.69	1.69	1.56	1.35	1.24	1.25	1.25	1.20	1.08	1.09	1.04	0.89
Turkey	0.08	0.15	0.26	0.42	0.63	1.23	1.51	1.50	1.43	1.34	1.43	1.30	1.30	1.55	1.50	1.67
United Kingdom	0.64	0.61	0.60	0.62	0.66	0.69	0.67	0.61	0.55	0.55	0.54	0.50	0.54	0.64	0.65	0.62
United States	1	1	1	1	1	1	1	1	1	1	1	1	1	1	1	1

Note: Data for EMU member countries are given in euros (EUR). Data for years prior to 1999 have been converted from national currencies into EUR by applying the irrevocable EUR/national currency conversion rates. The new Turkish lira was introduced on 1 January 2005, equivalent to 1 000 000 old Turkish lira.

Source: OECD Main Economic Indicators.

StatLink http://dx.doi.org/10.1787/888932802876

Annex Table A.2. **Purchasing power parities**

In national currency units per USD

	1996	1997	1998	1999	2000	2001	2002	2003	2004	2005	2006	2007	2008	2009	2010	2011
	1.32	1.32	1.30	1.30	1.31	1.33	1.34	1.35	1.37	1.39	1.41	1.43	1.43	1.46	1.53	1.56
	0.93	0.92	0.92	0.92	0.90	0.92	0.90	0.88	0.87	0.89	0.86	0.87	0.85	0.85	0.85	0.85
	0.91	0.91	0.92	0.92	0.89	0.89	0.87	0.88	0.90	0.90	0.88	0.89	0.87	0.86	0.86	0.87
	1.21	1.21	1.19	1.19	1.23	1.22	1.23	1.23	1.23	1.21	1.21	1.21	1.23	1.20	1.22	1.23
	265.70	272.35	274.63	277.39	284.03	288.40	295.76	306.64	320.56	333.69	363.42	372.43	365.24	376.15	399.64	402.19
epublic	11.95	12.70	13.88	14.14	14.21	14.23	14.32	14.04	14.29	14.32	14.03	13.94	14.26	13.91	14.21	13.87
k	8.44	8.43	8.39	8.47	8.41	8.47	8.30	8.54	8.40	8.59	8.32	8.23	8.01	7.89	7.87	7.82
	0.37	0.40	0.43	0.44	0.46	0.48	0.48	0.48	0.49	0.50	0.52	0.55	0.55	0.52	0.52	0.54
	1.00	1.00	1.00	1.00	0.99	1.01	1.00	1.01	0.98	0.98	0.95	0.94	0.92	0.91	0.92	0.93
	0.99	0.97	0.97	0.96	0.94	0.92	0.90	0.94	0.94	0.92	0.90	0.89	0.88	0.87	0.87	0.87
y	0.99	0.99	0.99	0.97	0.97	0.96	0.94	0.92	0.90	0.87	0.84	0.83	0.81	0.80	0.81	0.80
	0.60	0.63	0.66	0.68	0.68	0.67	0.66	0.69	0.70	0.71	0.70	0.72	0.70	0.70	0.71	0.71
	73.12	84.96	94.13	101.07	107.87	110.69	114.88	120.56	126.31	128.59	128.47	131.27	129.43	126.83	130.12	130.66
	74.95	74.39	77.20	79.68	84.30	88.96	91.34	94.52	94.25	99.08	107.17	113.05	117.42	127.75	135.55	141.61
	0.83	0.85	0.88	0.93	0.96	0.99	1.00	1.01	1.01	1.01	0.98	0.96	0.95	0.90	0.86	0.84
	2.98	3.16	3.34	3.50	3.44	3.42	3.46	3.63	3.53	3.72	3.83	3.74	3.87	4.02	4.02	4.02
	0.81	0.82	0.81	0.82	0.82	0.81	0.85	0.85	0.87	0.87	0.83	0.82	0.79	0.78	0.80	0.79
	170.60	168.52	166.58	162.04	154.75	149.46	143.77	139.69	134.41	129.55	124.72	120.31	116.85	115.04	111.36	106.83
	730.92	745.84	774.03	754.89	746.21	757.83	769.77	794.28	796.00	788.92	774.82	768.65	785.72	804.11	824.57	821.46
ɔurg	0.95	0.96	0.95	0.94	0.94	0.95	0.93	0.94	0.92	0.95	0.91	0.92	0.91	0.91	0.92	0.92
	3.76	4.35	4.97	5.63	6.10	6.31	6.55	6.82	7.22	7.13	7.19	7.38	7.47	7.71	7.93	8.19
nds	0.91	0.91	0.91	0.91	0.89	0.91	0.90	0.93	0.91	0.90	0.87	0.86	0.84	0.84	0.84	0.84
ıland	1.47	1.45	1.45	1.43	1.44	1.47	1.47	1.50	1.51	1.54	1.49	1.50	1.49	1.48	1.51	1.53
	9.05	9.08	9.38	9.33	9.13	9.18	9.11	9.12	8.99	8.90	8.69	8.77	8.75	8.92	9.01	8.88
	1.36	1.52	1.66	1.74	1.84	1.86	1.83	1.84	1.86	1.87	1.84	1.84	1.86	1.86	1.86	1.87
	0.66	0.67	0.69	0.70	0.70	0.71	0.71	0.71	0.72	0.68	0.66	0.66	0.65	0.64	0.64	0.63
ˈepublic	0.44	0.45	0.47	0.50	0.53	0.52	0.53	0.56	0.57	0.57	0.56	0.55	0.53	0.51	0.52	0.53
	0.43	0.46	0.48	0.51	0.53	0.57	0.59	0.62	0.61	0.61	0.61	0.63	0.63	0.64	0.64	0.63
	0.72	0.72	0.72	0.73	0.73	0.74	0.73	0.75	0.76	0.76	0.74	0.73	0.72	0.71	0.72	0.72
	9.25	9.29	9.37	9.29	9.13	9.35	9.35	9.34	9.11	9.38	9.08	8.88	8.77	8.94	9.03	8.95
ınd	1.94	1.89	1.88	1.87	1.85	1.84	1.77	1.78	1.75	1.74	1.66	1.60	1.55	1.53	1.52	1.46
	0.04	0.08	0.13	0.20	0.28	0.43	0.61	0.77	0.81	0.83	0.85	0.86	0.89	0.92	0.97	1.00
ingdom	0.64	0.63	0.64	0.65	0.64	0.63	0.63	0.64	0.63	0.64	0.63	0.65	0.65	0.65	0.66	0.68
tates	1	1	1	1	1	1	1	1	1	1	1	1	1	1	1	1

ta for EMU member countries are given in euros (EUR). Data for years prior to 1999 have been converted from national currencies into EUR by applying the irrevocable EUR/national currency conversion rates. The new Turkish lira ɩduced on 1 January 2005, equivalent to 1 000 000 old Turkish lira.

OECD Main Economic Indicators.

StatLink http://dx.doi.org/10.1787/888932802895

Annex Table A.3. **Gross domestic product**

USD millions

	1996	1997	1998	1999	2000	2001	2002	2003	2004	2005	2006	2007	2008	2009	2010	2011
Australia	435 142	436 552	390 896	427 121	410 985	391 165	435 278	558 108	677 132	759 392	814 331	979 958	1 052 284	1 010 453	1 285 475	1 521 578
Austria	234 494	207 102	213 234	211 985	191 261	191 251	208 046	252 804	289 763	306 554	323 793	375 370	415 800	383 543	376 838	417 656
Belgium	274 901	248 869	255 542	254 124	231 691	231 967	253 415	310 289	359 614	379 294	398 536	460 021	509 375	473 301	468 586	513 661
Canada	615 341	639 662	618 225	659 356	722 535	714 870	734 334	866 554	993 005	1 135 409	1 283 544	1 429 522	1 498 521	1 341 215	1 577 289	1 736 045
Chile	78 000	85 287	81 810	75 298	77 582	70 757	69 434	76 140	99 333	123 055	154 670	173 083	179 627	172 591	216 309	248 584
Czech Republic	64 907	59 461	63 866	62 164	58 800	64 368	78 422	95 289	113 976	130 052	148 345	180 511	225 449	197 218	198 929	217 027
Denmark	184 394	170 552	173 674	173 850	160 144	160 530	173 984	212 548	244 771	257 543	274 228	311 629	343 755	311 164	312 215	331 939
Estonia	4 725	5 035	5 591	5 700	5 704	6 224	7 336	9 797	11 957	13 978	16 739	22 012	23 875	19 114	19 097	22 154
Finland	128 742	123 425	129 590	130 129	121 280	124 364	135 515	163 518	187 983	196 786	207 206	246 342	273 044	239 331	235 258	263 011
France	1 572 573	1 421 172	1 467 893	1 454 261	1 320 737	1 335 316	1 455 592	1 784 161	2 043 916	2 147 559	2 247 645	2 584 647	2 842 934	2 619 115	2 549 028	2 773 032
Germany	2 435 065	2 148 989	2 177 444	2 127 872	1 878 440	1 876 696	2 011 509	2 412 921	2 710 741	2 780 500	2 892 375	3 326 712	3 637 941	3 297 917	3 284 474	3 600 833
Greece	137 328	134 873	134 854	138 900	126 208	129 552	146 408	191 983	226 646	241 313	261 116	305 166	342 529	321 725	299 103	298 733
Hungary	45 930	46 533	47 952	48 254	46 385	52 721	66 383	83 537	101 924	110 323	112 533	136 099	154 237	126 664	127 970	138 710
Iceland	7 331	7 423	8 292	8 742	8 697	7 923	8 907	10 970	13 252	16 287	16 651	20 427	16 832	12 113	12 570	14 026
Ireland	75 152	81 695	89 138	97 224	97 114	105 466	123 902	158 406	185 878	204 328	222 871	260 182	264 691	223 050	205 253	217 275
Israel	105 218	108 370	109 825	110 674	124 062	122 151	112 795	118 551	126 471	133 633	145 008	166 266	201 403	194 900	217 630	243 527
Italy	1 261 448	1 198 109	1 220 090	1 206 381	1 099 350	1 121 195	1 228 182	1 507 697	1 725 590	1 795 474	1 866 289	2 129 040	2 316 388	2 110 688	2 043 530	2 193 971
Japan	4 706 148	4 324 310	3 914 434	4 432 475	4 731 001	4 159 822	3 980 756	4 303 069	4 655 932	4 571 793	4 356 724	4 356 477	4 849 161	5 035 147	5 488 416	5 870 138
Korea	573 003	532 239	357 509	461 807	533 384	504 586	575 929	643 762	721 975	844 863	951 773	1 049 236	931 402	834 060	1 014 891	1 116 250
Luxembourg	20 516	18 451	19 350	21 156	20 184	20 154	22 634	29 027	33 896	37 854	42 400	51 358	57 996	51 935	52 983	59 475
Mexico	364 320	439 395	461 358	526 911	636 432	681 933	710 805	700 260	758 309	845 931	948 997	1 033 006	1 091 953	879 325	1 031 898	1 151 557
Netherlands	415 266	384 536	402 738	410 844	383 450	399 760	438 881	535 893	606 400	641 759	675 270	783 251	874 237	796 160	774 658	836 074
New Zealand	68 462	68 205	56 082	58 824	53 413	53 077	61 491	82 572	100 687	113 080	109 521	134 015	130 677	117 376	141 935	160 826
Norway	160 153	158 299	151 041	159 029	168 323	170 955	192 018	224 867	260 061	304 178	340 219	393 591	453 885	374 658	417 753	485 803
Poland	156 458	157 120	172 673	167 680	171 121	190 602	198 181	216 750	252 606	303 488	341 945	424 815	529 256	430 931	469 068	514 610
Portugal	121 061	116 260	122 641	126 235	116 805	120 063	132 610	161 204	184 337	192 836	201 069	231 944	252 916	234 033	227 197	237 371
Slovak Republic	21 105	21 310	22 369	20 518	20 377	21 044	24 538	33 289	42 207	47 878	55 558	74 939	94 144	87 215	86 504	95 914
Slovenia	21 332	20 310	21 849	22 266	19 963	20 559	23 195	29 297	33 956	35 903	38 806	47 389	54 771	49 383	46 851	50 239
Spain	623 493	572 638	599 437	616 960	577 896	607 497	687 979	879 867	1 038 635	1 136 623	1 231 934	1 442 686	1 599 688	1 455 639	1 380 109	1 476 882
Sweden	276 291	253 340	254 720	258 889	247 320	227 340	250 886	314 569	362 035	370 733	398 981	462 429	486 240	405 986	462 903	539 682
Switzerland	311 025	272 514	278 866	273 949	255 861	262 274	286 401	333 757	375 281	383 270	406 429	450 667	525 789	508 598	552 225	659 308
Turkey	248 217	258 417	270 012	249 038	264 537	195 304	232 103	303 187	390 932	484 277	530 343	648 599	731 180	614 554	732 533	777 283
United Kingdom	1 221 447	1 369 893	1 471 197	1 499 144	1 477 718	1 478 026	1 594 924	1 863 272	2 181 602	2 295 836	2 468 809	2 824 238	2 668 391	2 190 411	2 256 260	2 445 408
United States	7 783 900	8 278 900	8 741 000	9 301 000	9 898 800	10 233 900	10 590 200	11 089 300	11 797 800	12 564 300	13 314 500	13 961 800	14 219 300	13 898 300	14 419 400	14 991 300
OECD	24 752 888	24 369 246	24 505 192	25 798 760	26 257 561	26 053 413	27 252 976	30 557 235	33 908 600	35 906 079	37 799 159	41 477 425	43 849 671	41 017 813	42 985 137	46 219 882

Source: OECD Main Economic Indicators.

StatLink http://dx.doi.org/10.1787/888932802914

Annex Table A.4. **Total population**

Thousands

	1996	1997	1998	1999	2000	2001	2002	2003	2004	2005	2006	2007	2008	2009	2010	2011
Australia	18 417	18 606	18 812	19 036	19 270	19 531	19 768	20 009	20 250	20 542	20 871	21 261	21 728	22 129	22 446	22 769
Austria	7 959	7 968	7 977	7 992	8 012	8 042	8 082	8 118	8 169	8 225	8 268	8 301	8 337	8 363	8 388	8 421
Belgium	10 155	10 180	10 203	10 222	10 246	10 281	10 330	10 373	10 417	10 474	10 543	10 622	10 708	10 790	10 883	10 978
Canada	29 610	29 906	30 155	30 401	30 686	31 019	31 354	31 640	31 941	32 245	32 576	32 930	33 319	33 730	34 126	34 483
Chile	14 596	14 796	14 997	15 197	15 398	15 572	15 746	15 920	16 093	16 267	16 433	16 598	16 764	16 929	17 094	17 268
Czech Republic	10 315	10 304	10 295	10 283	10 273	10 224	10 201	10 202	10 207	10 234	10 267	10 323	10 430	10 492	10 517	10 497
Denmark	5 262	5 285	5 303	5 321	5 338	5 357	5 376	5 390	5 403	5 419	5 437	5 460	5 492	5 522	5 546	5 569
Estonia	1 425	1 406	1 393	1 379	1 372	1 367	1 361	1 356	1 351	1 348	1 345	1 342	1 341	1 340	1 340	1 340
Finland	5 125	5 140	5 154	5 166	5 176	5 188	5 201	5 213	5 228	5 246	5 266	5 289	5 313	5 339	5 363	5 387
France	59 589	59 795	60 011	60 315	60 725	61 163	61 605	62 038	62 491	62 958	63 393	63 781	64 133	64 476	64 824	65 176
Germany	81 896	82 052	82 029	82 087	82 188	82 340	82 482	82 520	82 501	82 464	82 366	82 263	82 120	81 875	81 757	81 779
Greece	10 709	10 777	10 835	10 883	10 918	10 950	10 988	11 024	11 062	11 104	11 149	11 193	11 237	11 283	11 308	11 310
Hungary	10 311	10 291	10 267	10 238	10 211	10 188	10 159	10 130	10 107	10 087	10 071	10 056	10 038	10 023	10 000	9 972
Iceland	269	271	274	277	281	285	288	289	293	296	304	311	319	319	318	319
Ireland	3 626	3 661	3 714	3 755	3 804	3 864	3 932	3 997	4 067	4 160	4 261	4 365	4 443	4 468	4 476	4 491
Israel	5 688	5 834	5 978	6 135	6 301	6 453	6 587	6 709	6 831	6 955	7 082	7 211	7 343	7 484	7 622	7 752
Italy	56 860	56 890	56 907	56 916	56 942	56 977	57 157	57 605	58 175	58 607	58 942	59 375	59 832	60 193	60 483	60 749
Japan	125 859	126 157	126 472	126 667	126 926	127 316	127 485	127 694	127 786	127 767	127 769	127 770	127 692	127 509	128 057	128 781
Korea	45 525	45 954	46 287	46 617	47 008	47 357	47 622	47 859	48 039	48 138	48 372	48 598	48 949	49 182	49 410	49 779
Luxembourg	415	420	425	431	437	442	447	452	459	466	473	481	489	498	508	519
Mexico	92 581	93 932	95 259	96 569	98 295	99 580	100 783	101 884	102 888	103 831	104 748	105 677	106 573	107 444	108 292	109 220
Netherlands	15 526	15 608	15 703	15 809	15 922	16 043	16 147	16 223	16 276	16 317	16 341	16 378	16 440	16 526	16 612	16 691
New Zealand	3 747	3 792	3 822	3 843	3 868	3 900	3 970	4 045	4 101	4 148	4 198	4 241	4 281	4 332	4 384	4 421
Norway	4 381	4 405	4 432	4 462	4 491	4 513	4 539	4 565	4 591	4 622	4 661	4 706	4 769	4 827	4 889	4 953
Poland	38 289	38 292	38 283	38 270	38 256	38 251	38 232	38 195	38 180	38 161	38 132	38 116	38 116	38 153	38 517	38 526
Portugal	10 058	10 091	10 129	10 172	10 226	10 293	10 368	10 441	10 502	10 549	10 584	10 608	10 622	10 633	10 637	10 651
Slovak Republic	5 374	5 384	5 391	5 396	5 401	5 380	5 379	5 379	5 382	5 387	5 391	5 397	5 406	5 418	5 430	5 441
Slovenia	1 990	1 986	1 982	1 984	1 989	1 992	1 995	1 996	1 997	2 001	2 008	2 019	2 022	2 042	2 049	2 053
Spain	39 479	39 583	39 722	39 927	40 264	40 721	41 314	42 005	42 692	43 398	44 068	44 874	45 593	45 930	46 073	46 125
Sweden	8 841	8 846	8 851	8 858	8 872	8 896	8 925	8 958	8 994	9 030	9 081	9 148	9 220	9 299	9 378	9 442
Switzerland	7 105	7 113	7 132	7 167	7 209	7 285	7 343	7 405	7 454	7 501	7 558	7 619	7 711	7 801	7 786	7 869
Turkey	60 671	61 582	62 464	63 364	64 252	65 133	66 008	66 873	67 723	68 566	69 395	70 215	71 095	72 050	73 003	73 950
United Kingdom	58 164	58 314	58 475	58 684	58 886	59 113	59 319	59 552	59 842	60 235	60 584	60 986	61 398	61 792	62 262	62 735
United States	269 714	272 958	276 154	279 328	282 398	285 225	287 955	290 626	293 262	295 993	298 818	301 696	304 543	307 240	309 774	312 040
OECD	1 119 531	1 127 578	1 135 285	1 143 149	1 151 840	1 160 242	1 168 445	1 176 684	1 184 755	1 192 741	1 200 754	1 209 209	1 217 817	1 225 427	1 233 553	1 241 456

Source: OECD Annual Labour Force Statistics.

StatLink http://dx.doi.org/10.1787/888932802933

Glossary

..	Data not available
2G	Second generation of mobile communications technology
3G	Third generation of mobile communications technology
ADSL	Asymmetric digital subscriber line
AFRINIC	African Network Information Centre
APNIC	Asia-Pacific Network Information Centre
ARIN	American Registry for Internet Numbers
AS (ASes)	Autonomous systems
ASEAN	Association of Southeast Asian Nations
ASN	Autonomous systems numbers
ATVoD	Association for Television on Demand
AV	Audio-visual
BB	Broadband
BGP	Border Gateway Protocol
BLS	Bureau of Labor Statistics (United States)
BRICS	Group of countries including Brazil, Russia, India, China and South Africa
CAGR	Compound annual growth rate (expressed as a percentage)
CAIDA	Cooperative Association for Internet Data Analysis
ccTLD	Country code top level domain
CDMA	Code division multiple access
CDN	Content delivery networks
CEPT	European Conference for Postal and Telecommunications Administrations
CIDR	Classless Inter-Domain Routing
CIS	Commonwealth of Independent States
CPE	Customer premises equipment
CPI	Consumer price index
CPP	Calling party-pays
CRT	Cathode ray tube
DBS	Direct broadcast satellite
DLNA	Digital Living Network Alliance
DNS	Domain name system
DOCSIS 3.0	Data over cable service interface specification
DSL	Digital subscriber lines
DTT	Digital terrestrial television
DTV	Digital television
DVB	Digital video broadcasting
DVB-H	Digital video broadcasting – handheld
DVB-T	Digital video broadcasting – television
EAO	European Audiovisual Observatory
EAP-SIM	Extensible Authentication Protocol SIM

EBOPS	Extended Balance of Payments Services Classification
EC	European Commission
ECLAC	UN Economic Commission for Latin America and the Caribbean
EDGE	Enhanced data rates for GSM evolution
ENUM	Electronic number mapping
EPG	Electronic programming guide
EPO	European Patent Office
ETNO	European Telecommunications Network Operators
EU	European Union
EAP-SIM	Extensible Authentication Protocol Method for GSM SIM
FCC	Federal Communications Commission (United States)
FTA	Free-to-air
FTP	File transfer protocol
FTTH	Fibre-to-the-home
FTTN	Fibre-to-the-node
FTTP	Fibre-to-the-premises
GDP	Gross domestic product
GFC	Global financial crisis
GFCF	Gross fixed capital formation
GPRS	GSM packet radio service
GSA	Global mobile Suppliers Association
GSM	Global system for mobile communications
GSMA	GSM Association
GST	Goods and services taxes
gTLD	Generic top level domain
HDTV	High-definition television
HE-AAC	High Efficiency Audio Coding
HFC	Hybrid fibre/coaxial networks
HICP	Harmonised indices of consumer prices
HLM	High-level meeting
HS	Harmonised system
HTML	Hypertext mark-up language
HTTP	Hypertext transfer protocol
IANA	Internet Assigned Numbers Authority
ICANN	Internet Corporation for Assigned Names and Numbers
ICT	Information and communication technology
IDN	Internationalised domain names
IEEE (802 Standards)	Institute of Electrical and Electronics Engineers
IETV	Internet-enabled televisions
IMSI	International mobile subscriber identity
IMT-2000	International Mobile Telecommunications 2000
IP	Internet protocol
IP-PBX	Internet protocol – private branch exchange
IPTV	Internet protocol television
IPv4	Internet protocol version 4
IPv6	Internet protocol version 6
IR	Internet registries

ISC	Internet System Consortium
ISDN	Integrated services digital network
ISO	International Organization for Standardization
ISP	Internet service provider
IT	Information technologies
ITCS	International trade by commodity statistics
ITU	International Telecommunication Union
IXP	Internet Exchange Points
KCC	Korea Communications Commission
Kbit/s	Kilobits per second (Kbps)
LACNIC	Latin American and Caribbean Internet Addresses Registry
LAN	Local area network
LBO	Local break-out
LCD	Liquid Crystal Display
LED	Light-emitting diode
LIR	Local Internet Registries
LRIC	Long-run incremental costs
LLU	Local loop unbundling
LTE	Long Term Evolution
M2M	Machine-to-machine
Mbit/s	Megabits per second (Mbps)
MDF	Main distribution frames
MiTT	Minutes of international telecommunication traffic
MMS	Multimedia messaging service
MNO	Mobile network operators
MoU	Minutes of use
MTR	Mobile termination rates
MVNO	Mobile virtual network operators
NAT	Network Address Translation
NDT	Network Diagnostic Tool
NFC	Near field communications
NGA	Next Generation Access networks
NIR	National Internet Registries
NRAs	National regulatory authorities
NVoD	Near video on demand
OCN	Open computer network
OFCOM	Office of Communications (United Kingdom)
OMB	United States' Office of Management Budget
OTT	Over the top
P2P	Peer-to-peer
PB	Petabytes
PBX	Private branch exchange
PC	Personal computer
PCB	Public call boxes
PCS	Personal communications service
PDA	Personal digital assistant
PPI	Producers price index
PPP	Purchasing power parities
PPV	Pay-per-view

PSB	Public service broadcasters
PSP	Public service publisher
PSTN	Public switched telecommunication network
PTO	Public telecommunications operator
PVR	Personal video recorder
QoS	Quality of service
R&D	Research and development
RIPE NCC	Réseaux IP Européens Network Co-ordination Centre
RIR	Regional Internet registry
RPP	Receiving party pays
S-DMB	Satellite digital media broadcasting
SDTV	Standard definition television
SIC	Standard industrial classification
SIM (card)	Subscriber identity module
SITC	Standard industrial trade classification
SMEs	Small and medium-sized enterprises
SMP	Significant market power
SMS	Short message service
SNA	Statistics of national accounts
SOE	State-owned enterprises
SSL	Secure sockets layer
TCP/IP	Transmission control protocol/Internet protocol
T-DAB	Terrestrial digital audio broadcasting
T-DMB	Terrestrial digital media broadcasting
TLCS	Television licensable content service
TLD	Top-level domain
TRAI	Telecom Regulatory Authority of India
TVHH	Television households
UMTS	Universal mobile telecommunications system
UNASUR	Union of South-American Nations
URL	Uniform resource locator
USD	United States Dollar
USO	Universal service obligations
VAT	Value-added tax
VDSL	Very high data rate digital subscriber line
VNI	Cisco's Visual Networking Index
VoBB	Voice over broadband
VoD	Video on demand
VoIP	Voice over Internet protocol
VoLTE	Voice over LTE
W-CDMA	Wideband code division multiple access
WAN	Wide Area Networks
WIDE	Widely integrated distributed environment
Wi-Fi	Wireless fidelity
WiMAX	Wireless interoperability for microwave access
WLAN	Wireless local area network
WLL	Wireless local loop

ORGANISATION FOR ECONOMIC CO-OPERATION AND DEVELOPMENT

The OECD is a unique forum where governments work together to address the economic, social and environmental challenges of globalisation. The OECD is also at the forefront of efforts to understand and to help governments respond to new developments and concerns, such as corporate governance, the information economy and the challenges of an ageing population. The Organisation provides a setting where governments can compare policy experiences, seek answers to common problems, identify good practice and work to co-ordinate domestic and international policies.

The OECD member countries are: Australia, Austria, Belgium, Canada, Chile, the Czech Republic, Denmark, Estonia, Finland, France, Germany, Greece, Hungary, Iceland, Ireland, Israel, Italy, Japan, Korea, Luxembourg, Mexico, the Netherlands, New Zealand, Norway, Poland, Portugal, the Slovak Republic, Slovenia, Spain, Sweden, Switzerland, Turkey, the United Kingdom and the United States. The European Commission takes part in the work of the OECD.

OECD Publishing disseminates widely the results of the Organisation's statistics gathering and research on economic, social and environmental issues, as well as the conventions, guidelines and standards agreed by its members.

ÉDITIONS OCDE, 2, rue André-Pascal, 75775 PARIS CEDEX 16
(93 2013 02 1P) ISBN 978-92-64-19459-5 – n° 60649 2013-02

EUW 90.00